Lecture Notes in Computer Science 5911

Commenced Publication in 1973
Founding and Former Series Editors:
Gerhard Goos, Juris Hartmanis, and Jan van Leeuwen

Editorial Board

David Hutchison, UK
Josef Kittler, UK
Alfred Kobsa, USA
John C. Mitchell, USA
Oscar Nierstrasz, Switzerland
Bernhard Steffen, Germany
Demetri Terzopoulos, USA
Gerhard Weikum, Germany

Takeo Kanade, USA
Jon M. Kleinberg, USA
Friedemann Mattern, Switzerland
Moni Naor, Israel
C. Pandu Rangan, India
Madhu Sudan, USA
Doug Tygar, USA

Advanced Research in Computing and Software Science

Subline of Lectures Notes in Computer Science

Subline Series Editors

Giorgio Ausiello, *University of Rome 'La Sapienza', Italy*
Vladimiro Sassone, *University of Southampton, UK*

Subline Advisory Board

Susanne Albers, *University of Freiburg, Germany*
Benjamin C. Pierce, *University of Pennsylvania, USA*
Bernhard Steffen, *University of Dortmund, Germany*
Madhu Sudan, *Microsoft Research, Cambridge, MA, USA*
Deng Xiaotie, *City University of Hong Kong*
Jeannette M. Wing, *Carnegie Mellon University, Pittsburgh, PA, USA*

Christophe Paul   Michel Habib (Eds.)

Graph-Theoretic Concepts in Computer Science

35th International Workshop, WG 2009
Montpellier, France, June 24-26, 2009
Revised Papers

 Springer

Volume Editors

Christophe Paul
CNRS - LIRMM
161 rue Ada
34392 Montpellier Cedex 5, France
E-mail: christophe.paul@lirmm.fr

Michel Habib
Université Paris Diderot
LIAFA
case 7014
75205 Paris Cedex 13
France
E-mail: habib@liafa.jussieu.fr

Library of Congress Control Number: 2009941869

CR Subject Classification (1998): G.2.2, I.2.8, E.1, G.1.6, C.2, J.2

LNCS Sublibrary: SL 1 – Theoretical Computer Science and General Issues

ISSN   0302-9743
ISBN-10   3-642-11408-3 Springer Berlin Heidelberg New York
ISBN-13   978-3-642-11408-3 Springer Berlin Heidelberg New York

This work is subject to copyright. All rights are reserved, whether the whole or part of the material is concerned, specifically the rights of translation, reprinting, re-use of illustrations, recitation, broadcasting, reproduction on microfilms or in any other way, and storage in data banks. Duplication of this publication or parts thereof is permitted only under the provisions of the German Copyright Law of September 9, 1965, in its current version, and permission for use must always be obtained from Springer. Violations are liable to prosecution under the German Copyright Law.

springer.com

© Springer-Verlag Berlin Heidelberg 2010
Printed in Germany

Typesetting: Camera-ready by author, data conversion by Scientific Publishing Services, Chennai, India
Printed on acid-free paper   SPIN: 12823185   06/3180   5 4 3 2 1 0

Preface

The 35th International Workshop on Graph-Theoretic Concepts in Computer Science (WG 2009) took place at Montpellier (France), June 24–26 2009. About 80 computer scientists from all over the world (Australia, Belgium, Canada, China, Czech Republic, France, Germany, Greece, Israel, Japan, Korea, The Netherlands, Norway, Spain, UK, USA) attended the conference.

Since 1975, it has taken place 20 times in Germany, four times in The Netherlands, twice in Austria, as well as once in Italy, Slovakia, Switzerland, the Czech Republic, France, Norway, and the UK. The conference aims at uniting theory and practice by demonstrating how graph-theoretic concepts can be applied to various areas in computer science, or by extracting new problems from applications. The goal is to present recent research results and to identify and explore directions of future research. The conference is well-balanced with respect to established researchers and young scientists.

There were 69 submissions. Each submission was reviewed by at least three, and on average four, Program Committee members. The Committee decided to accept 28 papers. Due to the competition and the limited schedule, some good papers could not be accepted.

The program also included excellent invited talks: one given by Daniel Král on "Algorithms for Classes of Graphs with Bounded Expansion," the other by David Eppstein on "Graph-Theoretic Solutions to Computational Geometry Problems." The proceedings contains two survey papers on these topics.

For the second year, the Best Student Paper was awarded. The recipients of the 2009 edition were Zhentao Li and Ignasi Sau for their paper "Graph Partitioning and Traffic Grooming with Bounded Degree Request Graph."

We wish to thank all those who contributed to the success of WG 2009, especially the authors for submitting high-quality papers, the referees and the Program Committee for their work, the speakers for their pedagogical talks and all the participants. We are grateful to all those who helped in the organization, especially the members and students of the ALGCO team of the LIRMM (Laboratoire d'Informatique, Robotique et Microélectronique de Montpellier), Céline Berger and Elisabeth Greverie. Special thanks to Alexandre Pinlou, who led all the local organization.

We should thank the EasyChair team for the service they offer, which saves to the Program Committee a lot of time. Finally, we would like to thank our sponsors: the LIRMM, Montpellier 2 University, the CNRS (Centre National de la Recherche Scientifique), the town of Montpellier and the Languedoc Roussillon district.

September 2009   Christophe Paul
                                                         Michel Habib

Conference Organization

The Tradition of WG

1975 U. Pape – Berlin, Germany
1976 H. Noltemeier – Göttingen, Germany
1977 J. Mühlbacher – Linz, Austria
1978 M. Nagl, H.J. Schneider – Castler Feuerstein, Germany
1979 U. Pape – Berlin, Germany
1980 H. Noltemeier – Bad Honnef, Germany
1981 J. Mühlbacher – Linz, Austria
1982 H.J. Schneider, H. Göttler – Neuenkirchen, Germany
1983 M. Nagl, J. Perl – Haus Ohrbeck near Onasbrück, Germany
1984 U. Pape – Berlin, Germany
1985 H. Noltemeier – Castle Schwanberg near Würzburg, Germany
1986 G. Tinhofer, G. Schmidt – Bernried near Munich, Germany
1987 H. Göttler, H.J. Schneider – Kloster Banz near Bamberg, Germany
1988 J. van Leeuwen – Amsterdam, The Netherlands
1989 M. Nagl – Castle Rolduc, The Netherlands
1990 R.H. Möhring – Berlin, Germany
1991 G. Schmidt, R. Berghammer – Fischbachau near Munich, Germany
1992 E.W Mayr – Wiesbaden-Naurod, Germany
1993 J. van Leeuwen – Utrecht, The Netherlands
1994 G. Tinhofer, E.W. Mayr, G. Schmidt – Herrsching near Munich, Germany
1995 M. Nagl – Aachen, Germany
1996 G. Ausiello, A. Marchetti-Spaccamela – Como, Italy
1997 R.H. Möhring – Berlin, Germany
1998 J. Hromkovič, O. Sýkora – Smolenice Castle, Slovak Republic
1999 P. Widmayer – Ascona, Switzerland
2000 D. Wagner – Konstanz, Germany
2001 A. Brandstädt, Boltenhagen near Rostock, Germany
2002 L. Kučera – Český Krumlov, Czech Republic
2003 H.L. Bodlaender – Elspeet, The Netherlands
2004 J. Hromkovič, M. Nagl – Bad Honnef, Germany
2005 D. Kratsch – Metz, France
2006 F.V. Fomin – Bergen, Norway
2007 A. Brandstädt, D. Kratsch, H. Müller – Dornburg near Jena, Germany
2008 H. Broersma, T. Erlebach – Durham, UK
2009 C. Paul, M. Habib – Montpellier, France

Program Chairs

Michel Habib (Co-chair)   LIAFA, Université Paris-Diderot, Paris, France
Christophe Paul (Co-chair)   CNRS - LIRMM - Université de Montpellier 2, France

Program Committee

Hans L. Bodlaender	Universiteit Utrecht, The Netherlands
Andreas Brandstädt	University of Rostock, Germany
Leizhen Cai	The Chinese University of Hong Kong, China
Feodor F. Dragan	Kent State University, USA
Jirí Fiala	Charles University, Prague, Czech Republic
Pinar Heggernes	University of Bergen, Norway
Michael Kaufmann	Universität Tübingen, Germany
Dieter Kratsch	University of Metz, France
Alberto Marchetti Spaccamela	Università di Roma La Sapienza, Italy
Ernst Wilhelm Mayr	Technische Universität München, Germany
Ross McConnell	Colorado State University, United States of America
Haiko Müller	University of Leeds, United Kingdom
Rolf Niedermeier	Friedrich-Schiller-Universität Jena, Germany
Sang-Il Oum	KAIST, Republic of Korea
Dimitrios M. Thilikos	National and Kapodistrian University of Athens, Greece
Ioan Todinca	LIFO - Université d'Orléans, France
Bernhard Westfechtel	Universität Bayreuth, Germany

Local Organization

Stéphane Bessy
Jean Daligault
Philippe Gambette
Émeric Gioan
Daniel Gonçalves
Chistophe Paul
Anthony Perez
Alexandre Pinlou (Chair)
Stéphan Thomassé

External Reviewers

Bampas, Evangelos
Bang-Jensen, Jørgen
Barat, Janos
Basavaraju, Manu
Bayer, Daniel
Bessy, Stéphane
Betzler, Nadja
Brandstadt, Andreas
Brass, Peter
Bui-Xuan, Binh-Minh
Chapelle, Mathieu
Chepoi, Victor
Chudnovsky, Maria
Corneil, Derek
Daligault, Jean
Dijk, Thomas C. van
Dom, Michael
Dorn, Britta
Dorn, Frederic
Durocher, Stephane
Effinger, Philip
Escoffier, Bruno
Fomin, Fedor
Fouquet, Jean-Luc
Fraigniaud, Pierre
Gaspers, Serge
Gerasch, Andreas
Geyer, Markus
Giannopoulou, Archontia
Gioan, Emeric
Golovach, Petr
Grandoni, Fabrizio
Guo, Jiong
Hoang, Chinh
Kaminski, Marcin
Kim, Seog-Jin
Knipe, David
Kolliopoulos, Stavros
Komusiewicz, Christian
Kral, Daniel
Lau, Lap Chi
Le, Van Bang

van Leeuwen, Erik Jan
Liedloff, Mathieu
Limouzy, Vincent
Lipshteyn, Marina
Mancini, Federico
Matamala, Martín
Mazoit, Frédéric
Meister, Daniel
Mnich, Matthias
de Montgolfier, Fabien
Moscardelli, Luca
Moser, Hannes
Nikolopoulos, Stavros
Nisse, Nicolas
Palios, Leonidas
Papadopoulos, Charis
Perez, Anthony
Pettie, Seth
Pinlou, Alexandre
Rao, Michael
Richerby, David
van Rooij, Johan M. M.
Rossmanith, Peter
Rue, Juanjo
Samal, Robert
Sau, Ignasi
Saurabh, Saket
Serna, Maria
Sritharan, Sri
Suchan, Karol
Telle, Jan Arne
Tittmann, Peter
Uetz, Marc
Uhlmann, Johannes
Vanherpe, Jean-Marie
Vaxès, Yann
Villanger, Yngve
Voloshin, Vitaly
Wanke, Egon
Xiang, Yang
Yamazaki, Koichi
Yeh, Roger

Table of Contents

Graph-Theoretic Solutions to Computational Geometry Problems
(Invited Talk) 1
    David Eppstein

Algorithms for Classes of Graphs with Bounded Expansion
(Invited Talk) 17
    Zdeněk Dvořák and Daniel Král'

A Graph Polynomial Arising from Community Structure
(Extended Abstract) 33
    Ilia Averbouch, Johann A. Makowsky, and Peter Tittmann

Fast Exact Algorithms for Hamiltonicity in Claw-Free Graphs 44
    Hajo Broersma, Fedor V. Fomin, Pim van 't Hof, and
    Daniël Paulusma

Maximum Series-Parallel Subgraph 54
    Gruia Călinescu, Cristina G. Fernandes, and Hemanshu Kaul

Low-Port Tree Representations 66
    Shiri Chechik and David Peleg

Fully Dynamic Representations of Interval Graphs 77
    Christophe Crespelle

The Parameterized Complexity of Some Minimum Label Problems 88
    Michael R. Fellows, Jiong Guo, and Iyad A. Kanj

Exact and Parameterized Algorithms for MAX INTERNAL SPANNING
TREE 100
    Henning Fernau, Serge Gaspers, and Daniel Raible

An Exact Algorithm for Minimum Distortion Embedding 112
    Fedor V. Fomin, Daniel Lokshtanov, and Saket Saurabh

Sub-coloring and Hypo-coloring Interval Graphs 122
    Rajiv Gandhi, Bradford Greening Jr., Sriram Pemmaraju, and
    Rajiv Raman

Parameterized Complexity of Generalized Domination Problems 133
    Petr A. Golovach, Jan Kratochvíl, and Ondřej Suchý

Connected Feedback Vertex Set in Planar Graphs 143
    Alexander Grigoriev and René Sitters

Logical Locality Entails Frugal Distributed Computation over Graphs
(Extended Abstract) 154
  Stéphane Grumbach and Zhilin Wu

On Module-Composed Graphs 166
  Frank Gurski and Egon Wanke

An Even Simpler Linear-Time Algorithm for Verifying Minimum
Spanning Trees 178
  Torben Hagerup

The k-Disjoint Paths Problem on Chordal Graphs 190
  Frank Kammer and Torsten Tholey

Local Algorithms for Edge Colorings in UDGs 202
  Iyad A. Kanj, Andreas Wiese, and Fenghui Zhang

Directed Rank-Width and Displit Decomposition 214
  Mamadou Moustapha Kanté and Michaël Rao

An Algorithmic Study of Switch Graphs 226
  Bastian Katz, Ignaz Rutter, and Gerhard Woeginger

Hardness Results and Efficient Algorithms for Graph Powers 238
  Van Bang Le and Ngoc Tuy Nguyen

Graph Partitioning and Traffic Grooming with Bounded Degree
Request Graph 250
  Zhentao Li and Ignasi Sau

Injective Oriented Colourings 262
  Gary MacGillivray, André Raspaud, and Jacobus Swarts

Chordal Digraphs 273
  Daniel Meister and Jan Arne Telle

A New Intersection Model and Improved Algorithms for Tolerance
Graphs 285
  George B. Mertzios, Ignasi Sau, and Shmuel Zaks

Counting the Number of Matchings in Chordal and Chordal Bipartite
Graph Classes 296
  Yoshio Okamoto, Ryuhei Uehara, and Takeaki Uno

Distance d-Domination Games 308
  Stephan Kreutzer and Sebastian Ordyniak

Cycles, Paths, Connectivity and Diameter in Distance Graphs 320
  Lucia Draque Penso, Dieter Rautenbach, and Jayme Luiz Szwarcfiter

Smallest Odd Holes in Claw-Free Graphs (Extended Abstract) 329
  Shimon Shrem, Michal Stern, and Martin Charles Golumbic

Finding Induced Paths of Given Parity in Claw-Free Graphs 341
  Pim van 't Hof, Marcin Kamiński, and Daniël Paulusma

Author Index 353

Graph-Theoretic Solutions to Computational Geometry Problems

David Eppstein

Computer Science Department, University of California, Irvine

Abstract. Many problems in computational geometry are not stated in graph-theoretic terms, but can be solved efficiently by constructing an auxiliary graph and performing a graph-theoretic algorithm on it. Often, the efficiency of the algorithm depends on the special properties of the graph constructed in this way. We survey the art gallery problem, partition into rectangles, minimum-diameter clustering, rectilinear cartogram construction, mesh stripification, angle optimization in tilings, and metric embedding from this perspective.

1 Introduction

Graph algorithms and computational geometry form separate communities with separate conferences such as the International Workshop on Graph-Theoretic Concepts in Computer Science and the ACM Symposium on Computational Geometry, respectively, but they also meet in broader algorithms conferences, and there has been much interplay between the research topics in the two areas.

Many classical graph algorithm problems have geometric analogues, algorithmic problems on graphs defined by a geometric input. In most cases, problems of this type can be solved directly by constructing the graph and then applying a general-purpose graph algorithm, but can be sped up by examining the graph algorithm's structure more closely and applying appropriate geometric data structures. A notable instance of this phenomenon is the Euclidean minimum spanning tree (the spanning tree of a complete graph in which the vertices represent points and edge lengths are Euclidean distances): by using a Delaunay triangulation in place of a complete graph, the quadratic time of a naive algorithm can be improved to $O(n\log n)$ [67]. Other work along these lines includes algorithms for Euclidean matching [1, 71] and bipartiteness testing [26].

Graph drawing, the visualization of graphs via geometric graph representations [19, 45, 59], forms another community represented by the annual International Symposium on Graph Drawing. Most work in the area concerns drawings in which a graph's vertices are represented as geometric points, disks, or polygons, and its edges are represented as line segments or curves. Researchers in this area seek algorithms that optimize mathematical stand-ins for their aesthetic quality and readability such as the number of crossings, the number of bends in non-straight edges, the angles formed by adjacent edges, the area of the drawing, or the amount of symmetry that the drawing displays; the interplay between these different measures provides much scope for research.

Geometric techniques have sometimes also been applied to solve problems that were originally defined in purely graph-theoretic terms. One instance concerns parametric minimum spanning trees, minimum spanning trees of graphs whose edge weights are linear functions of a parameter. Dey proved [18] that $O(mn^{1/3})$ different trees may be formed in this way from a parametric graph with m edges and n vertices, using the *crossing number inequality* that s simple plane curves with n shared endpoints have $\Omega(s^3/n^2)$ crossings [4, 50], while the best known lower bound, $\Omega(m\alpha(n))$, involves a reduction from another geometric problem, lower envelopes of line segments [22].

In this paper we survey connections between computational geometry and graph algorithms of yet another type: problems in computational geometry that, although not initially defined graph-theoretically, may be solved by constructing an auxiliary graph from the input, applying a purely graph-theoretic algorithm to this auxiliary graph, and translating the output of this algorithm back into geometric terms. We discuss problems including the art gallery problem, partition into rectangles, minimum-diameter clustering, rectilinear cartogram construction, mesh stripification, angle optimization in tilings, and metric embedding. The ordering of the problems is roughly chronological, and the selection of topics is idiosyncratic and (especially for the later problems) largely drawn from the author's own research rather than being comprehensive.

2 Art Gallery Problems

Most computer scientists are familiar both with Chvátal's *art gallery theorem* [13, 61] that every n-vertex simple polygon has a set of $\lfloor n/3 \rfloor$ *guard points* from which the whole polygon may be seen, and with Fisk's elegant graph-theoretic proof [36]. One begins the proof by adding a maximal set of non-crossing diagonals to the polygon, partitioning it into triangles. Graph-theoretically, the vertices, sides, and added diagonals of the polygon form a maximal outerplanar graph; the *weak dual* of this graph (the adjacency graph of the triangles, omitting the outer face) is a tree (Figure 1, center). Every maximal outerplanar graph may be colored with three colors, as may be proven by induction: the result is clearly true when the graph is a triangle, and any larger maximal planar graph may be colored by removing a leaf from the dual tree, coloring the remaining graph, restoring the leaf, and observing that the restored vertex has only two neighbors and therefore has a free color available. Each triangle has one vertex of each color, so each color class forms a valid guard set, and the smallest of the color classes has at most $\lfloor n/3 \rfloor$ vertices. This proof technique translates into an efficient algorithm: polygons may be triangulated in linear time [10], and the induction proof leads to a linear-time algorithm for 3-coloring maximal planar graphs.

A less well-known variant of the art gallery problem concerns *simple orthogonal polygons*, simple polygons all of whose sides are parallel to the coordinate axes. Simple orthogonal polygons need only $\lfloor n/4 \rfloor$ guards, as can be shown again by graph coloring. Every simple orthogonal polygon can be partitioned by diagonals into convex quadrilaterals [46]; the resulting tree of quadrilaterals can be viewed as a special type of *squaregraph*, a planar graph in which every bounded face has four sides and every vertex either belongs to the unbounded face or has at least four incident edges [12, 6]. Adding the two diagonals of each quadrilateral to a squaregraph forms a *kinggraph* [12] (Figure 1, right). As with maximal outerplanar graphs, the kinggraphs derived from simple orthogonal polygons may be shown to be 4-chromatic by removing leaves of the

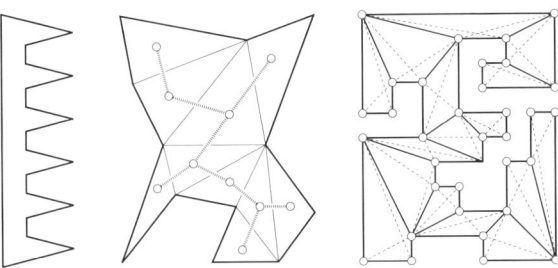

Fig. 1. The art gallery problem. Left: a comb polygon requiring $\lfloor n/3 \rfloor$ guards. Center: triangulating an input polygon produces a maximal outerplanar graph, the weak planar dual of which is a tree. Right: The kinggraph formed by adding diagonals (dashed) to a partition of an orthogonal polygon into convex quadrilaterals.

dual tree.[1] In any four-coloring, each quadrilateral has a vertex of each color, so each color class forms a guard set, and the smallest color class has at most $\lfloor n/4 \rfloor$ guards. This bound is tight: a comb-shaped orthogonal polygon requires $\lfloor n/4 \rfloor$ guards, just as a comb-shaped simple polygon requiring $\lfloor n/3 \rfloor$ guards shows that the bound for the standard art gallery problem is tight (Figure 1, left).

The biggest remaining open problem in art gallery theory concerns edge guards: how many edges must one select, from a simple polygon, so that every point of the polygon may be seen from some point on some selected edge? It is not clear, in this case, what graph one should define on the edges of the polygon to force the color classes of a coloring of the graph to form edge guard sets.

3 Partition into Rectangles

Many geometric algorithms take as input a complicated polygonal domain and cover or partition it using simpler shapes [65]; the partitions into triangles and quadrilaterals from the previous section are important examples. Another problem of this type concerns the partition into rectangles of an orthogonal polygon. The input polygon may have polygonal holes, unlike the simple orthogonal polygons of the previous section, and the rectangles are not required to meet edge-to-edge and vertex-to-vertex. The goal is to minimize the total number of rectangles in the partition (Figure 2).

Rectangular partitions have many applications. In VLSI design, it is necessary to decompose masks into the simpler shapes available in lithographic pattern generators [62], and similar mask decomposition problems also arise in DNA microarray design [39]. Rectangular partitions can simplify convolution operations in image processing [35] and can be used to compress bitmap images [11]. Closely related matrix decomposition problems have been applied to radiation therapy planning [21, 47], and rectangular partitions have also been used to design robot self-assembly sequences [51].

Define a *good diagonal* to be an axis-parallel line segment interior to the input polygon that connects two concave vertices of the polygon. As several authors independently

---

[1] More generally, every kinggraph is 4-chromatic, but a proof is beyond the scope of this survey.

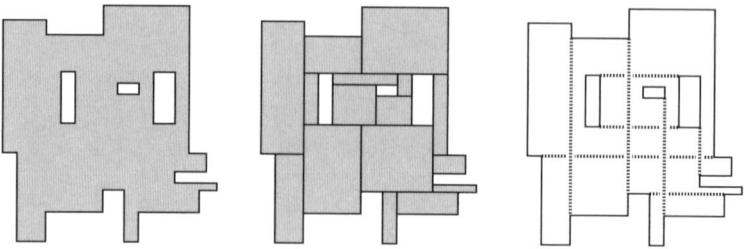

Fig. 2. Partitioning an orthogonal polygon (left) into the minimum number of rectangles (center). The right figure shows the axis-parallel diagonals that connect pairs of concave vertices; the rectangle partition problem may be solved by finding a maximum independent set in the bipartite intersection graph of these diagonals.

discovered [35, 55, 60], the minimum number of rectangles in a partition of a polygon with n vertices and h holes is $n/2 + h - g - 1$, where g is the maximum size of a set of disjoint good diagonals. To see this, consider the number of corners of rectangles in a partition, four times the number of rectangles. Let G be a maximal set of disjoint good diagonals that form a subset of the line segments in some partition of an orthogonal polygon, and define a *bad vertex* to be a nonconvex polygon vertex that is not an endpoint of G. Each polygon vertex forms at least one rectangle corner. Additionally, for each bad vertex v, let s be an interior line segment of the partition having v as an endpoint. Then either s crosses a segment in G, and the two corners formed by the crossing that are on the same side as v with respect to the crossed segment can be charged to v, or s ends at a non-vertex and the two corners formed at its other endpoint can be charged to v. This charging scheme shows that the number of rectangle corners in the partition is at least n plus twice the number of bad vertices. The number of nonconvex vertices in any orthogonal polygon with h holes is $n/2 + 2h - 2$, so the number of bad vertices is $n/2 + 2h - 2|G| - 2$, and a lower bound of $n/2 + h - |G| - 1$ rectangles follows. Conversely, if G is any set of disjoint good diagonals, a partition with exactly $n/2 + h - |G| - 1$ rectangles may be found by considering the bad vertices for G, one at a time, and extending a line segment from each bad vertex to the closest previously-added segment or polygon side. Thus, finding a partition into a minimum number of rectangles is equivalent to finding a maximum number of disjoint good diagonals.

The intersection graph of the good diagonals is bipartite: each horizontal diagonal intersects only vertical diagonals and vice versa. Therefore, finding the maximum number of disjoint good diagonals translates, in graph-theoretic terms, into finding a maximum independent set in a bipartite graph. By König's theorem [49], in any n-vertex bipartite graph the maximum independent set has size $n - M$, where M is the cardinality of a maximum matching; an independent set of this size may be found from a maximum matching by partitioning the vertices according to the lengths of the shortest alternating paths from an unmatched vertex to the given vertex, and including the vertices at even levels of this partition. Therefore, the maximum independent set of the intersection graph, the corresponding maximum set of disjoint good diagonals, and a partition into a minimum number of rectangles, may all be found in polynomial time.

A naive implementation of an algorithm that reduces the problem to a bipartite graph and then applies a general-purpose bipartite graph matching algorithm would solve the problem in time $O(n^{2.5})$, where n denotes the number of vertices of the input polygon: this is the time needed to apply the Hopcroft–Karp matching algorithm [41] to the intersection graph of the good diagonals, which may have $\Theta(n)$ vertices and $\Theta(n^2)$ edges. However, by using geometric range searching data structures to speed up the search for alternating paths within the matching algorithm, it is possible to improve the overall running time to $O(n^{3/2}\log n)$ [42, 53, 54].

4 Minimum Diameter Clustering

The *diameter* of a set of points is the maximum distance of any two of its points. The problem of finding low-diameter subsets of larger sets of input points [3, 27] may be formulated in several different ways: one may take as input a number k and produce as output the subset of k points with minimum diameter, one may take as input a number D and produce as output the largest subset with diameter at most D, or one may solve the *decision problem* in which D and k are both given and the task is to determine whether there exists a set of k points with diameter at most D. Since there are only $O(n^2)$ potential diameter values and n different values of k, an efficient algorithm for any one of these tasks leads to efficient algorithms for the other two tasks as well.

Finding the largest size k of a set with diameter at most a given value D has a direct translation to a graph theoretic problem, but the very directness of the translation means that it is unhelpful: it provides merely a restatement of the problem rather than conveying any new insight. We may scale the input point set so that $D = 2$; then what we seek is the largest clique in a *unit disk graph*, the intersection graph of unit disks centered at the points (Figure 3, left); note that a set of disks forming a clique need not have a common intersection point. Maximum cliques in unit disk graphs may be found in polynomial time, given a disk representation of the graph [15] but this is just a trivial restatement of the minimum diameter clustering problem; it is NP-hard to find a disk representation given only a graph-theoretic description of a unit disk graph [8].

However, just as in the rectangle partition problem, minimum diameter clustering may be reduced to a more basic graph-theoretic problem, maximum independent sets in bipartite graphs [3, 15]. Suppose that we know or can guess the two points p and q forming the diametral pair in a cluster. Then, all the other points of the cluster must be within the *lune* formed by intersecting the two circles with pq as radius, centered on p and q. The largest possible cluster within the lune is the maximum clique of the unit disk graph with disk centers inside this lune, or equivalently the maximum independent set of the complement G of this restricted unit disk graph. But G is bipartite: if the lune is bisected by line pq, then a point in G can be connected by an edge only to other points on the other side of the bisection line (Figure 3, right). Therefore, its maximum independent set may be found in polynomial time, as discussed in the previous section.

Based on this idea, one could test all $\Theta(n^2)$ pairs pq, find the bipartite graph derived from each pair and apply a bipartite matching algorithm to it, and return the best cluster found from all of these separate tests. However it is more efficient to use dynamic

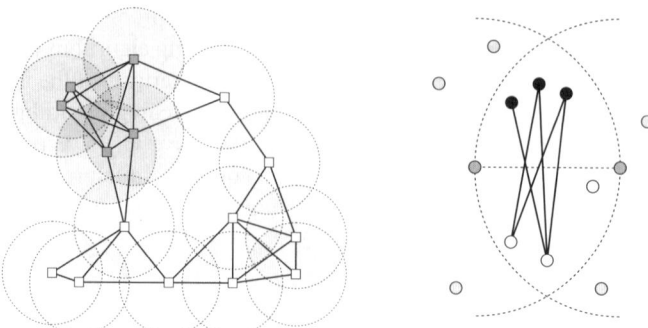

Fig. 3. Minimum diameter clustering. Left: a unit disk graph and its largest clique (the five darker circles in the upper left). Right: for any two specified points, the largest cluster having those two points as its diametral pair may be found as a maximum independent set of a bipartite graph in which the two sides of the bipartition are two halves of a lune defined by the two specified points.

graph algorithms to share work between multiple different matching problems [27]. Suppose we seek the size of the largest cluster for a given D. For each point p that could be an endpoint of a diametral pair, consider the lune defined by a segment of length D with p as one endpoint. If the defining segment rotates through an angle of 2π around p, the lune rotates with it, and the set of input points inside the lune undergoes a sequence of $O(n)$ discrete changes in which some point joins or leaves the set. After each change, one may update the maximum matching of the bipartite graph defined from the lune by a single alternating path search. Thus, the overall algorithm loops through all n possible choices of p, and performs a nested loop through the $O(n)$ set insertion and deletion operations defined by rotating a size-D lune around p. For each set update operation the algorithm performs an alternating path search to update a maximum matching and the maximum independent set in the bipartite graph defined from the lune, and when the nested loops terminate the algorithm returns the largest cluster found in each of these searches. As Aggarwal et al. [3] describe, each step of an alternating path search may be performed in logarithmic time with the aid of the circular hull data structure of Hershberger and Suri [40]. Therefore, each alternating path search takes time $O(n \log n)$, the sequence of $O(n)$ alternating path searches for a single choice of p takes time $O(n^2 \log n)$, and the overall clustering algorithm takes time $O(n^3 \log n)$ [27].

If k rather than D is given as input, the problem may be solved by a binary search among the $O(n^2)$ different distances defined by the input points, that checks for each distance whether it is the diameter of some k-point cluster. For this variant, the time may be further improved to $O(n \log n + k^2 n \log^2 k)$ by using a k-nearest-neighbor calculation to reduce the problem to $O(n/k)$ subproblems with $O(k)$ points per subproblem [27].

It appears to be open whether there exist polynomial time algorithms for solving the analogous minimum diameter clustering problems in higher dimensions (equivalently, finding maximum cliques in intersection graphs of unit balls in higher dimensions) or for finding a maximum clique in a unit disk graph when a geometric representation of the graph is not given.

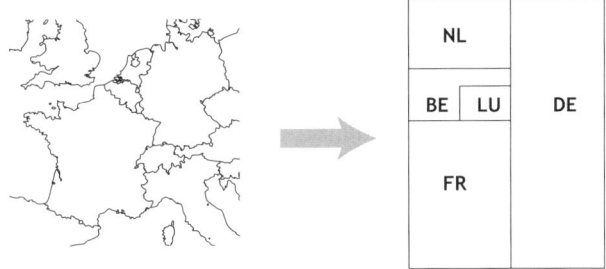

Fig. 4. A rectilinear cartogram of five countries in northwestern Europe. Map based on a CC-BY-SA licensed image by Brianski, Canuckguy, Zaparojdik, Madman2001, Roke, and Ssolbergj, online at http://commons.wikimedia.org/wiki/File:Blank_Map_of_Europe_-w_boundaries.svg.

5 Bend Minimization

A rectilinear cartogram [64,72,58] is a diagram in which geographic regions have been replaced by orthogonal polygons, with approximately the same shapes and in approximately the same positions with respect to each other as they hold geographically, but in which the areas of the regions have been distorted to represent numerical data about the regions unrelated to their physical areas. In introducing these diagrams in 1934, Raisz [64] wrote, "it should be emphasized that the statistical cartogram is not a map," and the stylization inherent in using orthogonal polygons helps perform this emphasis.

We have studied algorithms for constructing cartograms that can accomodate arbitrary area assignments [31] and in which the adjacencies between regions have desired orientations [30] but a more basic problem, constructing a cartogram in which the region shapes are as simple as possible, had already been solved in a different context, that of graph drawing [69]. Simple shapes such as rectangles aid the viewer in measuring and comparing areas and hence in understanding the data represented by a cartogram. We would like to find a cartogram that represents a given map with a minimum number of *bends* (corners interior to the boundary between two adjacent regions); for instance, in Figure 4 there is one bend, between Belgium and Luxembourg; this is optimal, as Luxembourg is surrounded by only three countries and hence must have at least one bend on its boundary. As Tamassia [69] shows, this problem of bend minimization can be solved in polynomial time by translating it into a network flow problem.

To form a flow problem that represents the bends of some given cartogram, construct a graph that has a single "circulation" vertex, a vertex for each region of the cartogram (including the exterior of the diagram as one of the regions), and a vertex for each junction where three or four regions of the cartogram meet (four regions at a junction may be needed to model places like the Four Corners in the southwest of the U.S. where four states meet, but five or more regions cannot meet at a single junction in an orthogonal cartogram). Include edges from the circulation vertex to each other vertex, between each two vertices that represent adjacent regions, and between two vertices that represent an incident junction-region pair. Label this graph with flow amounts, as follows: each junction vertex has four incoming units of flow from the circulation vertex, and sends one unit of flow to the vertices representing the regions in the four quadrants incident to the junction. For each bend in the cartogram, send one unit of flow from the region that has a convex vertex at the bend to the region that has a concave vertex. For each interior region having k junctions on its boundary, send $4 - 2k$ units of flow to the circulation vertex (or equivalently, if $4 - 2k < 0$, send $2k - 4$ units from the circulation vertex to the region vertex). And for an exterior region with k junctions on its boundary, send $2k + 4$ units of flow to the circulation vertex. The result can be shown to be a valid circulation: that is, for each vertex of the flow graph, the numbers of incoming and outgoing flow units are equal. For the junction vertices, this is clear because the four incoming units are balanced by the four quadrants into which a unit of flow is sent. An interior region with k junctions and no bends must form a rectangle, as a junction cannot form a concave corner; thus, it has four incident junction vertices that send a single unit of flow, and the remaining $k - 4$ incident junctions send two units of flow into the region, for a total of $2k - 4$ incoming units, balancing the flow to the circulation vertex. Each additional cave corner at a bend causes one unit of incoming flow from another region, but must be balanced by an additional convex corner; if that convex corner belongs to a bend, it provides a unit of outgoing flow, and if it belongs to a junction then that junction sends one fewer unit of incoming flow, in either case leading to the same total flow balance. A similar argument shows that the flow into and out of the exterior region vertex is also balanced, from which it follows that the flow must be balanced at the single remaining vertex, the circulation vertex.

Conversely, as Tamassia shows, one can assign costs and flow capacities to this network in order to force a minimum cost circulation to correspond to a minimum-bend drawing. Capacity constraints are needed to force the incoming flow to each junction vertex to be exactly four units; the edges between adjacent region vertices are assigned unit cost, and all other edges have zero cost. With these constraints and costs, the flow described above has cost equal to its number of bends. Conversely, any integer solution to the capacitated circulation problem can be translated into a drawing in which the total number of bends is equal to the cost of the circulation, so a cartogram with the minimum number of bends can be found from a minimum-cost circulation, which can be constructed in polynomial time. The flow graph is an *apex graph*: if one vertex, the circulation vertex, is removed, the rest is planar. Therefore, specialized techniques for finding flows in planar graphs may be used to speed up this algorithm [38, 43, 57].

6 Mesh Stripification

Stripification refers to the problem in computer graphics of partitioning a triangulated surface model of a three-dimensional object into *triangle strips*, sequences of triangles meeting edge-to-edge [5, 29, 34, 73]. Such a partition allows for fast rendering by requiring the coordinates of only one vertex per triangle to be transmitted to the graphics hardware; the other two vertex locations may be found from the previous triangle in the strip. Triangle strips aid in data compression of geometric models: a predicted location that aids in compressing the coordinates of each successive vertex may be found by extrapolation from the previous triangle in a strip [17]. Additionally, if a mesh can be

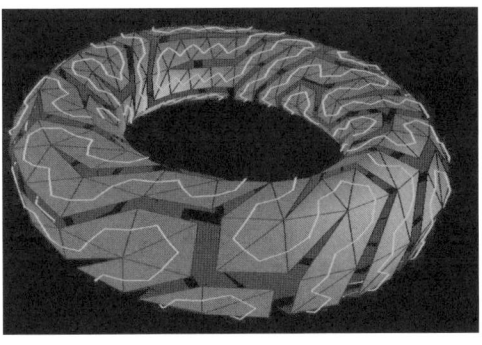 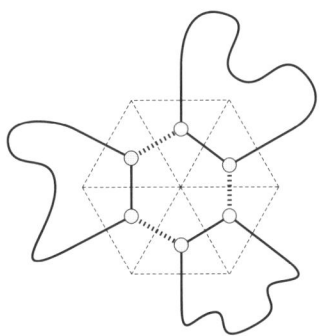

Fig. 5. Left: a single strip in a triangulated torus, rendered by M. Gopi, from [25]. Right: a local move at a vertex allows three cycles to be merged into a single cycle.

represented as a single cyclic triangle strip, as in Figure 5 (left), the structure of the strip may be used to cover the surface of the model with a space-filling curve [29], and contractions of the boundary edges of the strip can be used for topology-preserving simplifications of the model [20].

Finding such a strip may be viewed graph-theoretically, as finding a Hamiltonian cycle in the 3-regular dual graph of the triangulation, a graph with one vertex per triangle and one edge for each pair of adjacent triangles. However, this is problematic for two reasons. First, the Hamiltonian cycle problem is NP-complete and the known exponential-time algorithms for the problem [25, 44] are only capable of solving it within a reasonable time for models of at most a few hundred triangles. Second, and more importantly, not all triangulated models, even with spherical topology, have single triangle strips of this type. Tutte's counterexample to Tait's conjecture that 3-regular polyhedra are Hamiltonian [68, 70] dualizes to become a triangulated mesh that cannot be represented as a single cycle of triangles.

However, it is important to realize that the dual graph does not uniquely represent the geometry: we may change the triangulation, and hence change the graph, without changing the model's shape. The dual graph of any triangulated model is both 3-regular and bridgeless; therefore, by a theorem of Petersen [63], it has a perfect matching, which may be found efficiently [7]. The set of edges complementary to the matching forms a partition of the triangulation into a collection of disjoint cycles [29]; however, there may be more than one of these cycles. As we observed [29], in many cases a local move at a vertex of the triangulation, that swaps matched and unmatched edges connecting the triangles sharing that vertex, may reduce the number of cycles (Figure 5, right). If no such move is available, two adjacent triangles from two different cycles may be bisected across their shared edge, leading to a new triangulation of the same surface with the property that a cycle-reducing local move is available at the newly created vertex. By repeating this subdivision process, one eventually reaches a triangulation that covers a surface identical to the input model, but one that has a different dual graph than the input and that can be represented as a single strip. Although theoretically this could increase the total number of triangles by as much as a factor of $3/2$, in practice we saw at most a 2% increase.

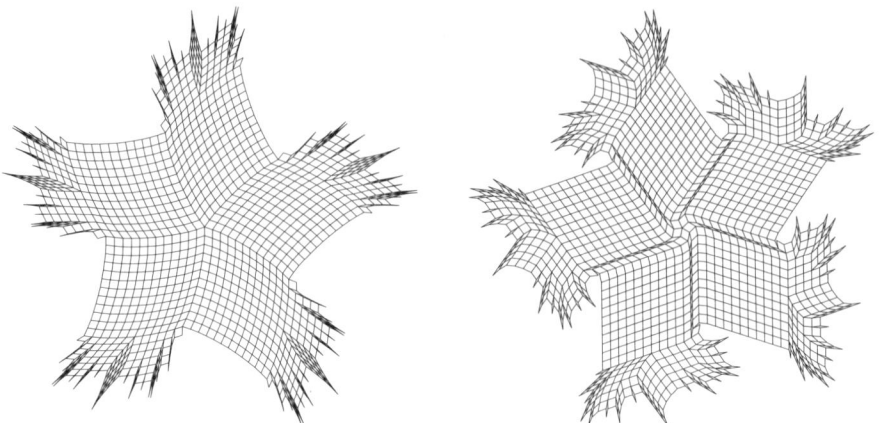

Fig. 6. Two drawings of the squaregraph dual to a 5-chromatic triangle-free circle graph described by Ageev [2], from [6]

7 Angle Optimization of Tilings

The planar dual of any arrangement of lines in the plane may be represented as a *tiling* of a convex subset of the plane by centrally symmetric convex polygons; this tiling has one polygon for each crossing point of the arrangement, which may be chosen to have unit-length edges perpendicular to the lines crossing at that point. De Bruijn [16] used this construction to form aperiodic tilings (including the rhombic Penrose tiling) from overlaid families of evenly-spaced parallel lines. More generally the dual of a *weak pseudoline arrangement* (a set of curves that are the image of a line under a homeomorphism of the plane, that have at most one point of intersection per pair of curves, and that cross each other at their intersection points), may be represented as a tiling of a nonconvex simple polygon by centrally-symmetric polygons, and every such tiling arises in this way [24, 28]; this provides a convenient method for visualizing squaregraphs, which are the duals of triangle-free hyperbolic line arrangements [6].

However, the tilings obtained directly from this construction may be hard to read due to having polygons with very sharp angles, as is true for instance for the tiling in Figure 6 (left). In unpublished work with Kevin Wortman [32], we considered the problem of finding a combinatorially equivalent tiling that maximizes the minimum angle in the tiling (the so-called angular resolution [9, 56]), such as the one in Figure 6 (right); we showed that this optimal tiling could be found by a parametric shortest path computation in an auxiliary graph derived from the input.

Our algorithm constructs a graph in which the edges have weights that are linear functions of a parameter λ that represents the angular resolution of the drawing. In a tiling of this type, one can define an equivalence relation on the sides of tiles in which opposite pairs of sides on the same side are equivalent; the equivalence classes form *zones* of line segments that are required to be parallel and to have the same length. Our graph has one vertex per zone, and an additional start vertex that has a zero-length edge to each other vertex; the distance from the start vertex to a zone's vertex represents the adjustment in angle for the segments in the zone between an initial tiling and the optimal tiling. The constraint that the angle between two sides of a polygon be at least λ can then be expressed by the existence of an edge between the vertices representing the zones containing the two sides, with length $\theta_i - \theta_j - \lambda$, where θ_i and θ_j are the angles formed by the two zones in the initial tiling. Similarly, the constraint that an interior angle be convex can be expressed by an edge with length $\pi + \theta_i - \theta_j$. With these vertices and edge lengths, it can be shown that there exists a tiling with angular resolution at least λ if and only if the result of substituting that value into each edge weight function is a graph with no negative cycles. Therefore, the optimal angular resolution is the largest value of λ giving no negative cycles. Due to the special form of the weights in the parametric graph (each weight is either constant or a constant minus λ) this parametric negative cycle detection problem can be solved in $O(n^3)$ time by an algorithm of Karp and Orlin [48]. The translation from the tiling angular resolution problem to the parametric negative cycle detection problem and back can be performed within the same time bound.

8 Metric Embedding into Stars

There has been a large amount of interest recently within the theoretical computer science community in problems of embedding unstructured metric spaces (which may be specified, for instance, as a distance matrix) into simpler and more highly constrained metrics, with low distortion [52]. Such embeddings may be used, for instance, in approximation algorithms: one can design an approximation algorithm for the constrained class of metrics, and apply it to arbitrary metrics, incurring the distortion of the embedding as a penalty factor in the approximation ratio of the algorithm. The construction of graph spanners [23] may also be seen as an instance of this type of problem: one wishes to approximate a metric describing the shortest paths in an arbitrary graph, by a more highly constrained metric of shortest paths in a sparse graph. Most work in this area has concentrated on finding embeddings that guarantee low but non-optimal distortion; however there has also been some work on finding the best possible embedding [33,37]. In particular, we describe here our work on finding optimal embeddings into star metrics, which (as in the angle optimization of the previous section) involves translating the problem into one of parametric negative cycle detection in an auxiliary graph.

Following [33], we define a *star metric* to be a metric space in which there exists a distinguished point h (the *hub*) of the metric) such that, for every two points p and q, h lies on a shortest path from p to q. Expressed algebraically: $d(p,q) = d(p,h) + d(h,q)$. Thus, in contrast to arbitrary n-point finite metric spaces which have $n(n-1)/2$ degrees of freedom, a star metric has only $n-1$ degrees of freedom. We wish to find a map from an arbitrary finite metric space (represented as a distance matrix D) to a star metric space (represented as a vector H of distances from each point to the hub) that minimizes the distortion of the embedding; we do not require that the hub be the image of any point in the input space. By scaling H, if necessary, we can assume without loss of generality that the map does not decrease any distance; thus, what we seek is a vector H satisfying the constraint that, for all p and q, $D[p,q] \leq H[p] + H[q]$, and minimizing the *dilation*

$$\delta = \max_{p,q} \frac{H[p] + H[q]}{D[p,q]}.$$

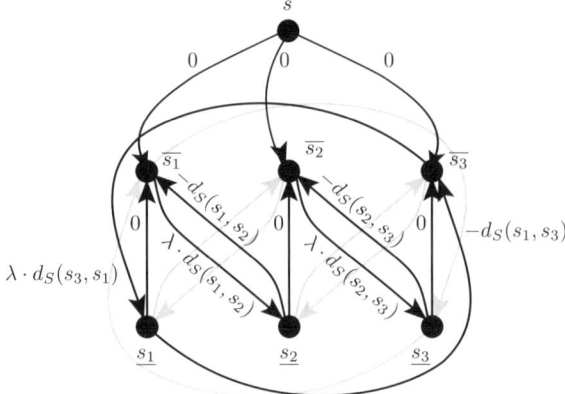

Fig. 7. The parametric network into which an optimal star embedding problem is translated. From [33].

This is a linear program: it may be rephrased as the problem of finding H and δ subject to the linear nondecreasing-distance constraints and the linear constraints that, for each p and q, $D[p,q]\delta - H[p] - H[q] \geq 0$, and, among all tuples of H and δ that satisfy the constraints, finding the tuple minimizing δ. However, it may be solved more efficiently, in strongly polynomial time, by transforming it into a parametric negative cycle detection problem.

Specifically, as our paper shows, we may find the optimal star embedding using an auxiliary graph defined from the input, having two vertices $\overline{p}$ and $\underline{p}$ for each metric space point p, as well as a special start vertex s. There is an edge with length zero from s to each vertex $\overline{p}$ and from each vertex $\underline{p}$ to the corresponding $\overline{p}$. In addition, for each pair of points p and q, there is an edge of length $-D[p,q]$ from $\underline{p}$ to $\overline{q}$ and another edge of length $\lambda D[p,q]$ from $\overline{p}$ to $\underline{q}$. As we show, the optimal dilation δ is the smallest value that can be assigned to λ such that the resulting graph has no negative cycles, and for this value of λ the distance $D[p]$ of p from the hub in the optimal embedding may be computed as half of the difference between the two distances from s to $\overline{p}$ and $\underline{p}$.

Thus, the star metric embedding problem may be solved optimally by a single parametric negative cycle detection calculation similar to that in the previous section. However, in this case the edge weight functions of the parametrized network no longer have the special form needed to apply the Karp–Orlin algorithm; instead, we developed a more general algorithm that solves any parametric negative cycle detection problem in strongly polynomial time. The basic idea of the algorithm is to use a matrix squaring method of Savage [66] to calculate the parametrized function representing the weights of paths between each pair of vertices; after i iterations of squaring a matrix whose entries represent these functions, they will correctly describe the lengths of the shortest path between each pair of vertices among paths constrained to have at most 2^i hops. After each matrix-squaring step, the entries of the matrix will not be linear functions but rather more general piecewise linear concave functions; we perform a binary search among the set of breakpoints of these functions to narrow down the range of values of λ within which the optimal value must lie, allowing us to simplify the functions to be linear once again. Each binary search step involves applying the Bellman-Ford algorithm to determine whether a specific value of λ causes the parametrized graph to have a negative cycle. The negative cycle detection algorithm resulting from this parametric search procedure has total running time $O(n^3 \log^2 n)$, which is therefore also the time for using this approach to solve the optimal star metric embedding problem.

9 Conclusions

As we have seen, a graph-theoretic point of view can be useful in many algorithmic problems that do not, initially, seem to have much to do with graphs. The graph problems occurring in the solutions of the geometry problems discussed here are mostly well-known and classical: they include the maximum independent set and maximum clique problems, maximum cardinality matching, shortest paths, and minimum cost flow. However, sometimes these problems occur in a somewhat different form than they have been studied elsewhere in the literature: the star metric embedding problem, for instance, required parametric negative cycle detection at a level of generality that had previously not been considered.

Special classes of graphs, and specialized algorithms that take advantage of these special classes, seem to be ubiquitous in this area. In the examples here, we have seen natural applications of trees, maximal outerplanar graphs, squaregraphs and kinggraphs, bipartite graphs, unit disk graphs, apex graphs, and 3-regular bridgeless graphs. The simple coloring algorithm for maximal outerplanar graphs described in the section on the art gallery theorem generalizes to chordal graphs (maximal outerplanar graphs are chordal) and to perfectly orderable graphs [14].

There are undoubtedly many more applications of graph-theoretic concepts in computational geometry waiting to be discovered, and it is important that the lines of communication remain open between the graph algorithm and computational geometry communities, so that computational geometers will know where to find the algorithms they need and so that graph algorithms researchers can focus their efforts on the problems and graph classes with the greatest benefit in geometric applications.

Acknowledgements. This work was supported in part by NSF grant 0830403 and by the Office of Naval Research under grant N00014-08-1-1015.

References

1. Agarwal, P.K., Efrat, A., Sharir, M.: Vertical decomposition of shallow levels in 3-dimensional arrangements and its applications. SIAM J. Comput. 29(3), 912–953 (2000)
2. Ageev, A.A.: A triangle-free circle graph with chromatic number 5. Discrete Mathematics 152, 295–298 (1996)
3. Aggarwal, A., Imai, H., Katoh, N., Suri, S.: Finding k points with minimum diameter and related problems. J. Algorithms 12(1), 38–56 (1991)
4. Ajtai, M., Chvátal, V., Newborn, M., Szemerédi, E.: Crossing-free subgraphs. In: Theory and Practice of Combinatorics. North-Holland Mathematics Studies, vol. 60, pp. 9–12 (1982)
5. Arkin, E.M., Held, M., Mitchell, J.S.B., Skiena, S.S.: Hamiltonian triangulations for fast rendering. The Visual Computer 12(9), 429–444 (1996)

6. Bandelt, H.-J., Chepoi, V., Eppstein, D.: Combinatorics and geometry of finite and infinite squaregraphs. Electronic preprint arxiv:0905.4537 (2009)
7. Biedl, T.C., Bose, P., Demaine, E.D., Lubiw, A.: Efficient algorithms for Petersen's matching theorem. J. Algorithms 38, 110–134 (2001)
8. Breu, H., Kirkpatrick, D.G.: Unit disk graph recognition is NP-hard. Computational Geometry Theory and Applications 9(1–2), 3–24 (1998)
9. Carlson, J., Eppstein, D.: Trees with convex faces and optimal angles. In: Kaufmann, M., Wagner, D. (eds.) GD 2006. LNCS, vol. 4372, pp. 77–88. Springer, Heidelberg (2007)
10. Chazelle, B.: Triangulating a simple polygon in linear time. Discrete & Computational Geometry 6(1), 485–524 (1991)
11. Cheng, Y., Iyengar, S.S., Kashyap, R.L.: A new method of image compression using irreducible covers of maximal rectangles. IEEE Trans. Software Engineering 14(5), 651–658 (1988)
12. Chepoi, V., Dragan, F., Vaxès, Y.: Center and diameter problem in planar quadrangulations and triangulations. In: Proc. 13th Annu. ACM–SIAM Symp. on Discrete Algorithms (SODA 2002), pp. 346–355. ACM Press, New York (2002)
13. Chvátal, V.: A combinatorial theorem in plane geometry. Journal of Combinatorial Theory, Series B 18, 39–41 (1975)
14. Chvátal, V.: Perfectly orderable graphs. In: Berge, C., Chvátal, V. (eds.) Topics in Perfect Graphs. Annals of Discrete Mathematics, vol. 21, pp. 63–68. North-Holland, Amsterdam (1984)
15. Clark, B.N., Colbourn, C.J., Johnson, D.S.: Unit disk graphs. Discrete Mathematics 86, 165–177 (1990)
16. de Bruijn, N.G.: Algebraic theory of Penrose's non-periodic tilings of the plane. Indagationes Mathematicae 43, 38–66 (1981)
17. Deering, M.: Geometry compression. In: Proc. 22nd Conf. Computer Graphics and Interactive Techniques (SIGGRAPH), pp. 13–20 (1995)
18. Dey, T.K.: Improved bounds for planar k-sets and related problems. Discrete & Computational Geometry 19(3), 373–382 (1998)
19. Di Battista, G., Eades, P., Tamassia, R., Tollis, I.G.: Graph Drawing: Algorithms for the Visualization of Graphs. Prentice-Hall, Englewood Cliffs (1998)
20. Díaz-Gutiérrez, P.: Using graph algorithms for geometry processing on surfaces. PhD thesis, Univ. of California, Irvine (2009)
21. Engel, K.: Optimal matrix-segmentation by rectangles. Discrete Applied Mathematics 157(9), 2015–2030 (2009)
22. Eppstein, D.: Geometric lower bounds for parametric matroid optimization. Discrete & Computational Geometry 20, 463–476 (1998)
23. Eppstein, D.: Spanning trees and spanners. In: Sack, J.-R., Urrutia, J. (eds.) Handbook of Computational Geometry, ch. 9, pp. 425–461. Elsevier, Amsterdam (2000)
24. Eppstein, D.: Algorithms for drawing media. In: Pach, J. (ed.) GD 2004. LNCS, vol. 3383, pp. 173–183. Springer, Heidelberg (2005)
25. Eppstein, D.: The traveling salesman problem for cubic graphs. J. Graph Algorithms and Applications 11(1), 61–81 (2007)
26. Eppstein, D.: Testing bipartiteness of geometric intersection graphs. ACM Trans. Algorithms 5(2), 15 (2009)
27. Eppstein, D., Erickson, J.: Iterated nearest neighbors and finding minimal polytopes. Discrete & Computational Geometry 11(3), 321–350 (1994)
28. Eppstein, D., Falmagne, J.-C., Ovchinnikov, S.: Media Theory. Springer, Heidelberg (2007)
29. Eppstein, D., Gopi, M.: Single-strip triangulation of manifolds with arbitrary topology. Eurographics Forum 23(3), 371–379 (2004); Proc. 25th Conf. Eur. Assoc. for Computer Graphics (EuroGraphics 2004)

30. Eppstein, D., Mumford, E.: Orientation-constrained rectangular layouts. In: Dehne, F., et al. (eds.) WADS 2009. LNCS, vol. 5664, pp. 266–277. Springer, Heidelberg (2009)
31. Eppstein, D., Mumford, E., Speckmann, B., Verbeek, K.A.B.: Area-universal rectangular layouts. In: Proc. 25th ACM Symp. Computational Geometry, pp. 267–276 (2009)
32. Eppstein, D., Wortman, K.: Optimal angular resolution for face-symmetric drawings. Electronic preprint arxiv:0907.5474 (2009)
33. Eppstein, D., Wortman, K.: Optimal embedding into star metrics. In: Dehne, F., et al. (eds.) WADS 2009. LNCS, vol. 5664, pp. 290–301. Springer, Heidelberg (2009)
34. Evans, F., Skiena, S.S., Varshney, A.: Optimizing triangle strips for fast rendering. In: Proc. 7th IEEE Conf. Visualization, pp. 319–326 (1996)
35. Ferrari, L., Sankar, P.V., Sklansky, J.: Minimal rectangular partitions of digitized blobs. Computer Vision, Graphics, and Image Processing 28(1), 58–71 (1984)
36. Fisk, S.: A short proof of Chvátal's watchman theorem. Journal of Combinatorial Theory, Series B 24(3), 374 (1978)
37. Fomin, F., Lokshtanov, D., Saurabh, S.: An exact algorithm for minimum distortion embedding. In: Paul, C., Habib, M. (eds.) WG 2009. LNCS, vol. 5911. Springer, Heidelberg (2009)
38. Garg, A., Tamassia, R.: A new minimum cost flow algorithm with applications to graph drawing. In: North, S.C. (ed.) GD 1996. LNCS, vol. 1190, pp. 201–216. Springer, Heidelberg (1997)
39. Hannenhalli, S., Hubbell, E., Lipshutz, R., Pevzner, P.A.: Combinatorial algorithms for design of DNA arrays. In: Chip Technology. Advances in Biochemical Engineering/Biotechnology, vol. 77, pp. 1–19. Springer, Heidelberg (2002)
40. Hershberger, J., Suri, S.: Finding tailored partitions. J. Algorithms 12(3), 431–463 (1991)
41. Hopcroft, J.E., Karp, R.M.: An $n^{5/2}$ algorithm for maximum matchings in bipartite graphs. SIAM J. Comput. 2(4), 225–231 (1973)
42. Imai, H., Asano, T.: Efficient algorithms for geometric graph search problems. SIAM J. Comput. 15, 478–494 (1986)
43. Imai, H., Iwano, K.: Efficient sequential and parallel algorithms for planar minimum cost flow. In: Asano, T., Imai, H., Ibaraki, T., Nishizeki, T. (eds.) SIGAL 1990. LNCS, vol. 450, pp. 21–30. Springer, Heidelberg (1990)
44. Iwama, K., Nakashima, T.: An improved exact algorithm for cubic graph TSP. In: Lin, G. (ed.) COCOON 2007. LNCS, vol. 4598, pp. 108–117. Springer, Heidelberg (2007)
45. Junger, M., Mutzel, P.: Graph Drawing Software. Springer, Heidelberg (2004)
46. Kahn, J., Klawe, M., Kleitman, D.: Traditional galleries require fewer watchmen. SIAM Journal on Algebraic and Discrete Methods 4(2), 194–206 (1983)
47. Kalinowski, T.: A dual of the rectangle-segmentation problem for binary matrices. Electronic J. Combinatorics 16(1), R89 (2009)
48. Karp, R.M., Orlin, J.B.: Parametric Shortest Path Algorithms with an Application to Cyclic Staffing. Technical Report OR 103-80, MIT Operations Research Center (1980)
49. Kőnig, D.: Gráfok és mátrixok. Matematikai és Fizikai Lapok 38, 116–119 (1931)
50. Leighton, T.: Complexity Issues in VLSI. Foundations of Computing Series. MIT Press, Cambridge (1983)
51. Li, G., Zhang, H.: A rectangular partition algorithm for planar self-assembly. In: Proc. IEEE/RSJ Int. Conf. Intelligent Robots and Systems, pp. 3213–3218 (2005)
52. Linial, N.: Finite metric spaces–combinatorics, geometry and algorithms. In: Proc. International Congress of Mathematicians, Beijing, vol. 3, pp. 573–586 (2002)
53. Lipski Jr., W.: Finding a Manhattan path and related problems. Networks 13, 399–409 (1983)
54. Lipski Jr., W.: An $O(n \log n)$ Manhattan path algorithm. Information Processing Letters 19, 99–102 (1984)
55. Lipski Jr., W., Lodi, E., Luccio, F., Mugnai, C., Pagli, L.: On two-dimensional data organization II. Fundamenta Informaticae 2, 245–260 (1979)

56. Malitz, S., Papakostas, A.: On the angular resolution of planar graphs. SIAM J. Discrete Mathematics 7, 172–183 (1994)
57. Miller, G.: Flow in planar graphs with multiple sources and sinks. SIAM J. Comput. 24(5), 1002–1017 (1995)
58. Mumford, E.: Drawing Graphs for Cartographic Applications. PhD thesis, Technische Universiteit Eindhoven (2008)
59. Nishizeki, T., Rahman, M.S.: Planar Graph Drawing. World Scientific, Singapore (2004)
60. Ohtsuki, T.: Minimum dissection of rectilinear regions. In: Proc. IEEE Int. Symp. Circuits and Systems, pp. 1210–1213 (1982)
61. O'Rourke, J.: Art Gallery Theorems and Algorithms. Oxford University Press, Oxford (1987)
62. Patel, K.: Computer-aided decomposition of geometric contours into standardized areas. Computer-Aided Design 9(3), 199–203 (1977)
63. Petersen, J.P.C.: Die theorie der regularen graphs. Acta Mathematica 15, 193–220 (1891)
64. Raisz, E.: The rectangular statistical cartogram. Geographical Review 24(2), 292–296 (1934)
65. Sack, J.-R., Urrutia, J.: Polygon decomposition. In: Sack, J.-R., Urrutia, J. (eds.) Handbook of Computational Geometry, pp. 491–518. Elsevier, Amsterdam (1999)
66. Savage, C.: Parallel Algorithms for Graph Theoretic Problems. PhD thesis, University of Illinois, Urbana-Champaign (1977)
67. Shamos, M.I., Hoey, D.: Closest-point problems. In: Proc. 16th IEEE Symp. Foundations of Computer Science, pp. 151–162 (1975)
68. Tait, P.G.: Listing's Topologie. Philosophical Magazine (5th ser.) 17, 30–46 (1884)
69. Tamassia, R.: On embedding a graph in the grid with the minimum number of bends. SIAM J. Comput. 16(3), 421–444 (1987)
70. Tutte, W.T.: On Hamiltonian circuits. Journal of the London Mathematical Society (2nd ser.) 21(2), 98–101 (1946); Reprinted in Scientific Papers, vol. II, pp. 85–98
71. Vaidya, P.M.: Geometry helps in matching. SIAM J. Comput. 18(6), 1201–1225 (1989)
72. van Kreveld, M., Speckmann, B.: On rectangular cartograms. Computational Geometry Theory and Applications 37(3), 175–187 (2007)
73. Xiang, X., Held, M., Mitchell, J.S.B.: Fast and effective stripification of polygonal surface models. In: Proc. Symp. Interactive 3D Graphics, pp. 71–78 (1999)

Algorithms for Classes of Graphs with Bounded Expansion

Zdeněk Dvořák* and Daniel Král'

Institute for Theoretical Computer Science (ITI),
Faculty of Mathematics and Physics, Charles University,
Malostranské náměstí 25, 118 00 Prague 1, Czech Republic
{rakdver,kral}@kam.mff.cuni.cz

Abstract. We overview algorithmic results for classes of sparse graphs emphasizing new developments in this area. We focus on recently introduced classes of graphs with bounded expansion and nowhere-dense graphs and relate algorithmic meta-theorems for these classes of graphs to their analogues for proper minor-closed classes of graphs, classes of graphs with bounded tree-width, locally bounded tree-width and locally excluding a minor.

1 Introduction

It is well-known that many hard problems are tractable for classes of graphs with restricted structure. A classical example of this phenomenon is the result of Courcelle [5] that every graph property that can be described by a monadic second order logic formula can solved in linear time for graphs with bounded tree-width. In particular, some NP-hard problems including graph coloring or vertex domination can be solved in linear time for graphs with bounded tree-width.

In this paper, we focus on algorithmic meta-theorems for classes of graphs whose structure is limited in some sense. To motivate the results we want to present, let us switch from the algorithmic to the structural point of view and look at the chromatic number. Graphs with bounded tree-width are degenerate and thus their chromatic number is bounded. Similarly, the chromatic number of planar graphs and more generally graphs that can be embedded on a fixed surface is bounded. Graphs with bounded tree-width, planar graphs and graphs that can be embedded on a fixed surface form minor-closed classes of graphs. A general experience says that most structural (and algorithmic) properties that hold both for classes of graphs with bounded tree-width and for classes of graphs embedded on a fixed surface are also true for classes of graphs excluding a fixed minor. The chromatic number being bounded is an example.

---

* Institute for Theoretical computer science is supported as project 1M0545 by Czech Ministry of Education. This work was also supported by the grant GACR 201/09/0197.

However, the chromatic number is not bounded only for classes of graphs excluding a fixed minor. Other classes of graphs with bounded chromatic number include graphs with bounded maximum degree or d-degenerate graphs (for fixed integer d). Since any graph is a minor of a cubic graph, these classes of graphs are clearly not minor-closed. A more tricky example of such class is the class of graphs obtained from planar graphs by adding at most two (not necessarily non-crossing) edges to each face. Still it turns out that some algorithmic properties of planar graphs also hold for the above mentioned graph classes.

Based on these examples, one would maybe guess that the only requirement we need is that the number of edges of a graph is bounded by the function linear in the number of its vertices, i.e., its average degree is bounded. This is however not sufficient since the average degree can be decreased by adding a sparse part to the graph (a set of isolated vertices being the simplest example, but one can easily think of more sophisticated ways which also preserve connectivity or other parameters). Similarly, the maximum average degree is not fine enough since subdividing each edge of an input graph decreases maximum average degree below four but most of the structural properties of an input graph are preserved. So, one needs a more robust structural parameter to capture the common properties of the above graph classes that are essential for the algorithmic results we are interested in.

A framework of classes of graphs with bounded expansion and a more general framework of classes of nowhere-dense graphs that have been introduced in a series of papers by Nešetřil and Ossona de Mendéz [19,20,21,22,23,24,25] seems to be the right one to be considered in this setting. In this paper, we will survey known structural and algorithmic results, including recent results of the authors and Thomas on decidability of first order logic properties, for classes of graphs with bounded expansion and classes of nowhere-dense graphs and relate these results to the earlier results for other graph classes. We will also provide proofs of some easier facts and those that are essential for algorithmic applications.

2 Definitions

In this section, we present definitions and notions important for our exposition. Though some of the notions we present are fairly standard, we decided to include them for the sake of completeness.

2.1 Graph Decompositions, Graph Minors

The graph minor project of Robertson and Seymour is one of the basic stones of modern graph theory. In this subsection, we recall some definitions and results from this area which we need in our further exposition.

A *tree-decomposition* of a graph G is a tree T whose vertices correspond to subsets of vertices of G, referred to as *bags*, and the following three properties hold:

1. every vertex of G is in at least one of the bags,

2. for every edge of G, there is a bag containing both its end-vertices, and
3. if a vertex v of G is contained in the bags associated with vertices u and u' of T, then v is contained in all the bags associated with the vertices on the path between u and u' in T.

The *order* of a tree-decomposition T is the maximum size of a bag associated to a vertex of T decreased by one. The *tree-width* $\mathrm{tw}(G)$ of a graph G is the minimum order of a tree-decomposition of G. Graphs with tree-width zero are edge-less and those with tree-width at most one are forests.

More restricted width parameter is the tree-depth. The *tree-depth* $\mathrm{td}(G)$ of a graph G is the minimum depth of a rooted tree T with the same vertex set as G that for every edge vv' of G, v is an ancestor of v' or v' is an ancestor of v. To fix our terminology, the *depth* of a rooted tree T is the maximum number of vertices in a path from the root to a vertex of T, e.g., the depth of the one-vertex rooted tree is one. Vertices on the path from a vertex v to the root are *ancestors* of v. Those vertices v' such that v is an ancestor of v' are *descendents* of v.

It is not hard to see that the tree-width of a graph G is bounded by its tree-depth decreased by one (consider the optimum tree T from the definition of the tree-depth, form bags as sets of vertices on the paths from the root to the leaves and associate them with vertices of a path in the order in which the leaves of T are visited during the depth-first search). On the other hand, the tree-depth of a graph is not bounded by any function of its tree-width (the tree-depth of the n-vertex path is $\lceil \log_2(n+1) \rceil$). In fact, the tree-depth of a graph G is proportional to the length ℓ of the longest path in G since $\lceil \log_2(\ell+2) \rceil \leq \mathrm{td}(G) \leq \binom{\ell+3}{2} - 1$. It also holds [3] that $\mathrm{td}(G) \leq (\mathrm{tw}(G)+1)\log_2(n+1)$ where n is the number of vertices of G.

An alternative definition of the tree-depth can be given by means of a vertex-coloring [26]. The *ranking number* of a graph G, as defined in [2], is the minimum number k of colors $1, \ldots, k$ needed to color the vertices of G such that any path joining two vertices of the same color contains a vertex with a bigger color. It can be shown that the tree-depth of a graph G is equal to its ranking number (to obtain the coloring, color the vertices of the tree T from the definition of the tree-depth based on their distance from the root, giving the root the largest color; to obtain a decomposition, proceed conversely).

A *minor* of a graph G is a graph obtained by deleting vertices and edges and contracting edges. Recall that the operation of contracting an edge e consists of removing e, identifying its end-vertices and deleting any loops and parallel edges that arise. A class $\mathcal{G}$ of graphs is *minor-closed* if every minor of a graph from $\mathcal{G}$ is also contained in $\mathcal{G}$. Examples of minor-closed classes of graphs include graphs embeddable in a fixed surface, graphs with tree-width at most k for an integer k, graphs with tree-depth at most k and many others. Proper minor-closed classes of graphs are *degenerate*, i.e., for every proper minor-closed class $\mathcal{G}$, there exists an integer k such that every graph $G \in \mathcal{G}$ is k-degenerate which means that G and each of its subgraphs has a vertex of degree at most k.

One of the main results in the graph minor series of Robertson and Seymour [30] asserts that every minor-closed class $\mathcal{G}$ of graphs has a finite list of obstructions, i.e., there exist $G_1, \ldots, G_k$ such that $G \in \mathcal{G}$ if and only if G does not contain any of the graphs $G_1, \ldots, G_k$ as a minor (these graphs are also called *obstructions*). E.g., the tree-width of a graph G is at most one if and only if G does not contain K_3 as a minor, and at most two if and only if G does not contain K_4 as a minor. The complete list containing four obstructions for graphs with tree-width at most three was given in [1]. A minor-closed class of graphs can contain graphs with arbitrary big tree-width (planar graphs being an example), but it is known that the tree-width of graphs in a minor-closed class $\mathcal{G}$ of graphs is bounded if and only if one of the obstructions for $\mathcal{G}$ is planar [29].

2.2 Local Parameters

First order logic graph properties are of localized nature as we discuss in Subsection 4.2. Because of this, graphs with locally restricted structure are important from the algorithmic point of view: classes of graphs with locally bounded tree-width were introduced by Eppstein [12] (using somewhat different notation) and classes of graphs locally excluding a minor were defined by Dawar, Grohe and Kreutzer [6].

Before we define these graph classes, we need to recall several definitions. If G is a graph and v is a vertex of G, then $N_d(v)$ is the *d-neighborhood* of v, i.e., the set of vertices of G at distance at most d from v. If A is a set of vertices of a graph G, then $G[A]$ is the subgraph of G *induced* by A, i.e., the subgraph with vertex set A that contains all edges of G with both end-vertices from A.

We say that a class $\mathcal{G}$ of graphs has *locally bounded tree-width* if there exists a function $f : \mathbb{N} \to \mathbb{N}$ such that the tree-width of $G[N_d(v)]$ is at most $f(d)$ for every graph $G \in \mathcal{G}$, every vertex v of G and every $d \geq 1$. Similarly, a class $\mathcal{G}$ of graphs *locally excludes a minor* if there exists an infinite sequence of graphs $H_1, H_2, \ldots$ such that for every graph $G \in \mathcal{G}$, every vertex v of G and every $d \geq 1$, the graph $G[N_d(v)]$ does not contain H_d as a minor.

Observe that every class of graphs with locally bounded tree-width locally excludes a minor. Similarly, every proper minor-closed class of graphs locally exclude a minor. We later define other locally restricted graph classes.

2.3 Grad and Expansion

We now present the framework of classes of graphs with bounded expansion and classes of nowhere-dense graph which was introduced by Nešetřil and Ossona de Mendéz in [24]. An *r-shallow minor* of a graph G is a graph obtained from G by removing some vertices and edges of G and contracting several vertex-disjoint subgraphs of radius at most r. Recall that the *radius* of a graph is the minimum r such that $G = G[N_r(v)]$ for some vertex v of G, i.e., every vertex of G is at distance at most r from v. If $\mathcal{G}$ is a class of graphs, then $\mathcal{G} \triangledown r$ is the class of all r-shallow minors of graphs contained in $\mathcal{G}$.

The edge-density of a graph G is $||G||/|G|$, i.e., the ratio of the number of edges of G and the number of its vertices. The *grad* $\triangledown_r(G)$ *with rank r* (greatest reduced average density) of a graph G is the maximum edge-density of an r-shallow minor of G. Observe that if $d \geq 2\triangledown_0(G)$, then a graph G is d-degenerate. A class $\mathcal{G}$ of graphs has *bounded expansion* if there exists a function $f : \mathbb{N} \to \mathbb{N}$ such that $\triangledown_r(G) \leq f(r)$ for every graph $G \in \mathcal{G}$ and every $r \geq 1$.

Let us give few examples of classes of graphs with bounded expansion. Since every proper minor-closed class $\mathcal{G}$ of graphs is degenerate, the grads of all ranks of graphs contained in $\mathcal{G}$ are bounded by a constant. Hence, all proper minor-closed classes of graphs have bounded expansion. Another example of a class of graphs with bounded expansion are graphs with bounded maximum degree: if G has maximum degree Δ, then $\triangledown_r(G) \leq \Delta(\Delta - 1)^r/2$. Hence, classes of graphs with bounded maximum degree have also bounded expansion. Another example is the class of graphs that can be embedded to the plane in such a way that each edge is crossed by at most one other edge; this class contains graphs with arbitrary large degrees and is not minor-closed. Other examples of classes of graphs with bounded expansion can be found in [27].

Analogously to already introduced definitions, a class $\mathcal{G}$ of graphs has *locally bounded expansion* if there exists a function $f : \mathbb{N} \times \mathbb{N} \to \mathbb{N}$ such that $\triangledown_r(G[N_d(v)]) \leq f(r, d)$ for every graph $G \in \mathcal{G}$, every vertex v of G and any two integers r and d. It can be shown that every class $\mathcal{G}$ of graphs with locally bounded expansion has almost bounded expansion in the following sense: for every $\varepsilon > 0$, there exist functions $f_r(n) : \mathbb{N} \to \mathbb{N}$ such that $f_r(n) \in O(n^\varepsilon)$ for every $r = 0, 1, \ldots$ and $\triangledown_r(G) \leq f_r(n)$ for every n-vertex graph $G \in \mathcal{G}$.

This leads us to the definition of nowhere-dense graphs. If $\mathcal{G}$ is a class of graphs and f a real-valued function on the set of all graphs, then

$$\limsup_{G \in \mathcal{G}} f(G)$$

is the supremum of all reals α such that there exists an infinite sequence of distinct graphs $G_1, G_2, \ldots$ from $\mathcal{G}$ with $\alpha = \lim_{k \to \infty} f(G_k)$. The trichotomy theorem of Nešetřil and Ossona de Mendéz [24] asserts the following:

Theorem 1. *For every infinite class $\mathcal{G}$ of graphs, the following holds:*

$$\lim_{r \to \infty} \limsup_{G \in \mathcal{G} \triangledown r} \frac{\log ||G||}{\log |G|} \in \{0, 1, 2\} \ . \tag{1}$$

Let us give the proof of this (at the first sight very surprising) theorem since it gives more insight into the structure of classes of graphs achieving each of the values of the limit.

Proof. If there exists a constant C such that every graph in $\mathcal{G}$ has at most C edges, then $\lim_{k \to \infty} \frac{\log ||G_k||}{\log |G_k|} = 0$ for every infinite sequence $G_1, G_2, \ldots$ of distinct graphs from $\mathcal{G} \triangledown r$ (the number of vertices of the graphs G_i must grow to the infinity but the number of their edges is bounded by C).

If there is no constant C bounding the number of edges of every graph in $\mathcal{G}$, proceed as follows: choose G_1 to be K_2, clearly, $K_2 \in \mathcal{G} \triangledown 0$. If $G_1, G_2, \ldots, G_k$ have already been fixed, choose G_k to be any graph of $\mathcal{G} \triangledown 0$ containing more edges than G_{k-1} and subject to this with the minimum number of vertices. Observe that $|G_k| \leq 2||G_k||$ for every k (otherwise, G_k contains an isolated vertex which contradicts our choice of G_k). It follows that

$$\liminf_{k \to \infty} \frac{\log ||G_k||}{\log |G_k|} \geq 1 \ .$$

Since $\mathcal{G} \triangledown 0 \subseteq \mathcal{G} \triangledown 1 \subseteq \cdots$, it follows that if the limit in (1) is not equal to zero, then the limit is at least one.

Assume now that the limit given in (1) is greater than 1 for $\mathcal{G}$. Hence, there exist r, $\varepsilon > 0$ and an infinite sequence of graphs $G_1, G_2, \ldots \in \mathcal{G} \triangledown r$ such that $||G_k|| \geq |G_k|^{1+\varepsilon}$. We now apply the following result from [8, Lemma 3.13]: for every $\varepsilon > 0$, there exist an integer d and $\delta > 0$ such that every n-vertex graph with average degree n^ε contains K_{n^δ} as a d-shallow minor. It follows that the class $\mathcal{G} \triangledown rd$ contains complete graphs of arbitrary order and the limit (1) is at least two. Since $||G|| \leq |G|^2/2$ for every graph G, the limit in (1) is at most two and the proof of the theorem is completed.

The classes $\mathcal{G}$ of graphs with the limit (1) equal to 0 or 1 are called classes of *nowhere-dense* graphs. It follows that every class of graphs with locally bounded expansion is a class of nowhere-dense graphs.

We finish this section with Figure 1 where the reader can find inclusions between graph classes we have introduced in this section.

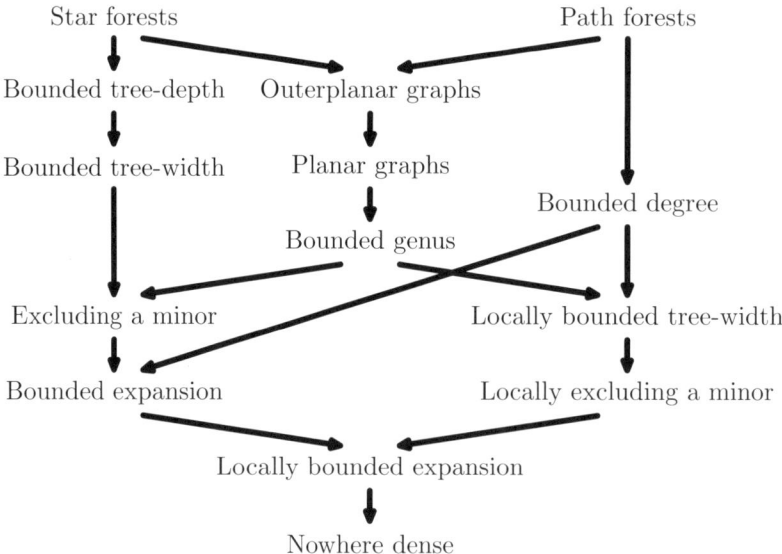

Fig. 1. Overview of inclusions between various graph classes

3 Structural Properties

In this section, we will introduce the notion of fraternal augmentations of orientations of graphs as defined by Nešetřil and Ossona de Mendéz in [20]. Since all the algorithms for classes of graphs with bounded expansion as well as classes of nowhere dense graphs are based on this notion, we decided to present it with full detail. The proofs follow the lines of those given in [20].

3.1 Orientations with Small In-Degree

Every graph $\mathcal{G}$ admits an orientation with maximum in-degree at most $\triangledown_0(G)$. The purpose is to develop a technique of augmenting orientations with small in-degrees preserving the fact that the grads remain small. To achieve this, we will need the following definition and lemma. Two vertices v and v' of a digraph[1] $\vec{G}$ are *k-reachable* for an integer k, if there exists a vertex w and oriented paths P and P' from v and v' to w of lengths ℓ and ℓ', respectively, such that $\ell + \ell' \leq k$. The paths P and P' form an (ℓ, ℓ')-*wedge* between v and v'.

We now state a key lemma for our further considerations.

Lemma 1. *There exist polynomials $P_k(x, y)$, $k = 1, 2, \ldots$, with the following properties. Let G be a graph and $\vec{G}$ an orientation of G with maximum in-degree Δ^-. If H_k is the graph with the vertex set of G and two vertices adjacent if they are k-reachable, then*

$$\triangledown_0(H_k) \leq P_k(\Delta^-, \triangledown_{k-1}(G)) .$$

Proof. The proof proceeds by induction on k. If $k = 1$, then the graphs H_1 and G are the same (observe that two vertices are 1-reachable if and only if they are adjacent in $\vec{G}$). Hence, $\triangledown_0(H_1) = \triangledown_0(G)$ and we can set $P_1(x, y) = y$.

Assume now that $k > 1$. Consider a proper vertex coloring of H_{k-1} with $2P_{k-1}(\Delta^-, \triangledown_{k-2}(G)) + 1$ colors (the existence of this coloring follows from the fact that H_{k-1} is $2P_{k-1}(\Delta^-, \triangledown_{k-2}(G))$-degenerate). Color now the arcs uv of $\vec{G}$ with pairs $[\alpha, \beta]$ of colors where α is the color of v and the color β is chosen in such a way that no two arcs coming to the same vertex have the same color. Since Δ^- choices of colors β suffice at each vertex, the arcs of $\vec{G}$ can be colored with at most $(2P_{k-1}(\Delta^-, \triangledown_{k-2}(G)) + 1)\Delta^-$ colors. Let K be this number of colors.

A *type* of an (ℓ, ℓ')-wedge formed by paths P and P' of lengths ℓ and ℓ' is the pair of two sequences of lengths ℓ and ℓ' formed by the colors of the arcs of P and P', respectively. Fix two integers ℓ and ℓ' such that $\ell + \ell' = k$ and $0 < \ell \leq \ell'$. Observe that the type of any (ℓ, ℓ')-wedge contains mutually distinct colors since the vertices with incoming arcs in an (ℓ, ℓ')-wedge have mutually distinct colors (they are $(k-1)$-reachable).

---

[1] We allow digraphs to have parallel arcs oriented in the opposite way. If we want to exclude parallel arcs, we will say that a digraph is simple.

Fix now two sequences σ and σ' of arc colors with lengths ℓ and ℓ' such that $\vec{G}$ contains an (ℓ, ℓ')-wedge of type $[\sigma, \sigma']$. Let F be the set of all arcs contained in an oriented path whose arcs are colored with the colors as in σ (respecting the order of the colors) and F' the set of all arcs contained in an oriented path whose arcs are colored as in σ'. Finally, F'' is the set of the arcs of F' that do not have the last color of σ'.

Now consider two paths P and Q of lengths ℓ with arc colors as in σ. Since the colors of all the $\ell+1$ vertices of P are mutually distinct as well as the colors of the $\ell+1$ vertices of Q and no vertex has two incoming arcs with the same color, P and Q are either vertex-disjoint or $P \cap Q$ is the initial sequence of both the paths. Hence, the arcs of F form (vertex-disjoint) out-branchings of depth ℓ in $\vec{G}$. The analogous reasoning also applies to F' and thus the arcs of F' form out-branchings of depth ℓ'.

Consider now two paths P and P' of lengths ℓ and ℓ' with arc colors as in σ and σ', respectively. Since only the pair of the first vertices of P and P' or the pair of last vertices (or both these pairs) can have the same color, either P and P' are vertex-disjoint, or they share their first vertices, or they share their last vertices, or they share both their first and last vertices. Hence, $F \cup F''$ form out-branchings rooted at their original vertices.

Let $\vec{G'}$ be the graph obtained from $\vec{G}$ by removing vertices not incident with arcs of $F \cup F''$ and contracting the out-branchings of $F \cup F''$. Since every leaf of any out-branching of $F \cup F''$ is at distance at most $\max\{\ell, \ell'-1\} \leq k-1$, the graph $\vec{G'}$ is a $(k-1)$-shallow minor of G (after disregarding the orientations of its arcs). If v and v' are k-reachable because of an (ℓ, ℓ')-wedge of type $[\sigma, \sigma']$, then v and v' are roots of out-branchings in $F \cup F''$ and they are adjacent after contracting these out-branchings (through the arc with the last color in σ'). We conclude that the edges between vertices v and v' that are k-reachable because of an (ℓ, ℓ')-wedge of type $[\sigma, \sigma']$ can be oriented in such a way that the in-degree of any vertex is at most $\nabla_{k-1}(G)$.

Ranging through all choices of $\ell + \ell' = k$ with $\ell > 0$ and $\ell' > 0$ and all choices of σ and σ', we obtain an upper bound of $2(k-1)K^k \nabla_{k-1}(G)$ on the number of incoming arcs added to H_k. If $\ell = 0$ and $\ell' = k$, then we just orient the new edges based on the direction of the paths they correspond to which increases the in-degree of each vertex by at most $(\Delta^-)^k$. Taking into account the edges present in H_{k-1}, we obtain that H_k has an orientation of its edges with maximum in-degree at most

$$\nabla_0(H_{k-1}) + (\Delta^-)^k + 2(k-1)K^k \nabla_{k-1}(G)$$

which is bounded by

$$2(k-1)((2P_{k-1}(\Delta^-, \nabla_{k-1}(G)) + 2)(\Delta^- + 1))^k \nabla_{k-1}(G).$$

The sought polynomial $P_k(x, y)$ can be set to be equal to $4(k-1)((P_{k-1}(x, y) + 1)(x+1))^k y$.

We now define a crucial notion of transitive fraternal augmentations. If $\vec{G}$ is a simple digraph, then the *transitive fraternal augmentation* of $\vec{G}$ is a simple digraph obtained from $\vec{G}$ by adding the following arcs:

1. **transitive arcs**: if uv and vw are arcs of $\vec{G}$, then the arc uw is added unless $\vec{G}$ already contains the arc uw or the arc wu.
2. **fraternal arcs**: if uv and $u'v$ are arcs of $\vec{G}$, then the arc uu' or the arc $u'u$ is added unless $\vec{G}$ already contains the arc uu' or the arc $u'u$.

Our aim is to add fraternal arcs (where it is possible to make a choice which arc to add) in such a way that the maximum in-degree of $\vec{G}$ does not increase significantly. To choose the fraternal arcs, we can apply Lemma 1. However, we would like to iterate the process and thus we need to have a bound on grads of the transitive fraternal augmentation. Such a bound is given in the following theorem from [20]:

Theorem 2. *There exists polynomial $Q_1(x,y), Q_2(x,y), \ldots$ with the following properties. Let G be a graph and $\vec{G}$ an orientation of G with maximum in-degree Δ^-. If H is the graph containing all the edges of the transitive fraternal augmentation of $\vec{G}$, then the following holds for every $r \geq 1$:*

$$\nabla_r(H) \leq Q_r(\Delta^-, \nabla_{2r+1}(G))$$

Proof. Let $V_1, \ldots, V_n$ be subsets of vertices of H such that the radius of $H[V_i]$ is at most r for every $i = 1, \ldots, n$. Let v_i be the center of $H[V_i]$. Consider the shortest distance tree T_i in $H[V_i]$ rooted at v_i and orient the edges of T_i in the direction from v_i. We now modify the simple digraph $\vec{G}$ in another digraph $\vec{G}'$ which need not be simple. If the arc uw of T_i corresponds to an edge of G, add the arc uw to $\vec{G}$. If the arc uw is a transitive edge corresponding to arcs wv and vu, add the arc vw. If the arc uw is a transitive edge corresponding to arcs uv and vw, no action is required. Finally, if the arc uw is a fraternal edge corresponding to arcs uv and wv, add arcs uv and vw.

Observe that the maximum in-degree of $\vec{G}'$ is at most $2\Delta^- + 1$: if an arc leading to v is added because of an arc uw of some T_i, then both u and w are in the same V_i and uv or wv is an arc of $\vec{G}$. Since the sets V_i are disjoint, at most Δ^- arcs leading to v can be added. The extra one in the estimate corresponds to an arc added because of the tree containing v.

If the subgraphs $H[V_i]$ and $H[V_j]$ are joined by an edge in H, then v_i and v_j are $2(r+1)$-reachable in $\vec{G}'$. In particular, the subgraph H' obtained from H by removing the vertices not contained in $V_1 \cup \cdots \cup V_n$ and contracting the subgraphs $H[V_i]$ is a subgraph of the graph H_{2r+2} as defined in Lemma 1. Consequently, $\nabla_0(H') \leq P_{2r+2}(2\Delta^-, \nabla_{2r+1}(G))$ and thus $\nabla_r(H) \leq P_{2(r+1)}(2\Delta^-, \nabla_{2r+1}(G))$.

Theorem 2 guarantees us that there is a choice of fraternal arcs to be added such that the maximum in-degree of the transitive fraternal augmentation of $\vec{G}$ is at most $(\Delta^-)^2 + 2Q_0(\Delta^-, \nabla_1(G))$. Moreover, since the grads of the transitive fraternal augmentations are bounded by polynomials in Δ^- and grads of G, the process can be iterated. In particular, we obtain the following corollaries. Note that since the existence of an orientation with small maximum in-degree is guaranteed by the fact that the grad with rank 0 is bounded, we can use a greedy algorithm to construct it.

Corollary 1. *Let $\mathcal{G}$ be a class of graphs with bounded expansion. There exist $\Delta_0, \Delta_1, \ldots$ such that every graph $G \in \mathcal{G}$ has an orientation $\vec{G}_0$ with maximum in-degree Δ_0 and $\vec{G}_i$ has a transitive fraternal augmentation $\vec{G}_{i+1}$ with maximum in-degree Δ_{i+1} for every $i \geq 0$. Moreover, for every i, $\vec{G}_i$ can be computed in time linear in the number of vertices of G.*

Corollary 2. *Let $\mathcal{G}$ be a class of nowhere dense graphs. For every $\varepsilon > 0$, there exist functions $f_i : \mathbb{N} \to \mathbb{N}$, $i = 0, 1, \ldots$, $f_i(n) \in O(n^\varepsilon)$, such that every n-vertex graph $G \in \mathcal{G}$ has an orientation $\vec{G}_0$ with maximum in-degree $f_0(n)$ and $\vec{G}_i$ has a transitive fraternal augmentation $\vec{G}_{i+1}$ with maximum in-degree $f_{i+1}(n)$ for every $i \geq 0$. Moreover, for every i, $\vec{G}_i$ can be computed in time $O(n^{1+\varepsilon})$ in the number of vertices of G.*

3.2 Low Tree-Width and Low-Tree-Depth Coloring

We now mention one structural result on a special type of vertex colorings of graphs which is important for algorithmic applications and is of independent interest. In [7], DeVos et al. established the existence of low tree-width color with bounded number of colors for proper minor-closed classes of graphs:

Theorem 3. *Let $\mathcal{G}$ be a proper minor-closed class of graphs. For every k, there exists K such that every graph $G \in \mathcal{G}$ has a vertex coloring with K colors such that any k' color classes, $1 \leq k' \leq k$, induce a subgraph of G with tree-width at most $k' - 1$.*

Theorem 3 was strengthened by Nešetřil and Ossona de Mendéz in [20] in two directions: first, the result holds for more general graph classes and second it guarantees the existence of low tree-depth colorings.

Theorem 4. *Let $\mathcal{G}$ be a class of graphs with bounded expansion. For every k, there exists K such that ever graph $G \in \mathcal{G}$ has a vertex coloring with K colors such that any k' color classes, $1 \leq k' \leq k$, induce a subgraph of G with tree-depth at most k'. Moreover, such a coloring can be constructed in linear time for any graph G from $\mathcal{G}$.*

Theorem 4 is implied by the following lemma; we provide its short proof for completeness.

Lemma 2. *For every $p \geq 1$ and $d \geq 1$, the following holds: if $\vec{G}_0$ is an orientation of G, $\vec{G}_1, \vec{G}_2, \ldots$ a series of its transitive fraternal augmentations and H a connected subgraph of G with tree-depth at most d, then $\vec{G}_{3pd}[V(H)]$ either contains a clique of order p or an out-branching T of depth at most p such that*

1. the end-vertices of every edge of H are joined by a directed path in T, and
2. if two vertices u and u' are joined by a directed path in T, then $\vec{G}_{3pd}[V(H)]$ contains the arc uu' or the arc $u'u$.

Proof. Fix p and let the proof proceed by induction on d. If $d = 1$, H is a single vertex and the claim clearly holds. Assume that $d > 1$ and let v be a vertex of H such that the tree-depth of each component of $H \setminus v$ is at most $d-1$. Let $V_1, \ldots, V_k$ be the vertex sets of the components of $H \setminus v$. By induction, $\vec{G}_{3pd-3p}[V_i]$ either contains a clique of order p or an out-branching T_i such that any edge of $G[V_i]$ joins a vertex with one of its ancestors and directed paths in T_i give rise to arcs in $\vec{G}_{3pd-3p}[V_i]$. If $\vec{G}_{3pd-3p}[V_i]$ for some i contains a clique of order p, then so does $\vec{G}_{3pd}[V(H)]$. Hence, we assume the existence of out-branchings T_i for all $i = 1, \ldots, k$.

Let r_i be the root vertex of T_i, $i = 1, \ldots, k$. We claim that r_i and v are adjacent in $\vec{G}_{3pd-2p-1}$: since H is connected, v is adjacent to one of the descendants of r_i in T_i, say w. Let $r_i w_1 \ldots w_\ell$ be the oriented path in T_i from r_i to $w = w_\ell$. Since the depth of T_i is at most p, $\ell \leq p-1$. Applying the transitive or the fraternal rule (depending on the orientation of the arc between v and w_ℓ), we obtain that $\vec{G}_{3pd-3p+1}[V_i]$ contains an arc between v and $w_{\ell-1}$. Repeating the argument, we get that the vertices v and r_i are adjacent in $\vec{G}_{3pd-3p+p-1}[V_i] = \vec{G}_{3pd-2p-1}$. Observe that we have actually proven that if v is adjacent to a vertex u of an out-branching T_i in $\vec{G}_{3pd-3p}$, then v is adjacent in $\vec{G}_{3pd-2p-1}$ to all the vertices on the path from r_i to u.

Let q be the first index such that the in-degree of v is the same in $\vec{G}_{3pd-2p-1+q}$ and $\vec{G}_{3pd-2p-1+q+1}$. If $q \geq p$, then the in-degree of v in $\vec{G}_{3pd-2p-1+p}$ is at least p and thus $\vec{G}_{3pd-p}$ contains a clique of order p (on the in-neighbors of v in $\vec{G}_{3pd-p-1}$). Consequently, $\vec{G}_{3pd}[V(H)]$ contains a clique of order p. Hence, we can assume that $q \leq p-1$.

Let W be the set of vertices w of H such that $\vec{G}_{3pd-2p-1+q}$ contains the arc wv and all the vertices on the path from the r_i to w in the out-branching T_i containing w are in-neighbors of v. Since $\vec{G}_{3pd-2p+q}$ contains an arc between any two vertices of W by the fraternity rule, $\vec{G}_{3pd-2p+q}[W \cup \{v\}]$ contains a directed Hamilton path, say $w_1, \ldots, w_\ell$. Observe that $w_\ell = v$ because of the choice of W.

Let $T'_1, \ldots, T'_{k'}$ be the out-branchings obtained from T_i by removing the vertices contained in W and let $r'_1, \ldots, r'_{k'}$ be their roots. Consider now the out-branching T in $\vec{G}_{3pd-2p+q}$ formed by the path $w_1 \ldots w_\ell$, the out-branchings $T'_1, \ldots, T'_{k'}$ and the arcs $w_i r'_j$ for $j = 1, \ldots, k'$ where i is the maximum index such that $\vec{G}_{3pd-2p+q}$ contains the arc $w_i r'_j$. Such an index i must exist since either r'_j is a root of one of the out-branchings $T_1, \ldots, T_k$ and thus $\vec{G}_{3pd-2p+q}$ contains the arc $w_\ell r'_j$ or W contains the in-neighbor of r'_j in one of the out-branchings $T_1, \ldots, T_k$. Hence, T is an out-branching contained in $\vec{G}_{3pd-2p+q}$.

We now verify that the end-vertices of every edge uu' of H are joined by a directed path in T. If $u = v$, then either $u' \in W$ (and thus the existence of the path follows) or u' is contained in one of the out-branchings $T'_1, \ldots, T'_{k'}$, say

T_j'. Since $r_j' \notin W$, T contains the arc vr_j' (here, we use that all the vertices between the root of T_i and a vertex of T_i adjacent to v are also adjacent to v in $\vec{G}_{3pd-2p-1} \subseteq \vec{G}_{3pd-2p+q}$). The case $u' = v$ is symmetric and thus we can assume that neither u nor u' is v.

If neither u nor u' is contained in W, then, by induction, they are contained in the same out-branching T_j' and are joined by a directed path in T. If both u and u' are contained in W, then they are clearly joined by a directed path in T since they are both contained in the path $w_1 \ldots w_\ell$. It remains to consider the case when $u \in W$ and $u' \notin W$. Let m the index such that $w_m = u$. Since u and u' are adjacent in G, they are contained in the same out-branching T_i. Further assume that u' is contained in an out-branching T_j'. By induction, $\vec{G}_{3pd-3p}$ contains either the arc ur_j' or the arc $r_j'u$. If the arc ur_j' is present in $\vec{G}_{3pd-3p}$, then r_j' is adjacent to a vertex $w_{m'}$ with $m' \geq m$ in T. If the arc $r_j'u$ is present in $\vec{G}_{3pd-3p}$, then $\vec{G}_{3pd-2p+q}$ contains an arc between r_j' and v since $\vec{G}_{3pd-2p-1+q}$ contains the arc uv. If $\vec{G}_{3pd-2p+q}$ contained the arc $r_j'v$, the choice of q would imply that $\vec{G}_{3pd-2p-1+q}$ also contained the arc $r_j'v$ which would imply that r_j' should have been included in W. Otherwise, $\vec{G}_{3pd-2p+q}$ contains the arc vr_j', thus the arc vr_j' is also contained in T and u and u' are joined by a directed path in T.

We have shown that the out-branching T satisfies that any two end-vertices of an edge of H are joined by a directed path in T. Since $q \leq p - 1$, T is an out-branching in $\vec{G}_{3pd-p}$. We claim that if $u_0, \ldots, u_m$ is a directed path in $\vec{G}_{3pd-p}$, then the vertices $u_0, \ldots, u_m$ form a clique in $\vec{G}_{3pd-p+m}$. Proceed by induction on m: if $m = 1$, there is nothing to prove. Otherwise, $\vec{G}_{3pd-p+m-1}$ contains a clique on the vertices $u_1, \ldots, u_m$. By the fraternity or transitivity rule, $\vec{G}_{3pd-p+m}$ contains an arc between u_0 and each of the vertices $u_1, \ldots, u_m$. Hence, the vertices $u_0, \ldots, u_m$ form a clique in $\vec{G}_{3pd-p+m}$. We conclude that if the depth of T is at least p, $\vec{G}_{3pd}[V(T)] = \vec{G}_{3pd}[V(H)]$ contains a clique of order p, and if the depth of T is less than p, then any two vertices joined by a directed path in T are adjacent in $\vec{G}_{3pd}$. The proof of the lemma is now finished.

4 Testing Graph Properties

In the final section of the paper, we want to focus on meta-algorithmic results for classes of graphs with restricted structure. Let us remark that the results we present in this section readily translate to relational structures by considering the concept of Gaifman graph. If R is a relational structure with a domain D, then the *Gaifman graph* of R is the graph with vertex set D where two distinct elements x and y of D are joined by an edge if R contains a relation including both x and y. For instance, if a graph G is viewed as a binary relational structure, then the Gaifman graph of G is G itself. Graph concepts we have introduced translate to relational structures by considering corresponding Gaifman graphs; e.g., the class of relational structures has bounded expansion, if the class of their Gaifman graphs has bounded expansion. The results we present further also hold for corresponding classes of relational structures under the assumption that the vocabulary is finite, i.e., the number of different types of relations is finite.

4.1 Σ_1-Properties

Analogously to Σ_1-formulas, which are first-order formulas with existential quantifiers only, a Σ_1-property is a property that can be described by Σ_1-formula. The easiest problem of this kind is testing the existence of a subgraph. Eppstein [10, 11] constructed a linear-time algorithm for deciding the existence of a fixed subgraph for planar graphs. He then extended his algorithm to minor-closed classes of graphs with locally bounded tree-width [12]. All these results were generalized to classes of graphs with bounded expansion by Nešetřil and Ossona de Mendéz in [21, 23]. In fact, they established a more general result on testing arbitrary Σ_1-properties:

Theorem 5. *Let Φ be a Σ_1-property and $\mathcal{G}$ a class of graphs with bounded expansion. There exists a linear time algorithm deciding Φ for graphs $G \in \mathcal{G}$.*

The main idea of the algorithm is that if Φ holds for $G \in \mathcal{G}$, then the witness assignment to variables can use at most k colors where k is the number of quantifiers of Φ. Hence, using Theorem 4, we can color vertices with K colors in such a way that any k colors induce a graph with tree-depth at most k. After finding the coloring (in linear time), the problem is reduced to deciding Φ for $\binom{K}{k}$ subgraphs of an input graph, each subgraph having tree-depth at most k (which can be solved, e.g., using the classical Courcelle's result mentioned at the beginning of the paper).

Following the lines of the above reasoning, we can obtain an analogous results for classes of nowhere-dense graphs, see [24] for further details. An algorithm is *almost linear*, if for any $\varepsilon > 0$ which is part of the input of the algorithm, the algorithm runs in time $O(n^{1+\varepsilon})$ where n is the number of vertices of an input graph.

Theorem 6. *Let Φ be a Σ_1-property and $\mathcal{G}$ a class of nowhere dense graphs. There exists an almost linear time algorithm deciding Φ for graphs $G \in \mathcal{G}$.*

Let us now focus on a particular case of Σ_1-properties, the existence of short paths between two vertices. Kowalik and Kurowski [17,18] designed a data structure with linear build-up time and constant query time answering the existence of a path of length at most d between two vertices of an input planar graph for a fixed integer d. In fact, they approach readily generalize to classes of graphs with bounded expansion. Let us sketch the main idea of the algorithm: let G be an input graph and consider the sequence of its transitive fraternal augmentations $\vec{G}_0, \ldots, \vec{G}_d$ as defined in Corollary 1. If two vertices u and v are joined by a path of length at most d, then they are either adjacent in $\vec{G}_d$ or they have a common in-neighbor in $\vec{G}_d$ (this can easily be proved by induction on d). Since the maximum in-degree of $\vec{G}_d$ is bounded, the existence of an edge joining the two vertices or the existence of their common in-neighbor can be done in constant time.

The data structure can be dynamized using the result of Brodal and Fagerberg [4] on maintaining orientations with small in-degrees of degenerate graphs, see [16]. The arguments readily translate to a setting of classes of graphs with bounded expansion, see [9]:

Theorem 7. *Let $\mathcal{G}$ be a class of graphs with bounded expansion and d a fixed integer. There exists a dynamic data structure for answering the existence of a path of length at most d in a graph $G \in \mathcal{G}$ with the following parameters:*

- *the data structure can be built in linear time,*
- *each query can be answered in constant time,*
- *an edge can be added to the represented graph in time $O(\log^d n)$ where n is its number of vertices, and*
- *an edge can be removed in constant time.*

4.2 First-Order Properties

We now address the complexity of deciding general first-order properties, i.e., those properties that can be described by formulas with quantifications over graph vertices only (quantification over sets of vertices is not allowed). As examples of first-order properties, we can mention deciding the existence of a dominating set of a fixed size or the existence of a vertex cover of a fixed size. First-order properties can always be decided in polynomial time (with degree depending on the property) but we are interested in fixed parameter results. The first result in this direction is the result of Seese [31] that every first-order property can be tested in linear time for any class of graphs with bounded maximum degree. The result is not that surprising after we realize that first-order properties are of very localized nature which is captured in the following classical result of Gaifman [15]:

Theorem 8. *Every first-order formula Φ for graphs is equivalent for some r to a Boolean combination of formulas of the form*

$$\exists x_1 \cdots \exists x_k \left(\wedge_{i \neq j} \mathrm{dist}(x_i, x_j) > 2r \bigwedge \wedge_{i=1,\ldots,k} \Phi_r(x_i) \right)$$

where each $\Phi_r(x_i)$ is r-local with respect to x_i, i.e., all quantifiers contained in $\Phi_r(x_i)$ have domain restricted to the r-neighborhood of x_i.

In the light of Theorem 8, deciding first-order properties for graphs with maximum degree Δ decomposes into a linear number of finite problems (the r-neighborhood of each vertex contains at most $\Delta(\Delta - 1)^{r-1}$ vertices) whose Boolean combination yields the result on whether the formula is satisfied for an input graph.

Frick and Grohe [13,14] extended this result by considering classes of graphs with locally bounded tree-width. They have shown that any first-order property can be decided in almost linear time for any class of graphs with locally bounded tree-width. In the particular case of planar graphs, they were able to obtain a linear time algorithm using a different covering algorithm. Their algorithm uses the following covering result of Peleg [28] which has applications in other algorithms in the area and thus we would like to mention it explicitly.

Lemma 3. *Let $k \geq 1$ be a fixed integer. There is an algorithm that given $r \geq 1$ and a graph G, outputs sets $A_1, \ldots, A_m$ of vertices of G such that*

- *for every vertex v of G, there exists A_i containing the r-neighborhood of v,*
- *every A_i is contained in the $2kr$-neighborhood of a vertex of G, and*
- *the sum $|A_1| + \cdots + |A_m|$ is at most $O(n^{1+1/k})$.*

The running time of the algorithm is linear in the sum of the numbers of edges contained in $G[A_i]$, $i = 1, \ldots, k$.

Another meta-theorem on graphs with locally restricted structure was obtained by Dawar, Grohe and Kreutzer [6] who showed that deciding first-order properties Φ is fixed-parameter tractable for classes of graphs locally excluding a minor, i.e., there exists a polynomial-time algorithm where the exponent does not depend on Φ. Nešetřil and Ossona de Mendéz [25] gave a linear time algorithm for deciding the existence of a dominating set of a fixed size for classes of graphs with bounded expansion. Their result indicates that the results we mention earlier could hold for classes of graphs with bounded expansion. This turns out to be true as proven by the authors and Thomas in [9]:

Theorem 9. *Let Φ be a first order formula and $\mathcal{G}$ a class of graphs with bounded expansion. There exists a linear-time algorithm deciding Φ for graphs $G \in \mathcal{G}$.*

Theorem 10. *Let Φ be a first order formula and $\mathcal{G}$ a class of nowhere dense graphs. There exists an almost linear time algorithm deciding Φ for graphs $G \in \mathcal{G}$.*

References

1. Arnborg, S., Proskurowski, A., Corneil, D.G.: Forbidden minors characterization of partial 3-trees. Discrete Math. 80, 1–19 (1990)
2. Bodlaender, H.L., Deogun, J.S., Jansen, K., Kloks, T., Kratsch, D., Müller, H., Tuza, Z.: Ranking of graphs. In: Mayr, E.W., Schmidt, G., Tinhofer, G. (eds.) WG 1994. LNCS, vol. 903, pp. 292–304. Springer, Heidelberg (1995)
3. Bodlaender, H.L., Gilbert, J.R., Hafsteinsson, H., Kloks, T.: Approximating treewidth, pathwidth, frontsize and shortest elimination tree. J. Algorithms 25, 1305–1317 (1996)
4. Brodal, G.S., Fagelberg, R.: Dynamic representations of sparse graphs. In: Dehne, F., Gupta, A., Sack, J.-R., Tamassia, R. (eds.) WADS 1999. LNCS, vol. 1663, pp. 342–351. Springer, Heidelberg (1999)
5. Courcelle, B.: The monadic second-order logic of graph I. Recognizable sets of finite graphs. Inform. and Comput. 85, 12–75 (1990)
6. Dawar, A., Grohe, M., Kreutzer, S.: Locally excluding a minor. In: Proc. LICS 2007, pp. 270–279. IEEE Computer Society Press, Los Alamitos (2007)

7. DeVos, M., Ding, G., Oporowski, B., Sanders, D.P., Reed, B., Seymour, P.D., Vertigan, D.: Excluding any graph as a minor allows a low tree-width 2-coloring. J. Combin. Theory Ser. B 91, 25–41 (2004)
8. Dvořák, Z.: Assymptotical structure of combinatorial objects, PhD thesis, Charles University (2007)
9. Dvořák, Z., Král', D., Thomas, R.: Deciding first order properties for nowhere dense graphs (manuscript) (2009)
10. Eppstein, D.: Subgraph isomorphism in planar graphs and related problems. In: Proc. SODA 1995, pp. 632–640. ACM&SIAM (1995)
11. Eppstein, D.: Subgraph isomorphism in planar graphs and related problems. J. Graph Algorithms Appl. 3, 1–27 (1999)
12. Eppstein, D.: Diameter and treewidth in minor-closed graph families. Algorithmica 27, 275–291 (2000)
13. Frick, M., Grohe, M.: Deciding first-order properties of locally tree-decomposable structures. In: Wiedermann, J., Van Emde Boas, P., Nielsen, M. (eds.) ICALP 1999. LNCS, vol. 1644, pp. 331–340. Springer, Heidelberg (1999)
14. Frick, M., Grohe, M.: Deciding first-order properties of locally tree-decomposable structures. J. ACM 48, 1184–1206 (2001)
15. Gaifman, H.: On local and non-local properties. In: Proc. Herbrand Symp. Logic Colloq. North-Holland, Amsterdam (1982)
16. Kowalik, L.: Adjacency queries in dynamic sparse graphs. Inf. Process. Lett. 102, 191–195 (2007)
17. Kowalik, L., Kurowski, M.: Oracles for bounded-length shortest paths in planar graphs. ACM Trans. Algorithms 2, 335–363 (2006)
18. Kowalik, L., Kurowski, M.: Short path queries in planar graphs in constant time. In: Proc. STOC 2003, pp. 143–148 (2003)
19. Nešetřil, J., Ossona de Mendéz, P.: First order properties of nowhere dense structures (manuscript) (2008)
20. Nešetřil, J., Ossona de Mendéz, P.: Grad and classes with bounded expansion I. Decompositions. European J. Combin. 29, 760–776 (2008)
21. Nešetřil, J., Ossona de Mendéz, P.: Grad and classes with bounded expansion II. Algorithmic aspects. European J. Combin. 29, 777–791 (2008)
22. Nešetřil, J., Ossona de Mendéz, P.: Grad and classes with bounded expansion III. Restricted graph homomorphism dualities. European J. Combin. 29, 1012–1024 (2008)
23. Nešetřil, J., Ossona de Mendéz, P.: Linear time low tree-width partitions and algorithmic consequences. In: Proc. STOC 2006, pp. 391–400 (2006)
24. Nešetřil, J., Ossona de Mendéz, P.: On nowhere dense graphs (manuscript) (2008)
25. Nešetřil, J., Ossona de Mendéz, P.: Structural properties of sparse graphs (manuscript) (2008)
26. Nešetřil, J., Ossona de Mendéz, P.: Tree depth, subgraph coloring and homomorphism bounds. European J. Combin. 27, 1022–1041 (2006)
27. Nešetřil, J., Ossona de Mendéz, P., Wood, D.: Characterizations and examples of graph classes with bounded expansion (manuscript) (2008)
28. Peleg, D.: Distance-dependent distributed directories. Info. Computa. 103, 270–298 (1993)
29. Robertson, N., Seymour, P.D.: Graph minors V: Excluding a planar graph. J. Combin. Theory Ser. B 41, 92–114 (1986)
30. Robertson, N., Seymour, P.D.: Graph Minors XX. Wagner's conjecture. J. Combin. Theory Ser. B 92, 325–357 (2004)
31. Seese, D.: Linear time computable problems and first-order descriptions. Math. Structu. Comput. Sci. 6, 505–526 (1996)

A Graph Polynomial Arising from Community Structure
(Extended Abstract)

I. Averbouch[1,*], J.A. Makowsky[1,**], and P. Tittmann[2]

[1] Department of Computer Science,
Technion–Israel Institute of Technology, 32000 Haifa, Israel
[2] Fachbereich Mathematik–Physik–Informatik
Hochschule Mittweida, Mittweida, Germany

Abstract. Inspired by the study of community structure in connection networks, we introduce the graph polynomial $Q(G; x, y)$, as a bivariate generating function which counts the number of connected components in induced subgraphs. We analyze the features of the new polynomial. First, we re-define it as a subset expansion formula. Second, we give a recursive definition of $Q(G; x, y)$ using vertex deletion, vertex contraction and deletion of a vertex together with its neighborhood, and prove a universality property. We relate $Q(G; x, y)$ to the universal edge elimination polynomial introduced by I. Averbouch, B. Godlin and J.A. Makowsky (2008), which subsumes other known graph invariants and graph polynomials, among them the Tutte polynomial, the independence and matching polynomials, and the bivariate extension of the chromatic polynomial introduced by K. Dohmen, A. Pönitz, and P. Tittmann (2003). Finally we show that the computation of $Q(G; x, y)$ is $\sharp\mathbf{P}$-hard, but Fixed Parameter Tractable for graphs of bounded tree-width and clique-width.

1 Introduction

1.1 Motivation: Community Structure in Networks

In the last decade stochastic social networks have been analyzed mathematically from various points of view. Understanding such networks sheds light on many questions arising in biology, epidemiology, sociology and large computer networks. Researchers have concentrated particularly on a few properties that seem to be common to many networks: the small-world property, power-law degree distributions, and network transitivity. For a broad view on the structure and dynamics of networks, see [30]. M. Girvan and M.E.J. Newman, [19], highlight another

---

[*] Partially supported by a grant of the Graduate School of the Technion–Israel Institute of Technology.
[**] Partially supported by a grant of the Fund for Promotion of Research of the Technion–Israel Institute of Technology and grant ISF 1392/07 of the Israel Science Foundation (2007-2010).

property that is found in many networks, the property of *community structure*, in which network nodes are joined together in tightly knit groups, between which there are only looser connections.

Motivated by [29], and the third author's involvement in a project studying social networks, we were led to study the graph parameter $q_{ij}(G)$, the number of vertex subsets $X \subseteq V$ with i vertices such that $G[X]$ has exactly j components. $q_{ij}(G)$, counts the number of degenerated communities which consist of i members, and which split into j isolated subcommunities.

The ordinary bivariate *generating function* associated with $q_{ij}(G)$ is the two-variable graph polynomial

$$Q(G;x,y) = \sum_{i=0}^{n} \sum_{j=0}^{n} q_{ij}(G) x^i y^j.$$

We call $Q(G;x,y)$ the *subgraph component polynomial* of G. The coefficient of y^k in $Q(G;x,y)$ is the ordinary generating function for the number of vertex sets that induce a subgraph of G with exactly k components.

1.2 $Q(G;x,y)$ as a Graph Polynomial

There is an abundance of graph polynomials studied in the literature, and slowly there is a framework emerging, [20,25,26], which allows to compare graph polynomials with respect to their ability to distinguish graphs, to encode other graph polynomials or numeric graph invariants, and their computational complexity. In this paper we study the subgraph component polynomial $Q(G;x,y)$ as a graph polynomial in its own right and explore its properties within this emerging framework.

Like the bivariate Tutte polynomial, see [9, Chapter 10], the polynomial $Q(G;x,y)$ has several remarkable properties. However, its distinguishing power is quite different from the Tutte polynomial and other well studied polynomials.

Our main findings are [1]:

– The Tutte polynomial satisfies a linear recurrence relation with respect to edge deletion and edge contraction, and is universal in this respect. $Q(G;x,y)$ also satisfies a linear recurrence relation, but with respect to three kinds of vertex elimination, and is universal in this respect. (Theorems 2 and 3).
– A graph polynomial in three indeterminates, $\xi(G;x,y,z)$, which satisfies a linear recurrence relation with respect to three kinds of edge elimation, and which is universal in this respect, was introduced in [4,5]. It subsumes both the Tutte polynomial and the matching polynomial. For line graph $L(G)$ of a graph G, we have $Q(L(G);x,y)$ is a substitution instance of $\xi(G;x,y,z)$ (Theorem 4).
– Distinguishing power of $Q(G;x,y)$ is incomparable with that of the Tutte polynomial, the characteristic polynomial and the bivariate chromatic polynomial introduced in [15] (Section 2).

---
[1] More results regarding $Q(G;x,y)$ are available in preprint [36].

- Also like for the Tutte polynomial, cf. [22], $Q(G; x_0, y_0)$ has the *Difficult Point Property*, i.e. it is $\sharp\mathbf{P}$-hard to compute for all fixed values of $(x_0, y_0) \in \mathbb{R}^2 - E$ where E is a semi-algebraic set of lower dimension (Theorem 5). In [26] it is conjectured that the Difficult Point Property holds for a wide class of graph polynomials, the graph polynomials definable in Monadic Second Order Logic. The conjecture has been verified for various special cases, [6,7,8].
- $Q(G; x, y)$ is fixed parameter tractable in the sense of [16] when restricted to graphs classes of bounded tree-width (Proposition 3) or even to classes of bounded clique-width (Proposition 4). For the Tutte polynomial, this is known only for graph classes of bounded tree-width, [2,28,31].

2 Distinguishing Power

We denote by $m(G; x) = \sum_i m_i(G) x^i$ be the matching polynomial with $m_i(G)$ the number of i-matchings of G, by $p(G; x)$ be the characteristic polynomial, by $\chi(G; x)$ the chromatic polynomial, by $T(G; x, y)$ the Tutte polynomial, and by $P(G; x, y)$ the bivariate chromatic polynomial introduced in [15].

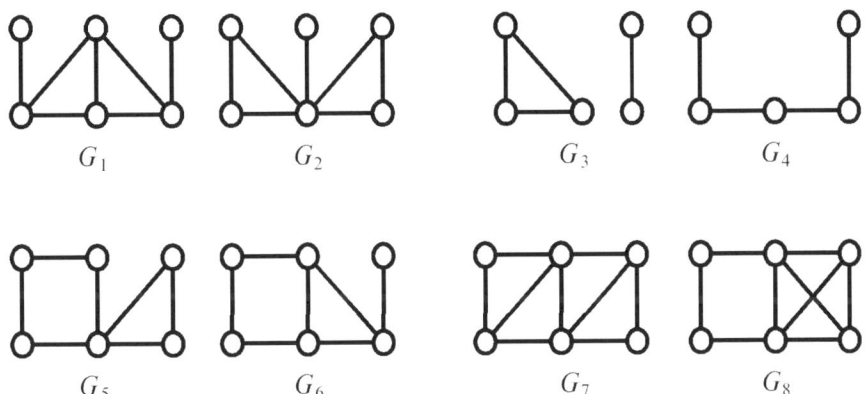

Fig. 1. Distinguishing power of $Q(G; x, y)$

Proposition 1. *For the graphs $G_i; i = 1, \ldots 6$ from Figure 1, and for P_4 and $K_{1,3}$ we have*

(1) $p(G_1; x) = p(G_2; x)$ *but* $Q(G_1; x, y) \neq Q(G_2; x, y)$.
(2) $m(G_3; x) = m(G_4; x)$ *but* $Q(G_3; x, y) \neq Q(G_4; x, y)$.
(3) $P(G_5; x, y) = P(G_6; x, y)$ *but* $Q(G_5; x, y) \neq Q(G_6; x, y)$.
(4) $T(P_4; x, y) = T(K_{1,3}; x, y)$ *but* $Q(P_4; x, y) \neq Q(K_{1,3}; x, y)$.

Proposition 2. *For the graphs G_7 and G_8 from Figure 1, we have: $Q(G_7; x, y) = Q(G_8; x, y)$ but $\chi(G_7; x) \neq \chi(G_8; x)$*

Corollary 1. *The distinguishing power of $Q(G;x,y)$ is incomparable neither with that of the chromatic polynomial nor with that of any generalization of the chromatic polynomial discussed above (Tutte polynomial, bivariate chromatic polynomial).*

Problem 1. *Are there simple graphs distinguished by $p(G;x)$ or $m(G;x)$, which are not distinguished by $Q(G;x,y)$?*

3   Subset Expansion and Definability in Logic

$Q(G;x,y)$ was defined as a generating function. Let us rewrite the definition of $Q(G;x,y)$ in a slightly different way. Instead of summation over the number of the used vertices i, and the number of induced connected components j, we shall sum over all the possible subsets of vertices:

$$Q(G;x,y) = \sum_{A\subseteq V} x^{|A|} y^{k(G[A])} = \sum_{A\subseteq V} \left(\prod_{v\in A} x\right) \left(\prod_{u\in F(A)} y\right). \tag{1}$$

where $k(G[A])$ is the number of connected components of the subgraph of G induced by A.

This is a *subset expansion formula*, a term coined in [37]. The relationship between recursive definitions of graph polynomials and the existence of subset expansion formulas has been studied from a logical point of view in [20]. Subset expansion formulas can often be used to show that a graph polynomial is definable in Monadic Second Order Logic, as studied in [23,26] . However, the exponent $k(G[A])$ in the left part of Equation (1) causes a problem. To remedy this, we use, like in [24], an auxiliary order $\prec$ over the vertices. Then we denote by $F(A)$ the subset of the *smallest* vertices in every respective connected component. Note that the result does not depend on the used auxiliary order.

Without having to go in the details of graph polynomials definable in Monadic Second Order Logic[2], Equation (1) shows that $Q(G;x,y)$ is a graph polynomial definable in Monadic Second Order Logic for graphs $G = (V,E)$ with universe V and a binary edge relation. Therefore all the theorems from [23,27] can be applied. In particular, the Feferman-Vaught-type theorems from [23] guarantee existence of reduction formulas like multiplicativity from Theorem 1 not only for the disjoint union or the join operation, but for a wide class of **MSOL**-definable operations. Also, a general theorem from [27] guarantees the existence of recurrence formulas, for a wide class of recursively defined families of graphs, as studied also in [33]. Among these we have the wheels W_n, the ladders L_n and the stars $Star_n$. It should not be difficult to compute the recurrence relations for these explicitly.

We shall exploit **MSOL**-definability also for our complexity analysis in Section 6.2.

---

[2] The interested reader can consult [17] for the use Monadic Second Order Logic in finite model theory, and [10] for its use in graph theory.

4 Recursive Definition and Universality

4.1 Recurrence Relation for Vertex Elimination

We turn now our attention to the investigation of properties of the subgraph polynomial that support its computation.

Theorem 1 (Multiplicativity). *Let $G = G_1 \sqcup G_1$ be the disjoint union of the graphs G_1 and G_2. Then*

$$Q(G; x, y) = Q(G_1; x, y) \cdot Q(G_2; x, y).$$

Theorem 2 (Recurrence relation). *Let $G = (V, E)$ be a graph and $v \in V$. Then the subgraph polynomial satisfies the decomposition formula*

$$Q(G; x, y) = Q(G - v; x, y) + x(y - 1)Q(G - N[v]; x, y) + xQ(G/v; x, y),$$

where three types of *vertex elimination* are used:

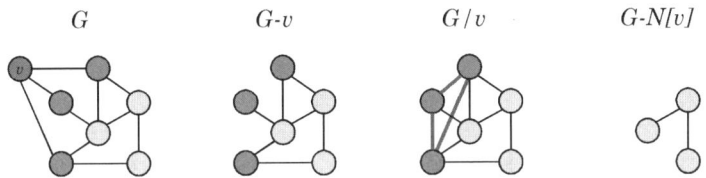

Fig. 2. Vertex elimination operations

Vertex deletion: For a given vertex $v \in V(G)$, let $G - v$ the graph obtained from G by removal of v and all edges that are incident to v. We call this operation *vertex deletion*.

Vertex extraction: Similarly, let $G - X$ be the graph obtained from G by removal of all vertices of the set $X \subseteq V$. Let $N(v)$ be the set of vertices that are adjacent to v in G (the neighborhood of v). We denote by $N[v]$ the *closed neighborhood* of a vertex v in G, i.e. the set of all vertices adjacent to v including v itself. The operation $G - N[v]$ is called *vertex extraction*.

Vertex contraction: A further special graph operation is needed here – the *contraction* of a vertex. That is the graph G/v obtained from G by removal of v and insertion of edges between all pairs of non-adjacent neighbor vertices of v.

The proof is available in [36].

4.2 The Universality Property of $Q(G; x, y)$

The vertex decomposition formula represented in Theorem 2 can be considered as a vertex equivalent to the well-known edge decomposition (deletion-contraction relations). Edge decomposition formulae of the form $f(G) = \alpha(e) f(G-e) + \beta(e) f(G/e)$ apply to the Tutte polynomial and derived graph invariants, for instance the number of spanning trees or the reliability polynomial. Indeed, it was shown by J.G. Oxley and D.J.A. Welsh, [35], that the Tutte polynomial is in a certain sense *universal*, meaning that all other graph invariants that satisfy edge decomposition formulae can be derived from the Tutte polynomial by substitution of variables. A textbook presentation is given in [9]. A general framework analyzing universality properties of graph polynomials is studied in [20].

It seems natural to ask for the most general vertex decomposition formula. Let us assume that we try to construct an ordinary generating function $f(G)$ that counts some type of vertex induced subgraphs with respect to the number of vertices. Which properties should such a function have? If the subgraphs in question are composed from subgraphs of the components then we can expect multiplicativity of f with respect to components of the graph. In order to assign the value $f(G)$ uniquely to a graph G by application of a decomposition formula as given in Theorem 2, certain initial values for the null graph and the empty graph have to be given. Therefore, we presuppose the following five properties of f:

(a) (Multiplicativity) If G_1 and G_2 are components of G then

$$f(G) = f(G_1) f(G_2).$$

(b) (Recurrence relation) Let $\alpha, \beta, \gamma \in \mathbb{R}$ and let v be a vertex of G, then

$$f(G) = \alpha f(G - v) + \beta f(G - N[v]) + \gamma f(G/v). \tag{2}$$

(c) (Initial condition) There exists $\delta \in \mathbb{R}$ such that $f(\emptyset) = \delta$ for the null graph $\emptyset = (\emptyset, \emptyset)$.
(d) (Initial condition) There exists $\varepsilon \in \mathbb{R}$ such that $f(E_1) = \varepsilon$ for a graph $E_1 = (\{v\}, \emptyset)$ consisting of one vertex.
(e) (Order invariance) The result of computing f has to be the same, irrespective of the order in which we apply the enabled computation steps, in particular, of the order of the vertices which we use to apply the relation (b).

In general we may choose $\alpha, \beta, \gamma, \delta, \varepsilon$ from a field of characteristic zero or from a ring. A graph invariant is *proper* if there are two graphs G_1 and G_2 with the same number of vertices such that $f(G_1) \neq f(G_2)$. In [36] it is proved that $Q(G; x, y)$ is universal among polynomials recursively defined using vertex deletion, vertex extraction and vertex contraction. More precisely:

Theorem 3 (Universality of $Q(G; x, y)$)

(1) For a graph polynomial $f(G; \alpha, \beta, \gamma, \delta, \varepsilon)$ to be proper and well-defined by conditions (a)-(d) we have $\alpha = 1$, $\delta = 1$ and $\varepsilon = 1 + \beta + \gamma$.
(2) There is a unique proper graph polynomial $U(G; \beta, \gamma)$ which is well-defined by conditions (a)-(d) and we have

$$Q(G;x,y) = U(G;x(y-1),x) \tag{3}$$

and

$$U(G;\beta,\gamma) = Q(G;\gamma,\frac{\beta}{\gamma}+1). \tag{4}$$

5 Vertex Eliminations vs Edge Elimination

We have seen in Theorem 3 that $Q(G;x,y)$ is universal among the polynomials defined recursively via deletion, extraction and contraction of vertices. In [4,5] the polynomial $\xi(G;x,y,z)$ was shown to be universal among the polynomials defined recursively via deletion, extraction and contraction of edges. In this section we will show the connection of $G(G;x,y)$ to the universal edge elimination polynomial $\xi(G;x,y,z)$.

The polynomial $\xi(G;x,y,z)$ generalizes both the Tutte and the matching polynomials, as well as the bivariate chromatic polynomial of [15]. We shall use the recursive decomposition of $\xi(G;x,y,z)$ from [5]:

$$\begin{aligned}
\xi(G;x,y,z) &= \xi(G-e;x,y,z) + y\xi(G/e;x,y,z) + z\xi(G\dagger e;x,y,z) \\
\xi(G_1 \sqcup G_2;x,y,z) &= \xi(G_1;x,y,z)\xi(G_2;x,y,z) \\
\xi(E_1;x,y,z) &= x \\
\xi(\emptyset) &= 1
\end{aligned} \tag{5}$$

where $G_1 \sqcup G_2$ denotes the disjoint union of graphs G_1 and G_2, and the three edge elimination operations are defined as follows:

Edge deletion: We denote by $G-e$ the graph obtained from G by simply removing the edge e.

Edge extraction: We denote by $G\dagger e$ the graph induced by $V \setminus \{u,v\}$ provided $e = \{u,v\}$. This operation removes also all the edges adjacent to e.

Edge contraction: We denote by G/e the graph obtained from G by unifying the endpoints of e.

Now we rewrite the decomposition of $Q(G;x,y)$ using Theorem 2.

$$\begin{aligned}
Q(G;x,y) &= Q(G-v;x,y) + xQ(G/v;x,y) + x(y-1)Q(G-N[v];x,y) \\
Q(G_1 \sqcup G_2;x,y) &= Q(G_1;x,y)Q(G_2;x,y) \\
Q(E_1;x,y) &= xy+1 \\
Q(\emptyset) &= 1
\end{aligned} \tag{6}$$

Theorem 4. *Let $G = (V,E)$ be a graph. Let $L(G) = (V_e, E_e)$ denote the line graph of G. Then the following equation holds:*

$$\xi(G;1,x,x(y-1)) = Q(L(G);x,y)$$

The proof is available in [36].

Problem 2. How does the distinguishing power of $\xi(G;x,y,z)$ compare to the distinguishing power of $Q(G;x,y)$?

6 Computational Complexity of $Q(G; x, y)$

6.1 Complexity of Evaluation

We deal with a problem of evaluation of $Q(-; x, y)$ at a given point $(x, y) \in \mathbb{Q}^2$ for arbitrary input graph G.

Theorem 5. *For every point $(x, y) \in \mathbb{Q}^2$, possibly except for the lines $xy = 0$, $y = 1$, $x = -1$ and $x = -2$, the evaluation of $Q(G; x, y)$ for an input graph G is $\sharp\mathbf{P}$-hard.*

C.Hoffmann in [21] showed the following:

Theorem 6 (Hoffmann 2008). *For every point $(x, y, z) \in \mathbb{Q}^3$, except possibly for the subsets $x = 0$, $z = -xy$, $(x, z) \in \{(1, 0), (2, 0)\}$ and $y \in \{-2, -1, 0\}$, the evaluation of $\xi(-; x, y, z)$ for an input graph G is $\sharp\mathbf{P}$-hard.*

Proof (Proof of Theorem 5:). We use Theorem 6 and our Theorem (4). Under the conditions of Theorem (4), Hoffmann's exception sets are mapped to the lines $xy = 0$, $y = 1$, $x = -1$ and $x = -2$. It follows that for every point $(x, y) \in \mathbb{Q}^2$ that does not lay on one of those lines, the polynomial $Q(-; x, y)$ is $\sharp\mathbf{P}$-hard to evaluate even for an input line graph $L(G)$. □

The evaluation of $Q(-; x, y)$ is polynomial time computable for $xy = 0$ and for $y = 1$. It remains open whether it is polynomial time computable for $x = -1$ and $x = -2$. One can also ask, whether there is some point $(x, y) \in \mathbb{Q}^2$, in which $Q(-; x, y)$ is hard to evaluate for general input graph, but easy for input line graph.

6.2 Parameterized Complexity

Here we discuss the computational complexity of $Q(G; x, y)$ for input graphs of bounded tree with, and for input graphs for bounded clique width. We do not need the exact definitions here. For background on tree-width the reader can consult [14]. Clique-width was defined in [12]. Both are discussed in [23].

Recall that the subgraph component polynomial is definable using the **MSOL**-formalism (formula (1)) with auxiliary order, while the result is order-independent. Hense, using a general theorem from [23,24], we have

Proposition 3. *$Q(G; x, y)$ is polynomial time computable on graphs of tree-width at most k where the exponent of the run time is independent of k.*

Moreover, applying the result of Courcelle, Makowsky and Rotics [11], combined with the results from [34], we have similar result for graphs of bouded clique width:

Proposition 4. *$Q(G; x, y)$ is polynomial time computable on graphs of clique-width at most k where the exponent of the run time is independent of k.*

The drawback of the general methods of [23,24] and [11], lies in the huge hidden constants, which make it practically unusable. However, an explicit dynamic algorithm for computing the polynomial $Q(G;x,y)$ on graphs of bounded tree-width, given the tree decomposition of the graph, where the constants are simply exponential in k, can be constructed along the same ideas as presented in [18,38]. For the graphs of bounded clique width, given the clique decomposition of the graph, we know an algorithm with constants doubly-exponential in k. It is open whether an algorithm with constants simply exponential in k exists. For a comparison of the complexity of computing graph polynomials on graphs classes of bounded clique-width, cf. [28].

7 Conclusions and Open Problems

We have shown that $Q(G;x,y)$ is a universal vertex elimination polynomial. We have looked at the graph polynomial $Q(G;x,y)$ from various angles and compared its behaviour and distinguishing power with the characteristic polynomial, the matching polynomial the Tutte polynomial and the universal edge elimination polynomial. We have not discussed the relationship of $Q(G;x,y)$ to other graph polynomials, such as the interlace polynomial, [1,3], or the many other graph polynomials listed in [26].

We have seen that $Q(G;x,y)$ distinguishes between graphs where these polynomials do not. For the chromatic polynomial and its generalizations we have also shown the opposite direction. We have not found cases where the matching and the characteristic polynomials do distinguish between graphs where $Q(G;x,y)$ does not. This is probably due to our lack of computerized tools for searching for such cases, cf. Problem 1. In Problem 2 we ask about comparing distinguishing power of $Q(G;x,y)$ and the universal edge elimination polynomial $\xi(G;x,y,z)$.

References

1. Aigner, M., van der Holst, H.: Interlace polynomials. Linear Algebra and Applications 377, 11–30 (2004)
2. Andrzejak, A.: Splitting formulas for Tutte polynomials. Journal of Combinatorial Theory, Series B 70(2), 346–366 (1997)
3. Arratia, R., Bollobás, B., Sorkin, G.B.: The interlace polynomial of a graph. Journal of Combinatorial Theory, Series B 92, 199–233 (2004)
4. Averbouch, I., Godlin, B., Makowsky, J.A.: The most general edge elimination polynomial (2007), http://uk.arxiv.org/pdf/0712.3112.pdf
5. Averbouch, I., Godlin, B., Makowsky, J.A.: An extension of the bivariate chromatic polynomial (submitted) (2008)
6. Bläser, M., Dell, H.: Complexity of the cover polynomial. In: Arge, L., Cachin, C., Jurdziński, T., Tarlecki, A. (eds.) ICALP 2007. LNCS, vol. 4596, pp. 801–812. Springer, Heidelberg (2007)
7. Bläser, M., Dell, H.: Complexity of the Bollobás-Riordan polynomia. exceptional points and uniform reductions. In: Hirsch, E.A., Razborov, A.A., Semenov, A., Slissenko, A. (eds.) Computer Science – Theory and Applications. LNCS, vol. 5010, pp. 86–98. Springer, Heidelberg (2008)

8. Bläser, M., Hoffmann, C.: On the complexity of the interlace polynomial. In: STACS, pp. 97–108 (2008)
9. Bollobás, B.: Modern Graph Theory. Springer, Heidelberg (1999)
10. Courcelle, B.: Graph Structure and Monadic Second Order Logic. Cambridge University Press, Cambridge (in preparation)
11. Courcelle, B., Makowsky, J.A., Rotics, U.: On the fixed parameter complexity of graph enumeration problems definable in monadic second order logic. Discrete Applied Mathematics 108(1-2), 23–52 (2001)
12. Courcelle, B., Olariu, S.: Upper bounds to the clique–width of graphs. Discrete Applied Mathematics 101, 77–114 (2000)
13. de Mier, A., Noy, M.: On graphs determined by their Tutte polynomials. Graphs and Combinatorics 20(1), 105–119 (2004)
14. Diestel, R.: Graph Decompositions, A Study in Infinite Graph Theory. Clarendon Press, Oxford (1990)
15. Dohmen, K., Pönitz, A., Tittmann, P.: A new two-variable generalization of the chromatic polynomial. Discrete Mathematics and Theoretical Computer Science 6, 69–90 (2003)
16. Downey, R.G., Fellows, M.F.: Parametrized Complexity. Springer, Heidelberg (1999)
17. Ebbinghaus, H., Flum, J.: Finite Model Theory. Springer, Heidelberg (1995)
18. Fischer, E., Makowsky, J.A., Ravve, E.V.: Counting truth assignments of formulas of bounded tree width and clique-width. Discrete Applied Mathematics 156, 511–529 (2008)
19. Girvan, M., Newman, M.E.J.: Community structure in social and biological networks. Proc. Natl. Acad. Sci. USA 99, 7821–7826 (2002)
20. Godlin, B., Katz, E., Makowsky, J.A.: Graph polynomials: From recursive definitions to subset expansion formulas (2008), http://uk.arxiv.org/pdf/0812.1364.pdf
21. Hoffmann, C.: A most general edge elimination polynomial–thickening of edges. arXiv:0801.1600v1 [math.CO] (2008)
22. Jaeger, F., Vertigan, D.L., Welsh, D.J.A.: On the computational complexity of the Jones and Tutte polynomials. Math. Proc. Camb. Phil. Soc. 108, 35–53 (1990)
23. Makowsky, J.A.: Algorithmic uses of the Feferman-Vaught theorem. Annals of Pure and Applied Logic 126(1-3), 159–213 (2004)
24. Makowsky, J.A.: Colored Tutte polynomials and Kauffman brackets on graphs of bounded tree width. Disc. Appl. Math. 145(2), 276–290 (2005)
25. Makowsky, J.A.: From a zoo to a zoology: Descriptive complexity for graph polynomials. In: Beckmann, A., Berger, U., Löwe, B., Tucker, J.V. (eds.) CiE 2006. LNCS, vol. 3988, pp. 330–341. Springer, Heidelberg (2006)
26. Makowsky, J.A.: From a zoo to a zoology: Towards a general theory of graph polynomials. Theory of Computing Systems 43, 542–562 (2008)
27. Makowsky, J.A., Fischer, E.: Linear recurrence relations for graph polynomials. In: Avron, A., Dershowitz, N., Rabinovich, A. (eds.) Pillars of Computer Science. LNCS, vol. 4800, pp. 266–279. Springer, Heidelberg (2008)
28. Makowsky, J.A., Rotics, U., Averbouch, I., Godlin, B.: Computing graph polynomials on graphs of bounded clique-width. In: Fomin, F.V. (ed.) WG 2006. LNCS, vol. 4271, pp. 191–204. Springer, Heidelberg (2006)
29. Newman, M.E.J.: Detecting community structure in networks. Eur. Phys. J.B. 38, 321–330 (2004)
30. Newman, M.E.J., Barabasi, A.L., Watts, D.: The Structure and Dynamics of Networks. Princeton University Press, Princeton (2006)

31. Noble, S.D.: Evaluating the Tutte polynomial for graphs of bounded tree-width. Combinatorics, Probability and Computing 7, 307–321 (1998)
32. Noy, M.: On graphs determined by polynomial invariants. TCS 307, 365–384 (2003)
33. Noy, M., Ribó, A.: Recursively constructible families of graphs. Advances in Applied Mathematics 32, 350–363 (2004)
34. Oum, S.: Approximating rank-width and clique-width quickly. In: Kratsch, D. (ed.) WG 2005. LNCS, vol. 3787, pp. 49–58. Springer, Heidelberg (2005)
35. Oxley, J.G., Welsh, D.J.A.: The Tutte polynomial and percolation. In: Bundy, J.A., Murty, U.S.R. (eds.) Graph Theory and Related Topics, pp. 329–339. Academic Press, London (1979)
36. Tittmann, P., Averbouch, I., Makowsky, J.A.: The Enumeration of Vertex Induced Subgraphs with Respect to the Number of Components (2009), http://uk.arxiv.org/pdf/0812.4147.pdf (To appear in the European Journal of Combinatorics, 2010)
37. Traldi, L.: A subset expansion of the coloured Tutte polynomial. Combinatorics, Probability and Computing 13, 269–275 (2004)
38. Traldi, L.: On the colored Tutte polynomial of a graph of bounded tree-width. Discrete Applied Mathematics 6, 1032–1036 (2006)

Fast Exact Algorithms for Hamiltonicity in Claw-Free Graphs

Hajo Broersma[1], Fedor V. Fomin[2], Pim van 't Hof[1], and Daniël Paulusma[1]

[1] Department of Computer Science, University of Durham,
Science Laboratories, South Road, Durham DH1 3LE, England*
{hajo.broersma,pim.vanthof,daniel.paulusma}@durham.ac.uk
[2] Institutt for informatikk, Universitet i Bergen,
Postboks 7803, 5020 Bergen, Norway
fomin@ii.uib.no

Abstract. The HAMILTONIAN CYCLE problem asks if an n-vertex graph G has a cycle passing through all vertices of G. This problem is a classic NP-complete problem. So far, finding an exact algorithm that solves it in $\mathcal{O}^*(\alpha^n)$ time for some constant $\alpha < 2$ is a notorious open problem. For a claw-free graph G, finding a hamiltonian cycle is equivalent to finding a closed trail (eulerian subgraph) that dominates the edges of some associated graph H. Using this translation we obtain two exact algorithms that solve the HAMILTONIAN CYCLE problem for the class of claw-free graphs: one algorithm that uses $\mathcal{O}^*(1.6818^n)$ time and exponential space, and one algorithm that uses $\mathcal{O}^*(1.8878^n)$ time and polynomial space.

1 Introduction

In this paper we study the well-known NP-complete decision problem HAMILTONIAN CYCLE (cf. [7]) that asks whether a graph G has a *hamiltonian cycle*, i.e., a cycle that passes through all vertices of G. The HAMILTONIAN CYCLE problem can be seen as a special case of the well-known TRAVELING SALESMAN problem. The input of the latter problem is a complete graph together with an edge weighting. The goal is to find a hamiltonian cycle of minimum total weight. Held & Karp [11] present a classic dynamic programming algorithm that solves the TRAVELING SALESMAN problem in $\mathcal{O}^*(2^n)$ time and $\mathcal{O}^*(2^n)$ space for graphs on n vertices. The $\mathcal{O}^*$-notation indicates that we suppress factors of polynomial order, and we use this notation throughout the paper. The slightly easier HAMILTONIAN CYCLE problem can be solved using $\mathcal{O}^*(2^n)$ time and polynomial space, as was shown by Karp [13] and independently by Bax [1]. It is a major and long outstanding open problem if the HAMILTONIAN CYCLE and the TRAVELING SALESMAN problem can be solved in $\mathcal{O}^*(\alpha^n)$ time for some constant $\alpha < 2$, even if the polynomial space restriction is dropped.

For some graph classes for which the HAMILTONIAN CYCLE, and consequently the TRAVELING SALESMAN problem, remains NP-complete, faster algorithms

---

* This work has been supported by EPSRC (EP/D053633/1).

have been designed. For planar graphs, the HAMILTONIAN CYCLE problem can be solved in $\mathcal{O}^*(c^{\sqrt{n}})$ for some constant c (cf. [19]). The TRAVELING SALESMAN problem can be solved in $\mathcal{O}^*(1.251^n)$ time for cubic graphs [12] and in $\mathcal{O}^*(1.890^n)$ time for graphs with maximum degree 4 [5]. Both algorithms use polynomial space. For graphs with maximum degree 4, an algorithm with time complexity $\mathcal{O}^*(1.733^n)$ is known [8], but this algorithm uses exponential space. More generally, Björklund et al. [2] present an algorithm that solves the TRAVELING SALESMAN problem in $\mathcal{O}^*((2-\epsilon)^n)$ for graphs with bounded degree, where $\epsilon > 0$ only depends on the maximum degree but not on the number of vertices. They show that this bound can be improved further for regular triangle-free graphs. These algorithms use exponential space. They also present an $\mathcal{O}^*((2-\epsilon)^n)$ time algorithm that uses polynomial space for bounded degree graphs in which the edges have bounded integer weights.

Our Results. We consider the class of claw-free graphs. This is a rich class containing, e.g., the class of line graphs and the class of complements of triangle-free graphs. It is also an intensively studied graph class, both within structural graph theory and within algorithmic graph theory; see [6] for a survey. The HAMILTONIAN CYCLE problem is NP-complete for claw-free graphs; the authors of [14] show that the problem is already NP-complete for 3-connected cubic planar claw-free graphs. We present two exact algorithms that solve the HAMILTONIAN CYCLE problem for claw-free graphs: our first algorithm uses $\mathcal{O}^*(1.6818^n)$ time and exponential space, and our second algorithm uses $\mathcal{O}^*(1.8878^n)$ time and polynomial space. Our techniques are based on a (known) transformation of the problem to the problem of finding a dominating closed trail in a graph and a new, more careful study of such trails. Hence, these techniques are different from the ones used in the already known algorithms, and as such may be of independent interest.

Preliminaries. All graphs in this paper are finite, undirected and without multiple edges and loops. For notation and terminology not defined in this paper we refer to [4]. Let $G = (V(G), E(G))$ be a graph. The *neighborhood* of a vertex v in G is denoted by $N_G(v) := \{w \in V(G) \mid vw \in E\}$, and $d_G(v) = |N_G(v)|$ denotes the *degree* of v. A *2-factor* of G is a spanning subgraph of G in which all vertices have degree 2. The subgraph of G induced by some $U \subseteq V$ is denoted by $G[U]$.

A graph is called *triangle-free* if it does not contain a subgraph isomorphic to the cycle on three vertices. A graph is called *claw-free* if it has no induced subgraph isomorphic to the *claw*, i.e., the four-vertex star $K_{1,3} = (\{u, a, b, c\}, \{ua, ub, uc\})$. Let G be a claw-free graph. Then, for each vertex v of G, the set of neighbors of v in G induces a subgraph with at most two components. If this subgraph has two components, both of them must be cliques. If the subgraph induced by $N_G(x)$ is connected but not complete, we can perform an operation called *local completion of G at x* by adding edges joining all pairs of nonadjacent vertices in $N_G(x)$.

The *line graph* of a graph H with edges $e_1, \ldots, e_p$ is the graph $L(H)$ with vertices $u_1, \ldots, u_p$ such that there is an edge between any two vertices u_i and u_j if and only if e_i and e_j share one end vertex in H. Note that $L(K_3) = L(K_{1,3}) = K_3$;

it is well-known that every connected line graph $F \neq K_3$ has a unique H with $F = L(H)$ (see e.g. [9]). We call H the *preimage graph* of F. For K_3 we let $K_{1,3}$ be its preimage graph. A graph is called *even* if all its vertices have even degree. A graph is called a *closed trail* (or *eulerian*) if it is a connected even graph. Let T be a closed trail in a graph H. If $V(H) \setminus V(T)$ is an independent set in H, then we say that T is a *dominating closed trail*, abbreviated DCT. Note that the latter means that every edge of H has at least one vertex in T, so in this context "dominating" means "edge-dominating". For any integer $k \geq 1$, a graph H is called *k-degenerate* if every non-empty subgraph of H has a vertex of degree at most k. We say that H is *k-ordered* if H allows a vertex ordering $\pi = v_1, \ldots, v_{|V(H)|}$ such that for $1 \leq i \leq |V(H)|$, $H[\{v_1, \ldots, v_i\}]$ is connected and v_i has at most k neighbors in $H[\{v_1 \ldots, v_i\}]$.

Paper organization. In Section 2 we translate the HAMILTONIAN CYCLE problem for claw-free graphs into the problem of finding a dominating closed trail in triangle-free graphs. In Section 3 we show that every graph with a spanning closed trail has a 2-degenerate 3-ordered spanning closed trail. We use this structural result in Section 4, where we present two exact algorithms for finding a dominating closed trail in a graph. Section 5 contains the conclusions and mentions some open problems.

2 The Two Exact Algorithms

Here we explain our two algorithms that solve the HAMILTONIAN CYCLE problem for a claw-free graph G on n vertices. For the first step we do not have to develop any new theory or algorithms, but can rely on the beautiful existing machinery from the literature.

Step 1: restrict to the preimage graph H of the closure of G

We recursively repeat the local completion operation, as long as this is possible. This way we obtain the *closure* $cl(G)$ of G. Ryjáček [17] showed that the closure of G is uniquely determined, i.e., that the ordering in which one performs the local completions does not matter. This means we can obtain $cl(G)$ in polynomial time. Ryjáček [17] also showed that G is hamiltonian if and only if $cl(G)$ is hamiltonian. Furthermore he showed that for any claw-free graph G there is a unique (triangle-free) graph H such that $L(H) = cl(G)$. We can obtain the preimage graph of a line graph in polynomial time (see e.g. [16]). Hence, we can efficiently compute the unique graph H with $L(H) = cl(G)$.

Step 2: find a DCT of H

Harary and Nash-Williams [10] showed that the line graph of any connected graph with at least three vertices is hamiltonian if and only if the graph itself contains a DCT. This result combined with the results from the previous step implies that G has a hamiltonian cycle if and only if H has a DCT. In Section 4 we present two exact algorithms for finding such a DCT in a graph with n

edges: one algorithm that uses $\mathcal{O}^*(1.6818^n)$ time and exponential space, and one algorithm that uses $\mathcal{O}^*(1.8878^n)$ time and polynomial space.

Step 3: translate the DCT of H back into a hamiltonian cycle of $cl(G)$

Suppose we have obtained a DCT T in Step 2. Then we construct a hamiltonian cycle of $cl(G)$ by traversing T, picking up the edges (corresponding to vertices in $cl(G)$) one by one and inserting dominated edges as soon as an end vertex of a dominated edge is encountered. For traversing T we use the polynomial-time algorithm that finds a eulerian tour in an even connected graph (cf. [4]).

Step 4: translate the hamiltonian cycle in $cl(G)$ to one in G

We can do this in polynomial time by using exactly the same method as described in [3]. There, we show how to translate a 2-factor of $cl(G)$ into a 2-factor of G. Since a hamiltonian cycle is a connected 2-factor we are done.

From the above it is clear that all steps except the third one can be performed in polynomial time. Hence, we have found the following.

Theorem 1. *The* HAMILTONIAN CYCLE *problem for a claw-free graph on n vertices can be solved in $\mathcal{O}^*(1.6818^n)$ time, using exponential space. It can also be solved in $\mathcal{O}^*(1.8878^n)$ time, using polynomial space.*

3 Closed Trails of Low Degeneracy and Ordering

A cycle C of a connected graph H is called *removable* if the graph $H - E(C)$ is connected and *non-separating* if $H - V(C)$ is connected. The following useful result is due to Thomassen and Toft [18].

Theorem 2 ([18]). *Any connected graph with minimum degree 3 has an induced non-separating cycle.*

Theorem 2 immediately yields the following result.

Corollary 1. *Any connected graph with minimum degree 3 has a removable cycle.*

Proof. Let H be a connected graph with minimum degree 3. By Theorem 2, H has an induced non-separating cycle C. Since $H - V(C)$ is connected, all vertices of $V(H) \setminus V(C)$ belong to the same component of $H - E(C)$. Since H has minimum degree 3 and C is an induced cycle, every vertex of C has a neighbor in $V(H) \setminus V(C)$. Hence $H - E(C)$ is connected, so C is removable. □

Using Corollary 1 we can prove the following theorem, which will help us to obtain the time complexity of the exact algorithms described in Section 4.

Theorem 3. *Every graph with a spanning closed trail contains a 2-degenerate 3-ordered spanning closed trail.*

Proof. We first show that every graph with a spanning closed trail contains a 2-degenerate spanning closed trail. Let H^* be a counterexample with $|E(H^*)|$ minimum. Let T be a spanning closed trail in H^*. We repeatedly remove vertices from T with degree at most 2 in T as long as possible. Let T' be the subgraph of T we obtain this way. Since H^* is a counterexample, T' is not empty. Let T_1 be a component of T'. Since T' has minimum degree at least 3, T_1 has a removable cycle C by Corollary 1. Then C is also a removable cycle of H^*, since H^* is a supergraph of T_1. This contradicts the minimality of $|E(H^*)|$.

So, every graph H with a spanning closed trail contains a 2-degenerate spanning closed trail T. Suppose T is not 3-ordered. We repeatedly remove vertices from T with degree at most 3 in T until T becomes disconnected. Let T' be the resulting (connected) subgraph of T. Since T is not 3-ordered, T' is not empty. Let U consist of all vertices of degree at most 3 in T'. By our procedure, every vertex of U is a cut-vertex of T', and since T is 2-degenerate, U is nonempty. Let $u \in U$ be such that $T'[V(T') \setminus \{u\}]$ contains a component D without vertices of U. Then all vertices of D have degree at least 3, contradicting the 2-degeneracy of T. □

4 Two Exact Algorithms for Finding a DCT

We present two exact algorithms for solving the following problem.

DOMINATING CLOSED TRAIL (DCT)
Instance: a connected graph H.
Question: does H have a dominating closed trail?

To solve the DCT problem for an instance H, both algorithms start by branching on vertices of low degree by the same branching procedure, explained in Section 4.1. This way both algorithms obtain a set of subproblems. Each subproblem has the original graph H as input. However, for some subset of edges of H it is already decided whether they will be included in or excluded from the dominating closed trail. Our first algorithm, described in Section 4.2, solves each of the subproblems using dynamic programming. Our second algorithm, described in Section 4.3, solves each of the subproblems by guessing the remaining edges of a possible dominating closed trail.

4.1 Branching on Vertices of Low Degree

Let $H = (V, E)$ be an instance of the DCT problem. We assign a so-called *parity label* $\ell(v) \in \{0, 1\}$ to each vertex v of H. Note that if H has a dominating closed trail T, then $d_T(v)$ is even for every $v \in V$. After all, a vertex is either not in T (i.e., $d_T(v) = 0$, in which case all of its neighbors must be in T), or a vertex has an even number of incident edges in T (since T is a closed trail). Hence we initially set $\ell(v) = 0$ for every $v \in V$.

The first stage of both algorithms consists of branching on vertices of degree at most d^*, thus creating a number of subproblems; more specifically, $d^* = 4$ for our first algorithm, and $d^* = 12$ for our second algorithm. The choice of these values of d^* is explained in the next sections. During the branching process, the size of the graphs under consideration decreases, and we might change the ℓ-labels of certain vertices.

Suppose v is a vertex of degree $d \leq d^*$ in H. If $\ell(v) = 0$ (respectively $\ell(v) = 1$), then the algorithm branches into 2^{d-1} subproblems, each subproblem corresponding to a possible way of choosing an even (respectively odd) number $0 \leq p \leq d$ of edges incident with v that are guessed to be in the dominating closed trail. We call the chosen edges *old trail edges*. For each choice W of old trail edges, we perform the following two operations:

1. set $\ell(w) := \ell(w) + 1 \pmod 2$ for every w with $vw \in W$;
2. delete v and all its d incident edges.

Repeat this procedure as long as the remaining graph contains a vertex of degree at most d^*. Let H' be the resulting graph. Then H' has minimum degree $d^* + 1$ and each vertex $u \in V(H')$ has some label $\ell(u) \in \{0, 1\}$. Let $E(H) = E(H') \cup R(H') \cup W(H')$, where $W(H')$ contains all old trail edges and $R(H')$ contains all other edges we removed from H. In the next stage, edges in $W(H')$ will be assumed to be in the dominating closed trail we are looking for, whereas edges in $R(H')$ will be assumed not to be in the dominating closed trail. Suppose $R(H')$ contains an edge $e = xy$ with $x, y \in V(H) \setminus V(H')$ such that both x and y are not incident with any old trail edge. Then e will not be dominated by any closed trail that we might discover in the next stage. Hence, we discard this subproblem. For the same reason, we also discard the subproblem if there is a vertex $v \in V(H) \setminus V(H')$ incident with an odd number of old trail edges. If these two cases do not occur, we keep the subproblem and call the tuple $(H', W(H'), \ell)$ a *stage-2 tuple*.

Lemma 1. *The branching phase of the algorithm creates $T(n_1) = \mathcal{O}^*(2^{\frac{d^*-1}{d^*} n_1})$ stage-2 tuples, where n_1 is the total number of edges deleted during this phase.*

Proof. Since for a vertex v of degree d we remove d edges and create 2^{d-1} subgraphs, we find $T(n_1) = 2^{d-1} \cdot T(n_1 - d)$, which yields $T(n_1) = \mathcal{O}^*(2^{\frac{d-1}{d} n_1})$. Since $d \leq d^*$, we end up with $\mathcal{O}^*(2^{\frac{d^*-1}{d^*} n_1})$ stage-2 tuples. □

We point out that the time complexity mentioned in Lemma 1 is $\mathcal{O}^*(1.6818^{n_1})$ if $d^* = 4$ and $\mathcal{O}^*(1.8878^{n_1})$ if $d^* = 12$.

4.2 An $\mathcal{O}^*(1.6818^n)$ Time Algorithm That Uses Exponential Space

Let $H = (V, E)$ be an input of the DCT problem. In case H has vertices of degree at most 4, we apply the branching procedure described in Section 4.1. Suppose that during the branching process n_1 edges were deleted (possibly $n_1 = 0$). Then, by Lemma 1, $\mathcal{O}^*(1.6818^{n_1})$ stage-2 tuples $(H', W(H'), \ell)$ have been created. Each of these stage-2 tuples will be processed using the dynamic programming procedure described below. If at least one of them leads to a dominating closed trail of H, then the algorithm outputs YES; the algorithm outputs NO otherwise.

Let $(H', W(H'), \ell)$ be a stage-2 tuple. We write $H' = (V', E')$. We output YES if $W(H')$ forms a dominating closed trail of H. If this is not the case, we enter the dynamic programming phase. In this procedure, we consider each $u \in V'$ and say that $(\{u\}, \ell(u))$ is an *option* if $u \in V'$ is incident with at least one old trail edge. Otherwise $(\{u\}, \ell(u))$ is not an option. Furthermore, $(\{u\}, \bar{\ell}(u))$ with $\bar{\ell}(u) = \ell(u) + 1 \pmod 2$ is not an option.

Suppose we know for all sets $S \subseteq V'$ of size at most k and all labelings $\ell' : S \to \{0, 1\}$ whether (S, ℓ') is an option or not. Then for each set $S \subseteq V'$ of size k, for each vertex $v \in V' \setminus S$, and for each $\{0, 1\}$-labeling ℓ' of $S \cup \{v\}$, we do as follows. Let p be the number of old trail edges incident with v. We consider every possible way of choosing $0 \leq q \leq 3$ edges incident with v and a vertex in S. The chosen edges will be referred to as *new trail edges*. For each choice N of new trail edges, we set $\ell'(x) := \ell'(x) + 1 \pmod 2$ for every $x \in S$ with $vx \in N$. We perform the following three tests.

(1) Check if (S, ℓ') is an option.
(2) Check if $p + q$ is even if $\ell'(v) = 0$ and odd if $\ell'(v) = 1$.
(3) If $q = 0$, check if there is a path from v to S in H only using old trail edges.

Only if tests (1), (2), (3) are all three affirmative, we say that $(S \cup \{v\}, \ell')$ is an option. If so, we also check whether

(4) each old trail edge allows a path to a vertex in $S \cup \{v\}$ that uses only old trail edges;
(5) each vertex x in $S \cup \{v\}$ has label $\ell'(x) = 0$ and each vertex $y \in V' \setminus (S \cup \{v\})$ incident with an old trail edge has label $\ell(y) = 0$;
(6) there is no edge $e = ab$ in H' for some $a, b \in V' \setminus (S \cup \{v\})$ such that both a and b are not incident with an old trail edge.

If the answers to tests (4), (5), (6) are all three affirmative, the algorithm concludes that H has a dominating closed trail (cf. Theorem 4) and returns YES. If no YES-answer has been returned and $k < |V'|$, the algorithm considers all sets $S \subseteq V'$ of size $k + 1$, all vertices $v \in V' \setminus S$ and all $\{0, 1\}$-labelings ℓ' of $S \cup \{v\}$. Otherwise, the algorithm outputs NO.

Theorem 4 (Correctness). *When run on a connected graph H, the algorithm returns* YES *if H has a dominating closed trail, and returns* NO *otherwise.*

Proof. Our algorithm only returns a YES-answer if it has found a stage-2 tuple $(H', W(H'), \ell)$ with some option (S, ℓ) for which tests (4), (5), (6) are all positive. In that case, let T be the subgraph of H consisting of all old trail edges in $W(H')$ plus all new trail edges that have been added between vertices of S. The dynamic programming, together with tests (3) and (4), ensures that T is connected. Tests (1), (2) and (5) together with the definition of a stage-2 tuple ensure that T is even, and (6) ensures that T is dominating. Hence, T is a dominating closed trail.

It remains to show that if H has a dominating closed trail, then the algorithm outputs YES. Suppose H has a dominating closed trail T. Due to Theorem 3 we may assume that T is 3-ordered. We show that our algorithm finds T, unless it finds another dominating closed trail of H first. Let V' consist of all vertices that are not removed in the branching procedure, so $V(H') = V'$ for the graph H' in every stage-2 tuple. Let T' be the subgraph of T with $V(T') = V(T) \cap V'$. Then there exists a stage-2 tuple $(H', W(H'), \ell)$ such that $W(H')$ is exactly the set of edges of T that are incident with at least one vertex in $V(T) \setminus V'$, and such that $\ell(v) = 0$ if $v \in V' \setminus V(T')$, and $\ell(v) = 0$ (respectively $\ell(v) = 1$) if $v \in V(T')$ and v is incident with an even (respectively odd) number of edges in $W(H')$. Since our algorithm considers all possible stage-2 tuples, it will detect tuple $(H', W(H'), \ell)$. As T is 3-ordered, each component of T' is 3-ordered. This means that our dynamic programming procedure based on the number of ways a vertex can be made adjacent to a set S with at most three edges will find a labeling ℓ' such that (T_i, ℓ') is an option for each component T_i of T. As these components are connected to each other via old trail edges, at some moment (T', ℓ) will be formed. Then tests (1)-(6) will all be successful and a YES-answer is returned. □

Below we give the overall running time of our algorithm.

Theorem 5 (Running time). *The algorithm runs in $\mathcal{O}^*(1.6818^n)$ time.*

Proof. We first prove that the dynamic programming procedure runs in $\mathcal{O}^*(3^p)$ time on any p-vertex graph. Let $H' = (V', E')$ be a graph on p vertices. There are $\binom{p}{k}$ sets $S \subseteq V'$ of cardinality k, each of those sets has 2^k possible labelings ℓ, and there are $\binom{k}{0} + \binom{k}{1} + \binom{k}{2} + \binom{k}{3} = \mathcal{O}(k^3)$ ways to attach a new vertex v to a subset of cardinality k by using at most 3 edges. Each of the tests (1)-(6) can be done in polynomial time. Hence the time complexity of this procedure is

$$\mathcal{O}^*\left(\sum_{k=1}^{p} \binom{p}{k} \cdot 2^k \cdot \mathcal{O}(k^3)\right) = \mathcal{O}^*(3^p).$$

Let H be an instance of the DCT problem having n edges. Suppose we repeatedly branch on vertices of degree at most $d^* = 4$, and suppose n_1 is the number of edges we delete during this branching phase. Then we obtain $\mathcal{O}^*(1.6818^{n_1})$ stage-2 tuples by Lemma 1. Let $(H', W(H'), \ell)$ be such a stage-2 tuple, where $H' = (V', E')$ is a graph of minimum degree 5 having $n_2 := n - n_1$ edges and, say, p vertices. As shown above, the dynamic programming procedure uses $\mathcal{O}^*(3^p)$ time. Since the minimum degree in H' is 5, we obtain $n_2 \geq 5p/2$, or equivalently $p \leq 2n_2/5$. Hence we can process each stage-2 tuple in time $\mathcal{O}^*(3^{\frac{2n_2}{5}}) = \mathcal{O}^*(1.5519^{n_2})$. This means that the overall running time of our algorithm on a graph H having $n = n_1 + n_2$ edges is

$$\mathcal{O}^*(1.6818^{n_1} \cdot 1.5519^{n_2}) = \mathcal{O}^*(1.6818^n).$$

If we choose $d^* \neq 4$, then the above upper bound is no longer guaranteed. □

4.3 An $\mathcal{O}^*(1.8878^n)$ Time Algorithm That Uses Polynomial Space

We describe our second algorithm in the proof of the following theorem.

Theorem 6. *The DCT problem for a graph H on n edges can be solved in $\mathcal{O}^*(1.8878^n)$ time, using polynomial space.*

Proof. Let H be an instance of the DCT problem with n edges. We execute the branching procedure described in Section 4.1, but this time we perform branching on vertices of degree at most $d^* = 12$. Suppose we delete n_1 edges during the branching process. By Lemma 1, this yields $\mathcal{O}^*(2^{11n_1/12}) = \mathcal{O}^*(1.8878^{n_1})$ stage-2 tuples $(H', W(H'), \ell)$, where each graph H' has p vertices of minimum degree 13 and $n_2 = n - n_1$ edges. Note that $n_2 \geq 13p/2$, or equivalently $p \leq 2n_2/13$.

If H has a dominating closed trail T, then T may be assumed to be 2-degenerate, due to Theorem 3. Let T' denote the (2-degenerate) subgraph of T that remains after the branching procedure; note that T' is a subgraph of some graph H'. A 2-degenerate graph on p vertices has at most $2p$ edges. This means that we only have to check in every H' for every possible subset of edges up to cardinality $2p$ whether this subset together with the old trail edges in $W(H')$ forms a dominating closed trail of H. Using Sterling's approximation $n_2! \approx n_2^{n_2} e^{-n_2} \sqrt{2\pi n_2}$ and the fact $p \leq 2n_2/13$, the total number of checks can be estimated as follows:

$$\sum_{k=1}^{2p} \binom{n_2}{k} \leq 2p \binom{n_2}{2p} \leq 2p \binom{n_2}{\frac{4n_2}{13}} = \mathcal{O}^*\left(\left(\frac{1}{\alpha^\alpha (1-\alpha)^{1-\alpha}}\right)^{n_2}\right),$$

where $\alpha = 4/13$, which leads to $\mathcal{O}^*(1.8539^{n_2})$ checks. Since each of them can be performed in polynomial time, the overall running time is

$$\mathcal{O}^*(1.8878^{n_1} \cdot 1.8539^{n_2}) = \mathcal{O}^*(1.8878^n).$$

If we choose $d^* \neq 12$, then the above upper bound is no longer guaranteed. It is clear that this algorithm only needs polynomial space. □

5 Conclusions

We presented two exact algorithms for the HAMILTONIAN CYCLE problem. Can we speed up these algorithms by making use of the triangle-freeness of the preimage graph? Another (more) interesting open problem is whether we can solve the TRAVELING SALESMAN problem for claw-free graphs in $\mathcal{O}^*(\alpha^n)$ time for some constant $\alpha < 2$. This requires some new ideas as our current approach that takes the closure of a graph and then makes a transformation to the domain of triangle-free graphs does not suffice. Can we find an $\mathcal{O}^*(\alpha^n)$ time algorithm that solves the HAMILTONIAN CYCLE problem for some constant $\alpha < 2$ for the class of bipartite graphs, or equivalently (cf. [15]) for the class of split graphs, or a superclass of split graphs such as the class of P_5-free graphs? As the HAMILTONIAN CYCLE problem is already NP-complete for chordal bipartite graphs [15], this question is interesting for that class as well. We can also try to design fast exact algorithms for superclasses of claw-free graphs such as $K_{1,4}$-free graphs.

References

1. Bax, E.T.: Inclusion and exclusion algorithm for the Hamiltonian path problem. Information Processing Letters 47, 203–207 (1993)
2. Björklund, A., Husfeldt, T., Kaski, P., Koivisto, M.: The travelling salesman problem in bounded degree graphs. In: Aceto, L., Damgård, I., Goldberg, L.A., Halldórsson, M.M., Ingólfsdóttir, A., Walukiewicz, I. (eds.) ICALP 2008, Part I. LNCS, vol. 5125, pp. 198–209. Springer, Heidelberg (2008)
3. Broersma, H.J., Paulusma, D.: Computing sharp 2-factors in claw-free graphs. In: Ochmański, E., Tyszkiewicz, J. (eds.) MFCS 2008. LNCS, vol. 5162, pp. 193–204. Springer, Heidelberg (2008)
4. Diestel, R.: Graph Theory, 2nd edn. Graduate Texts in Mathematics, vol. 173. Springer, Heidelberg (2000)
5. Eppstein, D.: The traveling salesman problem for cubic graphs. Journal of Graph Algorithms and Applications 11, 61–81 (2007)
6. Faudree, R., Flandrin, E., Ryjáček, Z.: Claw-free graphs—a survey. Discrete Mathematics 164, 87–147 (1997)
7. Garey, M.R., Johnson, D.S.: Computers and Intractability. W.H. Freeman and Co., New York (1979)
8. Gebauer, H.: On the number of hamilton cycles in bounded degree graphs. In: 4th Workshop on Analytic and Combinatorics (ANALCO 2008), SIAM, Philadelphia (2008)
9. Harary, F.: Graph Theory. Addison-Wesley, Reading (1969)
10. Harary, F., Nash-Williams, C.S.J.A.: On eulerian and hamiltonian graphs and line graphs. Canadian Mathematical Bulletin 8, 701–709 (1965)
11. Held, M., Karp, R.M.: A dynamic programming approach to sequencing problems. Journal of SIAM 10, 196–210 (1962)
12. Iwama, K., Nakashima, T.: An improved exact algorithm for cubic graph TSP. In: Lin, G. (ed.) COCOON 2007. LNCS, vol. 4598, pp. 108–117. Springer, Heidelberg (2007)
13. Karp, R.M.: Dynamic programming meets the principle of inclusion and exclusion. Operations Research Letters 1, 49–51 (1982)
14. Li, M., Corneil, D.G., Mendelsohn, E.: Pancyclicity and NP-completeness in planar graphs. Discrete Applied Mathematics 98, 219–225 (2000)
15. Müller, H.: Hamiltonian circuits in chordal bipartite graphs. Discrete Mathematics 156, 291–298 (1996)
16. Roussopoulos, N.D.: A max {m,n} algorithm for determining the graph H from its line graph G. Information Processing Letters 2, 108–112 (1973)
17. Ryjáček, Z.: On a closure concept in claw-free graphs. Journal of Combinatorial Theory, series B 70, 217–224 (1997)
18. Thomassen, C., Toft, B.: Non-separating induced cycles in graphs. Journal of Combinatorial Theory, Series B 31, 199–224 (1981)
19. Woeginger, G.J.: Open problems around exact algorithms. Discrete Applied Mathematics 156, 397–405 (2008)

Maximum Series-Parallel Subgraph

Gruia Călinescu[1,*], Cristina G. Fernandes[2,**], and Hemanshu Kaul[3]

[1] Department of Computer Science, Illinois Institute of Technology, Chicago, IL 60616, USA
calinescu@iit.edu
[2] Department of Computer Science, University of São Paulo, Rua do Matão, 1010, 05508-090 São Paulo, Brazil
cris@ime.usp.br
[3] Department of Applied Mathematics, Illinois Institute of Technology, Chicago, IL 60616, USA
kaul@iit.edu

Abstract. Consider the NP-hard problem of, given a simple graph G, to find a series-parallel subgraph of G with the maximum number of edges. The algorithm that, given a connected graph G, outputs a spanning tree of G, is a $\frac{1}{2}$-approximation. Indeed, if n is the number of vertices in G, any spanning tree in G has $n-1$ edges and any series-parallel graph on n vertices has at most $2n-3$ edges. We present a $\frac{7}{12}$-approximation for this problem and results showing the limits of our approach.

1 Introduction

The Maximum Series-Parallel Subgraph (MSP) problem is: given a simple graph G, find a series-parallel subgraph of G with the maximum number of edges. This problem is known to be NP-hard [3].

The algorithm that, given a connected graph G, outputs a spanning tree of G, is a $1/2$-approximation. Indeed, if n is the number of vertices in G, any spanning tree in G has $n-1$ edges and any series-parallel graph on n vertices has at most $2n-3$ edges. We present a $7/12$-approximation for this problem.

We apply a method, previously used for the Maximum Planar Subgraph problem [4], of producing a subgraph whose blocks (maximal 2-connected components) have a very simple structure. The way to produce such a subgraph also has similarities to some approximation algorithms for the Minimum Steiner Tree problem [1,6].

A novelty of this work is that we allow blocks to have unbounded size. Indeed, using only blocks of bounded size does not lead to an improvement (as we show later). This is a main difference to the works on Maximum Planar Subgraph and Minimum Steiner Tree [1,4,6]. A second difference, when compared to the

---

[*] Research supported in part by NSF grant CCF-0515088, and performed in part while on sabbatical at University of Wisconsin Milwaukee.
[**] Research supported in part by CNPq 312347/2006-5, 485671/2007-7, and 486124/2007-0.

Maximum Planar Subgraph algorithms, is that, to assure a good performance, our algorithm has to sometimes throw away or shrink previously selected blocks. We show ahead a family of examples that indicates that such an approach is necessary.

We call *spruces* the very simple series-parallel graphs that we admit as non-bridge blocks in the subgraph we produce. (We define spruces in the next subsection; a *bridge* consists of two adjacent vertices.) We prove that a subgraph whose non-bridge blocks are spruces, and with maximum number of edges among such subgraphs, achieves a ratio of 2/3, and this ratio is tight. Unfortunately, computing such a subgraph is NP-hard, as we also show. So our algorithm in fact computes only a large such subgraph. The ratio our algorithm achieves is 7/12, which happens to be the average between 1/2 and 2/3. This is a coincidence though, because our analysis compares directly the algorithm's output to an optimal solution.

In a related work, Cai [2] considered the variant of the problem where one is given a complete weighted graph, and wants to find a maximal series-parallel graph of minimum weight. He presented a 1.655-approximation for this variant when the input graph is a set of points in the plane with their distances as weights.

1.1 Preliminaries

Two edges of a multigraph are *parallel* if they have the same endpoints, and they are *series* edges if there is some vertex of degree two incident to both of them. A multigraph is *series-parallel* if it arises from a forest by repeated replacing edges by parallel or series edges [7].

All of our graphs are undirected and simple, unless otherwise specified. From the definition above, one can see that a maximal series-parallel graph can be constructed by the following procedure. Start with two adjacent vertices s and t, and then repeat the following: add one new vertex and make it adjacent to two existing adjacent vertices. (Such graphs are also called *2-trees* in the literature, and series-parallel graphs are also known as *partial 2-trees*.)

Based on the construction above, a *normalized* tree decomposition of a maximal series-parallel graph is built as follows (see Fig. 1 for an example). Start with one node with bag $\{s,t\}$, the root of our tree decomposition. We maintain the invariant that, for any edge of the series-parallel graph, there is exactly one node in the tree decomposition whose bag consists of the endpoints of the edge. Whenever a vertex z is added to the series-parallel graph, and made adjacent to existing adjacent vertices x and y, add to the tree decomposition three nodes: one with bag $\{x,y,z\}$, child of the node with bag $\{x,y\}$, and two "twin" children of this new node, with bags $\{x,z\}$ and $\{y,z\}$. In this tree decomposition, all even-level nodes have bags of size two, all odd-level nodes have bags of size three, and no leaf is in an odd level. For a normalized tree decomposition T of a maximal series-parallel graph H with $|V(H)| = n$, there are exactly $n-2$ odd-level nodes in T.

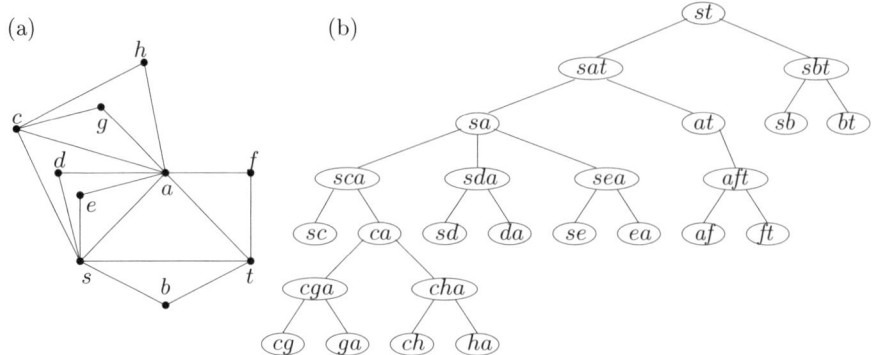

Fig. 1. (a) A maximal series-parallel graph, obtained by starting with the two adjacent vertices s and t, and then adding in order vertices a, b, c, d, e, f, g, h. (b) Its normalized tree decomposition.

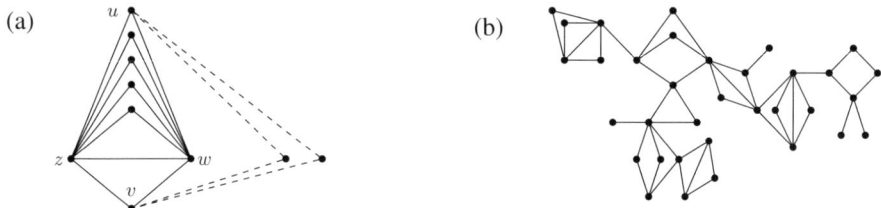

Fig. 2. (a) A graph with several spruces. (b) A connected spruce structure.

A *spruce* is a graph that has exactly two *base* vertices and at least one *tip* vertex, in which every tip vertex is adjacent to exactly the two base vertices. If the two base vertices are adjacent, the spruce is *complete*; otherwise it is incomplete. The *gain* of a spruce S is its cyclomatic number, and it is denoted $gain(S)$; this is the number of tips for complete spruces, and one less than the number of tips for incomplete spruces.

Fig. 2(a) depicts in solid lines a complete spruce with base vertices z and w, and six tip vertices including u and v. Another spruce contained in the same graph has base vertices u and v, and four tips including z and w; this second spruce is incomplete.

A *spruce cactus* is a graph such that each of its blocks is a spruce. A *spruce structure* is a graph each of whose blocks is a spruce or a bridge edge. See an example in Fig. 2(b).

Fact 1. *Spruce cactuses/structures are series-parallel graphs.*

We can view a spruce cactus as a collection of spruces — those giving the blocks of the spruce cactus. A spruce cactus is *well-behaved* if it is a collection of spruces that do not share tips. We define the *gain* of a spruce cactus to be its cyclomatic number.

Fact 2. *The gain of a spruce cactus equals the sum of the gains of its spruces.*

Before we proceed with the algorithm, we first elaborate on the need of spruces of unbounded size. First, if the input graph is a complete spruce with $n-2$ tips (and $2n-3$ edges), any approach which uses blocks of size bounded by, say, k, results in an output with gain at most $k-2$ and a total of $n+k-3$ edges. With n large and k fixed, this is only a $1/2$-approximation.

Our algorithm discards and shrinks selected spruces. Why one has to do this becomes clear from the following example, depicted in Fig. 3(a). The optimum has n vertices and $2n-3$ edges. It contains a spruce with base vertices x and y and circa $\sqrt{n}$ tips. For each of its tips v, there are two complete spruces, one with base vertices x and v, and the other with base vertices v and y, each with circa $\sqrt{n}/2$ tips. If an algorithm mistakenly (or greedily) selects the spruce with base vertices x and y, then it cannot add any more spruces and it ends up with circa $n+\sqrt{n}$ edges — asymptotically not better than a $1/2$-approximation.

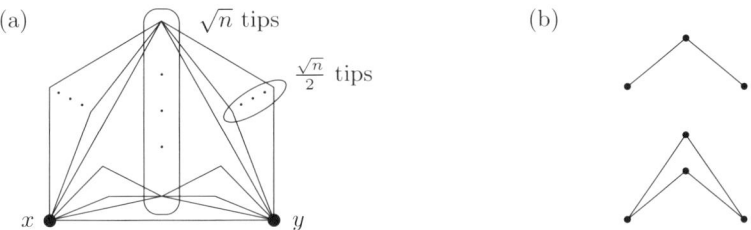

Fig. 3. (a) A graph where a naive greedy strategy that does not discard previously selected spruces fails to achieve a ratio better than $1/2$. (b) The only two types of degenerate spruces.

For the weighted version of our problem, the algorithm that returns a maximum weight spanning tree is a $1/2$-approximation. This follows from Lemma 3, which is also used in the analysis of our algorithm. Precisely, for any subgraph H' of an edge-weighted graph H, let $w(H')$ denote the sum of $w(e)$ for all e in $E(H')$. The proof of the next lemma follows closely that of Lemma 17 in [5].

Lemma 3. *Let F be a maximum weight forest in weighted simple series-parallel graph H. Then $w(H) \leq 2\,w(F)$, with the inequality being strict if $w(H) > 0$.*

Proof. We use the greedy algorithm to construct F, first sorting the edges of H into non-increasing order by weight. Let E_h be the set of the first h edges in this ordering, $1 \leq h \leq m$, where $m = |E(H)|$. By w_h we denote the weight of the hth edge in this ordering and we put $w_{m+1} = 0$. Starting with $F = \emptyset$, the greedy spanning tree algorithm scans the edges in the given order and adds an edge to F as long as it does not create any cycles.

Let F be the set of edges chosen by the greedy algorithm and let $F_h = E_h \cap F$. Then, by rearranging the terms,

$$w(F) = \sum_{h=1}^{m} |F_h|(w_h - w_{h+1}), \text{ and } w(H) = \sum_{h=1}^{m} |E_h|(w_h - w_{h+1}).$$

It is therefore enough to show that $|E_h| < 2|F_h|$ for $1 \leq h \leq m$. If this holds, of course $w(H) \leq 2w(F)$, and if $w_1 > 0$, the inequality is strict.

Choose an h such that $1 \leq h \leq m$. Let $p_1, p_2, \ldots, p_k$ be the number of vertices in the non-trivial components of F_h. Of course, $|F_h| = \sum_{z=1}^{k}(p_z - 1)$. Also note that $k \geq 1$, as F_h has at least one edge. Any edge of E_h must have its two endpoints in the same component of F_h. (Otherwise, the edge could have been selected by the greedy algorithm, merging two components of F_h.) Obviously this component is non-trivial. We associate each edge of E_h with the (non-trivial) component of F_h which contains both of its endpoints. The edges of E_h associated with a component of F_h are a subset of the edges of the graph induced in H by the vertices of this component. Thus, the number of edges associated with the zth non-trivial component is at most $2p_z - 3$, because this graph is series-parallel. But then, as $k \geq 1$, we have that $|E_h| \leq \sum_{z=1}^{k}(2p_z - 3) < \sum_{z=1}^{k} 2(p_z - 1) = 2|F_h|$. ∎

2 A Local Improvement Algorithm

We may assume the input graph G is connected. Our local improvement algorithm, when running on G, keeps a set Q of spruces in G that form a well-behaved spruce cactus. We abuse notation and sometimes think of Q as the spruce cactus it forms.

The algorithm uses a slightly modified notion of gain. (One could also get an approximation ratio higher than $1/2$ by only using gain in the algorithm, but we get a higher ratio.) For a spruce S, the *adjusted gain* of S is denoted by $\widehat{gain}(S)$, and is defined as $\widehat{gain}(S) = gain(S)$ if S is complete, and $\widehat{gain}(S) = gain(S) - 1$ if S is incomplete. We call a spruce *degenerate* if its adjusted gain is non-positive. See Fig. 3(b).

For each component C of Q, the algorithm keeps a weighted tree T_C whose vertex set is $V(C)$ and edge set is as follows. For each spruce S in C with base vertices x and y, and tips $v_1, v_2, \ldots, v_k$, there is an edge xy in T_C and edges xv_i for $i = 1, \ldots, k$. The weight of the edges is given as follows: $w(xy) = \widehat{gain}(S)$, and $w(xv_i) = 1$ for all i. Note that T_C is indeed a tree. For any two vertices x and y of C, let $index_Q(x, y)$ be an edge in T_C of minimum weight in the path in T_C from x to y. If x and y are in different components of Q, then let $index_Q(x, y)$ be undefined and consider its weight to be zero.

Let $v_1, v_2, \ldots, v_k$ be all vertices isolated in Q that are adjacent in G to both x and y. If $k \geq 1$, let $S_Q(x, y)$ be the spruce with base vertices x and y, tips $v_1, v_2, \ldots, v_k$, and the edge xy if it exists in G. Otherwise let $S_Q(x, y)$ be undefined.

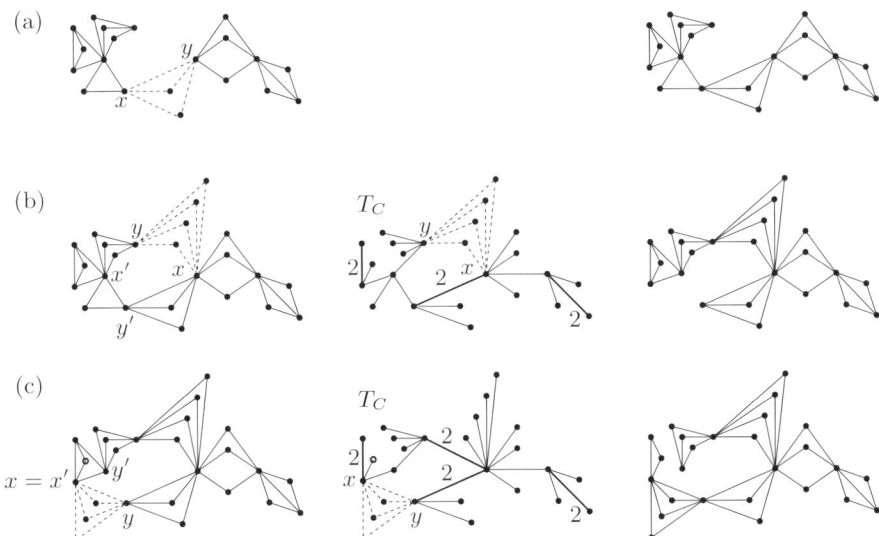

Fig. 4. Examples of local improvement, with $S_Q(x,y)$ given by the dashed lines in each case. (a) For such x and y, line 4 of the algorithm is executed resulting in Q as shown in the right. (b) For such x and y, line 7 of the algorithm is executed resulting in Q as shown in the right. The weighted tree T_C before the improvement is in the middle, with weights 1 except for those written in the figure. (c) For such x and y, line 7 and line 13 of the algorithm are executed resulting in Q as shown in the right.

The algorithm is shown in pseudocode later. We exemplify some of its cases in Fig. 4. Initially $Q = \emptyset$. The algorithm proceeds in iterations, each doing a local improvement. In each iteration, Q is updated as follows. If there are two vertices x and y of G for which $S_Q(x,y)$ is defined and $\widehat{gain}(S_Q(x,y)) > w(index_Q(x,y))$, then obtain a new Q' as follows, else go to the final phase. If $index_Q(x,y)$ is undefined, then let Q' be obtained from Q by adding $S_Q(x,y)$, and start a new iteration with Q' in the place of Q. Otherwise, let x' and y' be the endpoints of $index_Q(x,y)$, and C be the component of Q containing x, x', y, and y'. Let S' be the spruce in Q containing x' and y'. Note that such spruce exists by the construction of T_C. If x' and y' are the base vertices of S', then remove S' from Q and add $S_Q(x,y)$ to obtain Q'. Otherwise, by the construction of T_C, between x' and y' one is a base vertex of S', and the other is a tip of S'. Exchange x' and y' if needed so that x' is a base vertex of S'. Remove from S' the two edges incident to y'. If the resulting S' is degenerate or is a single edge, then remove S' from Q. Moreover, add $S_Q(x,y)$ to obtain Q', and start a new iteration with Q' in the place of Q.

Observe that, in this iterative part of the algorithm, we maintain the invariant that Q is a set of non-degenerate spruces that form a spruce cactus. Indeed, this follows by induction. It is enough to note that $\widehat{gain}(S_Q(x,y)) > 0$, and x and y are in different components, either from the start, or after we removed part or all of the spruce S' from Q.

The final phase consists of the following. Let Q now be the set of non-degenerate spruces produced by the iterative phase. Obtain a spanning connected subgraph of G from Q by adding bridges and let it be the output of the algorithm.

CONSTRUCT-SPRUCE-STRUCTURE (G)
1   $Q \leftarrow \emptyset$
2   **while** there are x and y such that $S_Q(x,y)$ is defined
        and $\widehat{gain}(S_Q(x,y)) > w(index_Q(x,y))$ **do**
3   **if** $index_Q(x,y)$ is undefined
4   **then** $Q \leftarrow Q \cup \{S_Q(x,y)\}$
5   **else** let x' and y' be the endpoints of $index_Q(x,y)$
6   let S' be the spruce in Q containing x' and y'
7               $Q \leftarrow Q \setminus \{S'\} \cup \{S_Q(x,y)\}$
8   **if** x' or y' is a tip of S'
9   **then** let z be one between x' and y' that is a tip of S'
10   let $\{e,f\}$ be the two edges of S' incident to z
11                      $S \leftarrow S' - \{e,f\}$
12   **if** S is not degenerate neither a single edge
13   **then** $Q \leftarrow Q \cup \{S\}$
14 add bridges to Q to obtain a connected spanning subgraph of G
15 **return** Q

2.1 Running Time Analysis

The main result of this section is the very technical Lemma 4 below, which shows that each iteration makes some "progress". Unfortunately, the definition of "progress" is not straightforward, for the following reason.

A natural measure of progress would be the gain of Q (that is, its cyclomatic number). If $gain(Q)$ increased in every iteration, then it would have been easy to conclude that the algorithm runs a polynomial number of iterations. However this is not the case, and a more careful analysis is required. Let us give some intuition in this paragraph. One can check that, in most of the cases, the gain of Q increases. Also, it never decreases and, in the iterations in which the gain of Q is maintained, the number of components increases — more components are helpful since more, or bigger, spruces become eligible to improve the current Q.

Define $\Phi(Q) = 3\,gain(Q) + c(Q)$, where $c(Q)$ is the number of components of Q when Q is seen as a spanning subgraph of G.

Lemma 4. *Every iteration of the algorithm increases the parameter Φ.*

From this lemma, whose proof we omit, we conclude that the number of iterations is polynomially bounded, because $\Phi(Q)$ is a non-negative integer and $gain(Q) \leq (2n-3) - (n-1) = n-2$, which means $\Phi(Q)$ is bounded by $3(n-2) + n = 4n-6$.

Also, each iteration can be easily implemented in polynomial time, as there are only $O(n^2)$ pairs x, y for which $S_Q(x,y)$ must be computed and, if possible, used in updating Q.

2.2 Approximation Ratio Analysis

Let m be the number of edges in the graph Q returned by the algorithm. Then
$$m = n - 1 + \sum_{S \in Q} gain(S).$$
Let A be an optimal solution for G and q be such that A has $2n - 3 - q$ edges. Thus, the algorithm achieves a ratio that is a constant greater than $1/2$ if

(i) $\sum_{S \in Q} gain(S)$ is at least a fraction of n, or
(ii) q is at least a fraction of n.

The analysis aims to prove that (i) or (ii) holds. Precisely, it will be shown that
$$6 \sum_{S \in Q} gain(S) + 3q \geq n - 2. \qquad (1)$$
From this, it is easy to derive the $7/12$ ratio:
$$m = n - 1 + \sum_{S \in Q} gain(S) \geq n - 1 + \tfrac{1}{6}(n - 2 - 3q) \geq \tfrac{7}{12}(2n - 3 - q).$$
The proof of Inequality (1) is not straightforward. We start by giving an overview. First we will derive a set M of spruces from A and prove that
$$\sum_{S \in M} \widehat{gain}(S) + 3q \geq n - 2.$$
This is done in Lemma 5, later. Then, to achieve Inequality (1), it remains to prove that
$$6 \sum_{S \in Q} gain(S) \geq \sum_{S \in M} \widehat{gain}(S). \qquad (2)$$
Consider Q to be the set of spruces when the algorithm finishes the iterations, and before the final phase (of adding bridges). Let t be the number of components of Q, and n' be the number of vertices in spruces of Q. Inequality (2) is a consequence of the following two inequalities:
$$4 \sum_{S \in Q} gain(S) \geq \sum_{S \in M} \widehat{gain}(S) - (n' - t),$$
which is given by Lemma 6, below, and
$$\sum_{S \in Q} gain(S) \geq \tfrac{1}{2}(n' - t),$$
which is given by Lemma 7.

In what follows, we present the description of the set M of spruces, and proceed to Lemmas 5, 6, and 7.

Let A^+ be a maximal series-parallel graph containing A. Call the edges of A^+ not in A of *missing* edges. As A^+ is maximal, it can be obtained by the incremental procedure described in the preliminaries. For each edge xy of A^+ for which this procedure added at least one new vertex adjacent to x and y, consider a spruce S^+_{xy} in A^+ that has x and y as base vertices, and as tips all the vertices adjacent to x and y that were added in the procedure. As an example, in Fig. 1(a), spruce S^+_{as} has a and s as base vertices, and tips c, d, e. Let S_{xy} be a maximal spruce of A contained in S^+_{xy}, if such a spruce exists. Let $M = \{M_1, M_2, \ldots, M_k\}$ be the set of all such spruces S_{xy}. First, note that the spruces in M do not share tips. Also,

Lemma 5. $\sum_{S \in M} \widehat{gain}(S) + 3q \geq n - 2$.

Proof. Observe that, as all S_{xy}^+ are complete, the sum of $gain(S_{xy}^+)$ for all x and y (for which S_{xy}^+ is defined) equals the cyclomatic number of A^+, which is $2n - 3 - (n-1) = n - 2$. Let us first argue that $\sum_{S \in M} gain(S) \geq n - 2 - 2q$. Indeed each missing edge e decreases the sum of $gain(S_{xy}^+)$ by at most two, because the edge e might appear in two spruces S_{xy}^+ (once as xy and once as an edge incident to a tip of S_{xy}^+). Note also that a spruce S_{xy}^+ for which S_{xy} is not a spruce corresponds to a term in the sum of $gain(S_{xy}^+)$ that will become zero or negative after these discounts, so it does not hurt to drop it from the sum. Finally, the sum $\sum_{S \in M} \widehat{gain}(S)$ is equal to the sum $\sum_{S \in M} gain(S)$ minus the number of incomplete spruces in M, which is bounded above by q. Therefore, the lemma holds. ∎

We proceed to Lemma 6.

Lemma 6. $4 \sum_{S \in Q} gain(S) \geq \sum_{S \in M} \widehat{gain}(S) - (n' - t)$.

Proof. For $i = 1, 2, \ldots, k$, let U_i be the set of tips of M_i that are in some spruce of Q. Let S_i be obtained from M_i after the removal of its tip vertices in U_i. Note that S_i might not be a spruce (it might be empty or a single edge). If S_i is a spruce, then $\widehat{gain}(S_i) = \widehat{gain}(M_i) - |U_i|$. To simplify, set $\widehat{gain}(S_i) = 0$ if S_i is not a spruce.

The proof of this lemma has two steps. The first one consists of the following simple observation. As $\sum_i |U_i| \leq n'$, we have that

$$\sum_{S \in M} \widehat{gain}(S) = \sum_i \widehat{gain}(M_i) \leq n' + \sum_i \widehat{gain}(S_i), \qquad (3)$$

because the spruces M_i do not share tips.

Let x and y be the base vertices of a spruce M_i from M. If x and y are in different components of Q, then S_i has to be a degenerate spruce or it is not a spruce (otherwise the algorithm would have included it in Q).

For each component C of Q, consider the following weighted simple graph $H = H_C$ on its set of vertices. For two vertices x and y in C that are the base vertices of a spruce S_i, the edge xy is present in H and it has weight $w(xy) = \widehat{gain}(S_i)$. Observe that H is a simple series-parallel graph. (It is a subgraph of A^+.)

Now, for the second step, let F_C be such a maximum weight forest in H. Recall that the algorithm constructs a weighted tree T_C on the same set of vertices; we treat the edges of T_C as distinct from the edges of F_C though both sets of edges have weight w. For each two vertices x and y with xy in F_C, there is a spruce S_i such that $w(xy) = \widehat{gain}(S_i)$. Now, the spruce $S_Q(x, y)$ was considered by the algorithm. Since Q is the set of spruces just before the final phase of the algorithm, $S_Q(x, y)$ was not added to Q and therefore $\widehat{gain}(S_Q(x, y)) \leq w(index_Q(x, y))$. Note that $\widehat{gain}(S_Q(x, y)) \geq \widehat{gain}(S_i)$ as all the tips of S_i, being isolated vertices in Q, are also in $S_Q(x, y)$. Thus, putting all this together, we have that $w(xy) = \widehat{gain}(S_i) \leq \widehat{gain}(S_Q(x, y)) \leq w(index_Q(x, y))$, for every x

and y such that $xy \in F_C$. But then, in the multigraph whose vertex set is C and the edge set is the disjoint union of $E(F_C)$ and $E(T_C)$, the tree T_C is a maximum weight tree [8]. Also, as F_C is a forest in this multigraph, we have that $w(F_C) \leq w(T_C)$.

Note that, for any spruce S in Q, the total weight of the edges of T_C obtained from S is $2\,gain(S)$, which holds both if S is complete or not. Let $\mathcal{C}$ be the collection of connected components of Q. Also, for C in $\mathcal{C}$, let Q_C be the (non-empty) set of spruces in C. By summing up for all spruces in Q, we obtain that

$$2 \sum_{C \in \mathcal{C}} gain(Q_C) = \sum_{C \in \mathcal{C}} w(T_C) \geq \sum_{C \in \mathcal{C}} w(F_C) \geq \frac{1}{2} \sum_{C \in \mathcal{C}} w(H_C) + \frac{1}{2} t,$$

where the last inequality comes from Lemma 3 and the fact that all weights are integers. Thus

$$2 \sum_{S \in Q} gain(S) = 2 \sum_{C \in \mathcal{C}} gain(Q_C) \geq \frac{1}{2} \sum_{i} \widehat{gain}(S_i) + \frac{1}{2} t.$$

and this, together with (3), implies the lemma. ∎

Now we proceed to Lemma 7.

Lemma 7. $\sum_{S \in Q} gain(S) \geq \frac{1}{2}(n' - t)$.

Proof. As in the previous proof, $\mathcal{C}$ is the collection of connected components of Q, and Q_C is the (non-empty) set of spruces in C, for C in $\mathcal{C}$. Let $n(C)$ be the number of vertices in C.

It is enough to prove that $gain(Q_C) \geq (n(C)-1)/2$ for all C in $\mathcal{C}$. So, consider a C in $\mathcal{C}$, and recall that Q does not have degenerate spruces. Let us prove by induction on the number of spruces in Q_C that $gain(Q_C) \geq (n(C)-1)/2$.

If Q_C has only one spruce S, then if S is complete, $n(S) = gain(S)+2$, and thus $gain(S) = n(S)-2 \geq (n(S)-1)/2$ because $n(S) \geq 3$. If S is incomplete, $n(S) = gain(S)+3$, and thus $gain(S) = n(S)-3 \geq (n(S)-1)/2$ because, as S is not degenerate, $n(S) \geq 5$.

Now suppose that Q_C has more than one spruce, and let S be a spruce in Q_C with at most one vertex in common with the others spruces in Q_C. (There is always one such spruce because Q_C is a spruce cactus.) Let C' be the connected subgraph of Q corresponding to the union of the spruces in $Q_{C'} = Q_C \setminus \{S\}$. By induction, $gain(Q_{C'}) \geq (n(C')-1)/2$. If S is complete, $n(C) = n(C') + gain(S) + 1$, and $gain(Q_C) = gain(Q_{C'}) + gain(S) \geq (n(C')-1)/2 + gain(S) = (n(C) - gain(S) - 2)/2 + gain(S) = (n(C) + gain(S) - 2)/2 \geq (n(C)-1)/2$, because $gain(S) \geq 1$. If S is incomplete, $n(C) = n(C') + gain(S) + 2$, and $gain(Q_C) = gain(Q_{C'}) + gain(S) \geq (n(C')-1)/2 + gain(S) = (n(C) - gain(S) - 3)/2 + gain(S) = (n(C) + gain(S) - 3)/2 \geq (n(C)-1)/2$, because $gain(S) \geq 2$, as S is non-degenerate. ∎

Having finished this proof, based on the discussion at the beginning of the subsection, we obtain the main result of the paper:

Theorem 1. *There is a polynomial-time $\frac{7}{12}$-approximation for Maximum Series-Parallel Subgraph.*

As an aside, observe that if we allowed the algorithm to include in Q the degenerate spruce which is a 4-cycle, then Lemma 7 would not hold anymore. Yet a weaker version of it would, with $1/3$ instead of $1/2$, and this would also lead to an approximation ratio greater than $1/2$. We introduced the adjusted gain concept specifically to forbid 4-cycles, so that Lemma 7 holds with $1/2$.

The analysis is tight. We will describe a family of graphs that proves this. Follow the description looking at Fig. 5. There is a graph G_k in this family for each even positive integer k. The graph G_k is the union of two edge-disjoint series-parallel graphs H_1 and H_2. The first one, H_1, is a path of length $8+k$, with a triangle on top of each of its edges (for a total of $7+k$ triangles and $3(7+k)$ edges). We call this path the *defining* path of H_1. In Fig. 5, the bottom edges form the defining path of H_1. The first 7 triangles on top of this path (shown by the darker edges) play a different role than the remaining k triangles. Call *top* the vertex in each of these triangles that is not on the defining path, and *round* the tops of the last k triangles plus the first and fourth top vertices. See the white circle vertices in Fig. 5. The final k vertices of the defining path are alternately named *square* and *triangular* vertices. The second and fifth top vertices are also square vertices, and the third and sixth are also triangular vertices. See Fig. 5. We will use these marks to describe the second graph.

The second graph, H_2, consists of three big spruces on the marked vertices of H_1, with a pair of new extra vertices per tip t, each of them adjacent to t and to one of the spruce base vertices. Each spruce is on one of the types of marked vertices in H_1. Let us now describe the first of the three big spruces, the one on the round vertices of H_1. This spruce has as base vertices the two first round vertices in H_1, and has as tips each of the other round vertices in H_1, for a total of k tips. In Fig. 5, this spruce is shown by the dotted edges, plus the "round" triangle with straight edges. For this triangle, we show also the two extra new vertices — the black small circle vertices, incident to the dashed edges.

The second big spruce is on the square vertices of H_1. Its base vertices are the two square top vertices, and its tips are the other $k/2$ square vertices of H_1. The third big spruce is defined similarly on the triangular vertices of H_1. This

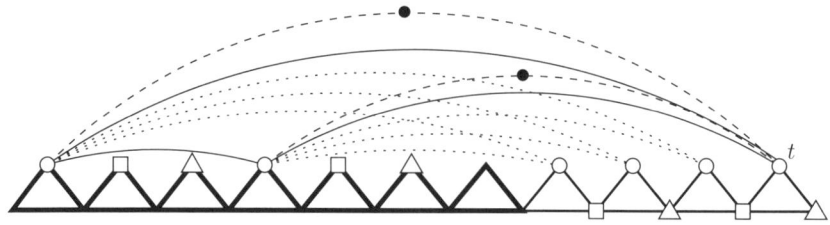

Fig. 5. Part of the graph G_4: the graph H_1 (the bottom path and the triangles on top of it), the first big spruce in H_2 (the subgraph induced by the white round vertices), and two extra vertices (the black small circle vertices)

completes the description of H_2, which, summarizing, consists of these three big spruces, plus the extra new vertices adjacent to the endpoints of the edges of these spruces incident to their tips. (In Fig. 5, we show only two of the extra vertices, the black small circle vertices.)

As we said, G_k consists of these two graphs H_1 and H_2. Note that both of them are indeed series-parallel. Thus, the number of edges in H_2, which is $(2k+1) + 2(k+1) + 8k = 12k+3$, is a lower bound on the size of a maximum series-parallel subgraph of G_k. Moreover, one can verify that our algorithm in the iterative phase can produce as Q the graph H_1, and output a graph with $|E(H_1)| + 4k = 3(7+k) + 4k = 7k+21$ edges. In this case, the ratio achieved is no more than $(7k+21)/(12k+3)$, which approaches $7/12$ as k gets large.

References

1. Berman, P., Ramaiyer, V.: Improved approximations for the Steiner tree problem. Journal of Algorithms 17, 381–408 (1994)
2. Cai, L.: On spanning 2-trees in a graph. Discrete Applied Mathematics 74(3), 203–216 (1997)
3. Cai, L., Maffray, F.: On the spanning k-tree problem. Discrete Applied Mathematics 44, 139–156 (1993)
4. Călinescu, G., Fernandes, C.G., Finkler, U., Karloff, H.: A better approximation algorithm for finding planar subgraphs. Journal of Algorithms 27(2), 269–302 (1998)
5. Călinescu, G., Fernandes, C.G., Karloff, H., Zelikovski, A.: A new approximation algorithm for finding heavy planar subgraphs. Algorithmica 36(2), 179–205 (2003)
6. Robins, G., Zelikovsky, A.: Improved steiner tree approximation in graphs. In: Proceedings of the Tenth ACM-SIAM Symposium on Discrete Algorithms (SODA), pp. 770–779 (2000)
7. Schrijver, A.: Combinatorial Optimization, vol. A. Springer, Heidelberg (2003), http://homepages.cwi.nl/~lex/files/dict.ps
8. Tarjan, R.E.: Data Structures and Networks Algorithms. Society for Industrial and Applied Mathematics (1983)

Low-Port Tree Representations*

Shiri Chechik and David Peleg

Department of Computer Science and Applied Mathematics,
The Weizmann Institute of Science, Rehovot, 76100 Israel
{shiri.chechik,david.peleg}@weizmann.ac.il

Abstract. Consider an n-node undirected graph $G(V, E)$ with a pre-assigned port numbering for the outgoing edges of each node. The port numbers assigned to a node u of degree $\deg(u)$ are $\{0, 1, \ldots, \deg(u) - 1\}$. In certain contexts it is necessary to maintain a directed spanning tree of G, in which case each node needs to remember the port number leading to its parent. Hence the cost of a spanning tree T is the total number of bits the nodes need to store in order to remember T. This paper addresses the question of asymptotically bounding the cost of the optimal tree, as a function of the graph size. A tight upper bound of $O(n)$ is established on this cost, thus improving on the best previously known bound of $O(n \log \log n)$ [6] and proving the conjecture raised therein. This is achieved by presenting a polynomial time algorithm for constructing a spanning tree T of cost $O(n)$ for a given general graph G with an arbitrary port labeling.

1 Introduction

Many distributed applications make use of pre-defined network representations for guaranteeing efficient performance. These representations are often designed to be as compact as possible (cf. [16]) in terms of their memory requirements. Two notable examples are compact routing schemes [1,3,4,7,17,18,19,20], which commonly rely on sparse network representations such as partitions, covers and decompositions, and informative labeling schemes for a variety of applications, e.g., [5,6,8,9,10,11,12,14,15].

An important special case of a pre-defined network representation is that of a spanning tree. A number of well-known algorithms for basic distributed operations, such as broadcast, convergecast and graph exploration (cf. [2,13,16]), are based on maintaining a spanning tree for the network and using it for efficient communication.

The problem of compact port-based representations for spanning trees was introduced in [6], which focused on the problem of providing, during a pre-processing stage, a compact local encoding of a spanning tree for a given graph. In particular, it raised the question of the existence of encodings in which the average number of bits stored at each node is constant.

---
* Supported by a grant from the Israel Science Foundation.

More precisely, the following problem was considered in [6]. Consider an n-node undirected graph $G(V, E)$. Each node u has a pre-assigned port number for each of its outgoing edges, with the port number of the edge connecting it to the neighbor v denoted $\text{Port}(u, v)$. Denoting the degree of the node u by $\deg(u)$, the port numbers assigned to u's ports are $\{0, \ldots, \deg(u) - 1\}$, or more formally, the port number $\text{Port}(u, v)$ is in $\{0, \ldots, \deg(u) - 1\}$, where $\text{Port}(u, v_1) \neq \text{Port}(u, v_2)$ for every two distinct neighbors v_1 and v_2 of u. The port numbers are not necessarily symmetric, i.e., it could be that $\text{Port}(u, v) \neq \text{Port}(v, u)$. The network nodes are required to maintain a directed spanning tree of G, with each node required to remember the port number leading to its parent. For a port number p, denote by $\omega(p)$ the number of bits required to encode p using the standard binary representation for integers. Formally,

$$\omega(p) = \begin{cases} 1, & \text{if } p = 0, \\ \lfloor \log p \rfloor + 1, & \text{if } p \geq 1. \end{cases}$$

The cost of a tree T is the total number of bits the nodes need to remember, denoted $\text{Cost}(T, G)$. Formally,

$$\text{Cost}(T, G) = \sum_{v \in V,\ v \neq r(T)} \omega(\text{Port}(v, parent(v, T))),$$

where $r(T)$ is the root of the tree T and $parent(v, T)$ is the parent of v in the tree T. Define $\text{Cost}(G)$ to be $\min\{\text{Cost}(T, G)\}$, where the minimum is taken over all spanning trees T of G. The question of constructing a spanning tree T minimizing $\text{Cost}(T, G)$ for a given graph G, hence also determining $\text{Cost}(G)$, was shown in [6] to enjoy a polynomial time algorithm.

In this paper, we are interested in bounding the asymptotic behavior of $\text{Cost}(G)$ as a function of the graph size. Define $\text{Cost}(n)$ to be $\max\{\text{Cost}(G)\}$, where the maximum is taken over all n-node graphs G. An upper bound of $O(n \log \log n)$ on $\text{Cost}(n)$ was shown in [6], by presenting a polynomial time algorithm constructing a spanning tree T of cost $O(n \log \log n)$ for a given n-node graph G. It was conjectured in [6] that the actual bound is $\text{Cost}(n) = \Theta(n)$. In fact, a tight upper bound of $O(n)$ is proved therein for the special cases of complete graphs with arbitrary labeling and of arbitrary graphs with symmetric port labeling. In what follows, we confirm the above conjecture for arbitrary graphs and arbitrary assignments of port numbers, by establishing a tight upper bound of $O(n)$ on $\text{Cost}(n)$. This is achieved by presenting a polynomial time algorithm for constructing a spanning tree T of cost $O(n)$ for a given general graph G with an arbitrary port labeling.

2 Construction of a Low Port Tree with Cost $O(n)$

In this section we present an algorithm for constructing a spanning tree with cost $O(n)$ for a given graph $G(V, E)$.

2.1 Escape-Paths

We start by defining some basic notions. For a subtree T of G, we say that the edge (u, w) is an *exit edge* of T if $u \in V(T)$ and $w \notin V(T)$, and the node u is an *exit node* of T if it has an exit edge of T. A tree T rooted at $r(T)$ is *well-oriented* if $r(T)$ is an exit node of T.

For a directed subtree T of G rooted at $r(T)$, a path $P = (v_1, \ldots, v_k)$ is said to be a *escape-path* of T if

1. $v_1 = r(T)$,
2. $v_1, \ldots, v_k \in V(T)$,
3. $(v_i, v_{i+1}) \in E$ for every $1 \leq i \leq k-1$,
4. v_k is an exit node (i.e., it has a neighbor $z \notin V(T)$).

Note that the edges of the escape-path P need not be in T.
Define the following order relation on paths in G. Given a path $P_1 = (v_1, \ldots, v_k)$ and a path $P_2 = (z_1, \ldots, z_r)$, we say that the path P_2 is *lighter* than P_1 if either $r < k$ or $r = k$ and $\text{Port}(z_{r-1}, z_r) \leq \text{Port}(v_{k-1}, v_k)$. Notice that this lightness relation is transitive. For a subtree T of G, a path $\mathcal{P}$ is said to be a *lightest escape-path* of T if it is an escape-path of T and no other escape-path of T is lighter than $\mathcal{P}$.

2.2 Outline of the Algorithm

The algorithm consists of two phases, a *preprocessing phase* and a *tree construction phase*. Our approach in the construction process of the spanning tree is based on starting with individual nodes and merging them gradually by taking a small well-oriented tree T and hanging it on another subtree $\hat{T}$ by adding an exit edge from $r(T)$ to $\hat{T}$. Whenever this process succeeds, it leads to a low cost spanning tree. The complications arise once the process encounters some small subtree T that is not well-oriented, i.e., whose root has no exit edge. In this case, the algorithm tries to re-orient the subtree T by finding a lightest escape-path of T and reversing the relevant edges in that path.

The construction process proceeds in iterations, where in each iteration i the algorithm chooses a subtree T_i and applies this process on it. For the sake of a clearer description of the algorithm and its analysis, let us keep a record of this merging process by remembering, for each iteration i, the basic small subtree T_i from which we started and the escape-path from the original root $r(T_i)$ to the selected exit node (which acts as the new root of T_i after the reversal). The preprocessing stage deals with selecting such escape-paths.

The preprocessing phase. Preprocessing consists of two stages. In the first, Stage S1, the algorithm chooses for each T_i a lightest escape-path $\mathcal{P}_i$ in a naive way. By the analysis given in [6], these paths yields a tree of cost $O(n \log \log n)$. In the second stage of the preprocessing phase, Stage S2, the algorithm builds cheaper paths that lead to the desired cost of $O(n)$.

Stage S1 of the preprocessing phase consists of $n-1$ iterations. The algorithm maintains a *forest* F, namely, a collection of (vertex disjoint) subtrees whose union spans V. Initially, the forest F contains n subtrees, each consisting of a single vertex. In each iteration i, the algorithm chooses the smallest subtree T_i in the collection and merges it with another subtree in the collection. At the end of this stage, the forest F consists of a single tree spanning the entire graph. In addition, the algorithm also keeps record of the trees $T_1, \ldots, T_{n-1}$ chosen during the $n-1$ iterations. Note that these trees are not necessarily disjoint; it is possible that $V(T_i) \subset V(T_j)$ for some $i < j$, although partial overlaps may not occur.

Let us now describe in detail the process of transforming and merging the subtree T_i, in iteration i of Stage S1. The process consists of two main steps. In the first step, the algorithm identifies a lightest escape-path $\mathcal{P}_i$ of T_i, ending at an exit node v_k. (If the root $r(T_i)$ itself is an exit node, then $\mathcal{P}_i$ consists of the single node $v_1 = r(T_i)$.) The algorithm then transforms T_i into a well-oriented tree T'_i on the same set of vertices. (If $r(T_i)$ is an exit node then no change is needed, i.e., $T'_i = T_i$ and $r(T'_i) = r(T_i)$.) In the second step, the algorithm looks at the set of exit edges of $r(T'_i)$, selects the exit edge $(r(T'_i), z)$ of minimum $\mathsf{Port}(r(T'_i), z)$, sets $Out_i \leftarrow z$, and lets $\hat{T} \in F$ be the subtree containing z. The algorithm then merges the subtrees T'_i and $\hat{T}$ into a subtree $\tilde{T}$ by adding the edge $(r(T'_i), z)$, removes T_i and $\hat{T}$ from F and adds the merged tree $\tilde{T}$ instead. For convenience, the formal description of the above process is broken into two procedures (presented in Figure 1): Procedure Transform, performing the first step, and Procedure Merge, performing the second.

Procedure Transform(G, T_i)

1. $T'_i \leftarrow T_i$
2. Find a lightest escape-path of T_i, $\mathcal{P}_i = (v_1, \ldots, v_k)$.
3. For every $1 \leq r \leq k-1$, add the edge (v_r, v_{r+1}) of $\mathcal{P}_i$ to T'_i.
   In turn, for $2 \leq r \leq k$, remove from T'_i the (unique) outgoing edge of v_r in T'_i, (v_r, w_r).
4. Return the transformed tree and the escape-path, $(T'_i, \mathcal{P}_i)$.

Procedure Merge(G, T_i, v_k, T'_i, F)

1. $Out_i \leftarrow$ node z outside T'_i with edge to v_k of minimal $\mathsf{Port}(v_k, z)$.
2. Let $\hat{T} \in F$ be the tree in F that contains Out_i.
3. $\tilde{T} \leftarrow (V(T'_i) \cup V(\hat{T}), E(T'_i) \cup E(\hat{T}))$.
4. Connect v_k to Out_i in $\tilde{T}$.
5. Remove T_i and $\hat{T}$ from F and add the merged $\tilde{T}$ instead.
6. Return (Out_i, F).

Fig. 1. The procedure for merging the tree T_i with the tree $\hat{T}$

The transformation process, performed by Procedure Transform, operates as follows. Let $\mathcal{P}_i = (v_1, v_2, \ldots, v_k)$ be a lightest escape-path of T_i. Set $T'_i \leftarrow T_i$. For every $1 \leq j \leq k-1$, add the edge (v_j, v_{j+1}) of $\mathcal{P}_i$ to T'_i. In turn, for $2 \leq j \leq k$, remove from T'_i the (unique) outgoing edge of v_j in T'_i, (v_j, w_j). Figure 2 illustrates this transformation process.

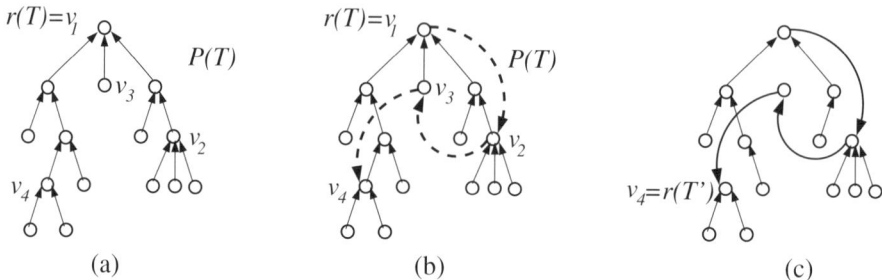

Fig. 2. (a) The tree T. (b) The lightest escape-path $\mathcal{P} = (v_1, v_2, v_3, v_4)$ of T (dashed). The node v_4 has an exit edge (not shown in the figure). (c) The tree T', with v_4 as its root. T' is obtained from T by erasing the edges that connected v_2, v_3 and v_4 to their parents in T, and replacing them by the edges of $\mathcal{P}$.

In the second stage of the preprocessing phase, Stage S2, the algorithm examines the chosen trees T_i and their escape-paths $\mathcal{P}_i$ in reverse order, from T_{n-1} to T_1, and tries to find *shortcut* paths $\mathcal{P}^*_i$ using the nodes in the previously selected shortcut paths $\mathcal{P}^*_j$ for $j > i$.

Loosely speaking, when considering a tree T_i, the algorithm treats all nodes in the tree T_i that participate in some shortcut path $\mathcal{P}^*_j$ for some $j > i$ as nodes outside the tree T_i. It then tries to use these new outer nodes in finding a shortcut path $\mathcal{P}^*_i$ of lower cost than the escape-path $\mathcal{P}_i$. The definition of the *shortcut* path is similar to that of the escape-path, with a small modification: instead of the requirement that the last node in the path has a neighbor outside of $V(T_i)$, we require that the last node has an edge to a node that participates in $\mathcal{P}^*_j$ for some $j > i$. The *lightest shortcut path* is the analogue of the lightest escape-path, i.e., a path $\mathcal{P}^*$ is said to be a *lightest* shortcut-path of T if it is a shortcut-path of T and no other shortcut-path of T is lighter than $\mathcal{P}^*$. If the algorithm finds a shortcut-path lighter than the original escape-path $\mathcal{P}_i$, then it replaces the escape-path $\mathcal{P}_i$ with the lightest shortcut-path $\mathcal{P}^*_i$.

Hence iteration i of Stage S2 of the preprocessing phase proceeds as follows. Consider the subtree T_i for $1 \leq i \leq n-1$. The algorithm examines all shortcut-paths $(v_1, \ldots, v_k)$ of the tree T_i such that v_k has a neighbor z that belongs to some shortcut path $\mathcal{P}^*_j$ for $j > i$. If the algorithm finds a shortcut-path lighter than $\mathcal{P}_i$ then it sets $\mathcal{P}^*_i$ to be the lightest such path $(v_1, \ldots, v_k)$ and Out_i to the node z that belongs to some shortcut path $\mathcal{P}^*_j$ for $j > i$ with minimal $\mathsf{Port}(v_k, z)$.

The preprocessing phase is described formally in Figure 3. After the preprocessing phase, the algorithm keeps a set of subtrees $T_1, \ldots, T_{n-1}$, a set of shortcut paths $\mathcal{P}^*_1, \ldots, \mathcal{P}^*_{n-1}$ and a set of out nodes $Out_1, \ldots, Out_{n-1}$.

Procedure Find_Paths(G)

1. $F \leftarrow \{(\{v\}, \emptyset) \mid v \in V\}$
2. For $i = 1, \ldots, n-1$ do: /* Stage S1 */
   (a) Let T_i be the smallest-size tree in F.
   (b) Invoke $(T'_i, \mathcal{P}_i) \leftarrow$ Transform(G, T_i).
   (c) Invoke $(Out_i, F) \leftarrow$ Merge(G, T_i, v_k, T'_i, F).
3. $S \leftarrow \emptyset$
4. For $i = n-1, \ldots, 1$ do: /* Stage S2 */
   (a) Consider $\mathcal{P}_i = (v_1, \ldots, v_k)$.
   (b) If there exists a shortcut-path $\mathcal{P}^*_i$ (using the nodes in S) lighter than $\mathcal{P}_i$ then do:
      i. Set $\mathcal{P}^*_i$ to be the lightest shortcut path $(z_1, \ldots, z_r)$ of T_i using the nodes in S.
      ii. Set Out_i to be the node z in S with an edge (z_r, z) of minimal Port(z_r, z).
   (c) Else, set $\mathcal{P}^*_i \leftarrow \mathcal{P}_i$.
   (d) $S \leftarrow S \cup V(\mathcal{P}^*_i)$.
5. Return $(T_1, \ldots, T_{n-1}, \mathcal{P}^*_1, \ldots, \mathcal{P}^*_{n-1}, Out_1, \ldots, Out_{n-1})$

Fig. 3. The procedure for finding the transformation paths for all subtrees

Procedure Tree_Cons$(G, T_1, \ldots, T_{n-1}, \mathcal{P}^*_1, \ldots, \mathcal{P}^*_{n-1}, Out_1, \ldots, Out_{n-1})$

1. $\mathcal{T} \leftarrow (V, \emptyset)$
2. For $i = 1, \ldots, n-1$ do:
   (a) Consider $\mathcal{P}^*_i = (v_1, \ldots, v_k)$.
   (b) For every $1 \leq j \leq k-1$, add the edge (v_j, v_{j+1}) to $\mathcal{T}$. In turn, for $2 \leq j \leq k$, remove from $\mathcal{T}$ the (unique) outgoing edge of v_j in $\mathcal{T}$, (v_j, w_j).
   (c) Add to $\mathcal{T}$ an edge from v_k to Out_i.
3. Return $\mathcal{T}$.

Fig. 4. The procedure for constructing a tree with total cost $O(n)$

Main(G)

1. Invoke $(T_1, \ldots, T_{n-1}, \mathcal{P}^*_1, \ldots, \mathcal{P}^*_{n-1}, Out_1, \ldots, Out_{n-1}) \leftarrow$ Find_Paths(G).
2. Invoke $\mathcal{T} \leftarrow$ Tree_Cons$(G, T_1, \ldots, T_{n-1}, \mathcal{P}^*_1, \ldots, \mathcal{P}^*_{n-1}, Out_1, \ldots, Out_{n-1})$.

Fig. 5. Constructing a spanning tree with cost $O(n)$

The tree construction phase. The tree construction phase consists of $n-1$ iterations. The algorithm maintains a subgraph $\mathcal{T}$ which is first initialized to be $\mathcal{T} \leftarrow (V, \emptyset)$. In each iteration i, the algorithm does the following. Let $\mathcal{P}_i^* = (v_1, \ldots, v_k)$. For every $1 \leq j \leq k-1$, the algorithm adds the edge (v_j, v_{j+1}) to $\mathcal{T}$. In turn, for $2 \leq j \leq k$, it removes from $\mathcal{T}$ the (unique) outgoing edge of v_j in $\mathcal{T}$, (v_j, w_j). It then adds to $\mathcal{T}$ an edge from v_k to Out_i. In the end, it returns $\mathcal{T}$.

The tree construction phase is presented formally in Figure 4. The main algorithm is given in Figure 5.

3 Analysis

Let $\mathcal{T}$ be the final tree returned by the algorithm. Denote by $\mathcal{S}$ the set of all edges added to $\mathcal{T}$ at some iteration in procedure Tree_Cons. Notice that $E(\mathcal{T}) \subseteq \mathcal{S}$, where $E(\mathcal{T})$ is the set of edges in the final tree $\mathcal{T}$. It could be that $E(\mathcal{T}) \neq \mathcal{S}$, as some of the edges in $\mathcal{S}$ might be removed in later iterations. We show that the total cost of all edges in $\mathcal{S}$, denoted by $\text{Cost}(\mathcal{S})$, is $O(n)$, which implies that $\text{Cost}(\mathcal{T}, G) = O(n)$. For the analysis, we partition the edges that were added to $\mathcal{S}$ into two subsets, E_{out} and E_{esc}, and bound separately the total cost of edges in each subset by $O(n)$, thus the total cost of all edges in $\mathcal{S}$ is also $O(n)$.

For a subtree T_r with $\mathcal{P}_r^* = (v_1, \ldots, v_k)$, all edges (v_i, v_{i+1}) for $1 \leq i \leq k-1$ belong to the subset E_{esc} and are referred to as *escape-edges*. The edge from the node v_k to the node Out_r is called an *out-edge* and it belongs to the subset E_{out}.

Consider an edge (x, y), and let $c = \text{Port}(x, y)$. Notice that there are exactly c neighbors w of x that are "cheaper" than y, i.e., such that $\text{Port}(x, w) < \text{Port}(x, y)$. For the sake of the analysis, we employ the following *charging rule* on escape-edges. If the algorithm selects the escape-edge (x, y) to the constructed tree $\mathcal{T}$ at some iteration of Procedure Tree_Cons, then each such cheaper node w incurs a charge of 1 upon adding (x, y) to $\mathcal{T}$.

For each node w, we are interested in identifying the subtrees T_i that cause w to incur a charge. Formally, we say that T_i is a *charging subtree* of w if w incurs a charge upon adding the shortcut path $\mathcal{P}_i^* = (v_1, \ldots, v_k)$, namely, there exists some $1 \leq j \leq k-1$ such that the escape-edge (v_j, v_{j+1}) is more expensive than (v_j, w), or in other words, $\text{Port}(v_j, v_{j+1}) > \text{Port}(v_j, w)$. Denote by $M(w)$ the set of charging subtrees of w. In the analysis we prove that the algorithm guarantees that every node w has at most one charging subtree, namely, $|M(w)| \leq 1$.

For each node $w \in V$, denote by $C(w)$ the *charge count* of w, namely, the number of escape-edges for which w incurred a charge. Formally,

$$C(w) = \#\{v \mid \text{Port}(v, u) > \text{Port}(v, w) \text{ and } (v, u) \in E_{esc}\}.$$

We now show that the shortcut-paths chosen by the algorithm are disjoint.

Lemma 1. *The paths $\mathcal{P}_i^*$ for $1 \leq i \leq n-1$ are disjoint.*

Proof: The proof is straightforward from the construction of $\mathcal{P}_i^*$. Consider some T_i and T_j where $1 \leq j < i \leq n-1$. Assume, towards contradiction, that the paths $\mathcal{P}_i^*$ and $\mathcal{P}_j^*$ intersect, i.e., there exists a node v such that $v \in \mathcal{P}_i^*$ and $v \in \mathcal{P}_j^*$. Letting $\mathcal{P}_j^* = (z_1, \ldots, z_r, v, \ldots)$, notice that the path $P = (z_1, \ldots, z_r)$ is shorter than $\mathcal{P}_j^*$. Moreover, after iteration i of step 4 of Procedure Find_Paths, the node v was added to S. Hence when looking for the lightest shortcut path for the tree T_j in iteration j of step 4 of Procedure Find_Paths, the algorithm should have chosen P as $\mathcal{P}_j^*$; contradiction. ∎

The following two lemmas establish that the resulting subgraph $\mathcal{T}$ is a tree.

Lemma 2. *After iteration i of step 2 of Procedure Tree_Cons, all cycles in $\mathcal{T}$ contain an edge (z, w) such that $z \in \mathcal{P}_r^*$ for some $r > i$.*

Proof: By induction on i. For $i = 1$, after the first iteration of Procedure Tree_Cons $\mathcal{T}$ contains only one edge and the claim is trivial. Assume the claim holds for every $j < i$, and consider iteration i of step 2 of Procedure Tree_Cons. By the inductive hypothesis, in the beginning of iteration i all cycles in $\mathcal{T}$ contain a node z such that $z \in \mathcal{P}_r^*$ for some $r \geq i$.

Consider $\mathcal{P}_i^* = (v_1, \ldots, v_k)$ and consider a cycle C that contains an edge (z, w) such that $z \in \mathcal{P}_i^*$. Note that during iteration i of step 2(b) of Procedure Tree_Cons, the outgoing edge (z, w) is removed, and therefore the subgraph $\mathcal{T}$ no longer contains the cycle C. In addition, for all edges (v_j, v_{j+1}) that are added to $\mathcal{T}$ in iteration i of step 2 of Procedure Tree_Cons, the outgoing edge of v_{j+1} is removed, so clearly no cycle is created. The only exception is for the edge from v_k to Out_i. If $Out_i \notin T_i$, then again no cycle is closed. If $Out_i \in T_i$ then it must be that $Out_i \in \mathcal{P}_r^*$ for some $r > i$. It follows that the claim also holds for iteration i. ∎

Corollary 1. *The final tree $\mathcal{T}$ does not contain cycles.*

Lemma 3. *The number of edges in $\mathcal{T}$ in the end of procedure Tree_Cons is $n-1$.*

Proof: In step 2(b) of Procedure Tree_Cons, the number of edges that are added to $\mathcal{T}$ is equal to the number of edges that are removed from $\mathcal{T}$. In step 2(c) of Procedure Tree_Cons, one edge is added to $\mathcal{T}$. Therefore, one edge is added to $\mathcal{T}$ in each iteration of procedure Tree_Cons. As the procedure has $n - 1$ iterations overall, the number of edges in $\mathcal{T}$ in the end of procedure Tree_Cons is $n - 1$. ∎

Corollary 1 and Lemma 3 directly yield the following.

Lemma 4. *The subgraph $\mathcal{T}$ is a tree.*

We now show that each node z has at most one charging subtree.

Lemma 5. *The charging trees of every $z \in V$ satisfy $|M(z)| \leq 1$.*

Proof: Consider a node $z \in V$. Assume, towards contradiction, that both $T_{r_1}, T_{r_2} \in M(z)$, and without loss of generality assume that $r_1 > r_2$. Let $\mathcal{P}_{r_1}^* = (v_1, \ldots, v_{k_1})$ and $\mathcal{P}_{r_2}^* = (z_1, \ldots, z_{k_2})$. As $T_{r_1}, T_{r_2} \in M(z)$, it must be that z has an edge to both v_i and z_j, and moreover, $\text{Port}(v_i, z) < \text{Port}(v_i, v_{i+1})$

and $\text{Port}(z_j, z) < \text{Port}(z_j, z_{j+1})$ for some $1 \leq i \leq k_1 - 1$ and $1 \leq j \leq k_2 - 1$. Notice that the path $P = (z_1, \ldots, z_j, z)$ is lighter than the path $\mathcal{P}^*_{r_2}$ and also z has an edge to $v_i \in \mathcal{P}^*_{r_1}$. Moreover, after iteration r_1 in step 4 of Procedure Find_Paths, the node v_i was added to S. When looking for a shortcut for the tree T_{r_2} in iteration r_2 in step 4 of Procedure Find_Paths, the algorithm was supposed to choose P (or some other path lighter than P) as $\mathcal{P}^*_{r_2}$; contradiction. ∎

We now turn to the cost analysis of the resulting tree $\mathcal{T}$.

Lemma 6. *The charging count of every $z \in V$ satisfies $0 \leq C(z) \leq 3$.*

Proof: Consider some node $z \in V$. By Lemma 5, $|M(z)| \leq 1$. This means that z incurs a charge only on one subtree. It remains to show that when z incurs a charge on some subtree T_i, that charge is at most 3. Assume $T_i \in M(z)$. Let $\mathcal{P}_i = (v_1, \ldots, v_k)$ be the lightest escape-path generated by procedure Transform in the transformation process of T_i. Since $\mathcal{P}_i$ is a shortest path from v_1 to v_k in $G(T_i)$, we have that z has at most three neighbors in $\mathcal{P}_i$, otherwise the procedure could have used z to get a shorter path between v_1 and v_k (this is due to the fact that if z is adjacent to nodes $v_{l_1}, \ldots, v_{l_t}$ on $\mathcal{P}_i$, then $(v_1, \ldots, v_{l_1}, z, v_{l_t}, \ldots, v_k)$ is an alternate path between v_1 and v_k, and if $t \geq 4$ then this alternate path is necessarily shorter than the original.). When updating $\mathcal{P}_i$ to a lighter path $\mathcal{P}^*_i = (z_1, \ldots, z_r)$, again there can be at most three nodes among $z_1, \ldots, z_r$ that have an edge to z. Thus $C(z) \leq 3$. ∎

For the analysis, we partition the overall cost of $\mathcal{S}$ into

$$\text{Cost}(\mathcal{S}) = C_{out} + C_{esc},$$

where

$$C_{out} = \sum_{(x,y) \in E_{out}} \omega(\text{Port}(x,y))$$

and

$$C_{esc} = \sum_{(x,y) \in E_{esc}} \omega(\text{Port}(x,y)).$$

Lemma 7. *Consider some subtree T_i for $1 \leq i \leq n-1$. Let $\mathcal{P}^*_i = (v_1, \ldots, v_k)$ and $z = Out_i$. Then $\text{Port}(v_k, z) < |T_i|$.*

Proof: The proof is straightforward from the definition of Out_i. By definition, Out_i is a node z outside T_i (or a node in T_i that participate in some shortcut path $\mathcal{P}^*_j$ for some $j > i$) with minimal $\text{Port}(v_k, z)$. So in the worst case, $\text{Port}(v_k, z) = |T_i| - 1$. ∎

Lemma 8. $C_{out} = O(n)$.

Proof: For each subtree T_i, the algorithm adds at most one edge to E_{out}, and by Lemma 7 the cost of this edge is at most $\lfloor \log |T_i| \rfloor + 1$. There are $n-1$ such subtrees T_i. In each iteration in step 2 of Procedure Find_Paths, the algorithm chooses the smallest subtree in the forest F and merges it with another subtree. As initially F contains n subtrees of size 1, there are at least $n/2$ iterations with subtrees of size 1, and at least $n/4$ subsequent iterations with subtrees of size at most 2, and so on. It follows that the total cost is

$$C_{out} \leq \sum_{i=1}^{\log n} \frac{n}{2^i} \cdot (i+1) = O(n) .$$ ∎

Lemma 9. $C_{esc} = O(n)$.

Proof: Notice that exactly $\text{Port}(x,y)$ nodes incur a charge for each edge $e = (x,y) \in E_{esc}$, and therefore

$$\sum_{(x,y) \in E_{esc}} \text{Port}(x,y) = \sum_{v \in V} C(v) \leq 3n ,$$

where the last inequality follows by Lemma 6. It follows that

$$C_{esc} = \sum_{\substack{(x,y) \in E_{esc} \\ \text{Port}(x,y)=0}} 1 + \sum_{\substack{(x,y) \in E_{esc} \\ \text{Port}(x,y)>0}} (\lfloor \log \text{Port}(x,y) \rfloor + 1) \leq n + \sum_{(x,y) \in E_{esc}} \text{Port}(x,y)$$

$$\leq 4n = O(n).$$ ∎

Finally, Lemmas 8 and 9 yield the desired bound.

Lemma 10. $\text{Cost}(\mathcal{T}, G) = O(n)$.

References

1. Abraham, I., Gavoille, C., Malkhi, D., Nisan, N., Thorup, M.: Compact name-independent routing with minimum stretch. In: Proc. 16th ACM Symp. on Parallel Algorithms & Architectures (SPAA), pp. 20–24 (2004)
2. Attiya, H., Welch, J.: Distributed Computing: Fundamentals, Simulations and Advanced Topics. McGraw-Hill, New York (1998)
3. Awerbuch, B., Bar-Noy, A., Linial, N., Peleg, D.: Compact distributed data structures for adaptive network routing. In: Proc. 21st ACM Symp. on Theory of Computing, pp. 230–240 (1989)
4. Awerbuch, B., Peleg, D.: Routing with polynomial communication-space trade-off. SIAM J. on Discrete Math. 5, 151–162 (1992)
5. Cohen, R., Fraigniaud, P., Ilcinkas, D., Korman, A., Peleg, D.: Label-guided graph exploration by a finite automaton. In: Caires, L., Italiano, G.F., Monteiro, L., Palamidessi, C., Yung, M. (eds.) ICALP 2005. LNCS, vol. 3580, pp. 335–346. Springer, Heidelberg (2005)
6. Cohen, R., Fraigniaud, P., Ilcinkas, D., Korman, A., Peleg, D.: Labeling Schemes for Tree Representation. In: Pal, A., Kshemkalyani, A.D., Kumar, R., Gupta, A. (eds.) IWDC 2005. LNCS, vol. 3741, pp. 13–24. Springer, Heidelberg (2005)
7. Fraigniaud, P., Gavoille, C.: Routing in trees. In: Orejas, F., Spirakis, P.G., van Leeuwen, J. (eds.) ICALP 2001. LNCS, vol. 2076, pp. 757–772. Springer, Heidelberg (2001)

8. Gavoille, C., Peleg, D., Pérennes, S., Raz, R.: Distance labeling in graphs. In: Proc. 12th ACM Symp. on Discrete Algorithms, pp. 210–219 (2001)
9. Kannan, S., Naor, M., Rudich, S.: Implicit Representation of Graphs. SIAM J. on Descrete Math. 5, 596–603 (1992)
10. Katz, M., Katz, N.A., Korman, A., Peleg, D.: Labeling Schemes for Flow and Connectivity. SIAM J. Computing 34, 23–40 (2004)
11. Korman, A., Kutten, S., Peleg, D.: Proof Labeling Schemes. In: Proc. 24th ACM Symp. on Principles of Distributed Computing, PODC (2005)
12. Korman, A., Peleg, D., Rodeh, Y.: Labeling Schemes for Dynamic Tree Networks. Theory of Computing Systems (Special Issue of STACS 2002) 37, 49–75 (2004)
13. Lynch, N.: Distributed Algorithms. Morgan Kaufmann, San Francisco (1995)
14. Peleg, D.: Proximity-preserving labeling schemes and their applications. In: Widmayer, P., Neyer, G., Eidenbenz, S. (eds.) WG 1999. LNCS, vol. 1665, pp. 30–41. Springer, Heidelberg (1999)
15. Peleg, D.: Informative labeling schemes for graphs. In: Nielsen, M., Rovan, B. (eds.) MFCS 2000. LNCS, vol. 1893, pp. 579–588. Springer, Heidelberg (2000)
16. Peleg, D.: Distributed Computing: A Locality-Sensitive Approach. SIAM, Philadelphia (2000)
17. Peleg, D., Upfal, E.: A tradeoff between size and efficiency for routing tables. J. ACM 36, 510–530 (1989)
18. Santoro, N., Khatib, R.: Labelling and implicit routing in networks. The Computer Journal 28, 5–8 (1985)
19. Thorup, M., Zwick, U.: Compact routing schemes. In: Proc. 13th ACM Symp. on Parallel Algorithms & Architectures (SPAA), pp. 1–10 (2001)
20. van Leeuwen, J., Tan, R.B.: Routing with compact routing tables. In: Rozenberg, G., Salomaa, A. (eds.) The Book of L, pp. 259–273. Springer, Heidelberg (1986)

Fully Dynamic Representations of Interval Graphs

Christophe Crespelle

LIP6, CNRS - Université Paris 6
christophe.crespelle@lip6.fr

Abstract. We present a fully dynamic algorithm that maintains three different representations of an interval graph: a minimal interval model of the graph, the PQ-tree of its maximal cliques, and its modular decomposition. After each vertex or edge modification (insertion or deletion), the algorithm determines whether the new graph is an interval graph in $O(n)$ time, and, in the positive, updates the three representations within the same complexity.

1 Introduction

In this paper, we are interested in the *dynamic recognition and representation problem* for the class of interval graphs. For a family $\mathcal{F}$ of graphs, this problem is to maintain a characteristic representation of dynamically changing graphs as long as the modified graph belongs to $\mathcal{F}$ [3,4,6,8,15,16]. The input of the problem is a graph $G \in \mathcal{F}$ with its representation and a modification which is one of the following: inserting or deleting a vertex (along with the edges incident to it), inserting or deleting an edge. After any modification, the algorithm determines whether the new graph belongs to $\mathcal{F}$ and, in the positive, updates the chosen representation.

Related works. The seminal paper for the recognition of interval graphs [1] solved the problem in linear time by introducing a data structure called PQ-tree. The algorithm of [1] is not dynamic: even though the consecutiveness constraints of each vertex are added one by one, the maximal cliques of the graph need to be computed in advance. The algorithm of [11] also considers the vertices arriving one by one and updates the PQ-tree. But in order to achieve a linear complexity, the ordering on the vertices is not arbitrary and must be precomputed statically. On the opposite, the algorithm of [7] is truly incremental on vertices. In the worst case, the cost of a vertex insertion may be up to $\Theta(n)$. But unfortunately, as mentioned by the author, the data structure he uses does not allow to treat vertex deletion, while our algorithm is able to do so, within the same worst case time complexity. For edge modifications, [9] designed a fully dynamic algorithm that runs in $O(n \log n)$ time per operation. Here, we lower this complexity to $O(n)$.

Our results. Our algorithm treats the insertion of a vertex in an interval graph in a truly dynamic manner and is the first one also treating the deletion of a vertex; both operations being handled in $O(n)$ time. We also lower the complexity of the best dynamic algorithm for edges [9] from $O(n \log n)$ to $O(n)$ per operation, insertion or deletion. In addition, we do not only deal with the recognition problem but also maintain, within the same complexity, three useful representations of the graph: a minimal interval model, the PQ-tree and the modular decomposition.

Beside our algorithmic results, we give new insight into the structure of interval graphs by showing strong connections between the PQ-tree and the modular decomposition of an interval graph. It should also be noted that Theorem 3 gives a characterisation of the neighbourhood of a vertex in an interval graph. Complete proofs of all results presented here can be found in [2].

2 Preliminaries

Every graph considered here will be finite, undirected, loopless and simple. Throughout the paper, V denotes the vertex set of graph G and E its edge set; we write $G = (V, E)$. n stands for $|V|$ and an edge between x and y is denoted indifferently xy or yx. The neighbourhood of a vertex $x \in V$ is denoted $N(x)$ and its non-neighbourhood $\overline{N}(x)$. $\mathcal{K}(G)$ is the set of maximal cliques of G. A vertex x is *simplicial* in G iff its neighbourhood is a clique. A subset $S \subsetneq V$ of vertices is *uniform* wrt. vertex $x \in V \setminus S$ if $S \subseteq N(x)$ (S is *full*) or $S \subseteq \overline{N}(x)$ (S is *hollow*). If S is not hollow, S is *linked*, and *mixed* if S is neither hollow nor full. When there is no confusion, we omit to mention the vertex x referred to. For a rooted tree T and a node u of T, we denote $parent(u)$ for the parent of u in T, $\mathcal{C}(u)$ for its set of children, $Anc(u)$ for the ancestors of u in T ($u \in Anc(u)$), and T_u for the subtree of T rooted at u. We sometimes identify the tree and its set of nodes by denoting $u \in T$. For a linear ordering σ, we denote $min(\sigma)$ and $max(\sigma)$ for respectively the first and last element of σ.

Interval graphs. An interval model of a graph G is a set $\mathcal{I}$ of intervals of the real line along with a mapping from V to $\mathcal{I}$ such that two vertices of G are adjacent iff their corresponding intervals intersect. Interval graphs are the graphs that admit such a model. In all the models considered in the following, intervals will be closed and will have integer bounds (the class remains the same under this restriction). Associating with each vertex the two integer bounds of its corresponding interval yields an efficient data structure providing adjacency testing in constant time. Interval graphs are well known to be *chordal*, that is, they do not contain any induced cycle of length ≥ 4. One of their nicest characterisations is the following.

Theorem 1. [5] *A graph G is an interval graph iff its maximal cliques can be linearly ordered such that, for every vertex x of G, the maximal cliques containing x occur consecutively.*

Such an ordering of the maximal cliques is called a *consecutive ordering* of G (or $\mathcal{K}(G)$). Numbering the maximal cliques with their rank in a consecutive ordering σ and assigning to each vertex x of G the interval of σ consisting of the cliques containing x results in a model of G. The minimal models are precisely those that can be obtained this way.

It is shown in [1] that all the consecutive orderings of the maximal cliques of a graph G can be represented by an $O(|\mathcal{K}(G)|)$-space structure called PQ-tree. The PQ-tree of G, denoted T^c, is a rooted tree whose leaves are the maximal cliques of G. Its internal nodes are labeled P (*degenerate nodes*) or Q (*prime nodes*). Any Q-node q is assigned two linear orderings, denoted σ_q and $\bar{\sigma}_q$, on the set of its children, $\bar{\sigma}_q$ being the reverse order of σ_q. A *solidification* of a PQ-tree T, is an assignation, to each node u of T, of a linear ordering on its children: any linear ordering if u is a P-node, σ_u or $\bar{\sigma}_u$ if u is a Q-node. The *frontier* of a solidification s is the prefix order of the leaves of T resulting from a depth first search where the children of a given node $u \in T$ are explored in the order defined by s. A result of [1] states that the frontier is a one to one mapping from the set of solidifications of T^c onto the set of consecutive orderings of G.

Modular decomposition. The reader which is not familiar with the basic notions of modular decomposition theory such as *module, strong module, maximal strong module* (whose set is denoted $\mathcal{MSM}(G)$) and *prime graph* may refer to [13].

For a module M of G, we define the quotient graph $G/M = G[(V \setminus M) \cup \{a\}]$, where $a \in M$ is called the *representative vertex* of M. Similarly, for a family $\mathcal{P}$ of pairwise disjoint modules, we define the quotient graph $G/\mathcal{P}$ by choosing a representative vertex for each module in $\mathcal{P}$.

The modular decomposition tree of G is denoted T^m, its leaves are the vertices of G and a node $p \in T^m$ represents the strong module of G, denoted $V(p)$, which is the set of leaves of T^m_p. The children of a node p of T^m are the maximal strong modules of $G[V(p)]$. To each node p of T^m, we associate its quotient graph $G_p = G[V(p)]/\mathcal{MSM}(G[V(p)])$. From the well-known modular decomposition theorem, the quotient G_p is either a stable set, then p is labeled *parallel*, or a clique, then p is labeled *series*, or a prime graph, then p is labeled *prime*. The parallel and series nodes are also called degenerate nodes. We will need the following lemma.

Lemma 1. *Let G and H be interval graphs, and x a vertex of G. $G_{x \leftarrow H}$ is an interval graph iff: **(i)** x is simplicial; or **(ii)** H is a clique.*

3 Three Representations of Interval Graphs

Minimal interval models. of an interval graph G consist of a consecutive ordering σ of G stored as a list. Each cell contains its position in the list and each vertex of G is assigned two pointers (possibly the same) toward the cells representing the first and the last (wrt. σ) maximal clique of G containing x.

The size of such a structure is $O(n + |\mathcal{K}(G)|) = O(n)$ as $|\mathcal{K}(G)| \leq n-1$ for any interval graph.

The PQ-representation. is essentially the same structure as the MPQ-tree introduced in [10]. In the classic PQ-tree, the maximal clique corresponding to a leave of T^c is stored by the list of its vertices, which results in an $O(n+m)$ space structure, while the number of nodes in the PQ-tree is only $O(n)$. In the PQ-representation, the vertices of G are stored in the internal nodes of T^c (thanks to the pointers defined below) instead of being stored in its leaves. Let u be a node of T^c, we denote $\mathcal{K}_{T^c}(u)$ for the maximal cliques of G corresponding to the leaves of T_u^c. For a subset $S \subseteq V$ of vertices, we denote $\mathcal{K}(S)$ for the set of maximal cliques of G containing S; and for a singleton we denote $\mathcal{K}(x)$ instead of $\mathcal{K}(\{x\})$. For a vertex $x \in V$, we denote e_x for the least common ancestor of the leaves of T^c corresponding to the maximal cliques of G containing x.

Lemma 2. *[11] For any vertex x of an interval graph G, exactly one of the two following conditions holds: (i) $\mathcal{K}(x) = \mathcal{K}_{T^c}(e_x)$, or (ii) e_x is a prime node and $\exists (u_1, u_2) \in (\mathcal{C}(e_x))^2 \setminus \{(min(\sigma_{e_x}), max(\sigma_{e_x}))\}, u_1 <_{\sigma_{e_x}} u_2$ and $\mathcal{K}(x) = \bigcup_{u_1 \leq v \leq u_2} \mathcal{K}_{T^c}(v)$.*

When (ii) is satisfied, we denote e_x^1 and e_x^2 for the children u_1 and u_2 of e_x. The PQ-representation of an interval graph G, denoted $PQ(G)$, is made of T^c and the set of vertices of G, where each vertex x stores a *primary pointer* toward e_x, and two *secondary pointers* toward resp. e_x^1 and e_x^2 when x satisfies (ii). These pointers encode which maximal cliques of G (i.e. the leaves of T^c) contain x. Since the number of nodes in T^c is $O(n)$ and since each vertex of G stores at most three pointers, it follows that the total size of the PQ-representation is $O(n)$.

Notation 1. *(cf. Fig 1) Let ρ be the root of T^c. For each node u of T^c, we define the following sets:*
$X_u = \{y \in V \mid e_y = u \text{ and } y \text{ has no secondary pointers}\}$
$Y_u = \{y \in V \mid e_y = u \text{ and } y \text{ has secondary pointers toward the children of } u\}$
$u^* = \{y \in V \mid e_y \in T_u^c\}$
$\Delta_u = \begin{cases} \{y \in Y_{\hat{u}} \mid e_y^1 \leq_{\sigma_{\hat{u}}} u \leq_{\sigma_{\hat{u}}} e_y^2\} & \text{if } u \neq \rho \quad \text{(where } \hat{u} = parent(u)) \\ \emptyset & \text{if } u = \rho \end{cases}$
$B_u = \bigcup_{v \in Anc(u)} X_v \cup \Delta_v$

Note that, by definition, if u is degenerate then $Y_u = \emptyset$, and if $parent(u)$ is degenerate then $\Delta_u = \emptyset$. B_u is the set of vertices that belong to all the maximal cliques corresponding to the leaves of T_u^c, and u^* is the set of vertices that are involved only in those cliques. The maximal clique of G corresponding to a leaf $f \in T^c$ is precisely B_f.

The MD-representation. of an arbitrary graph G, denoted $MD(G)$, is its modular decomposition tree T^m along with the quotient graphs G_p of its prime nodes p and a mapping from the vertices of G_p onto $\mathcal{C}(p)$. In the case where G

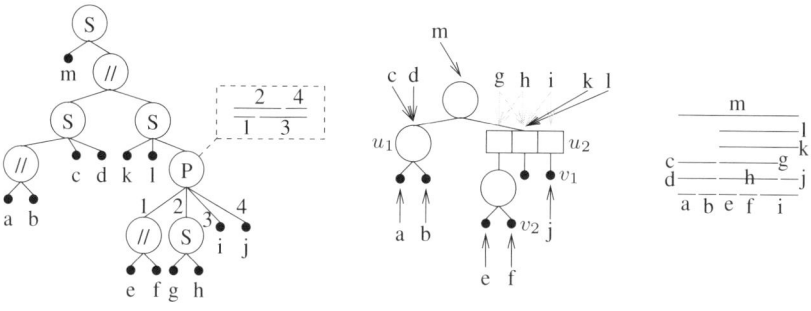

Fig. 1. Three representations of an interval graph. In the PQ-representation, the degenerate nodes are represented by circles and the prime nodes by rectangles. The primary pointers are black, while the secondary pointers are grey. The primary pointers of the vertices that have secondary pointers are not represented. We have $u_1^* = \{a, b, c, d\}$, $X_{u_1} = \{c, d\}$, $X_{u_2} = \{k, l\}$, $Y_{u_2} = \{g, h, i\}$, $B_{v_2} = \{f, g, h, k, l, m\}$ and $\Delta_{v_1} = \{i\}$.

is an interval graph, the quotient graphs are stored as minimal interval models. As the model of a prime node p takes $O(|\mathcal{C}(p)|)$ space, it yields an $O(n)$ space representation of G.

3.1 Linear-Time Equivalence of PQ-Representation and MD-Representation

The equivalence is based on the fact that the PQ-representation of an interval graph is quite well structured regarding its strong modules.

Theorem 2. *Let G be an interval graph. M is a non-trivial strong module of G iff $|M| > 1$ and there exists some node $u \in T^c$ satisfying one of the two following conditions:*

1. $M = u^*$ or $M = u^* \setminus X_u$; or
2. u is prime and $\exists u_1, u_2 \in \mathcal{C}(u), M = \{y \in Y_u \mid e_y^1 = u_1 \text{ and } e_y^2 = u_2\}$.

Theorem 2 justifies that the MD-representation can be obtained from the PQ-representation as follows. Hereinafter, we refer to this process as the PQ-MD-transformation. In a bottom-up manner, for each node u of T^c, we compute $T^m(G[u^*])$. For an internal node u, whose set of children is $\{u_1, \ldots, u_k\}$, we have to consider three different cases:

1. If $X_u = \emptyset$ and u is a degenerate node, the root $\tilde{u}$ of $T^m(G[u^*])$ is a parallel node whose children are the roots of the trees $T^m(G[u_1^*]), \ldots, T^m(G[u_k^*])$.
2. If $X_u = \emptyset$ and u is a prime node, the root $\tilde{u}$ of $T^m(G[u^*])$ is a prime node. The interval model of $G_{\tilde{u}}$ is made with the list Z of children of u ordered by σ_u. The set of simplicial vertices of $G_{\tilde{u}}$ is the set $\mathcal{S} = \{v \in \mathcal{C}(u) \mid v^* \neq \emptyset\}$. For any $v \in \mathcal{S}$, the root of $T^m(G[v^*])$ is made a child of $\tilde{u}$ and its corresponding vertex in $G_{\tilde{u}}$ has two pointers toward the cell of v in Z. The non-simplicial

vertices of $G_{\tilde{u}}$ are the classes $\bar{y}$ of vertices $y \in Y_u$ having the same secondary pointers. The child $\tilde{v}$ of $\tilde{u}$ corresponding to $\bar{y}$ is a series node (or a leaf if $|\bar{y}| = 1$) whose children are the vertices of $\bar{y}$. The pointers of $\tilde{v}$ toward Z are the same as the pointers of any $y \in \bar{y}$.
3. If $X_u \neq \emptyset$, we first ignore the vertices of X_u and build $T^m(G[u^* \setminus X_u])$ like described above. Then, we introduce a new series node whose children are the leaves representing vertices of X_u and the root of $T^m(G[u^* \setminus X_u])$.

Processing the leaves of T^c takes $O(n)$ time. For a degenerate node u, treatment 1 takes $O(|\mathcal{C}(u)|)$ time. In the processing of a prime node, the difficult operation is to find the equivalence classes $\bar{y}$ in Y_u. To that purpose, we can bucket sort the vertices $y \in Y_u$ with the rank of e_y^1 in Z as primary key, and the rank of e_y^2 as secondary key. As we sort $|Y_u|$ elements having values between 1 and $|\mathcal{C}(u)|$, this takes $O(|Y_u| + |\mathcal{C}(u)|)$ time. It follows that the processing time of a prime node is $O(|Y_u| + |\mathcal{C}(u)|)$. Finally, treatment 3 takes $O(|X_u|)$ time. Thus, the total computation time of $T^m(G)$ is $O(\sum_{u \in T^c} |X_u| + \sum_{u \in T^c} |Y_u| + \sum_{u \in T^c} |\mathcal{C}(u)|) = O(n)$.

For lack of space, we do not detail the converse operation that gives the PQ-representation from the MD-representation. It leans on a bottom-up search of T^m, similar to the one of T^c, which also runs in $O(n)$ time.

4 The Dynamic Algorithm

Since our three representations are $O(n)$ time equivalent, and since we want to get an $O(n)$ time algorithm, we can focus on maintaining only one of them and get the others within the same complexity. We chose to concentrate on showing how to maintain the MD-representation. However, when they are more convenient, we will also use the other representations and the equivalence between them.

Since edge modifications can be handled by one vertex deletion followed by one vertex insertion, we will not specifically consider them.

For lack of space, we do not present the deletion algorithm but only give its general idea. When the parent $u \in T^m$ of the vertex x to be deleted is degenerate, we simply remove the leaf corresponding to x and make some local cleaning of the tree, exactly as in [3]. When u is prime, we use the algorithm of [12] that computes the PQ-tree from an interval model, and we use the PQ-MD-transformation to get the updated MD-representation. We now concentrate on vertex insertion.

4.1 Focus on the Key Node

[3] showed that, for an arbitrary graph, the modifications of T^m under vertex insertion are located in the subtree of T^m rooted at the *insertion node* w_m, defined further. Here, we introduce the *key node* w that plays the same role in T^c and we show that, in order to determine whether $G + x$ is an interval graph, we can restrict our attention to $G[w^*] + x$.

From now on, x will be a vertex to be inserted in an interval graph G, and we denote G' for $G + x$. We will say that a node $u \in T^m$ (resp. T^c) is uniform, mixed, full, hollow or linked (see definitions p. 78) referring to the set $V(u)$ (resp. u^*). A node $u \in T^c$ is *saturated* iff u^* and B_u are full.

Definition 1. *[3] A node $u \in T^m$ is proper iff either u is uniform wrt. x, or u is a mixed node with a (unique) mixed child f such that $V(f) \cup \{x\}$ is a module of $G'[V(u) \cup \{x\}]$. The insertion node, denoted w_m, is the least common ancestor of the non-proper nodes of T^m.*
Node w_m is said to be cut iff w_m has no mixed child and either w_m is prime and has a child f such that $V(f) \cup \{x\}$ is a module of $G'[V(w_m) \cup \{x\}]$, or w_m is degenerate.

In the following, we do not consider the trivial case where V is uniform. Moreover, from now on, we only consider the case where w_m is not cut and the neighbourhood of $V(w_m)$ is a clique: the other cases are easy to deal with. We adopt the following definition.

Definition 2. *The key node w is the node of T^c such that $V(w_m) = w^*$ or $V(w_m) = w^* \setminus X_w$.*

Note that in Condition 2 of Theorem 2, the neighbourhood of M is not a clique. Thus, in the present case, $V(w_m)$ satisfies Condition 1 of Theorem 2, which ensures the existence of node w. It is straightforward to see that $V(w_m) \cup \{x\}$ is a module of $G + x$, and since w^* is a module of G and $V(w_m) \subseteq w^*$, it follows that $w^* \cup \{x\}$ is a module of $G + x$. Furthermore, since the neighbourhood of $V(w_m)$ in G is a clique, the neighbourhood of $w^* \cup \{x\}$ in $G + x$ is a clique. Then, the lemma below follows from Lemma 1.

Lemma 3. *$G + x$ is an interval graph iff $G[w^*] + x$ is an interval graph.*

4.2 Dynamic Characterisation of Interval Graphs

In this section, we characterise the insertions of a vertex x in an interval graph G that result in an interval graph. We start with the definitions and notations we use in our characterisation.

Definition 3. *Let u be a prime node of T^c and $v \in \mathcal{C}(u)$. v satisfies the left (resp. right) property iff $\forall y \in Y_u \cap N(x), e_y^2 \geq v$ (resp. $e_y^1 \leq v$).*

Notation 2. *If the saturated children of u form an interval of σ_u, we denote I_u for this interval and l_u (resp. r_u), if it exists, for the child of u immediately preceding (resp. following) I_u.*

Lemmas 4 to 7 give some necessary conditions for $G + x$ to be an interval graph, and Theorem 3 states that they are also sufficient. We omitted the proofs of the Lemmas since they are too technical to be sketched within the space limitation. As a general hint, we can say that their statement widely lean on the fact that deleting x in all the cliques of a consecutive ordering σ' of G' and removing the obtained cliques that are not maximal in G results in a consecutive ordering of G. Roughly speaking, this implies that, in σ', the maximal cliques of the new graph G' appear in an order that somehow respects the constraints previously imposed by the nodes of T^c.

Lemma 4. *If $G + x$ is an interval graph, any node $u \in T_w^c \setminus \{w\}$ has at most one mixed child, and w has at most two mixed children. Furthermore, for all $u \in T_w^c$, if B_u is not full, then u has at most one linked child.*

The other necessary conditions for G' to be an interval graph apply only to prime nodes of T_w^c.

Lemma 5. *If $G + x$ is an interval graph, for all prime nodes $u \in T_w^c$, the set of saturated children of u is an interval I_u of σ_u. And if $I_u \neq \varnothing$, then any node $v_1 \in C(u) \setminus (I_u \cup \{l_u, r_u\})$ is hollow.*

Lemma 6. *If $G + x$ is an interval graph, any prime node $u \neq w$ of T_w^c satisfies one of the following conditions:*

1. *B_u is full and $I_u \neq \varnothing$; and, up to reversing σ_u, $max(\sigma_u) \in I_u$ and l_u satisfies the left property.*
2. *B_u is full and $I_u = \varnothing$, or B_u is not full; and, up to reversing σ_u, $max(\sigma_u)$ satisfies the left property and the nodes of $C(u) \setminus \{max(\sigma_u)\}$ are hollow.*

Lemma 7. *If $G + x$ is an interval graph and if w is a prime node, it satisfies one of the following conditions:*

1. *B_w is full and $I_w \neq \varnothing$; and l_w and r_w satisfy respectively the left and right property.*
2. *B_w is full and $I_w = \varnothing$; and, one of the two following conditions holds:*
   (a) there exist two consecutive elements l and r in σ_w, with $l <_{\sigma_w} r$, that satisfy respectively the left and right property, and the nodes of $C(w) \setminus \{l, r\}$ are hollow, and $\Delta_l \cap \Delta_r \subseteq N(x)$; or
   (b) up to reversing σ_w, $max(\sigma_w)$ satisfies the left property and the nodes of $C(w) \setminus \{max(\sigma_w)\}$ are hollow.
3. *B_w is not full; and, up to reversing σ_w, $max(\sigma_w)$ satisfies the left property and the nodes of $C(w) \setminus \{max(\sigma_w)\}$ are hollow.*

Theorem 3. *$G + x$ is an interval graph iff the conditions of Lemmas 4 to 7 are satisfied.*

Sketch of proof. If the conditions of Lemmas 4 to 7 are satisfied, we can build a consecutive ordering of $G[w^*] + x$. To that purpose, we first build, for every full node u, a consecutive ordering of $G[u^*] + x$. Then, inductively, in a bottom up traversal of T_w^c, we build, for every mixed node $u \in T_w^c \setminus \{w\}$, a consecutive ordering of $G[u^*] + x$ st. the last maximal clique contains x. There are several cases to be considered. We cannot discuss each of them but we detail, as an example, the case where u satisfies Cond. 1 of Lemma 6 and l_u is mixed and B_{l_u} is full. In this case we obtain a consecutive ordering σ' of $G[u^*] + x$ by appending, in the order defined by σ_u, the consecutive orderings of $G[v^*]$ of nodes $v <_{\sigma_u} l_u$, the consecutive ordering of $G[l_u^*] + x$ built previously in the induction, and the consecutive orderings of $G[v^*] + x$ of nodes $v >_{\sigma_u} l_u$. Moreover, for any $v \in \mathcal{C}(u)$, we add the vertices of $\Delta_v \cup X_u$ to the cliques of $G[v^*]$ (or $G[v^*] + x$ if $v \geq_{\sigma_u} l_u$). Since we use a consecutive ordering of $G[l_u^*] + x$ whose last clique contains x, the cliques of σ' containing x form an interval. Once we get the consecutive orderings related to the mixed children of w, as w satisfies the conditions of Lemma 7, a last induction step allow us to obtain, in a similar way, a consecutive ordering of $G[w^*] + x$. □

4.3 Overview of the Algorithm and Complexity

The first step of our algorithm collects some information about T^m and T^c, and finds the key node w. The second step checks whether T_w^c satisfies the conditions of Lemmas 4 to 7, that is whether $G + x$ is an interval graph. In the positive, the third step updates $MD(G)$ by building $MD(G')$.

Marking step. We first determine for each node of T^m whether it is full, mixed or hollow by a well-known bottom-up marking process of the tree (see [14]), in $O(n)$ time. Then, we find the insertion node w_m by following a path from the root to w, while the visited node u is proper, we visit its unique mixed child (see [3]); the first non-proper node found is w_m. As we mentionned, the cases where w_m is cut or where the neighbourhood of $V(w_m)$ is not a clique are easy to deal with. We now describe the algorithm in the opposite case. Thanks to the correspondence between T^c and T^m, we find the key node w and determine for each node of T^c whether it is full, mixed or hollow. Finally, a simple top-down search of T^c allows us to determine for each node u whether B_u is full, mixed or hollow. The first step runs in $O(n)$ time.

Testing step. The conditions of Lemma 4 can be tested in $O(|\mathcal{C}(u)|)$ time by a simple search of $\mathcal{C}(u)$. The difficulty of checking the conditions of Lemma 5 is to decide whether the saturated children of u form an interval. To that purpose, we determine the set $S = \{v \in \mathcal{C}(u) \mid \Delta_v \subseteq N(x)\}$. We first bucket sort the vertices of $y \in Y_u \cap \overline{N}(x)$ by increasing e_y^1. As $1 \leq e_y^1 \leq |\mathcal{C}(u)|$, it takes $O(|\mathcal{C}(u)|)$ time. Examining the vertices of $Y_u \cap \overline{N}(x)$ in this order, we are able to maintain a partition of the children v of u such that Δ_v contains none of the vertices $y \in Y_u \cap \overline{N}(x)$ examined so far; each set of this partition being an interval of $\mathcal{C}(u)$. At the end of the routine, we obtain a partition of S. Then, checking the conditions of Lemma 5 becomes easy. It can be done in $O(|\mathcal{C}(u)| + |Y_u|)$ time. For a child v of a prime node u, it is easy to check whether v satisfies the left or right property by scanning Y_u. It follows that the only difficulty in checking the conditions of Lemmas 6 and 7 is to check Cond. 2a of Lemma 7. Let $w_1 = \min_{\sigma_w} \{e_y^2 \mid y \in Y_w \cap N(x)\}$. The children of w satisfying the left property are exactly the nodes $v \leq_{\sigma_w} w_1$. In the same way, we find the children of w satisfying the right property. Then, the couples (f, l) st. f and l resp. satisfy the left and right property define an interval of σ_w. The same technique as the one used to check the conditions of Lemma 5 determines the couples (f,l) such that $\Delta_f \cap \Delta_l \subseteq N(x)$. Hence, Cond. 2a of Lemma 7 can be tested in $O(|\mathcal{C}(w)|+|Y_w|)$ time. Finally, since all the conditions can be tested for a node u in $O(|\mathcal{C}(u)|+|Y_u|)$ time, we can determine whether $G+x$ is an interval graph in $O(n)$ time.

Insertion step. If $G+x$ is not an interval graph, then the algorithm stops. Otherwise, the MD-representation is updated. Since $V(w_m) \cup \{x\}$ is a strong module of $G' = G+x$ (cf. [3]), we can obtain $MD(G')$ by replacing node w_m of $MD(G)$ with the root of $MD(G[V(w_m)]+x)$. In order to get $MD(G[V(w_m)]+x)$ we first compute an interval model of $G[w^*]+x$. In the proof of Theorem 3, it is shown how to build a consecutive ordering of $G[w^*]+x$ by a bottom-up traversal of T_w^c. At each step, we concatenate the orderings computed for the children of the current node u, and we assign pointers to the vertices of $X_u \cup Y_u$; this takes $O(|\mathcal{C}(u)|+|X_u|+|Y_u|)$ time. Thus, the whole processing of T_w^c takes $O(n)$ time. Once we get an interval model of $G[w^*]+x$ we can easily extract a model of $G[V(w_m)]+x$, and thanks to the algorithm of [12] that computes the PQ-tree from an interval model, we get $PQ(G[V(w_m)]+x)$ in $O(n)$ time. The PQ-MD-transformation (see p.81) provides us with $MD(G[V(w_m)]+x)$ within the same complexity. Thus, the total computation time of $MD(G')$ is $O(n)$.

Acknowledgements

The author thanks Christophe Paul for useful discussions on the subject.

References

1. Booth, K.S., Lueker, G.S.: Testing for the consecutive ones property, interval graphs, and graph planarity using PQ-tree algorithms. J. Comput. Syst. Sci. 13(3), 335–379 (1976)
2. Crespelle, C.: Dynamic representations of interval graphs (manuscript) (2009), http://www-npa.lip6.fr/~crespell/publications/DynInt_long.pdf
3. Crespelle, C., Paul, C.: Fully dynamic algorithm for recognition and modular decomposition of permutation graphs. Algorithmica,
 http://www.springerlink.com/content/u6x0054g3h348810/;
 Ext. abs. in WG 2005
4. Crespelle, C., Paul, C.: Fully dynamic recognition algorithm and certificate for directed cographs. Discrete Applied Mathematics 154(12), 1722–1741 (2006); Ext. abs. in WG 2004
5. Gilmore, P.C., Hoffman, A.J.: A characterization of comparability graphs and of interval graphs. Canad. J. Math. 16, 539–548 (1964)
6. Hell, P., Shamir, R., Sharan, R.: A fully dynamic algorithm for recognizing and representing proper interval graphs. SIAM J. Comput. 31(1), 289–305 (2002)
7. Hsu, W.-L.: On-line recognition of interval graphs in $O(m+nlogn)$ time. In: Combinatorics and Computer Science, pp. 27–38 (1996)
8. Ibarra, L.: Fully dynamic algorithms for chordal graphs. In: SODA, pp. 923–924 (1999)

9. Ibarra, L.: A fully dynamic algorithm for recognizing interval graphs using the clique-separator graph. Tech. Report DCS-263-IR, Dept. of Computer Science, University of Victoria (2001)
10. Korte, N., Möhring, R.H.: Transitive orientation of graphs with side constraints. In: Nagl, M. (ed.) WG 1995. LNCS, vol. 1017, pp. 143–160. Springer, Heidelberg (1995)
11. Korte, N., Möhring, R.H.: An incremental linear-time algorithm for recognizing interval graphs. SIAM J. Comput. 18, 68–81 (1989)
12. McConnell, R.M., de Montgolfier, F.: Algebraic operations on PQ trees and modular decomposition trees. In: Kratsch, D. (ed.) WG 2005. LNCS, vol. 3787, pp. 421–432. Springer, Heidelberg (2005)
13. Möhring, R.H., Radermacher, F.J.: Substitution decomposition for discrete structures and connections with combinatorial optimization. Annals of Discrete Mathematics 19, 257–356 (1984)
14. Muller, J.H., Spinrad, J.P.: Incremental modular decomposition algorithm. JACM 36(1), 1–19 (1989)
15. Nikolopoulos, S.D., Palios, L., Papadopoulos, C.: A fully dynamic algorithm for the recognition of p_4-sparse graphs. In: Fomin, F.V. (ed.) WG 2006. LNCS, vol. 4271, pp. 256–268. Springer, Heidelberg (2006)
16. Tedder, M., Corneil, D.G.: An optimal, edges-only fully dynamic algorithm for distance-hereditary graphs. In: Thomas, W., Weil, P. (eds.) STACS 2007. LNCS, vol. 4393, pp. 344–355. Springer, Heidelberg (2007)

The Parameterized Complexity of Some Minimum Label Problems

Michael R. Fellows[1], Jiong Guo[2], and Iyad A. Kanj[3]

[1] The University of Newcastle, Callaghan, NSW 2308, Australia
michael.fellows@newcastle.edu.au
[2] Institut für Informatik, Friedrich-Schiller-Universität Jena, Ernst-Abbe-Platz 2, D-07743 Jena, Germany
guo@minet.uni-jena.de
[3] School of Computing, DePaul University, 243 S. Wabash Avenue, Chicago, IL 60604, USA
ikanj@cs.depaul.edu

Abstract. We study the parameterized complexity of several minimum label graph problems, in which we are given an undirected graph whose edges are labeled, and a property Π, and we are asked to find a subset of edges satisfying property Π that uses the minimum number of labels. These problems have a lot of applications in networking. We show that all the problems under consideration are W[2]-hard when parameterized by the number of used labels, and that they remain W[2]-hard even on graphs whose pathwidth is bounded above by a small constant. On the positive side, we prove that most of these problems are FPT when parameterized by the solution size, that is, the size of the sought edge set. For example, we show that computing a maximum matching or an edge dominating set that uses the minimum number of labels, is FPT when parameterized by the solution size. Proving that some of these problems are FPT is nontrivial, and requires interesting and elegant algorithmic methods that we develop in this paper.

1 Introduction

In this paper we consider several *minimum label graph problems* that are defined as follows:

Input: A graph $G = (V, E)$ whose edges are associated with labels or colors specified by a function $\mathcal{C} : E \rightarrow C$, where C denotes the set of labels (also referred to as colors in this paper), a graph property Π, and an integer d.
Output: A set $E' \subseteq E$ such that the subgraph of G consisting of the set of edges in E' satisfies Π, and the number of labels/colors used by the edges in E' is at most d.

Minimum label problems have been extensively studied in the last few years. These problems are motivated by applications from telecommunication networks, electrical networks, and multi-modal transportation networks. For example, in communication networks, there are different types of communication media, such as optic fiber, cable, microwave, and telephone line. A communication node may communicate with different nodes by choosing different types of communication media. Given a set of communication network nodes, the problem of finding a connected communication network using as few types of communication media (i.e., labels/colors) as possible is exactly the MINIMUM LABEL SPANNING TREE problem, in which the property Π is the property of being a spanning tree of G (see [5,13] for more details). Among the minimum label problems that have been extensively studied, we mention the MINIMUM LABEL SPANNING TREE problem [1,2,3,5,8,10,13,14,17,18,19], the MINIMUM LABEL PATH problem [2,4,8,16,20] (where Π is the property of being a path between two designated vertices), the MINIMUM LABEL CUT problem [9,20] (where Π is the property of being a cut between two designated vertices), and the MINIMUM LABEL PERFECT MATCHING problem [11] (where Π is the property of being a perfect matching).

The previous work on minimum label problems mainly dealt with determining the classical complexity of these problems and studying their approximabilty. Some of the previous work, however, dealt with developing exact algorithms for these problems. For example, Broersma et al. [2] devised two exact algorithms for the MINIMUM LABEL PATH and MINIMUM LABEL CUT problems with running time
$O(n \cdot \min\{|C|^{d(s,t)}, 2^{|C|}\})$ and $O(n^2 \cdot |C|!)$, respectively, where C denotes the set of labels (colors), and $d(s,t)$ denotes the distance between the two designated vertices s and t.

In the current paper we study the parameterized complexity of several minimum label graph problems, with respect to two natural parameters: the number of used labels d, and the size of the solution $|E'|$. The problems under consideration are: MINIMUM LABEL SPANNING TREE (MLST), MINIMUM LABEL HAMILTONIAN CYCLE (MLHC) (where Π is the property of being a Hamiltonian cycle), MINIMUM LABEL CUT (MLC), MINIMUM LABEL EDGE DOMINATION SET (MLEDS) (where Π is the property of being an edge dominating set, that is, every edges in $E \setminus E'$ shares at least one endpoint with some edge in E'), MINIMUM LABEL PERFECT MATCHING (MLPM), MINIMUM LABEL MAXIMUM MATCHING (MLMM) (where Π is the property of being a maximum matching of G), and MINIMUM LABEL PATH (MLP).

From some of the NP-hardness reductions for the above problems, we can derive parameterized intractability results with respect to the parameter d; for example, the NP-hardness reduction for MINIMUM LABEL SPANNING TREE shows that this problem is W[2]-hard [10]. In this paper, we strengthen these intractability results by showing that, even on graphs whose pathwidth is at most a small constant, when parameterized by the number of used labels d, these problems remain W[2]-hard. These results are interesting, as very few natural parameterized problems are known to be (parameterized) intractable on graphs with bounded pathwidth. When parameterized by the solution size $|E'|$, we show that, with the only exceptions of MINIMUM LABEL PATH and MINIMUM LABEL CUT, which are W[1]-hard, all other problems are fixed-parameter tractable (on general graphs). Showing that some of these problems are FPT is non-trivial, and requires elegant algorithmic methods that we develop in this paper.

All the hardness results will be presented in Section 2, while Section 3 contains all the fixed-parameter tractability results.

For the background and terminologies on graphs, we refer the reader to West [15], and for that on parameterized complexity, we refer the reader to Downey and Fellows' book [7].

2 Parameterized Hardness Results

First, we show that even on graphs whose pathwidth is at most a small constant, all the considered minimum label problems are W[2]-hard, when parameterized by the number of used labels d. These results are very interesting since there are few problems that are known to be W-hard on graphs of bounded pathwidth.

Theorem 2.1. *Parameterized by the number of used labels d:*

- MINIMUM LABEL EDGE DOMINATING SET *and* MINIMUM LABEL MAXIMUM MATCHING *are W[2]-hard on trees of pathwidth at most 1;*
- MINIMUM LABEL SPANNING TREE *and* MINIMUM LABEL PATH *are W[2]-hard on graphs with pathwidth at most 2;*
- MINIMUM LABEL CUT *and* MINIMUM LABEL PERFECT MATCHING *are W[2]-hard on graphs with pathwidth at most 3; and,*
- MINIMUM LABEL HAMILTONIAN CYCLE *is W[2]-hard on graphs with pathwidth at most 5.*

Proof. All the corresponding FPT-reductions are from the W[2]-hard HITTING SET (HS) problem, defined as follows. Given a ground set S, a collection $\mathcal{L}$ of subsets of S, and a nonnegative integer k, decide if there exists a subset S' of S of cardinality at most k, such that every subset in $\mathcal{L}$ has a non-empty intersection with S'. We only give one FPT-reduction showing that MINIMUM LABEL SPANNING TREE (MLST) is W[2]-hard. The reductions for the other problems are similar.

For a given instance of HS, we construct a graph G where, for each subset c in $\mathcal{L}$, there is a star consisting of a root vertex and $|c|$ leaves. The edges in this star are labeled with the elements of c. Then, we connect the leaves of this star by a path whose edges have the same label x, where $x \notin S$. Finally, we connect all root vertices by a path whose edges have the same label x. Clearly, the resulting graph has pathwidth 2. Observe that every size-d solution of the HS-instance corresponds to a solution of the resulting MLST-instance using $d+1$ labels, and vice versa. This gives the W[2]-hardness of MLST.

By reductions from MULTICOLORED CLIQUE, we can show the following theorem:

Theorem 2.2. *Parameterized by the solution size $|E'|$:*
- MINIMUM LABEL CUT *is W[1]-hard on graphs with pathwidth at most 4, and*
- MINIMUM LABEL PATH *is W[1]-hard on graphs with pathwidth at most 2.*

3 Fixed-Parameter Tractability Results

Parameterized by the solution size, MINIMUM LABEL SPANNING TREE, MINIMUM LABEL PERFECT MATCHING, and MINIMUM LABEL HAMILTONIAN CYCLE are all fixed-parameter tractable, since the instance size is bounded by a function of the parameter. However, it requires much more effort to show that MINIMUM LABEL MAXIMUM MATCHING (MLMM) and MINIMUM LABEL EDGE DOMINATING SET (MLEDS) are fixed-parameter tractable with respect to the same parameter. We note that we are mainly concerned with establishing the fixed-parameter tractability of MLMM and MLEDS. Consequently, the running time of the parameterized algorithms developed in this paper is not very practical, and can definitely be improved much further.

3.1 Minimum Label Maximum Matching (MLMM)

Let (G, k) be an instance of MLMM, where k is the size of a maximum matching in G. We denote by $e(G)$ and $n(G)$ the number of edges and vertices, respectively, in G. Let M be a maximal matching in G, $I = V(G) \setminus V(M)$, and note that I is an independent set in G. We denote by $G[M]$ the subgraph of G induced by the endpoints of the edges in M.

The algorithm is a search-tree based algorithm: it starts by growing a set of partial solutions, i.e., matchings, into an optimal solution, i.e., a maximum matching that uses the minimum number of colors. To do so, the algorithm branches on some vertices and edges in G to decide whether they belong to an optimal solution or not. Since the branching will consider all possibilities, we will maintain the invariant that at least one partial solution, among all partial solutions we keep, can be extended to an optimal solution. The algorithm can be split into three stages, each trying to simplify the resulting instance further by possibly performing more branchings.

Stage 1 Let M_{opt} be an optimal solution that we are trying to compute. For every edge e in $G[M]$ we branch as follows.

- e in M_{opt}: in this case we include e, decrement k by 1, and remove e and its endpoints from the graph. We also record that the color $\mathcal{C}(e)$ is used in the optimal solution.
- e is not in M_{opt}: in this case we simply remove e, that is, we set $G := G - e$.

For every remaining vertex v in $G[M]$ we branch as follows.

- v in M_{opt}: in this case we keep v in the graph.
- v is not in M_{opt}: in this case we remove v by setting $G := G - v$.

Let S be the set of remaining vertices in $G[M]$, and note that since all the edges in $G[M]$ have been removed during the branching, S is an independent set. Moreover, assuming that our partial solution (branching) is valid (i.e., leads to an optimal solution), every vertex in S must be an endpoint of an edge in the optimal solution M_{opt}. Without loss of generality, and since the parameter k can only decrease during the branching, we will denote the resulting parameter by k; this will simplify the notation in the remaining discussion. Assuming that our branching is valid, we have the following observation.

Observation 3.1. *The following are true:*

(a) $|S| = k$, and hence,
(b) for every $u \in I$, $deg(u) \leq k$.

Let $B = (S, I)$ be the resulting bipartite graph from G after the branching. The remaining task amounts to computing a matching with the minimum number of colors that matches S into I—and hence has size k, under the constraint that some of the colors have been used.

Analysis of the number of partial solutions enumerated in Stage 1
Since $|M| \leq k$, the number of vertices in $G[M]$ is at most $2k$, and the number of edges in $G[M]$ is at most $\binom{2k}{2} = k(2k - 1)$.

The branching in Stage 1 can be implemented as follows. For each $i = 0, \ldots, k$, we choose a matching of size i from the edges in $G[M]$ to be included in M_{opt}. For each of the remaining at most $(2k - 2i)$ vertices in $G[M]$, we branch on it as indicated above, thus creating at most 2^{2k-2i} partial solutions. Therefore, the number of partial solutions enumerated in Stage 1 is bounded above by:

$$\sum_{i=0}^{k} \binom{k(2k-1)}{i} 2^{2k-2i} = 4^k \sum_{i=0}^{k} \binom{k(2k-1)}{i} 1/4^i \tag{1}$$

$$\leq 4^k \binom{k(2k-1)}{k} \sum_{i=0}^{k} 1/4^i \tag{2}$$

$$\leq 4^k \cdot (e(2k-1))^k \cdot O(1) \tag{3}$$

$$\leq 4^k \cdot (2ek)^k \cdot O(1) = O((8ek)^k).$$

Inequality (2) is justified by the fact that the coefficient $\binom{k(2k-1)}{k}$ is the largest coefficient in the summation. Inequality (3) uses the fact that $\binom{n}{k} \leq (en/k)^k$, where e is the base of the natural logarithm (for instance, see [6]). It follows that the number of partial solutions enumerated in Stage 1 is $O((8ek)^k)$.

Stage 2 Given the bipartite graph $B = (S, I)$ and the parameter k, we try in this stage to simplify the instance further by performing more branching. We say that a matching is *monochromatic* if all its edges have the same color. If M' is a monochromatic matching, we denote by $\mathcal{C}(M')$ the color of the edges in M'.

We would like to partition S into groups such that all the vertices in the same group are matched in M_{opt} by a monochromatic matching of a distinct color. For this purpose we try all possible partitions of S. For a fixed partition of S into ℓ groups $S_1, \ldots, S_\ell$, we work under the assumption that the vertices in each group are matched by a monochromatic matching in M_{opt} of a distinct color (with respect to the colors of the other groups). Clearly, there exists at least one partition of S for which this working hypothesis is true, namely the one induced by the color classes in M_{opt}.

Let $S_1, \ldots, S_\ell$ be a fixed partition of S into ℓ nonempty groups, where $1 \leq \ell \leq k$ is an integer. It is possible that a group S_i uses the color of an edge that was added to a partial solution in Stage 1. Therefore, for each (possibly empty) subset C_{used} of the set of colors of the edges added in Stage 1, we try all one-to-one mappings from C_{used} to $\{S_1, \ldots, S_\ell\}$. Fix such a mapping. Then some groups in $\{S_1, \ldots, S_\ell\}$ have been assigned colors, and hence the colors of the monochromatic matchings sought for these groups are fixed. Clearly, since we are trying all possible assignments of the used colors to the groups, there will be an assignment that corresponds to that of M_{opt}, and we are safe.

Let S_i, $i \in \{1, \ldots, \ell\}$, be a group. If S_i has a preassigned color, let c_i be this color and define $\mathcal{M}_i = \{M_i \mid M_i$ is a monochromatic matching that matches S_i into I and $\mathcal{C}(M_i) = c_i\}$. Otherwise, the color of S_i is undetermined yet, and in this case define $\mathcal{M}_i = \{M_i \mid M_i$ is a monochromatic matching that matches S_i into $I\}$.

Let $h(k)$ be a function of k to be determined later, and let S_i, $i \in \{1, \ldots, \ell\}$, be a group. We perform more branching to simplify the instance as follows.

If $|\mathcal{M}_i| \leq h(k)$, we branch on every matching in $\mathcal{M}_i$ as the matching that matches S_i in M_{opt}. For each branch corresponding to a matching M_i in $\mathcal{M}_i$, we add the edges in M_i to the potential solution, decrement k by $|S_i|$, remove the vertices in $V(M_i)$ from the graph, and remove every edge whose color is $\mathcal{C}(M_i)$ from the graph (such an edge can no longer be used). Since we are trying all possible matchings M_i in $\mathcal{M}_i$, we are safe.

If the total number of colors used by the matchings in $\mathcal{M}_i$ is at most $h(k)$, we branch by trying all possible colors appearing in $\mathcal{M}_i$ to determine the color used in M_{opt} to match S_i (this color has to be one of the colors in $\mathcal{M}_i$). For each such color c, we remove all the edges in $\mathcal{M}_i$ whose colors are different from c. Again, since we are branching on all possible colors in $\mathcal{M}_i$, we are safe.

If all the edges of the matchings in $\mathcal{M}_i$ have the same color, and if there exists a vertex v in S_i with at most $h(k)$ edges incident on it in the matchings in $\mathcal{M}_i$, we branch on which edge in a matching in $\mathcal{M}_i$ matches v in M_{opt}. For each branch corresponding to an edge e_v, we add e_v to the potential solution, remove the endpoints of e_v from the graph, and decrement k by 1. We can now assume the following.

Assumption 3.2. *For each $i \in \{1, \ldots, \ell\}$:*

(i) $|\mathcal{M}_i| > h(k)$.
(ii) *Either the number of colors appearing in $\mathcal{M}_i$ is more than $h(k)$, or it is exactly 1.*

(iii) If $\mathcal{M}_i$ has exactly one color appearing in it, then every vertex in S_i has more than $h(k)$ edges that are incident on it in the matchings in $\mathcal{M}_i$.

In the next stage we show how, given the above assumption, we can easily compute a solution to the resulting instance.

Analysis of the number of partial solutions enumerated in Stage 2
Let c_{used} be the number of colors used in Stage 1. The number of partitions of S into ℓ groups is at most $\ell^{|S|} \leq \ell^k$. For each partition of S into ℓ groups, where $\ell \geq c_{used}$, we map the colors used in a one-to-one fashion to a subset of the ℓ groups. There are at most $\ell!/(\ell - c_{used})! \leq \ell!$ such mappings. Therefore, the total number of partitions of S in which some of the ℓ groups (exactly c_{used} many groups among them) have been assigned the used colors is at most $\sum_{\ell=1}^{k} \ell^k \ell! \leq k^{k+1} k!$.

Now for each S_i, $i \in \{1, \ldots, \ell\}$, we compute at most $h(k) + 1$ monochromatic matchings $M_i \in \mathcal{M}_i$. To do so, we iterate over each color c, and compute at most $h(k) + 1$ monochromatic matchings of color c. For a fixed color c, we consider the subgraph of B consisting only of the edges incident on S_i whose color is c. Note that each matching in this subgraph that matches S_i into I is a maximum matching. It was shown in [12] how, after computing a maximum matching in a bipartite graph, every other maximum matching can be computed in linear time in the number of vertices of the subgraph, per matching. Therefore, computing at most $h(k) + 1$ monochromatic matchings of color c that match S_i into I can be done in time $O(e(G)\sqrt{n(G)} + n(G)h(k))$. As a matter of fact, since whenever we fix a color c for a group S_i we only look at the edges of color c incident on the vertices in S_i, and since we totally compute at most $h(k) + 1$ matchings incident on the vertices in S_i, computing at most $h(k) + 1$ monochromatic matchings (regardless of the color) incident on the vertices of S_i can be done in time $O(e(G)\sqrt{n(G)} + n(G)h(k))$. Since there are at most k groups, computing the sets $\mathcal{M}_i$, $i = 1, \ldots, \ell$, can be done in time $O(ke(G)\sqrt{n(G)} + kh(k)n(G))$.

To make the graph B satisfy the statements in Assumption 3.2, we do the following. After computing the set $\mathcal{M}_i$ for each group S_i as indicated above, we check if $|\mathcal{M}_i| \leq h(k)$. If it is, we branch on every monochromatic matching in $\mathcal{M}_i$. For each such matching M_i, we remove the endpoints of the edges in M_i, and hence the group S_i from the graph, and decrease the parameter by $|S_i|$. Since we are branching on every monochromatic matching in $\mathcal{M}_i$, we are safe. Since there are at most $h(k)$ matchings in $\mathcal{M}_i$, and at most k groups S_i, the total number of enumerations is at most $h(k)^k$.

Now we can assume that the cardinality of each set $\mathcal{M}_i$ is at least $h(k) + 1$.

If there is a set $\mathcal{M}_i$ such that the total number of colors appearing in it is at most $h(k)$, then we branch by trying every color in $\mathcal{M}_i$ as the color used to match S_i in M_{opt}. For each such color c, we remove all the edges incident on S_i whose color is different from c, and we remove every edge whose color is c but is not incident on a vertex in S_i. The total number of enumerations is again at most $h(k)^k$.

Finally, if we have a set $\mathcal{M}_i$ such that all the matchings in this set have the same color c, then for every vertex v (if any) in S_i whose degree in $\mathcal{M}_i$ is at most $h(k)$, we branch on which edge in $\mathcal{M}_i$ is used to match v in M_{opt}. For each edge in $\mathcal{M}_i$ incident on v, we remove the endpoints of the edge from the graph and decrement k by 1. Since we are trying all possible edges incident on such a vertex v, we are safe. The total number of enumerations in this case is at most $h(k)^k$ (there are at most k vertices in S).

We can now assume that B satisfies the statements in Assumption 3.2. The total number of enumerations incurred to make B satisfy the statements in Assumption 3.2 is at most $h(k)^k \cdot h(k)^k \cdot h(k)^k = h(k)^{3k}$.

It follows that the number of partial solutions enumerated in Stage 2 is bounded above by the number of partitions of S, multiplied by the number of enumerations to make B satisfy the statements in Assumption 3.2. From the above discussion, it follows that the number of partial solutions enumerated in Stage 2 is $O(k^{k+1}k! + h(k)^{3k})$.

Stage 3 Given an instance $B = (S, I)$ and a parameter k such that S is partitioned into $S_1, \ldots, S_\ell$, where each set $\mathcal{M}_i$ associated with S_i, for $i = 1, \ldots, \ell$, satisfies the statements of Assumption 3.2, the following theorem asserts that, in time $O(k^3)$, we can compute a matching M' that matches S into I, and such that the set of edges in M' incident on S_i is a monochromatic matching whose edges are edges from the matchings in $\mathcal{M}_i$. The proof is omitted for lack of space.

Theorem 3.1. *Let $h(k) \geq k^2 + k$. Assuming that each $\mathcal{M}_i$, $i = 1, \ldots, \ell$, satisfies Assumption 3.2, then in time $O(k^3)$ we can compute a matching M' that matches S into I, such that the set of edges in M' incident on S_i, for $i = 1, \ldots, \ell$, is a monochromatic matching whose edges are edges from the matchings in $\mathcal{M}_i$.*

Analysis of the running time of Stage 3
By Theorem 3.1, computing the matching M' takes $O(k^3)$ time.

Putting all together. The correctness of the algorithm follows from the fact that it is enumerating all possible branchings. For each possible branching, either we reject the instance, or we end up computing a maximum matching that uses a certain number of colors. The maximum matching we output at the end is the maximum matching with the minimum number of colors. The running time of the algorithm is bounded by the number of partial solutions enumerated, multiplied by the running time spent along each enumeration (path in the search tree). The number of partial solutions we enumerate is the product of those enumerated in Stage 1 ($O((8ek)^k)$) and Stage 2 ($O(k^{k+1}k! + h(k)^{3k})$), which is $O((8e)^k \cdot k^{7k})$ after choosing $h(k) = k^2 + k$. Along each path in the search tree we end up processing the graph G, which takes linear time in its number of vertices and edges, computing a maximum matching in G, which takes $O(e(G)\sqrt{n(G)})$, and computing the sets $\mathcal{M}_i$ in Stage 2, which takes $O(ke(G)\sqrt{n(G)} + k^3 n(G))$. Therefore, the running time of the algorithm is $O((8e)^k \cdot k^{7k+3} e(G)\sqrt{n(G)})$.

Theorem 3.2. MINIMUM LABEL MAXIMUM MATCING *is FPT when parameterized by the size of the maximum matching in the graph.*

3.2 Minimum Label Edge Dominating Set (MLEDS)

The ideas used by the algorithm are similar in flavor to those used for the MLMM problem. Therefore, we will omit some details to avoid repetition. We start with the following easy observation.

Observation 3.3. *Let M be a matching in G, and let Q be an edge dominating set of G. Then $|Q| \geq |M|/2$.*

Let (G, k) be an instance of MLEDS. Let M be a maximal matching in G, $I = V(G) \setminus V(M)$, and note that I is an independent set in G. If $|M| > 2k$, then by Observation 3.3, G does not have an edge dominating set of size at most k, and we can reject the instance (G, k). Therefore, we may assume henceforth that $|M| \leq 2k$.

Similar to what we did for the MLMM problem, we will branch on the edges and vertices in M to determine which ones contribute to a solution Q_{opt}, which is an edge dominating set of G of size at most k that uses the minimum number of colors (if such a solution exists).

For an edge $e \in G[M]$, we branch on e as follows. If e is decided to be in Q_{opt}, we set $G = G - e$, decrement k by 1, mark all the edges incident on e in the graph as dominated, and label both endpoints of e with the label "IN_{used}" to indicate that they are in Q_{opt}, and are endpoints of some edge that is already decided to be in Q_{opt}. (We will use the label "IN" later to indicate that the vertex is decided to be in Q_{opt} but has no incident edge that was decided to be in Q_{opt} yet.) We also indicate that the color of e has been used by storing it in a set of colors C_{used}. On the other hand, if e is decided not be in Q_{opt}, we set $G = G - e$.

For a vertex $v \in G[M]$ whose status has not been determined yet by the above branching (i.e., v does not have the label IN_{used}), we branch on v as follows. If v is decided to be an endpoint of an edge in Q_{opt}, we label v as IN, and mark every edge incident on v as dominated. If v is decided not be an endpoint of an edge in Q_{opt}, we label it as OUT.

Note that since I is an independent set in G, every edge in G must be dominated by an edge in Q_{opt} with at least one endpoint in $G[M]$. In particular, this is true for every edge in $G[M]$. Therefore, after branching on the edges and vertices in $G[M]$, we need to check that, for every edge $e \in G[M]$ that was decided not to be in Q_{opt} and subsequently removed from G, at least one of its endpoints has label IN or IN_{used}. If this is not the case, then the partial solution that we have enumerated is not valid, and we reject it.

Noting that after the above branching all the edges of $G[M]$ were removed from G, we end up with a bipartite graph $B = (S, I)$, where S consists of the vertices in $G[M]$. Every vertex in S has one of the following labels: (1) IN_{used} indicating that the vertex is an endpoint of a known edge which was determined to be in Q_{opt}, (2) IN indicating that the vertex is the endpoint of some edge in Q_{opt} but this edge has not been determined yet, and (3) OUT indicating that the vertex is not an endpoint of an edge in Q_{opt}. The edges in B have one of two possible types: (1) dominated, those are the edges with at least one endpoint of label IN_{used} or IN, and (2) not dominated, and those are the edges whose endpoint in S is of label OUT.

Since we are trying all possible branches for the edges and vertices in $G[M]$, we are safe. The number of partial solutions enumerated by the branching can be upper bounded in a similar fashion to that in Stage 1 of the algorithm for MLMM. The only difference here is that the number of edges in the maximal matching M is at most $2k$, and hence, the number of vertices in $G[M]$ is at most $4k$, and consequently the number of edges in $G[M]$ is at most $2k(4k-1)$. Using a similar analysis to that in Stage 1 of MLMM, we obtain that the number of partial solutions enumerated by the above branching is at most $(128ek)^k$.

Now given the instance $B = (S, I)$, and the parameter k (without loss of generality), we will branch further to simplify the instance. First, observe that since the number of edges in Q_{opt} is at most k, the number of vertices in S that are labeled with IN_{used} or IN is at most $2k$ (otherwise, we reject the enumeration).

Observation 3.4. *For every vertex w in I, the number of edges incident on w whose endpoint in S is labeled with IN_{used} or IN is at most $2k$.*

Note that, for every edge $e = \{u, v\}$ where $u \in S$ has label OUT, e needs to be dominated by an edge incident on v; therefore, the vertex v must be an endpoint of some edge in Q_{opt}. Since the number of edges in Q_{opt} is at most k, and B is bipartite, there can be at most k vertices in I that are neighbors of vertices in S of label OUT; let I_{in} be the set of such vertices. Since (by Observation 3.4) every vertex in I has at most $2k$ edges incident on it whose endpoint in S is labeled IN_{used} or IN, we can branch on every such edge incident on a vertex in I_{in} to determine if the edge is in Q_{opt} or not. For each such edge, if the edge is decided to be in Q_{opt}, we include the edge in the solution, label both its endpoints IN_{used}, we remove the edge, decrement k by 1, and update C_{used} appropriately; if the edge is decided not be in Q_{opt}, we simply remove it. After this branching, we check that for every vertex in I_{in} at least one of the edges incident on it was decided to be in Q_{opt}; otherwise, we reject the branch. The number of partial solutions generated by this branching is at most $(2k)^k$.

After branching on the edges incident on the vertices in I_{in} and removing them, the vertices in I_{in} and the vertices in S of label OUT can be removed. Every remaining vertex in S is either of label IN_{used} or IN.

Since a vertex in S of label IN_{used} is an endpoint of an edge already in Q_{opt}, every edge incident on a vertex in IN_{used} is dominated. Therefore, if for every vertex of label IN in S we determine one of its incident edges to be in Q_{opt}, we obtain an edge dominating set of B. On the other hand, our branching stipulates that from every vertex in S of label IN we must determine at least one edge incident on it to be in Q_{opt}. Therefore, our problem reduces to picking for every vertex of label IN in S exactly one edge incident on it, so that the total number of colors used is minimized. To do so, we first remove the vertices of label IN_{used} from S, since no edge incident on any of them needs to be considered. At this point S should have at most k vertices; otherwise, we can reject. Then for every color c in C_{used}, and for every vertex v of label IN in S, if there is an edge of color c incident on v, we include e in the solution, decrement k, and remove the vertex from B. (Note that edges whose color is in C_{used} are gained for free.)

After this step, every vertex in S is of label IN, and there is no edge incident on any vertex in S whose color appears in C_{used}. To compute a set of edges of minimum colors, such that for every vertex in S exactly one edge in this set is incident on it, we try each partition of S into ℓ groups, $\ell \in \{1, \ldots, k\}$, such that all vertices in the same group are incident on edges of the same color in Q_{opt} (as we did in Stage 2 of the MLMM problem). For each such partition and each group in this partition, we find a color c such that every vertex in this group is incident on an edge of color c. If such a choice is not possible for some group, then we reject the partition.

At the end, we end up with an edge dominating set for G. We output the edge dominating set of G of size at most k that uses the minimum number of colors, over all partial solutions generated from all branches.

Since S has at most k vertices at this stage, the total number of partitions of S is at most k^{k+1}.

It follows that the total number of partial solutions enumerated by the algorithm is $O((128ek)^k \cdot (2k)^k \cdot k^{k+1}) = O((256e)^k k^{3k+1})$. For each such partial solution we need to process the graph G during the branching, which takes time $O(n(G) + e(G))$. Therefore, the running time of the algorithm is $O((256e)^k k^{3k+1}(n(G) + e(G)))$.

Theorem 3.3. MINIMUM LABEL EDGE DOMINATING SET *is FPT when parameterized by the size of the edge dominating set.*

4 Concluding Remarks

In this paper, we considered some minimum label graph problems. We showed that, when parameterized by the number of used labels, most of these problems are intractable, even on graphs of bounded pathwidth.

On the other hand, we showed that most of these problems become parameterized tractable when parameterized by the solution size.

We note that, recently, there has been a lot of interest in studying structured graph problems, such as problems on colored graphs, due to their applications in various fields such as networking and computational biology. (The convex recoloring problem is such an example in computational biology.) While these problems are practically very important, they are often computationally hard due to the structural requirement on the solution sought. Therefore, it is both natural and interesting to study whether these problems remain intractable with respect to different parameters, such as the number of colors, the pathwidth/treewidth of the graph, the solution size, or even with respect to more restrictive parameters, such as the vertex cover or the max leaf number. This paper follows this line of research.

Finally, it is interesting to study the parameterized complexity of other minimum label graph problems that have practical applications. A good candidate would be the Minimum Label Feedback Arc Set problem on directed graphs.

References

1. Broersma, H., Li, X.: Spanning trees with many or few colors in edge-colored graphs. Discussiones Mathematicae Graph Theory 17(2), 259–269 (1997)
2. Broersma, H., Li, X., Woeginger, G., Zhang, S.: Paths and cycles in colored graphs. Australasian Journal on Combinatorics 31, 299–311 (2005)
3. Brüggemann, T., Monnot, J., Woeginger, G.: Local search for the minimum label spanning tree problem with bounded color classes. Operations Research Letters 31(3), 195–201 (2003)
4. Carr, R., Doddi, S., Konjevod, G., Marathe, M.: On the red-blue set cover problem. In: Proc. 11th ACM-SIAM SODA, pp. 345–353 (2000)
5. Chang, R., Leu, S.: The minimum labeling spanning trees. IPL 63(5), 277–282 (1997)
6. Cormen, T., Leiserson, C., Rivest, R., Stein, C.: Introduction to Algorithms, 2nd edn. McGraw-Hill Book Company, Boston (2001)
7. Downey, R.G., Fellows, M.R.: Parameterized complexity. Monographs in Computer Science. Springer, New York (1999)
8. Hassin, R., Monnot, J., Segev, D.: Approximation algorithms and hardness results for labeled connectivity problems. Journal of Combinatorial Optimization 14(4), 437–453 (2007)
9. Jha, S., Sheyner, O., Wing, J.: Two formal analyses of attack graphs. In: Proc. 15th IEEE Computer Security Foundations Workshop, pp. 49–63 (2002)
10. Krumke, S., Wirth, H.: On the minimum label spanning tree problem. IPL 66(2), 81–85 (1998)
11. Monnot, J.: The labeled perfect matching in bipartite graphs. IPL 96(3), 81–88 (2005)
12. Uno, T.: Algorithms for enumerating all perfect, maximum and maximal matchings in bipartite graphs. In: Leong, H.-V., Jain, S., Imai, H. (eds.) ISAAC 1997. LNCS, vol. 1350, pp. 92–101. Springer, Heidelberg (1997)
13. Voss, S., Cerulli, R., Fink, A., Gentili, M.: Applications of the pilot method to hard modifications of the minimum spanning tree problem. In: Proc. 18th MINI EURO Conference on VNS (2005)
14. Wan, Y., Chen, G., Xu, Y.: A note on the minimum label spanning tree. IPL 84(2), 99–101 (2002)
15. West, D.: Introduction to graph theory. Prentice Hall Inc., Upper Saddle River (1996)
16. Wirth, H.: Multicriteria Approximation of Network Design and Network Upgrade Problems. PhD thesis, Department of Computer Science, Universität Würzburg, Germany (2005)
17. Xiong, Y.: The Minimum Labeling Spanning Tree Problem and Some Variants. PhD thesis, Graduate School of the University of Maryland, USA (2005)
18. Xiong, Y., Golden, B., Wasil, E.: A one-parameter genetic algorithm for the minimum labeling spanning tree problem. IEEE Transactions on Evolutionary Computation 9(1), 55–60 (2005)
19. Xiong, Y., Golden, B., Wasil, E.: Worst case behavior of the MVCA heuristic for the minimum labeling spanning tree problem. Operations Research Letters 33(1), 77–80 (2005)
20. Zhang, P., Tang, L., Zhao, W., Cai, J., Li, A.: Approximation and hardness results for label cut and related problems (manuscript) (2007)

Exact and Parameterized Algorithms for MAX INTERNAL SPANNING TREE*

Henning Fernau[1], Serge Gaspers[2], and Daniel Raible[1]

[1] Univ. Trier, FB 4—Abteilung Informatik, D-54286 Trier, Germany
{fernau,raible}@uni-trier.de
[2] LIRMM – Univ. of Montpellier 2, CNRS, 34392 Montpellier, France
gaspers@lirmm.fr

Abstract. We consider the $\mathcal{NP}$-hard problem of finding a spanning tree with a maximum number of internal vertices. This problem is a generalization of the famous HAMILTONIAN PATH problem. Our dynamic-programming algorithms for general and degree-bounded graphs have running times of the form $\mathcal{O}^*(c^n)$ ($c \leq 3$). The main result, however, is a branching algorithm for graphs with maximum degree three. It only needs polynomial space and has a running time of $\mathcal{O}(1.8669^n)$ when analyzed with respect to the number of vertices. We also show that its running time is $2.1364^k n^{\mathcal{O}(1)}$ when the goal is to find a spanning tree with at least k internal vertices. Both running time bounds are obtained via a Measure & Conquer analysis, the latter one being a novel use of this kind of analysis for parameterized algorithms.

1 Introduction

Motivation. In this paper we investigate the following problem:

MAX INTERNAL SPANNING TREE (MIST)
Given: A graph $G = (V, E)$ with n vertices and m edges.
Task: Find a spanning tree of G with a maximum number of internal vertices.

MIST is a generalization of the well-studied HAMILTONIAN PATH problem: find a path in a graph such that every vertex is visited exactly once. Clearly, such a path, if it exists, is also a spanning tree, namely one with a maximum number of internal vertices. Whereas the running time barrier of 2^n has not been broken for general graphs, there are faster algorithms for cubic graphs (using only polynomial space). It is natural to ask if for the generalization, MIST, this can also be obtained.

A second issue is to find an algorithm for MIST with a running time of the form $\mathcal{O}^*(c^n)$. [1] The naïve approach gives only an upper bound of $\mathcal{O}^*(2^m)$. A

---
* Supported by a PPP grant between DAAD (Germany) and NFR (Norway).
[1] $f(n) = \mathcal{O}^*(g(n))$ if $f(n) \leq p(n) \cdot g(n)$ for some polynomial $p(n)$.

possible application could be the following scenario. Consider cities which should be connected with water pipes. The possible connections between them can be represented by a graph G. It suffices to compute a spanning tree T for G. In T we may have high degree vertices that have to be implemented by branching pipes which cause turbulences and therefore pressure may drop. To minimize the number of branching pipes one can equivalently compute a spanning tree with the smallest number of leaves, leading to MIST. Vertices representing branching pipes should not be of arbitrarily high degree, motivating us to investigate MIST on degree-restricted graphs.

Previous Work. It is well-known that the more restricted problem, HAMILTONIAN PATH, can be solved within $\mathcal{O}(n^2 2^n)$ steps and exponential space. This result has been independently obtained by Bellman [1], and Held and Karp [6]. The TRAVELING SALESMAN problem (TSP) is very closely related to HAMILTONIAN PATH. Basically, the same algorithm solves this problem, but there has not been any improvement on the running time since 1962. The space requirements have, however, been improved and now there are $\mathcal{O}^*(2^n)$ algorithms needing only polynomial space. In 1977, Kohn et al. [9] gave an algorithm based on generating functions with a running time of $\mathcal{O}(2^n n^3)$ and space requirements of $\mathcal{O}(n^2)$ and in 1982 Karp [8] came up with an algorithm which improved storage requirements to $\mathcal{O}(n)$ and preserved this run time by an inclusion-exclusion approach.

Eppstein [4] studied TSP on cubic graphs. He could achieve a running time of $\mathcal{O}(1.260^n)$ using polynomial space. Iwama and Nakashima [7] could improve this to $\mathcal{O}(1.251^n)$. Björklund et al. [2] considered TSP with respect to degree-bounded graphs. Their algorithm is a variant of the classical 2^n-algorithm and the space requirements are therefore exponential. Nevertheless, they showed that for a graph with maximum degree d there is a $\mathcal{O}^*((2-\epsilon_d)^n)$-algorithm. In particular for $d = 4$ there is a $\mathcal{O}(1.8557^n)$- and for $d = 5$ a $\mathcal{O}(1.9320^n)$-algorithm.

MIST was also studied with respect to parameterized complexity. The (standard) parameterized version of the problem is parameterized by k, and asks whether G has a spanning tree with at least k internal vertices. Prieto and Sloper [11] proved a $\mathcal{O}(k^3)$-vertex kernel for the problem showing $\mathcal{FPT}$-membership. In [12] the kernel size has been improved to $\mathcal{O}(k^2)$ and in [5] to $3k$. Parameterized algorithms for MIST have been studied in [3,5,12]. Prieto and Sloper [12] gave the first FPT algorithm, with running time $2^{4k \log k} \cdot n^{\mathcal{O}(1)}$. This result was improved by Cohen et al. [3] who solve a more general directed version of the problem in time $49.4^k \cdot n^{\mathcal{O}(1)}$. The current fastest algorithm has running time $8^k \cdot n^{\mathcal{O}(1)}$ [5].

Salamon [14] studied the problem considering approximation. He could achieve a $\frac{7}{4}$-approximation. A $2(\Delta - 2)$-approximation for the node-weighted version is a by-product. Cubic and claw-free graphs were considered by Salamon and Wiener [13] introducing algorithms with approximation ratios $\frac{6}{5}$ and $\frac{3}{2}$, respectively.

Our Results. This paper gives two algorithms:

(a) A dynamic-programming algorithm solving MIST in time $\mathcal{O}^*(3^n)$. We extend this algorithm and show that for any degree-bounded graph a running time of $\mathcal{O}^*((3-\epsilon)^n)$ with $\epsilon > 0$ can be achieved. To our knowledge this is the first algorithm for MIST with a running time bound of the form $\mathcal{O}^*(c^n)$.[2]

(b) A polynomial-space branching algorithm solving the maximum degree 3 case in time $\mathcal{O}(1.8669^n)$. We also analyze the same algorithm from a parameterized point of view, achieving a running time of $2.1364^k n^{\mathcal{O}(1)}$ to find a spanning tree with at least k internal vertices (if possible). The latter analysis is novel in a sense that we use a potential function analysis—Measure & Conquer—in a way that, to our knowledge, is much less restrictive than any previous analysis for parameterized algorithms that were based on the potential function method.

Notions and Definitions. We consider only simple undirected graphs $G = (V, E)$. The *neighborhood* of a vertex $v \in V$ in G is $N_G(v) := \{u \mid \{u, v\} \in E\}$ and its *degree* is $d_G(v) := |N_G(v)|$. The *closed neighborhood* of v is $N_G[v] := N_G(v) \cup \{v\}$ and for a set $V' \subseteq V$ we let $N_G(V') := \left(\bigcup_{u \in V'} N_G(u)\right) \setminus V'$. We omit the subscripts of $N_G(\cdot)$, $d_G(\cdot)$, and $N_G[\cdot]$ when the graph is clear from the context. A *subcubic graph* has maximum degree at most three. For a (partial) spanning tree $T \subseteq E$ let $I(T)$ be the set of its internal (non-leaf) vertices and $L(T)$ the set of its leaves. An *i-vertex* u is a vertex with $d_T(u) = i$ with respect to some spanning tree T, where $d_H(u) := \{\{u, v\} \mid \{u, v\} \in H\}$ for any $H \subseteq E$. The *tree-degree* of some $u \in V(T)$ is $d_T(u)$. We also speak of the T-degree $d_T(v)$ when we refer to a specific spanning tree. A *Hamiltonian path* is a sequence of pairwise distinct vertices $v_1, \ldots, v_n$ from V such that $\{v_i, v_{i+1}\} \in E$ for $1 \leq i \leq n-1$.

The Problem on General Graphs. By means of Dynamic Programming and the help of an upper bound on the number of connected vertex-subsets of degree bounded graph (shown by [2]) we show the next statement.

Lemma 1. *MIST can be solved in time $\mathcal{O}^*(3^n)$ and for graphs with maximum degree Δ, MIST can be solved in time $\mathcal{O}^*(3^{(1-\epsilon_\Delta)n})$ with $\epsilon_\Delta > 0$.*

2 Subcubic Maximum Internal Spanning Tree

2.1 Observations

Let t_i^T denote the number of vertices u such that $d_T(u) = i$ for a spanning tree T. Then the following proposition can be proved by induction on $n_T := |V(T)|$.

Proposition 1. *In any spanning tree T, $2 + \sum_{i \geq 3}(i - 2) \cdot t_i^T = t_1^T$.*

Due to Proposition 1, MIST on subcubic graphs boils down to finding a spanning tree T such that t_2^T is maximum. Every internal vertex of higher degree would also introduce additional leaves.

---

[2] Before the camera-ready version of this paper was prepared, Nederlof [10] came up with a polynomial-space $\mathcal{O}^*(2^n)$ algorithm for MIST on general graphs, answering a question in a preliminary version of this paper.

Lemma 2. *[11] An optimal solution T_o to* MAX INTERNAL SPANNING TREE *is a Hamiltonian path or the leaves of T_o are independent.*

The proof of Lemma 2 shows that if T_o is not a Hamiltonian path and there are two adjacent leaves, then the number of internal vertices can be increased. In the rest of the paper we assume that T_o is not a Hamiltonian path due to the next lemma which uses the $\mathcal{O}(1.251^n)$ algorithm for HAMILTONIAN CYCLE on subcubic graphs [7] as a subroutine.

Lemma 3. HAMILTONIAN PATH *can be solved in time $\mathcal{O}(1.251^n)$ on subcubic graphs.*

Lemma 4. *Let T be a spanning tree and $u, v \in V(T)$ two adjacent vertices with $d_T(u) = d_T(v) = 3$ such that $\{u, v\}$ is not a bridge. Then there is a spanning tree $T' \supset (T \setminus \{\{u, v\}\})$ with $|I(T')| \geq |I(T)|$ and $d_{T'}(u) = d_{T'}(v) = 2$.*

Proof. By removing $\{u, v\}$, T is separated into two parts T_1 and T_2. The vertices u and v become 2-vertices. As $\{u, v\}$ is not a bridge, there is another edge $e \in E \setminus E(T)$ connecting T_1 and T_2. By adding e we lose at most two 2-vertices. Then let $T' := (T \setminus \{\{u, v\}\}) \cup \{e\}$ and it follows that $|I(T')| \geq |I(T)|$. □

2.2 Reduction Rules

Let $E' \subseteq E$. Then, $\partial E' := \{\{u, v\} \in E \setminus E' \mid u \in V(E')\}$ are the edges outside E' that have a common end point with an edge in E' and $\partial_V E' := V(\partial E') \cap V(E')$ are the vertices that have at least one incident edge in E' and another incident edge not in E'. During the algorithm we will maintain an acyclic subset of edges F which will be part of the final solution. The following invariant will always be true: $G[F]$ consists of a tree T and a set P of *pending tree edges (pt-edges)*. Here a pt-edge $\{u, v\} \in F$ is an edge with one end point u of degree 1 and the other end point $v \notin V(T)$. $G[T \cup P]$ will always consist of $1 + |P|$ components. Next we present several reduction rules. The order in which they are applied is crucial: Before a rule is applied the preceding ones were carried out exhaustively.

Cycle: Delete any edge $e \in E$ such that $E(T) \cup \{e\}$ has a cycle.
Bridge: If there is a bridge $e \in \partial E(T)$, then add e to F.
Deg1: Let $u \in V \setminus V(F)$ with $d(u) = 1$. Then add its incident edge to F.
Pending: If a a vertex v is incident to $d_G(v) - 1$ pt-edges, then remove them.
ConsDeg2: If there are edges $\{v, w\}, \{w, z\} \in E \setminus E(T)$ such that $d_G(w) = d_G(z) = 2$, then delete $\{v, w\}, \{w, z\}$ from G and add the edge $\{v, z\}$ to G.
Deg2: If there is an edge $\{u, v\} \in \partial E(T)$ such that $u \in V(T)$ and $d_G(u) = 2$, then add $\{u, v\}$ to F.
Attach: If there are edges $\{u, v\}, \{v, z\} \in \partial E(T)$ such that $u, z \in V(T)$, $d_T(u) = 2$, $1 \leq d_T(z) \leq 2$, then delete $\{u, v\}$. See Fig. 1(a)
Attach2: If there is a vertex $u \in \partial_V E(T)$ with $d_T(u) = 2$ and $\{u, v\} \in E \setminus E(T)$ such that v is incident to a pt-edge, then delete $\{u, v\}$.
Special: If there are two edges $\{u, v\}, \{v, w\} \in E \setminus F$ with $d_T(u) \geq 1$, $d_G(v) = 2$, and w is incident to a pt-edge, then add $\{u, v\}$ to F. See Fig. 1(b).

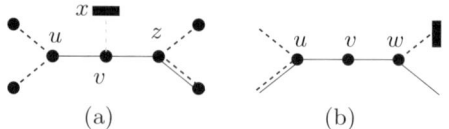

Fig. 1. Light edges may be not present. Double edges (dotted or solid, resp.) refer to edges which are either T-edges or not, resp. Edges attached to oblongs are pt-edges.

We mention that **ConsDeg2** can create double edges. In this case simply delete one of them which is not in F (at most one can be part of F).

Lemma 5. *The reduction rules stated above are sound.*

Proof. Let $T_o \supset F$ be a spanning tree of G with a maximum number of internal vertices. The correctness of the first five reduction rules is easily verified.

Deg2. Since the preceding reduction rules do not apply, we have $d_G(v) = 3$ and there is one incident edge, say $\{v,z\}$, $z \neq u$, that is not pending. Assume u is a leaf in T_o. Define another spanning tree $T'_o \supset F$ by setting $T'_o = (T_o \cup \{\{u,v\}\}) \setminus \{v,z\}$. Since $|I(T_o)| \leq |I(T'_o)|$, T'_o is also optimal.

Attach. If $\{u,v\} \in E(T_o)$ then $\{v,z\} \notin E(T_o)$ due to the acyclicity of T_o and as T is connected. Then by exchanging $\{u,v\}$ and $\{v,z\}$ we obtain a solution T'_o with at least as many 2-vertices.

Attach2. Suppose $\{u,v\} \in E(T_o)$. Let $\{v,p\}$ be the pt-edge and $\{v,z\}$ the third edge incident to v (that must exist and is not pending, since **Pending** did not apply). Since **Bridge** did not apply, $\{u,v\}$ is not a bridge. Firstly, suppose $\{v,z\} \in E(T_o)$. Due to Lemma 4, there is also an optimal solution $T'_o \supset F$ with $\{u,v\} \notin E(T'_o)$. Secondly, assume $\{v,z\} \notin E(T_o)$. Then $T' = (T_o \setminus \{\{u,v\}\}) \cup \{\{v,z\}\}$ is also optimal as u has become a 2-vertex.

Special. Suppose $\{u,v\} \notin E(T_o)$. Then $\{v,w\}, \{w,z\} \in E(T_o)$ where $\{w,z\}$ is the third edge incident to w. Let $T'_o := (T_o \setminus \{\{v,w\}\}) \cup \{\{u,v\}\}$. In T'_o, w is a 2-vertex and hence T' is also optimal. □

2.3 The Algorithm

The algorithm we describe here is recursive. It constructs a set F of edges which are selected to be in every spanning tree considered in the current recursive step. The algorithm chooses edges and considers all relevant choices for adding them to F or removing them from G. It selects these edges based on priorities chosen to optimize the running time analysis. Moreover, the set F of edges will always be the union of a tree T and a set of edges P that are not incident to the tree and have one end point of degree 1 in G (pt-edges). We do not explicitly write in the algorithm that edges move from P to T whenever an edge is added to F that is incident to both an edge of T and an edge of P. To maintain the connectivity of T, the algorithm explores edges in the set $\partial E(T)$ to grow T.

If $|V| > 2$ every spanning tree T must have a vertex v with $d_T(v) \geq 2$. Thus initially the algorithm creates an instance for every vertex v and every possibility that $d_T(v) \geq 2$. Due to the degree constraint there are no more than $4n$ instances. After this initial phase, the algorithm proceeds as follows.

1. Carry out each reduction rule exhaustively in the given order.
2. If $\partial E(T) = \emptyset$ and $V \neq V(T)$, then G is not connected and does not admit a spanning tree. Ignore this branch.
3. If $\partial E(T) = \emptyset$ and $V = V(T)$, then return T.
4. Select $\{a,b\} \in \partial E(T)$ with $a \in V(T)$ according to the following priorities (if such an edge exists):
   a) there is an edge $\{b,c\} \in \partial E(T)$,
   b) $d_G(b) = 2$,
   c) b is incident to a pt-edge, or
   d) $d_T(a) = 1$.
   Recursively solve the two instances where $\{a,b\}$ is added to F or removed from G respectively, and return the spanning tree with most internal vertices.
5. Otherwise, select $\{a,b\} \in \partial E(T)$ with $a \in V(T)$. Let c, x be the other two neighbors of b. Recursively solve three instances where
   (i) $\{a,b\}$ is removed from G,
   (ii) $\{a,b\}$ and $\{b,c\}$ are added to F and $\{b,x\}$ is removed from G, and
   (iii) $\{a,b\}$ and $\{b,x\}$ are added to F and $\{b,c\}$ is removed from G.
   Return the spanning tree with most internal vertices.

By a Measure & Conquer analysis taking into account the degrees of the vertices, their number of incident edges that are in F, and to some extent the degrees of their neighbors, we obtain the following result.

Theorem 1. MIST *can be solved in time* $\mathcal{O}(1.8669^n)$ *on subcubic graphs.*

Let us provide the measure we use in the analysis. Let $D_2 := \{v \in V \mid d_G(v) = 2, d_F(v) = 0\}$, $D_3^\ell := \{v \in V \mid d_G(v) = 3, d_F(v) = \ell\}$ and $D_3^{2*} := \{v \in D_3^2 \mid N_G(v) \setminus N_F(v) = \{u\} \text{ and } d_G(u) = 2\}$. Then the measure we use is

$$\mu(G) = \omega_2 \cdot |D_2| + \omega_3^1 \cdot |D_3^1| + \omega_3^2 \cdot |D_3^2 \setminus D_3^{2*}| + |D_3^0| + \omega_3^{2*} \cdot |D_3^{2*}|$$

with the weights $\omega_2 = 0.3193$, $\omega_3^1 = 0.6234$, $\omega_3^2 = 0.3094$ and $\omega_3^{2*} = 0.4144$. The proof of the theorem uses the following result.

Lemma 6. *None of the reduction rules increase μ for the given weights.*

Let $\Delta_3^0 := \Delta_3^{0*} := 1 - \omega_3^1$, $\Delta_3^1 := \omega_3^1 - \omega_3^2$, $\Delta_3^{1*} := \omega_3^1 - \omega_3^{2*}$, $\Delta_3^2 := \omega_3^2$, $\Delta_3^{2*} := \omega_3^{2*}$ and $\Delta_2 = 1 - \omega_2$. We define $\tilde{\Delta}_3^i := \min\{\Delta_3^i, \Delta_3^{i*}\}$ for $1 \leq i \leq 2$, $\Delta_m^\ell = \min_{0 \leq j \leq \ell}\{\Delta_3^j\}$, and $\tilde{\Delta}_m^\ell = \min_{0 \leq j \leq \ell}\{\tilde{\Delta}_3^j\}$.

Proof. (of Theorem 1) As the algorithm deletes edges or moves edges from $E \setminus F$ to F, cases 1–3 do not contribute to the exponential function in the running time of the algorithm. It remains to analyze cases 4 and 5, which we do now. Note that after applying the reduction rules exhaustively, we have that for all $v \in \partial_V E(T)$, $d_G(v) = 3$ (**Deg2**) and for all $u \in V$, $d_P(u) \leq 1$ (**Pending**).

4.(a) Obviously, $\{a,b\}, \{b,c\} \in E \setminus E(T)$, and there is a vertex d such that $\{c,d\} \in E(T)$; see Figure 2(a). We have $d_T(a) = d_T(c) = 1$ due to the reduction rule **Attach**. We consider three cases.

$d_G(b) = 2$. When $\{a,b\}$ is added to F, **Cycle** deletes $\{b,c\}$. We get an amount of ω_2 and ω_3^1 as b drops out of D_2 and c out of D_3^1 (**Deg2**). Also a will be removed from D_3^1 and added to D_3^2 which amounts to a reduction of at least $\tilde{\Delta}_3^1$. When $\{a,b\}$ is deleted, $\{b,c\}$ is added to $E(T)$ (**Bridge**). By a symmetric argument we get a reduction of $\omega_2 + \omega_3^1 + \tilde{\Delta}_3^1$ as well. In total this yields a $(\omega_2 + \omega_3^1 + \tilde{\Delta}_3^1, \omega_2 + \omega_3^1 + \tilde{\Delta}_3^1)$-branch.

$d_G(b) = 3$ and there is one pt-edge incident to b. Adding $\{a,b\}$ to F decreases the measure by $\tilde{\Delta}_3^1$ (from a) and $2\omega_3^1$ (deleting $\{b,c\}$, then **Deg2** on c). By Deleting $\{a,b\}$ we decrease μ by $2\omega_3^1$ and by $\tilde{\Delta}_3^1$ (from c). This amounts to a $(2\omega_3^1 + \tilde{\Delta}_3^1, 2\omega_3^1 + \tilde{\Delta}_3^1)$-branch.

$d_G(b) = 3$ and no pt-edge is incident to b. Let $\{b,z\}$ be the third edge incident to b. In the first branch the measure drops by at least $\omega_3^1 + \tilde{\Delta}_3^1$ from c and a (**Deg2**), 1 from b (**Deg2**). In the second branch we get $\omega_3^1 + \Delta_2$. Observe that we also get an amount of at least $\tilde{\Delta}_m^1$ from $q \in N_T(a) \setminus \{b\}$ if $d_G(q) = 3$. If $d_G(q) = 2$ we get ω_2. It results a $(\omega_3^1 + \tilde{\Delta}_3^1 + 1, \omega_3^1 + \Delta_2 + \min\{\omega_2, \tilde{\Delta}_m^1\})$-branch.

Note that from this point on, for all $u, v \in V(T)$ there is no $z \in V \setminus V(T)$ with $\{u,z\}, \{z,v\} \in E$.

4.(b) As the previous case does not apply, the other neighbor c of b has $d_T(c) = 0$, and $d_G(c) \geq 2$ (**Pending**), see Figure 2(b). Additionally, observe that $d_G(c) = 3$ due to **ConsDeg2** and that $d_P(c) = 0$ due to **Special**. We consider two subcases.

$d_T(a) = 1$. When we add $\{a,b\}$ to F, then $\{b,c\}$ is also added due to **Deg2**. The reduction is at least $\tilde{\Delta}_3^1$ from a, ω_2 from b and Δ_3^0 from c. When $\{a,b\}$ is deleted, $\{b,c\}$ becomes a pt-edge. There is $\{a,z\} \in E \setminus E(T)$ with $z \neq b$, which is subject to a **Deg2** reduction rule. We get at least ω_3^1 from a, ω_2 from b, Δ_3^0 from c and $\min\{\omega_2, \tilde{\Delta}_m^1\}$ from z. This is a $(\tilde{\Delta}_3^1 + \Delta_3^0 + \omega_2, \omega_3^1 + \Delta_3^0 + \omega_2 + \min\{\omega_2, \tilde{\Delta}_m^1\})$-branch.

$d_T(a) = 2$. Similarly, we obtain a $(\Delta_3^{2*} + \omega_2 + \Delta_3^0, \Delta_3^{2*} + \omega_2 + \Delta_3^0)$-branch.

4.(c) In this case, $d_G(b) = 3$ and there is one pt-edge attached to b, see Figure 2(c). Note that $d_T(a) = 2$ can be ruled out due to **Attach2**. Thus, $d_T(a) = 1$. Let $z \neq b$ be such that $\{a,z\} \in E \setminus E(T)$. Due to the priorities, $d_G(z) = 3$. We distinguish between the cases where c, the other neighbor of b, is incident to a pt-edge or not.

$d_P(c) = 0$. First suppose $d_G(c) = 3$. Adding $\{a,b\}$ to F allows a reduction of $2\Delta_3^1$ (due to case 4.(b) we can exclude Δ_3^{1*}). Deleting $\{a,b\}$ implies that we get a reduction from a and b of $2\omega_3^1$ (**Deg2** and **Pending**). As $\{a,z\}$ is added to F we reduce $\mu(G)$ by at least $\tilde{\Delta}_3^1$ as the state of z changes. Now due to **Pending** and **Deg1** we include $\{b,c\}$ and get Δ_3^0 from c. We have at least a $(2\Delta_3^1, 2\omega_3^1 + \tilde{\Delta}_3^1 + \Delta_3^0)$-branch.

If $d_G(c) = 2$ we consider the two cases for z also. These are $d_P(z) = 1$ and $d_P(z) = 0$. The first entails $(\omega_3^1 + \Delta_3^{1*}, 2\omega_3^1 + \tilde{\Delta}_3^1 + \omega_2 + \tilde{\Delta}_m^2)$.

Note that when we add $\{a,b\}$ we trigger **Attach2**. The second is a $(\Delta_3^1 + \Delta_3^{1*}, 2\omega_3^1 + \Delta_3^0 + \omega_2 + \tilde{\Delta}_m^2)$-branch.

$d_P(c) = 1$. Let $d \neq b$ be the other neighbor of c that does not have degree 1. When $\{a,b\}$ is added to F, $\{b,c\}$ is deleted by **Attach2** and $\{c,d\}$ becomes a pt-edge (**Pending** and **Deg1**). The changes on a incur a measure decrease of Δ_3^{1*} and those on b,c a measure decrease of $2\omega_3^1$. When $\{a,b\}$ is deleted, $\{a,z\}$ is added to F (**Deg2**) and $\{c,d\}$ becomes a pt-edge by two applications of the **Pending** and **Deg1** rules. Thus, the decrease of the measure is at least $3\omega_3^1$ in this branch. In total, we have a $(\Delta_3^{1*} + 2\omega_3^1, 3\omega_3^1)$-branch here.

4.(d) Now, $d_G(b) = 3$, b is not incident to a pt-edge, and $d_T(a) = 1$. See Figure 2(c). There is also some $\{a,z\} \in E \setminus E(T)$ such that $z \neq b$. Note that $d_T(z) = 0$, $d_G(z) = 3$ and $d_P(z) = 0$. Otherwise either **Cycle** or cases 4.(b) or 4.(c) would have been triggered. From the addition of $\{a,b\}$ to F we get $\Delta_3^1 + \Delta_3^0$ and from its deletion ω_3^1 (from a via **Deg2**), Δ_2 (from b) and at least Δ_3^0 from z and thus, a $(\Delta_3^1 + \Delta_3^0, \omega_3^1 + \Delta_2 + \Delta_3^0)$-branch.

5. See Figure 2(d). The algorithm branches in the following way: 1) Delete $\{a,b\}$, 2) add $\{a,b\}, \{b,c\}$, and delete $\{b,x\}$, 3) add $\{a,b\}, \{b,x\}$ and delete $\{b,c\}$. Due to **Deg2**, we can disregard the case when b is a leaf. Due to Lemma 4 we also disregard the case when b is a 3-vertex. Thus by branching in this manner we find at least one optimal solution.

The reduction in the first branch is at least $\omega_3^2 + \Delta_2$. We get an additional amount of ω_2 if $d(x) = 2$ or $d(c) = 2$ from **ConsDeg2**. In the second we have to consider also the vertices c and x. There are exactly three situations for $h \in \{c,x\}$ $\alpha)$ $d_G(h) = 2$, $\beta)$ $d_G(h) = 3$, $d_P(h) = 0$ and $\gamma)$ $d_G(h) = 3$, $d_P(h) = 1$. We will only analyze branch 2) as 3) is symmetric. We first get a reduction of $\omega_3^2 + 1$ from a and b. We reduce μ due to deleting $\{b,x\}$ by: $\alpha)$ $\omega_2 + \tilde{\Delta}_m^2$, $\beta)$ Δ_2, $\gamma)$ $\omega_3^1 + \tilde{\Delta}_m^2$. Next we examine the amount by which μ will be decreased by adding $\{b,c\}$ to F. We distinguish between the cases α, β and γ: $\alpha)$ $\omega_2 + \tilde{\Delta}_m^2$, $\beta)$ Δ_3^0, $\gamma)$ $\tilde{\Delta}_3^1$.

For $h \in \{c,x\}$ and $W \in \{\alpha, \beta, \gamma\}$ let 1_W^h be the indicator function which is set to one if we have situation W at vertex h. Otherwise it is zero. Now the branching tuple can be stated the following way :

$(\omega_3^2 + \Delta_2 + (1_\alpha^x + 1_\alpha^c) \cdot \omega_2,$

$\omega_3^2 + 1 + 1_\alpha^x \cdot (\omega_2 + \tilde{\Delta}_m^2) + 1_\beta^x \cdot \Delta_2 + 1_\gamma^x \cdot (\omega_3^1 + \tilde{\Delta}_m^2) + 1_\alpha^c \cdot (\omega_2 + \tilde{\Delta}_m^2) + 1_\beta^c \cdot \Delta_3^0 + 1_\gamma^c \cdot \tilde{\Delta}_3^1),$

$\omega_3^2 + 1 + 1_\alpha^c \cdot (\omega_2 + \tilde{\Delta}_m^2) + 1_\beta^c \cdot \Delta_2 + 1_\gamma^c \cdot (\omega_3^1 + \tilde{\Delta}_m^2) + 1_\alpha^x \cdot (\omega_2 + \tilde{\Delta}_m^2) + 1_\beta^x \cdot \Delta_3^0 + 1_\gamma^x \cdot \tilde{\Delta}_3^1)$

The amount of $(1_\alpha^x + 1_\alpha^c) \cdot \omega_2$ comes from possible applications of **ConsDeg2**.

Observe that every instance created by branching is smaller than the original instance in terms of μ. Together with Lemma 6 we see that every step of the algorithm only decreases μ. Now if we evaluate the upper bound for every given branching tuple for the given weights we can conclude that MIST can be solved in time $\mathcal{O}^*(1.8669^n)$ on subcubic graphs. □

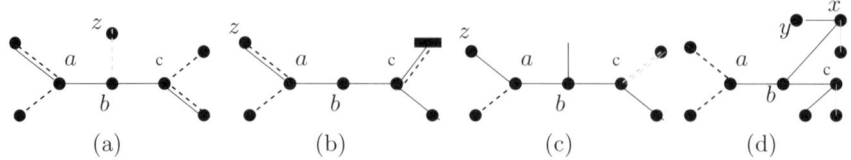

Fig. 2. Light edges may be not present. Double edges (dotted or solid, resp.) refer to edges which are either T-edges or not, resp. Edges attached to oblongs are pt-edges.

Let us now turn to a parameterized analysis of the algorithm. For general graphs, the smallest known kernel has size $3k$. This can be easily improved to $2k$ for subcubic graphs.

Lemma 7. *MIST on subcubic graphs has a $2k$-kernel.*

Applying the algorithm of Theorem 1 on this kernel for subcubic graphs would lead to a running time of $3.4854^k n^{\mathcal{O}(1)}$. However, we can achieve a faster parameterized running time by applying a Measure & Conquer analysis which is customized to the parameter k. We would like to put forward that our use of the technique of Measure & Conquer for a parameterized algorithm analysis goes beyond previous work as our measure is not restricted to differ from the parameter k by just a constant. We first demonstrate our idea with a simple analysis.

Theorem 2. *Deciding whether a subcubic graph has a spanning tree with at least k internal vertices can be done in time $2.7321^k n^{\mathcal{O}(1)}$.*

Proof. Consider the algorithm described earlier, with the only modification that the parameter k is adjusted whenever necessary (for example, when two pt-edges incident to the same vertex are removed), and that the algorithm stops and answers Yes whenever T has at least k internal vertices. Note that the assumption that G has no Hamiltonian path can still be made due to the $2k$-kernel of Lemma 7: the running time of the Hamiltonian path algorithm is $1.251^{2k} n^{\mathcal{O}(1)} = 1.5651^k n^{\mathcal{O}(1)}$. The running time analysis of our algorithm relies on the following measure: $\kappa := \kappa(G, F, k) := k - \omega \cdot |X| - |Y|$
where $X := \{v \in V \mid d_G(v) = 3, d_T(v) = 2\}$, $Y := \{v \in V \mid d_G(v) = d_T(v) \geq 2\}$ and $\omega = 0.45346$. Let $U := V \setminus (X \cup Y)$. Note that a vertex which has already been decided to be internal, but that still has an incident edge in $E \setminus T$, contributes a weight of $1 - \omega$ to the measure. Or equivalently, such a vertex has been only counted by ω. None of the reduction and branching rules increases κ and we have that $0 \leq \kappa \leq k$ at any time of the execution of the algorithm. By a simple case analysis and evaluating the branching factors, the proof follows. □

This analysis can be improved by also measuring vertices of degree 2 and vertices incident to pt-edges differently.

Theorem 3. *Deciding whether a subcubic graph has a spanning tree with at least k internal vertices can be done in time $2.1364^k n^{\mathcal{O}(1)}$.*

Table 1. Analysis of the branching for the running time of Theorem 3

	add	delete	branching tuple
Case 4.(a), $d_G(b) = 2$			
	$a : U \to X$ $b : Z \to U$ $c : U \to Y$	symmetric	$(1 + \omega_1 - \omega_2, 1 + \omega_1 - \omega_2)$
Case 4.(a), $d_G(b) = 3$, b is incident to a pt-edge			
	$a : U \to X$ $b : W \to Y$ $c : U \to Y$	symmetric	$(2 + \omega_1 - \omega_3, 2 + \omega_1 - \omega_3)$
Case 4.(a), $d_G(b) = 3$, b is not incident to a pt-edge			
	$a : U \to X$ $b : U \to Y$ $c : U \to Y$	$a : U \to Y$ $b : U \to Z$	$(2 + \omega_1, 1 + \omega_2)$
Case 4.(b), $d_T(a) = 1$			
	$a : U \to X$ $b : Z \to Y$	$a : U \to Y$ $b : Z \to U$ $c : U \to W$	$(1 + \omega_1 - \omega_2, 1 + \omega_3 - \omega_2)$
Case 4.(b), $d_T(a) = 2$			
	$a : X \to Y$ $b : Z \to Y$	$a : X \to Y$ $b : Z \to U$ $c : U \to W$	$(2 - \omega_1 - \omega_2, 1 - \omega_1 - \omega_2 + \omega_3)$
Case 4.(c)			
	$a : U \to X$ $b : W \to X$	$a : U \to Y$ $b : W \to Y$ $c : U \to W$	$(2\omega_1 - \omega_3, 2)$
Case 4.(d)			
	$a : U \to X$	$a : U \to Y$ $b : U \to Z$	$(\omega_1, 1 + \omega_2)$
Case 5, $d_G(x) = d_G(c) = 3$ and there is $q \in (X \cap (N(x) \cup N(c))$, w.l.o.g. $q \in N(c)$			
	$a : X \to Y$ $b : U \to Y$ (2nd branch) $q : X \to Y$	$a : X \to Y$ $b : U \to Z$	$(2 - \omega_1, 3 - 2\omega_1, 1 - \omega_1 + \omega_2)$
Case 5, $d_G(x) = d_G(c) = 3$			
	$a : X \to Y$ $b : U \to Y$ $c/x : U \to Z$ There are 3 branches; 2 of them (add) are symmetric.	$a : X \to Y$ $b : U \to Z$	$(1 - \omega_1 + \omega_2, 2 - \omega_1 + \omega_2, 2 - \omega_1 + \omega_2)$
Case 5, $d_G(x) = 2$ or $d_G(c) = 2$ and			
	$a : X \to Y$ $b : U \to Y$ When $\{a, b\}$ is deleted, **ConsDeg2** additionally decreases k by 1 and removes a vertex of Z.	$a : X \to Y$ $b : U \to Z$	$(2 - \omega_1, 2 - \omega_1, 2 - \omega_1)$

We consider a more detailed measure:
$\kappa := \kappa(G, F, k) := k - \omega_1 \cdot |X| - |Y| - \omega_2 |Z| - \omega_3 |W|$, where

- $X := \{v \in V \mid d_G(v) = 3, d_T(v) = 2\}$ is the set of vertices of degree 3 that are incident to exactly 2 edges of T,
- $Y := \{v \in V \mid d_G(v) = d_T(v) \geq 2\}$ is the set of vertices of degree at least 2 that are incident to only edges of T,
- $W := \{v \in V \setminus (X \cup Y) \mid d_G(v) \geq 2, \exists u \in N(v) \text{ st. } d_G(u) = d_F(u) = 1\}$ is the set of vertices of degree at least 2 that have an incident pt-edge, and
- $Z := \{v \in V \setminus W \mid d_G(v) = 2, N[v] \cap (X \cup Y) = \emptyset\}$ is the set of degree 2 vertices that do not have a vertex of $X \cup Y$ in their closed neighborhood, and are not incident to a pt-edge.

We immediately set $\omega_1 := 0.5485, \omega_2 := 0.4189$ and $\omega_3 := 0.7712$. Let $U := V \setminus (X \cup Y \cup Z \cup W)$. We first have to show that the algorithm can be stopped whenever the measure drops to 0 or less.

Lemma 8. *Let $G = (V, E)$ be a connected graph, k be an integer and $F \subseteq E$ be a set of edges that can be partitioned into a tree T and a set of pending edges P. If none of the reduction rules applies to this instance and $\kappa(G, F, k) \leq 0$, then G has a spanning tree $T^* \supseteq F$ with at least k internal nodes.*

We also show that reducing an instance does not increase its measure.

Lemma 9. *Let (G', F', k') be an instance resulting from the application of a reduction rule to an instance (G, F, k). Then, $\kappa(G', F', k') \leq \kappa(G, F, k)$.*

Proof. (of Theorem 3) Table 1 outlines how vertices a, b, and their neighbors move between U, X, Y, Z, and W in the branches where an edge is added to F or deleted from G in the different cases of the algorithm. For each case, the worst branching tuple is given.

The tight branching numbers are found for cases 4.(b) with $d_T(a) = 2$, 4.(c), 4.(d), and 5. with all of b's neighbors having degree 3. The respective branching numbers are $(2 - \omega_1 - \omega_2, 1 - \omega_1 - \omega_2 + \omega_3)$, $(2\omega_1 - \omega_3, 2)$, $(\omega_1, 1 + \omega_2)$, and $(1 - \omega_1 + \omega_2, 2 - \omega_1 + \omega_2, 2 - \omega_1 + \omega_2)$. They all equal 2.1364. □

3 Conclusion and Future Research

We have shown that MAX INTERNAL SPANNING TREE can be solved in time $\mathcal{O}^*(3^n)$. In a preliminary version of this paper we asked if MIST can be solved in time $\mathcal{O}^*(2^n)$ and also expressed our interest in polynomial space algorithms for MIST. These questions have been settled very recently by Nederlof [10] by providing a $\mathcal{O}^*(2^n)$ polynomial-space algorithm for MIST which is based on the principle of Inclusion-Exclusion and on a new concept called "branching walks".

This paper focuses on algorithms for MIST that work for the degree-bounded case, in particular, for subcubic graphs. The main novelty is a Measure & Conquer approach to analyze our algorithm from a parameterized perspective (parameterizing by the solution size). We are not aware of many examples where this was successfully done without cashing the obtained gain at an early stage, see [15]. More examples in this direction would be interesting to see.

A related problem is the generalization to directed graphs: Find a directed tree, which consist of directed paths form the root to the leaves with as few leaves as possible. Which results can be carried over to the directed case?

Acknowledgment. We would like to thank Alexey A. Stepanov for useful discussions in the initial phase of this paper.

References

1. Bellman, R.: Dynamic programming treatment of the Travelling Salesman Problem. J. Assoc. Comput. Mach. 9, 61–63 (1962)
2. Björklund, A., Husfeldt, T., Kaski, P., Koivisto, M.: The travelling salesman problem in bounded degree graphs. In: Aceto, L., Damgård, I., Goldberg, L.A., Halldórsson, M.M., Ingólfsdóttir, A., Walukiewicz, I. (eds.) ICALP 2008, Part I. LNCS, vol. 5125, pp. 198–209. Springer, Heidelberg (2008)
3. Cohen, N., Fomin, F.V., Gutin, G., Kim, E.J., Saurabh, S., Yeo, A.: Algorithm for finding k-Vertex Out-trees and its application to k-Internal Out-branching problem. In: COCOON 2009. Springer, Heidelberg (2008) (to appear)
4. Eppstein, D.: The Traveling Salesman problem for cubic graphs. J. Graph Algorithms Appl. 11(1), 61–81 (2007)
5. Fomin, F.V., Gaspers, S., Saurabh, S., Thomassé, S.: A linear vertex kernel for Maximum Internal Spanning Tree. ArXiv Report CoRR abs/0907.3473 (2009)
6. Held, M., Karp, R.M.: A dynamic programming approach to sequencing problems. J. Soc. Indust. Appl. Math. 10, 196–210 (1962)
7. Iwama, K., Nakashima, T.: An improved exact algorithm for cubic graph TSP. In: Lin, G. (ed.) COCOON 2007. LNCS, vol. 4598, pp. 108–117. Springer, Heidelberg (2007)
8. Karp, R.M.: Dynamic programming meets the principle of inclusion-exclusion. Inf. Process. Lett. 1(2), 49–51 (1982)
9. Kohn, S., Gottlieb, A., Kohn, M.: A generating function approach to the Traveling Salesman Problem. In: Proceedings of the 1977 ACM Annual Conference (ACM 1977), pp. 294–300. Association for Computing Machinery (1977)
10. Nederlof, J.: Fast polynomial-space algorithms using Möbius inversion: Improving on Steiner Tree and related problems. In: Albers, S., et al. (eds.) ICALP 2009, Part I. LNCS, vol. 5555, pp. 713–725. Springer, Heidelberg (2009)
11. Prieto, E., Sloper, C.: Either/or: Using vertex cover structure in designing FPT-algorithms—the case of k-internal spanning tree. In: Dehne, F., Sack, J.-R., Smid, M. (eds.) WADS 2003. LNCS, vol. 2748, pp. 474–483. Springer, Heidelberg (2003)
12. Prieto, E., Sloper, C.: Reducing to independent set structure – the case of k-internal spanning tree. Nord. J. Comput. 12(3), 308–318 (2005)
13. Salamon, G., Wiener, G.: On finding spanning trees with few leaves. Inf. Process. Lett. 105(5), 164–169 (2008)
14. Salamon, G.: Approximation algorithms for the maximum internal spanning tree problem. In: Kučera, L., Kučera, A. (eds.) MFCS 2007. LNCS, vol. 4708, pp. 90–102. Springer, Heidelberg (2007)
15. Wahlström, M.: Algorithms, Measures and Upper Bounds for Satisfiability and Related Problems. PhD thesis, Linköpings universitet, Sweden (2007)

An Exact Algorithm for Minimum Distortion Embedding*

Fedor V. Fomin, Daniel Lokshtanov, and Saket Saurabh

Department of Informatics, University of Bergen, N-5020 Bergen, Norway
{fedor.fomin,daniello,saket.saurabh}@ii.uib.no

Abstract. Let G be an unweighted graph on n vertices. We show that an embedding of the shortest path metric of G into the line with minimum distortion can be found in time $5^{n+o(n)}$. This is the first algorithm breaking the trivial $n!$-barrier.

1 Introduction

Given an undirected graph G with the vertex set $V(G)$ and the edge set $E(G)$, the *graph metric* of G is $M(G) = (V(G), D_G)$, where the distance function D_G is the shortest path distance between u and v for every pair of vertices $u, v \in V(G)$. Given a graph metric M and another metric space M' with distance functions D and D', a mapping $f : M \to M'$ is called an *embedding* of M into M'. The mapping f has *contraction* c_f and *expansion* e_f if for every pair of points p, q in M,

$$D(p,q) \le D'(f(p), f(q)) \cdot c_f,$$

and

$$D(p,q) \cdot e_f \ge D'(f(p), f(q))$$

respectively. We say that f is *non-contracting* if c_f is at most 1. A non-contracting mapping f has *distortion* d if e_f is at most d.

In this paper we provide an exact algorithm for the following fundamental problem: For a given graph G, find a minimum distortion embedding of the graph metric of G into the line. In this case the metric space M' is $\mathbb{R}^1$ and D' is the Euclidean distance. A simple algorithm is to try all possible permutations of the vertex set. Each permutation corresponds to an embedding where the distance between two consecutive vertices on the line is equal to the shortest path distance between them. The running time this algorithm is $O(n!n)$ and to the best of our knowledge, no faster exact algorithm for any kind of embedding problem was known prior to our work.

The problem of finding an embedding with low distortion between metric spaces is a fundamental mathematical problem [8,10] that has been studied intensively. Embedding a graph metric into a simple low-dimensional metric space like the real line has proved to be a useful algorithmic tool in various fields. A long list of applications given in [7] includes approximation algorithms for graph and network problems, such as sparsest cut, minimum bandwidth, low-diameter decomposition and optimal group Steiner trees,

---
* Partially supported by the Research Council of Norway.

and on-line algorithms for metrical task systems and file migration problems. The algorithmic issues of metric embeddings has recently begun to develop [1,2,3,9]. For example, Bǎdoiu et al. [1] describe approximation algorithms and hardness results for embedding general metrics into the line. In particular they show that the minimum distortion for a line embedding is hard to approximate up to a factor polynomial in n, even for weighted trees where the ratio of maximum/minimum weights is bounded by a polynomial in n. For the case of unweighted graphs, it was shown by Bǎdoiu et al. [2] that there is a constant $a > 1$, such that it is NP-hard to compute an a-approximation of the minimum distortion of an embedding into the line. Bǎdoiu et al. also provided an exact algorithm for computing an embedding with distortion at most d in time $n^{O(d)}$. For $d = \Omega(n)$ the running time of such an algorithm is $n^{O(n)}$. Fellows et al. [6] studied the parameterized complexity of metric embeddings and proved that embedding into the line and more generally, into trees with bounded vertex degrees, is fixed parameter tractable when parameterized by the distortion. For embedding a graph metric into the line the running time of the algorithm described in [6] is $O(nd^4(2d+1)^{2d})$, which also does not break the barrier of $n!$ when $d = \Theta(n)$.

It is worth to mention the resemblance between the problem of embedding into the line and the BANDWIDTH MINIMIZATION problem. In the BANDWIDTH MINIMIZATION problem the objective is for a given graph G to find a bijective mapping $f : V(G) \to \{1,\ldots,n\}$, for which the *bandwidth*, that is $b = \max_{(u,v) \in E(G)} |f(u) - f(v)|$, is minimized. Observe that the only difference between the two problems is that in the BANDWIDTH MINIMIZATION problem we demand $1 \le |f(p)-f(q)|$ for every pair of vertices while the non-contraction constraint in our embedding problem is $D(p,q) \le |f(p) - f(q)|$.

The BANDWIDTH MINIMIZATION problem is one of the test-bed problems in the area of moderately exponential time algorithms and has been studied intensively. Trying all possible permutations of the vertex set yields a simple $O(n!n)$ time algorithm while the known algorithms for the problem with running time $O(c^n)$ are far from straightforward. The $O(n!)$-barrier was broken by Feige and Kilian [5] who gave an algorithm with running time $O(10^n n^{O(1)})$. This result was subsequently improved by Cygan and Pilipczuk [4] down to $O(5^n n^{O(1)})$.

Despite the similarities between low distortion embedding into the line and bandwidth, the non-contraction constraint makes the algorithmic complexity of the two problems significantly different. A striking example is that the parameterized version of the BANDWIDTH MINIMIZATION problem is one of the hardest problems in Parameterized Complexity, while low distortion embedding into the line is fixed parameter tractable [6]. Thus, it is not surprising that a direct transmission of the ideas for the BANDWIDTH MINIMIZATION problem to low distortion embeddings does not work. Nevertheless, our approach is still based on the approaches from [4,5], especially the initial and final parts of our algorithm. However, to handle non-contraction need a non-trivial additional link connecting these parts.

2 Preliminaries

Let G be an undirected graph with vertex set $V(G)$ and edge set $E(G)$. We denote the number of vertices by n. For u and $v \in V(G)$, $D_G(u,v)$ is the shortest path distance between u and v in G. For a subset $V' \subseteq V(G)$, by $G[V']$ we mean the subgraph of G induced by V'. An *integer interval* is a set $\{x, x+1, \ldots, y-1, y\}$ of integers appearing consecutively. An embedding of a graph G into the line is a function $f : V(G) \to \mathbb{R}$. The *distortion* of an embedding f is $\max_{u,v \in V(G)} \frac{|f(u)-f(v)|}{D_G(u,v)}$. An embedding is called *non-contracting* if $|f(u) - f(v)| \geq D_G(u,v)$ for every pair u, v of vertices. If f is non-contracting we say that a vertex u *pushes* vertex v if $D_G(u,v) = |f(u) - f(v)|$. For an embedding f, let $v_1, v_2, \ldots, v_n$ be an ordering of the vertices such that $f(v_1) < f(v_2) < \cdots < f(v_n)$. We say that f is *pushing* if v_i pushes v_{i+1}, for each $1 \leq i \leq n-1$.

A *partial embedding* of G into the line is a function $f' : V' \to \mathbb{R}$ for some subset V' of V. For a partial embedding f' with domain V', let $v'_1, v'_2, \ldots, v'_{n'}$ be an ordering of V' such that $f'(v'_1) < f'(v'_2) < \cdots < f(v'_{n'})$. We say that f' is *pushing* if v'_i pushes v'_{i+1}, for each $1 \leq i \leq n'-1$. The distortion of a pushing partial embedding f' is $\max_{uv \in E(G[V'])} |f'(u) - f'(v)|$.

3 Exact Algorithm for Distortion

In this section we give an exact algorithm for the following problem.

Given an input graph G with the vertex set $V(G)$ and the edge set $E(G)$, find a mapping $f : V(G) \to \mathbb{R}^+$ such that for all $u, v \in V(G)$, $|f(u) - f(v)| \geq D_G(u,v)$ and the function

$$dist(G) = \max_{u,v \in V(G)} \frac{|f(u) - f(v)|}{D_G(u,v)}$$

is minimized.

In order to reduce the search space we apply a simple lemma proved in [6] on minimum distortion embedding of graphs into the line.

Lemma 1 ([6])

- *If G can be embedded into the line with distortion d, then there is a pushing embedding of G into the line with distortion d. Furthermore, every pushing embedding of G into the line is non-contracting.*
- *Let f be a pushing embedding of a connected graph G into the line with distortion at most d. Then $D(v_{i-1}, v_i) \leq d$ for every $1 \leq i \leq n$.*

Lemma 1 implies that it is sufficient to look for an optimal pushing embedding. Notice that a pushing embedding with $f(v_1) = 0$ maps every vertex to an integer coordinate. Therefore we can without loss of generality restrict ourselves to functions $f : V(G) \to \{0, \ldots, dn\}$. We also assume that our input graph G is *connected*, because otherwise some pair of vertices have infinite distance between them and hence there is no non-contracting embedding of G into the line.

We now present an algorithm that decides whether there is an embedding of distortion at most d for the input graph G. It is well known that any graph G with n vertices can be embedded into the line with distortion at most $2n-1$ [2]. Thus, if we want to find the minimum d such that there is an embedding of G into the line with distortion at most

d it is sufficient to try all values between 1 and $2n - 1$ for d. Next we describe the three main components of our algorithm and show how to combine them in order to obtain an algorithm running in time $5^{n+o(n)}$ and using $2^{n+o(n)}$ space. The first and third part of our algorithm go along the lines of the known algorithms for BANDWIDTH [5,4]. While these two parts are sufficient to compute bandwidth, in order to solve our problem we need an intermediate divide and conquer step to bridge the first and last part.

3.1 Fixing an Assignment into Buckets

The algorithm loops over all possible distributions of the vertices into "buckets" on the integer line. The remaining two steps of the algorithm deal with finding an optimal embedding that agrees with the distribution made in the first step. Formally, we are looking for a pushing embedding $f : V(G) \to \{0, \ldots, dn\}$. A *bucket assignment* is a function $h : V(G) \to \{0, \ldots, n\}$ and an embedding $f : V(G) \to \{0, \ldots, dn\}$ of G *agrees with h* if for every vertex v of G we have $h(v) = \lfloor \frac{f(v)}{d+1} \rfloor$. For $i \geq 0$, the i-th bucket of h (or the i-th bucket for short) is $\mathcal{B}_i = \{(d + 1)i, \ldots, (d + 1)(i + 1) - 1\}$ and the content of the i-th bucket is $V_i = \{v : h(v) = i\}$.

The outer loop of the algorithm goes over a set of bucket assignments such that if there is a pushing embedding $f : V(G) \to \{0, \ldots, dn\}$ with distortion at most d then some h we have looped over agrees with f. We guess a vertex v such that $h(v) = 0$ and fix a spanning tree T of G with r_T as root. Once $h(p)$ has been determined for the parent p of a node u in T, we loop over all possible values of $h(u)$. If h is to agree with some pushing embedding $f : V(G) \to \{0, \ldots, dn\}$ with distortion at most d we have that $h(u) = h(p) - 1$, $h(u) = h(p)$ or $h(u) = h(p) + 1$ and that $h(u) \geq 0$. Since we have at most 3 possibilities for the placement of each vertex the outer loop needs only to go over at most $n \cdot 3^n$ different bucket assignments h.

3.2 Dealing with Many Buckets

In this section and Section 3.3, we provide an algorithm which given an initial bucket assignment h, decides whether there is a pushing embedding f of the input graph into the line with distortion at most d that agrees with h.

Our algorithm EXACT-DIST solves a slight modification of the problem. Input to this problem is a graph G, an integer d, a bucket assignment h, an interval $J = \{x, x + 1, \ldots, y\}$ of integers and a function $g : V' \to \{0, \ldots, dn\}$ for some subset V' of $V(G)$. Let $\mathcal{B}_J = \bigcup_{j \in J} \mathcal{B}_j$ and $V_J = \bigcup_{j \in J} V_j$. The algorithm determines whether there is a partial pushing embedding $f : V_J \to \mathcal{B}_J$ with distortion at most d such that f agrees with h and $f(v) = g(v)$ for all vertices in $V' \cap V_J$. To solve the original problem we make a call to EXACT-DIST(G, d, h, J, g) where $J = \{0, \ldots, n\}$ and the domain V' of g is empty. Before commencing with the algorithm, we perform a "sanity check". That is, given h check whether it is even remotely feasible that f can exist. We verify that h satisfies the following properties.

- For every i, $|V_i| \leq d + 1$.
- Similarly, for every edge uv, $|h(u) - h(v)| \leq 1$.

EXACT-DIST(G, d, h, J, g)
(Here d is the distortion, h is the fixed bucket assignment, $J = \{x, \ldots, y\}$ is the set of indices of buckets and g is a partial embedding of some of the vertices in the graph.)

1. If the size of $|J| > \frac{n}{\log^2 n}$ then find a bucket V_j of the kind described in Lemma 2 else go to Step 3.
2. Enumerate all possible pushing partial embeddings $g_j : V_j \to \mathcal{B}_j$ of distortion at most d. For every such g_j:
   – Assign $g'(v) = g_j(v)$ if $v \in V_j$ and $g'(v) = g(v)$ if v is in the domain of g. Let $J_1 = \{x, \ldots, j-1, j\}$ and $J_2 = \{j, j+1, \ldots y\}$. Recursively solve the subproblems EXACT-DIST(G, d, h, J_1, g') and EXACT-DIST(G, d, h, J_2, g'). Return "YES" if both recursive calls return "YES".
3. In this case solve the problem using Lemma 5 of Section 3.3.

Fig. 1. Description of the Algorithm

Indeed, if some of these cases do not hold, there is no embedding f with distortion d that agrees with h and we can immediately answer "NO". At all later stages of the algorithm we assume that h satisfies these properties. An outline of the algorithm without these preliminary steps is given in Figure 1. In Section 3.3 we will give an algorithm which implements Step 3 in time $2^n \cdot n^{O(b)}$ time, where $b = |J|$ is the number of buckets considered.

The idea behind the algorithm is as follows. When the number of buckets $|J|$ is large, our algorithm follows a divide-and-conquer approach and if the number of buckets is "small", that is roughly $n/\log^2 n$, we do dynamic programming. To deal with the large number of buckets we look for a "small balanced separator" to branch on. The first step of algorithm EXACT-DIST is based on the following lemma.

Lemma 2. *Let h be a bucket assignment and let $J = x, x+1, \ldots, y$ be an integer interval such that $\frac{n}{\log^2 n} < |J|$. Then there exists $j \in I = \{\frac{3x+y}{4} + 1, \ldots, \frac{x+3y}{4}\}$ such that $|V_j| \leq 2\log^2 n$.*

Proof. The proof follows from an averaging principle. For the sake of contradiction, let us assume that for every $j \in I$, $|V_j| > 2\log^2 n$. Then the total number of elements in the buckets V_j with $j \in I$ is at least

$$\sum_{j \in I} |V_j| > 2\log^2 n \cdot \frac{|J|}{2} > 2\log^2 n \cdot \frac{n}{2\log^2 n} = n.$$

But the sets V_j are disjoint, and thus the sum does not exceed $|V(G)| = n$, which is a contradiction. □

If $|J|$ is at least $n/\log^2 n$, the algorithm picks a bucket $\mathcal{B}_j$ and branches on all possible ways to lay out V_j in $\mathcal{B}_j$. After this the problem breaks up into two independent subproblems (G, d, h, J_1, g') and (G, d, h, J_2, g'), see Figure 1. We argue that the two subproblems are indeed independent. Let f be a pushing partial embedding of V_J into $\mathcal{B}_J$ with distortion at most d such that f agrees with h and coincides with g. This means that f restricted to V_j is a pushing partial embedding of V_j into $\mathcal{B}_j$. We choose g_j to coincide with f on V_j and define $g'(v) = g_j(v)$ if $v \in V_j$ and $g'(v) = g(v)$ if v is in the domain of g, just as in step 2 of algorithm EXACT-DIST. If $J = \{x, \ldots, y\}$ then $J_1 = \{x, \ldots, j\}$ and $J_2 = \{j, \ldots, y\}$. Now f restricted to J_1 is a pushing partial embedding from V_{J_1} to $\mathcal{B}_{J_1}$ while f restricted to J_2 is a pushing partial embedding from V_{J_2} to $\mathcal{B}_{J_2}$.

In the other direction, let f_1 and f_2 be pushing partial embeddings from V_{J_1} to $\mathcal{B}_{J_1}$ and from V_{J_2} to $\mathcal{B}_{J_2}$ respectively, agreeing with h and coinciding with g'. Since $J = J_1 \cup J_2$ and $J_1 \cap J_2 = \{j\}$ we can choose f to be the partial embedding from V_J to $\mathcal{B}_J$ that coincides with both f_1 and f_2. Since both f_1 and f_2 are pushing partial embeddings, so is f. Since every edge with both endpoints in V_J has both endpoints either in V_{J_1} or in V_{J_2} and both f_1 and f_2 have distortion at most d, so does f.

Let $T(n,b)$ be the time required by algorithm EXACT-DIST on a n-vertex graph G with $|J| = b$. Let $T^*(n)$ be the time required by algorithm EXACT-DIST on a n-vertex graph G and with $|J| < n/\log^2 n$. An analysis of step 1 and 2 of algorithm EXACT-DIST yields the following recurrence.

$$T(n,b) = \begin{cases} \binom{d+1}{2\log^2 n}(2\log^2 n)! \cdot 2T\left(n, \frac{3b}{4}\right) & \text{if } b > \frac{n'}{\log^2 n} \\ T^*(n) & \text{otherwise.} \end{cases}$$

Thus, since $b \le n$ we have $T(n,b) \le 2^{O(\log \frac{n}{n/\log^2 n})} \cdot T^*(n) = 2^{o(n)} \cdot T^*(n)$. In Section 3.3 we show how to implement the last step of algorithm EXACT-DIST to run in time $2^n n^{O(b)}$ which is at most $2^n \cdot 2^{o(n)}$ since $b \le n/\log^2 n$. This yields a $2^{n+o(n)}$ runtime bound for algorithm EXACT-DIST and a $6^{n+o(n)}$ bound for deciding whether G can be embedded into the line with distortion at most d. In Section 3.4 we will show that the running time of our algorithm in fact is bounded by $5^{n+o(n)}$.

3.3 Dealing with Few Buckets

In this section we give an algorithm which given an initial bucket assignment h, a partial assignment g, and an integer interval $J = \{x, \ldots, y\}$ with $|J| = b < n/\log^2 n$ decides whether there is a pushing partial embedding $f : V_J \to \mathcal{B}_J$ with distortion at most d, agreeing with h and coinciding with partial assignment g. Our algorithm runs in time and space $2^n n^{O(b)}$.

The number of slots in J, that is positions in the line to where vertices can be mapped, is at most $b \cdot (d + 1)$. Thus there could be many slots with no vertex mapped to them. We start our algorithm by guessing for every $j \in J$ the leftmost non-empty slot in each bucket $\mathcal{B}_j$ and a vertex from V_j to be placed there. Naturally, if the layout of a bucket $\mathcal{B}_j$ with $j \in J$ has already been determined by g our guesses must be consistent with this. For every $j \in J$, let t_j denote the vertex guessed to be placed leftmost in bucket j. Also let l_j denote the position guessed for t_j. After having made the guess we modify the problem at hand—we now look for a pushing partial embedding $f : V_J \to \mathcal{B}_J$ with distortion at most d, agreeing with h, coinciding with g such that for every bucket $\mathcal{B}_j$ with $j \in J$, the leftmost vertex mapped to $\mathcal{B}_j$ is t_j, which is mapped to l_j. The number of possible guesses is bounded by $(d+1)^b n^b$.

We choose the ordering $\pi_1, \pi_2, \ldots, \pi_{|\mathcal{B}_J|}$ of the entries of $\mathcal{B}_J$ such that for every $i < j$ we have that $\pi_i \bmod (d+1) \leq \pi_j \bmod (d+1)$ and such that if $\pi_i \bmod (d+1) = \pi_j \bmod (d+1)$ then $\frac{\pi_i}{d+1} \leq \frac{\pi_j}{d+1}$. For example, if $J = 3, 4, 5$ and $d = 4$, then

$$\pi_1, \ldots, \pi_{15} = 15, 20, 25, 16, 21, 26, 17, 22, 27, 18, 23, 28, 19, 24, 29.$$

We call the ordering $\pi_1, \ldots, \pi_{|\mathcal{B}_J|}$ the *bucket order* of $\mathcal{B}_j$. Next we define the notion of a state.

Definition 1 *A state ζ is a quadruple (P, Q, R, p), where $P \subseteq V_J$, $Q \subseteq P$ is a set of vertices containing at most one vertex from each V_j such that if $V_j \cap P \neq \emptyset$ then $V_j \cap Q \neq \emptyset$ and $t_j \in P$, $R \subseteq \mathcal{B}_J$, is a set of integers containing at most one integer from each bucket $\mathcal{B}_j$ and $p \leq |J|$ is a non-negative integer.*

Let us observe that the number of states is at most $2^n \times n^{|J|} \times (d+1)^{|J|} \times |\mathcal{B}_j|$. If $Q \cap V_j \neq \emptyset$, then define q_j to be the vertex in $Q \cap V_j$. If $R \cap \mathcal{B}_j \neq \emptyset$ let r_j be the integer in $R \cap \mathcal{B}_j$. Next we define what it means for a state to be feasible:

Definition 2 *A state is called* feasible *if there exists a partial embedding f assigning the vertices of P to the first p positions in the bucket order such that the following condition hold:*

1. *For any edge uv with $u \in P$ and $v \in P$, $|f(u) - f(v)| \leq d$, f agrees with h and coincides with g.*
2. *If $V_j \cap P \neq \emptyset$, then $f(t_j) = l_j$ and $f(q_j) = r_j$. There is no vertex $v \in V_j \cap P$ such that $f(v) < l_j$ or $f(v) > r_j$.*
3. *For any bucket V_j with $j \in J$, if $x, y \in V_j$, $f(x) < f(y)$ and no vertex is mapped by f to the interval $\{f(x) + 1, f(y) - 1\}$, then $f(y) - f(x) = D_G(x, y)$;*
4. *If $j \in J$ and j is not the largest element of J, $V_j \subseteq P$ and $V_{j+1} \cap P \neq \emptyset$, then $f(l_{j+1}) - f(r_j) = D_G(l_{j+1}, r_j)$.*

The idea is to go through the slots in J one by one in the bucket order and for each of them determine which vertex (if any) gets mapped by f to this slot. The number p denotes the position in the bucket order that we have reached. The set P corresponds to the set of vertices that have already been placed. For every $j \in J$, t_j and q_j denote the vertices placed leftmost and rightmost in $\mathcal{B}_j$ respectively. Also l_j and r_j denotes the position of t_j and q_j in $\mathcal{B}_j$. Now we define the notion of a state succeeding another state.

Definition 3 *Let $\zeta_1 = (P_1, Q_1, R_1, p)$ and $\zeta_2 = (P_2, Q_2, R_2, p+1)$ be two states. We say that ζ_2 succeeds ζ_1 if the following holds.*

- *Either $P_2 = P_1$, or $P_2 = P_1 \cup \{v\}$.*
- *If $P_1 = P_2$, then $Q_1 = Q_2$ and $R_1 = R_2$.*
- *If $P_2 = P_1 \cup \{v\}$ and $v \in V_j$, then $j = \lfloor \frac{\pi_{p+1}}{d+1} \rfloor$ and*
    1. *If $v \in t_j$, then $l_j = \pi_{p+1}$. If $g(v)$ is defined then $g(v) = \pi_{p+1}$.*
    2. *$Q_2 = (Q_1 \setminus \{q_j\}) \cup \{v\}$ and $R_2 = (R_1 \setminus \{r_j\}) \cup \{\pi_{p+1}\}$.*
    3. *If $V_j \cap P_1 \neq \emptyset$ then $\pi_{p+1} - r_j = D_G(v, q_j)$.*
    4. *If j is not the largest element of J then $l_{j+1} - \pi_{p+1} \geq D(v, t_{j+1})$.*

5. If $j \in J$ and j is not the largest element of J, $V_j \subseteq P_2$ and $V_{j+1} \cap P_2 \neq \emptyset$ then $f(l_{j+1}) - f(v) = D_G(l_{j+1}, v)$.
6. If j is not the smallest element of J then $N(v) \cap V_{j-1} \cap P_2 = \emptyset$.

We now proceed to prove an observation that will be helpful for the correctness proof.

Lemma 3. *Let $\zeta_1 = (P_1, Q_1, R_1, p)$ be a feasible state and $\zeta_2 = (P_2, Q_2, R_2, p+1)$ be a state that succeeds ζ_1. Then ζ_2 is feasible.*

Proof. Since $\zeta_1 = (P_1, Q_1, R_1, p)$ is feasible there is a partial embedding f satisfying points 1 – 4 in Definition 2. If $P_1 = P_2$ then f satisfies the points 1 – 4 for ζ_2 as well. If $P_2 \neq P_1$ then $P_2 \setminus P_1$ contains a single vertex v. Let f' be a partial embedding assigning the vertices of P_2 to the first $p+1$ positions in the bucket order such that f' and f coincide and $f'(v) = \pi_{p+1}$. By point 1 of the definition of succession f' agrees with h and coincides with g. Since v has no neighbour in $P_1 \cap V_{j-1}$ it follows that for any edge uw with $u \in P_2$ and $w \in P_2$, $|f(u) - f(w)| \leq d$. Also, f' satisfies point 2 of definition 2 because π_{p+1} is the rightmost position in $P_2 \cap \mathcal{B}_j$. Furthermore f' satisfies point 3 of Definition 2 by point 3 of Definition 3. Finally f' satisfies point 4 of Definition 2 by point 5 of Definition 3. □

Now we are ready to prove the main lemma of the section which allows us to obtain the desired result.

Lemma 4. *There is a pushing partial embedding $f : V_J \to \mathcal{B}_J$ with distortion at most d such that f agrees with h, coincides with g and such that for every $j \in J$, $f(t_j) = l_j$ and no other vertex in V_j is mapped before t_j by f if and only if there exists sequence of states $\zeta_1, \zeta_1, \ldots, \zeta_{|\mathcal{B}_J|}$ such that (a) $\zeta_1 = (\emptyset, \emptyset, \emptyset, 0)$; (b) ζ_{i+1} succeeds ζ_i for all $i \in \{1, \ldots, |\mathcal{B}_J| - 1\}$; and (c) $\zeta_{|\mathcal{B}_J|} = (V_J, X, Y, |\mathcal{B}_J|)$.*

Proof. Let $f : V_J \to \mathcal{B}_J$ be a pushing partial embedding with distortion at most d such that f agrees with h, coincides with g and such that for every $j \in J$, $f(t_j) = l_j$ and no other vertex in V_j is mapped before t_j by f. With the help of f we define the sequence of feasible states as follows. For every $p \leq |\mathcal{B}_J|$, P is the set of vertices f maps to $\pi_0, \ldots, \pi_p$, Q is the set of vertices in P such that for every j such that $P \cap V_j \neq \emptyset$, Q contains exactly one vertex q_j, f maps all vertices in $P \cap V_j$ to the left of q_j. Finally R is the set of positions that f maps the vertices of Q. The construction of the sequence of states implies that $\zeta_1 = (\emptyset, \emptyset, \emptyset, 0)$, ζ_{i+1} succeeds ζ_i for all $i \in \{1, \ldots, |\mathcal{B}_J| - 1\}$ and that $\zeta_{|\mathcal{B}_J|} = (V_J, X, Y, |\mathcal{B}_J|)$.

For the reverse direction suppose that we have sequence of feasible states $\zeta_1, \zeta_1, \ldots, \zeta_{|\mathcal{B}_J|}$ such that $\zeta_1 = (\emptyset, \emptyset, \emptyset, 0)$; (b) ζ_{i+1} succeeds ζ_i for all $i \in \{1, \ldots, |\mathcal{B}_J| - 1\}$; and (c) $\zeta_{|\mathcal{B}_J|} = (V_J, X, Y, |\mathcal{B}_J|)$. Since $\zeta_1 = (\emptyset, \emptyset, \emptyset, 0)$ is feasible Lemma 3 implies that $\zeta_{|\mathcal{B}_J|} = (V_J, X, Y, |\mathcal{B}_J|)$ is feasible as well. The definition of feasibility guarantees the existence of the desired f, concluding the proof. □

Finally, we ready to proceed with the lemma used for the analysis of Step 3.

Lemma 5. *There is an algorithm that for given G, d, h, J, g and T decides whether there is a pushing partial embedding $f : V_J \to \mathcal{B}_J$ with distortion at most d such that f agrees with h, coincides with g and such that for every $j \in J$, $f(t_j) = l_j$ and no other vertex in V_j is mapped before t_j by f in time and space $2^n \cdot n^{O(|J|)}$.*

Proof. As we observed already, the number of states is at most $2^n \times n^{|J|} \times (d+1)^{|J|} \times |\mathcal{B}_J| \le 2^n \cdot n^{O(|J|)}$. The algorithm decides the existence of f by applying Lemma 4. The algorithm starts in the state $(\emptyset, \emptyset, \emptyset, 0)$ and does breadth first search on the graph where vertices are the states and there is a directed edge from a state ζ_i to a state ζ_j if ζ_j succeeds ζ_i. We do not keep this graph explicitly and rather generate the vertices of this graph as and when required in our breadth first search. Whenever we are at state ζ we can find all possible successor states in polynomial time. By Lemma 4 there is a required embedding f if and only if there is a path from $(\emptyset, \emptyset, \emptyset, 0)$ to $(V_J, X, Y, |\mathcal{B}_J|)$. Our algorithm needs $2^n \cdot n^{O(|J|)}$ space to keep track of the set of states visited by the breadth first search algorithm. Since the number of states is bounded by $2^n \cdot n^{O(|J|)}$ and the number of successors of a state is at most $d+2$ the number of vertices and edges in the state graph is upper bounded by $2^n \cdot n^{O(|J|)}$. Hence the algorithm takes $2^n \cdot n^{O(|J|)}$ time and space. □

Observe that applying Lemma 5 together with the analysis presented for Algorithm EXACT-DIST over the previous section yields a running time bound of $6^{n+o(n)}$. In fact, our algorithm runs in time $5^{n+o(n)}$. The next section is devoted to proving this.

3.4 A Refined Analysis

In this section we prove that the total number of states ever produced by our algorithm is $5^{n+o(n)}$. Since the running time of the algorithm is proportional to the number of states we generate up to a subexponential factor, this implies that algorithm EXACT-DIST runs in time $5^{n+o(n)}$. The proof of the following lemma is essentially an adaptation of the running time analysis given in [4] and will appear in the full version of the paper.

Lemma 6. *The algorithm described in the previous sections runs in time $5^{n+o(n)}$.*

We conclude with the following theorem.

Theorem 1. *There is an algorithm that given a graph G on n vertices constructs a non-contracting embedding of the shortest path metric generated by G into the line with minimum distortion in time $5^{n+o(n)}$ and space $2^{n+o(n)}$.*

4 Concluding Remarks and Open Problems

In this paper we have provided the first single vertex exponential time algorithm for computing a minimum distortion embedding of a graph metric into the line. This result gives rise to many challenging questions.

How fast is it possible to compute a minimum distortion embedding of a graph G into the metric of another graph H? Is there a $2^{O(|V(G)|)}$ time algorithm for this problem, or can one show that this is impossible up to some complexity theoretic assumption? How does the problem behave if the host graph H is a tree? Even when H is a binary tree, this does not seem to be an easy problem. At a first glance it would seem that our algorithm should be directly extendable to find a minimum distortion embedding of a graph G into a given cycle C. However this does not look to be easy and we leave it as an open problem whether finding a minimum distortion embedding of a graph G into a given cycle C can be done in $2^{O(|V(G)|)}$ time.

We believe that the world of embeddings provides a lot of challenges to the area of moderately exponential time algorithms and is worth to be explored. We hope that our result will lead to further investigation of the combinatorially challenging field of embeddings within the framework of moderately exponential time algorithms.

References

1. Bădoiu, M., Chuzhoy, J., Indyk, P., Sidiropoulos, A.: Low-distortion embeddings of general metrics into the line. In: Proceedings of the 37th Annual ACM Symposium on Theory of Computing (STOC), ACM, pp. 225–233 (2005)
2. Bădoiu, M., Dhamdhere, K., Gupta, A., Rabinovich, Y., Räcke, H., Ravi, R., Sidiropoulos, A.: Approximation algorithms for low-distortion embeddings into low-dimensional spaces. In: Proceedings of the 16th Annual ACM-SIAM Symposium on Discrete Algorithms (SODA), pp. 119–128. SIAM, Philadelphia (2005)
3. Badoiu, M., Indyk, P., Sidiropoulos, A.: Approximation algorithms for embedding general metrics into trees. In: Proceedings of the 18th Annual ACM-SIAM Symposium on Discrete Algorithms (SODA), pp. 512–521. ACM, SIAM (2007)
4. Cygan, M., Pilipczuk, M.: Faster Exact Bandwidth. In: Broersma, H., Erlebach, T., Friedetzky, T., Paulusma, D. (eds.) WG 2008. LNCS, vol. 5344, pp. 101–109. Springer, Heidelberg (2008)
5. Feige, U.: Coping with the NP-hardness of the graph bandwidth problem. In: Halldórsson, M.M. (ed.) SWAT 2000. LNCS, vol. 1851, pp. 10–19. Springer, Heidelberg (2000)
6. Fellows, M.R., Fomin, F.V., Lokshtanov, D., Losievskaja, E., Rosamond, F.A., Saurabh, S.: Distortion is Fixed Parameter Tractable. In: Albers, S., et al. (eds.) ICALP 2009. LNCS, vol. 5555, pp. 463–474. Springer, Heidelberg (2009)
7. Gupta, A., Newman, I., Rabinovich, Y., Sinclair, A.: Cuts, trees and l_1-embeddings of graphs. Combinatorica 24, 233–269 (2004)
8. Indyk, P.: Algorithmic applications of low-distortion geometric embeddings. In: Proceedings of the 42nd IEEE Symposium on Foundations of Computer Science (FOCS), pp. 10–33. IEEE, Los Alamitos (2001)
9. Kenyon, C., Rabani, Y., Sinclair, A.: Low distortion maps between point sets. In: Proceedings of the 36th Annual ACM Symposium on Theory of Computing (STOC), pp. 272–280. ACM Press, New York (2004)
10. Linial, N.: Finite metric-spaces—combinatorics, geometry and algorithms. In: Proceedings of the International Congress of Mathematicians, Beijing, vol. III, pp. 573–586. Higher Ed. Press (2002)

Sub-coloring and Hypo-coloring Interval Graphs[*]

Rajiv Gandhi[1], Bradford Greening Jr.[1], Sriram Pemmaraju[2], and Rajiv Raman[3]

[1] Department of Computer Science, Rutgers University-Camden, Camden, NJ 08102
rajivg@camden.rutgers.edu
[2] Department of Computer Science, University of Iowa, Iowa City, Iowa 52242
sriram@cs.uiowa.edu
[3] Max-Planck Institute for Informatik, Saarbrücken, Germany
rraman@mpi-inf.mpg.de

Abstract. In this paper, we study the *sub-coloring* and *hypo-coloring* problems on interval graphs. These problems have applications in job scheduling and distributed computing and can be used as "subroutines" for other combinatorial optimization problems. In the sub-coloring problem, given a graph G, we want to partition the vertices of G into minimum number of sub-color classes, where each sub-color class induces a union of disjoint cliques in G. In the hypo-coloring problem, given a graph G, and integral weights on vertices, we want to find a partition of the vertices of G into sub-color classes such that the sum of the weights of the heaviest cliques in each sub-color class is minimized. We present a "forbidden subgraph" characterization of graphs with sub-chromatic number k and use this to derive a 3-approximation algorithm for sub-coloring interval graphs. For the hypo-coloring problem on interval graphs, we first show that it is NP-complete, and then via reduction to the max-coloring problem, show how to obtain an $O(\log n)$-approximation algorithm for it.

1 Introduction

Given a graph $G = (V, E)$, a *k-sub-coloring* of G is a partition of V into *sub-color classes* $V_1, V_2, \ldots, V_k$; a subset $V_i \subseteq V$ is called a sub-color class if it induces a union of disjoint cliques in G. Figure 1(a) shows a 2-sub-coloring of a graph, with the black vertices forming one sub-color class and the white vertices the other. The smallest k for which a graph has a k-sub-coloring is called the *sub-chromatic number* of G, and is denoted $\chi_s(G)$. The *sub-coloring problem* [1,4,5,8,22] seeks to find a partition of vertices of G into the smallest number of sub-color classes. Clearly, any proper coloring of G is also a sub-coloring, since any proper color class can be viewed as the disjoint union of size-1 cliques; hence, $\chi_s(G) \leq \chi(G)$. Of course, the sub-chromatic number can be much smaller than the chromatic number (e.g., consider a large clique). Figure 1(b) shows a graph

---

[*] Part of this work was done when the first author was visiting the University of Iowa. Research partially supported by Rutgers University Research Council Grant and by NSF award CCF-0830569.

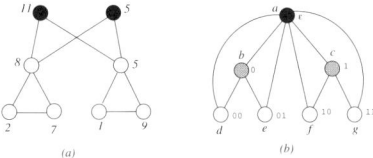

Fig. 1. (a) shows a 2-sub-coloring of a graph whose chromatic number is 3. If the numbers next to the vertices are taken to be vertex-weights then the white-black coloring is a hypo-coloring of cost $17 + 11 = 28$. (b) shows a graph whose chromatic number and sub-chromatic number are both 3. This is an example of $BC(3)$, a binary clique of order 3. The binary string vertex-labels are useful in defining the family of binary cliques (see Section 2).

G with $\chi(G) = \chi_s(G) = 3$. To see that $\chi_s(G) \geq 3$, observe that each of the 2-paths induced by $\{b, d, e\}$ and $\{c, f, g\}$ require 2 sub-color classes. Furthermore, if the subgraph induced by $\{b, c, d, e, f, g\}$ is colored using 2 sub-colors, then vertex a cannot be added to either of the sub-color classes because each of the sub-color classes will contain at least one vertex from $\{b, d, e\}$ and at least one vertex from $\{c, f, g\}$. We call the graph shown in Figure 1(b) a *binary clique of order 3*, denoted $BC(3)$. Later (in Section 2) we define the family, $BC(k)$, $k \geq 1$, of order k binary cliques and show that the presence of an induced binary clique is an obstacle to having a small sub-chromatic number, in the sense that $\chi_s(BC(k)) \geq k$.

Given a graph $G = (V, E)$, and a vertex weight function $w : V \to \mathbb{N}$, the *hypo-coloring problem* [6] seeks to find a partition of the vertices of G into sub-color classes such that the sum of the weights of the heaviest cliques in each sub-color class is minimized. In other words, if $V_1, V_2, \ldots, V_k$ are the sub-color classes of a hypo-coloring solution, then the cost of the solution is $\sum_{i=1}^{k} \max_{K \subseteq V_i} w(K)$, where each K is a clique in the sub-color class V_i and $w(K)$ is the sum of the weights of vertices in K. Figure 1(a) shows a hypo-coloring of a vertex-weighted graph with cost $17 + 11 = 28$.

Our Contribution. This paper studies the approximability of sub-coloring and hypo-coloring on interval graphs. On the positive side, we present (in Section 2) a 3-approximation algorithm for sub-coloring interval graphs. This is the first constant-factor approximation algorithm for the problem. We also present an $O(\log n)$-approximation algorithm for the hypo-coloring problem, via reduction to the max-coloring problem [20]. In fact, we get an $O(\log n)$-approximation for hypo-coloring on a variety of graph classes including perfect graphs, unit disk graphs, circle graphs, etc., all of which admit constant-factor approximation algorithms for the max-coloring problem [19]. On the negative side, we show (in Section 3) that hypo-coloring on interval graphs is NP-complete.

It is worth noting here that the complexity status of sub-coloring on interval graphs is unknown. Results due to Broersma et al. [5] imply that there is an $n^{O(\log n)}$-time algorithm for the sub-coloring problem on interval graphs. The existence of a subexponential time algorithm for the problem makes it unlikely that sub-coloring on interval graphs is NP-complete. Given this, it is worth further exploring the possibility that the problem has a polynomial-time algorithm.

1.1 Applications

The sub-coloring and hypo-coloring problems have a variety of applications to job scheduling, distributed computing, and combinatorial optimization. We sketch some of these applications next.

Combinatorial Optimization: The sub-coloring problem can be used as a subroutine in solving some combinatorial optimization problems on graphs, where solving the problem on a clique is easier than solving the problem on general graphs. This approach has been used for example in approximation algorithms for the *maximum feasible subsystem problem(MFS)*[1]. In [7], Elbassioni et al. study the MFS problem where the marix A is a consecutive-ones matrix, and x is restricted to be non-negative. They give an algorithm for solving the problem on a clique, and then use this result in conjunction with a sub-coloring to obtain approximation algorithms for this problem.

Job Scheduling: The hypo-coloring problem arises in the problem of *batch scheduling* jobs in conflict. In a batch scheduling environment with conflicts, we are given jobs $\mathcal{J} = \{J_1, \cdots, J_n\}$ with processing times p_j, and a conflict graph with vertices as the jobs, and an edge representing conflict, where jobs that are in conflict can not be scheduled simultaneously. A schedule that minimizes the makespan of the schedule (with an arbitrarily large number of machines) has been studied as the *max-coloring* problem [20]. When the conflict graph is an *interval graph*, the problem reduces to the max-coloring problem on interval graphs. Now suppose that in a batch we have one job with a *large* processing time, and other jobs with very *small* processing times, then all the machines except for the machine processing this large job are idle until the batch completes. We can get an improved schedule if in each batch the jobs form a union of disjoint cliques. Thus, each clique can be run on the same machine sequentially, while the next set of jobs is scheduled once this batch of jobs completes. Such schedules can be seen as batch schedules with a kind of *backfilling* [21,15].

Distributed Computing: The sub-coloring problem on general graphs is also motivated by the *network decomposition* problem in distributed computing [3,17]. A vertex partition $V_1, V_2, \ldots, V_k$ of a graph $G = (V, E)$ induces a *cluster graph* with vertex set $\{1, 2, \ldots, k\}$ and edges $\{i, j\}$ iff there is an edge in G between some $u \in V_i$ and $v \in V_j$. The network decomposition problem seeks to find a vertex partition $V_1, V_2, \ldots, V_k$ of G such that each *cluster* $G[V_i]$ has small diameter and the cluster graph has small chromatic number. For example, Awerbuch et al. [3] present a deterministic, distributed algorithm running in $O(n^{\epsilon(n)})$ time for

---

[1] The maximum feasible subsytem problem is the following. Given a system $l \leq Ax \leq b$, which is infeasible, the goal is to find a solution vector x that satisfies the maximum number of inequalites.

computing a $(n^{\epsilon(n)}, n^{\epsilon(n)})$-network decomposition[2] where $\epsilon(n) = \sqrt{\log\log n}/\sqrt{\log n}$. Using this they obtain the first, deterministic, sublinear distributed algorithms for several classical problems in distributed computing. If we restrict the diameter of each V_i to 1, i.e., a clique, then the network decomposition problem is equivalent to the sub-coloring problem since each proper color class of the cluster graph is a disjoint union of cliques from the input graph. Lately, this approach to designing fast distributed algorithms has become very popular in wireless networks [12]. This motivates the study of sub-coloring on geometric intersection graphs such as *unit disk graphs (UDGs)* or *disk graphs*, used to model wireless networks.

1.2 Related Work

The sub-chromatic number of a graph was introduced by Mynhardt and Broere [4,16], and studied as *spot-coloring* by Hartman [11]. Achlioptas [1] proved that F-free coloring[3] is NP-hard even when F is any graph with at least 3 vertices. By setting $F = P_3$ we get that the sub-coloring problem on general graphs is NP-hard. Fiala et al. [8] showed that F-free coloring is NP-hard even for triangle-free planar graphs with maximum degree 4, while giving polynomial time algorithms for sub-coloring on cographs and graphs of bounded tree-width. Stacho [22] has shown that sub-coloring on chordal graphs is NP-complete. Broersma et al. [5] study algorithmic and combinatorial aspects of sub-coloring on various classes of graphs. Specifically, they show that when G is chordal $\chi_s(G)$ is $\Theta(\log n)$. They also show that for any constant r, there is a polynomial time algorithm to compute a sub-coloring of interval graphs that have sub-coloring $\leq r$. However, they do not consider the problem of obtaining an approximation algorithm for sub-coloring on general interval graphs.

Motivated by the problem of batch scheduling conflicting jobs, de Werra et al. [6] introduced the hypo-coloring problem. The authors give a polynomial time algorithm for graphs with maximum degree 2, and for forests with bounded maximum degree. They also show that the problem is NP-hard on bipartite graphs and triangle-free planar graphs.

A problem that seems related to hypo-coloring is the *max-coloring problem*. Given a graph $G = (V, E)$ and a weight function $w : V \to \mathbb{N}$ the problem is to find a proper vertex coloring $C_1, C_2, \ldots, C_k$ of G that minimizes $\sum_{i=1}^{k} \max_{v \in C_i} w(v)$. Note that the special case of this problem in which $w(v) = 1$ for all $v \in V$ is simply the problem of coloring graphs using fewest colors. Pemmaraju et al. [20] show that the max-coloring problem on interval graphs is NP-complete and give a 2-approximation algorithm. In Section 3 we study the relation between the optimal solutions for max-coloring and hypo-coloring and provide a reduction

---
[2] A $(c(n), d(n))$-network decomposition is one in which the cluster graph chromatic number is bounded above by $c(n)$ and the diameter of each cluster is bounded about by $d(n)$.

[3] For a graph $G = (V, E)$, an F-free coloring is a partition of the vertex set of G such that in each color class, the vertices do not have F as an induced subgraph.

from hypo-coloring to max-coloring that leads to an $O(\log n)$-approximation to hypo-coloring.

2 A 3-Approximation for Sub-coloring Interval Graphs

This section presents an algorithm that takes as input an interval graph $G = (V, E)$ and returns a partition into sub-color classes $S_1, S_2, \ldots, S_k$, such that $k \leq 3 \cdot \chi_s(G)$. We start by first establishing a lower bound on $\chi_s(G)$, for any graph G. A *complete binary tree of order* k, $k \geq 1$, denoted $CBT(k)$, is a rooted tree with vertex set $\{0,1\}^{k-1}$ and edge set $\{\{\alpha, \alpha 0\}, \{\alpha, \alpha 1\} \mid \alpha \in \{0,1\}^{k-2}\}$. The *root* of the tree is the vertex labeled ε, the empty string. A *binary clique of order* k, denoted $BC(k)$, is obtained from $CBT(k)$ by adding edges $\{\alpha, \beta\}$ where β is a strict prefix of α. Figure 1(b) shows $BC(3)$ along with the binary string labels for the vertices. The edges $\{a, d\}$, $\{a, e\}$, $\{a, f\}$, and $\{a, g\}$ were added in going from $CBT(3)$ to $BC(3)$.

Lemma 1. *If a graph G (not necessarily an interval graph) contains $BC(k)$ as an induced subgraph, then $\chi_s(G) \geq k$.*

Proof. (Sketch) The proof follows by induction on k. The base case can be easily seen to hold. Assuming the inductive hypothesis for $k' \leq k-1$, if $BC(k)$ can be sub-colored with $k-1$ or fewer colors, this leads to a contradiction.

The 3-approximation algorithm that we will present next has two main phases. Suppose that G is the input interval graph and A is the set of intervals corresponding to this graph. In the first phase (partitioning phase), we partition the intervals in A into subsets $S_1, S_2, \ldots, S_k$ and show that G contains an induced binary clique of order k, implying via Lemma 1 that $\chi_s(G)$ is lower bounded by the size of the partition. In the second phase (coloring phase) of the algorithm we sub-color each S_i using at most 3 colors, and using a different set of colors for each subclass. This yields a sub-coloring with $3k \leq 3\chi_s(G)$ colors.

2.1 Partitioning Phase

Our partitioning procedure takes as input a collection A of intervals. An interval I in A is said to be *internal* if it completely contains (in the geometric sense) two disjoint intervals I_1 and I_2. Any interval that is not internal is called *external*. The partitioning algorithm (shown in Figure 2) simply peels of "layers" of external intervals. Note that every non-empty collection of intervals has a non-empty subset of external intervals.

Lemma 2. *If $\text{PARTITION}(A)$ returns $S_1, S_2, \ldots, S_k$, then G, the interval graph corresponding to A, contains an induced $BC(k)$.*

Proof. We will prove by induction that for any j, $1 \leq j \leq k$, and any interval $I \in S_j$, I is the root of a binary clique H of order j contained entirely within the

PARTITION(A)

1    $A_0 \leftarrow A;\ k \leftarrow 0$
2   **while** $(A_k \neq \emptyset)$ **do**
3        $k \leftarrow k + 1$
4        $S_k \leftarrow$ intervals that are external in A_{k-1}
5        $A_k \leftarrow A_{k-1} \setminus S_k$
6   **return** $S_1, S_2, \ldots, S_k$

Fig. 2. Partitioning algorithm that takes a set A of intervals as input and returns a partition $S_1, S_2, \ldots, S_k$

intervals in $B_j = S_1 \cup S_2 \cup \cdots \cup S_j$ and furthermore all intervals in H are entirely contained (in the geometric sense) in I. The base case ($j = 1$) is trivially true. Consider any interval $I \in S_j$. There are two disjoint intervals I_1 and I_2 in S_{j-1} that are completely contained in I. Otherwise, I would have been a member of S_{j-1}. By the inductive hypothesis, I_1 is the root of H_1, a binary clique of order $j-1$ and I_2 is the root of H_2, a binary clique of order $j-1$. Furthermore, all intervals in H_i, $i = 1, 2$ are contained in I_i. This implies that H_1 and H_2 are disjoint and also that interval I has edges to all the intervals in H_1 and in H_2. The graph induced by I and the intervals in H_1 and H_2 is a binary clique of order j, with root I, and with all intervals contained within I.

After partitioning A into subsets $S_1, S_2, \ldots, S_k$, we "color" the intervals in each subset S_i as follows. For the rest of this subsection let S denote an arbitrary S_i. We start by choosing the left-most maximal clique in S; call this M_1. Let $I_1 \in M_1$ be the interval with the right-most right endpoint, and let N_1 be the set of intervals not in M_1 that are completely contained within I_1. We then remove intervals in $M_1 \cup N_1$ from S and if $S \neq \emptyset$, we repeat the process and compute M_2 and N_2 and so on. Once this process terminates, for some $k \geq 1$, we have partitioned S into subsets $M_1, M_2, \ldots, M_k$ and $N_1, N_2, \ldots, N_k$. We then "color" all intervals in $N_1 \cup N_2 \cup \cdots \cup N_k$ using color C_2 and alternately "color" the intervals in $M_1, M_2, \ldots, M_k$, using color C_1 and C_0. The pseudocode for this algorithm is given in Figure 3.

2.2 Analysis

Observe that after the 3COLOR algorithm finishes processing a subset S, each interval in S is assigned to exactly one of 3 subsets, C_0, C_1, or C_2. We now show that each C_j, $j = 0, 1, 2$, is a union of disjoint cliques. This suffices to prove that C_0, C_1, C_2 is a valid 3-sub-coloring of S.

Lemma 3. *C_2 is a union of disjoint cliques.*

Proof. Let N_x and N_y be any two particular but arbitrary cliques in C_2. Without loss of generality, let $x < y$. Note that N_x is the set of intervals in C_2 which are completely contained in interval I_x and N_y is the set of intervals in C_2 which are

3COLOR(S)

```
1    k ← 0
2    C_0 ← C_1 ← C_2 ← ∅
3    while (S ≠ ∅) do
4        k ← k + 1
5        M_k ← leftmost maximal clique
6        I_k ← interval in M_k with the rightmost right endpoint
7        N_k ← intervals not in M_k that are completely contained in I_k
8        S ← S \ (M_k ∪ N_k)
9        if k is even then
10           C_0 ← C_0 ∪ M_k
11       else
12           C_1 ← C_1 ∪ M_k
13       C_2 ← C_2 ∪ N_k
14   return C_0, C_1, C_2
```

Fig. 3. Sub-Coloring algorithm that takes a subset S of intervals produced by the partition algorithm and computes a 3-sub-coloring of S

completely contained in interval I_y. If I_x and I_y do not overlap, then clearly there are no overlaps between intervals in N_x and intervals in N_y. However, if I_x and I_y do overlap, then it must be that $y = x + 1$. Now assume for contradiction that interval $J \in N_x$ overlaps with interval $K \in N_y$. Since J is completely contained in I_x, and K overlaps J, clearly K also overlaps I_x. This means that K belongs to M_y (the leftmost maximal clique after all intervals in M_x are removed), and hence cannot belong to N_y, a contradiction. □

Lemma 4. C_1 (C_0, respectively) is a union of disjoint cliques.

Proof. Let M_x and M_y be any two particular but arbitrary cliques in C_1 (C_0, respectively) and assume for contradiction that there are intervals in M_x which overlap intervals in M_y. Without loss of generality assume that $x < y$. Let $J \in M_x$ be an interval which overlaps with an interval $K \in M_y$. Recall that $I_x \in M_x$ is the interval having rightmost right endpoint in M_x, therefore I_x also overlaps K. This implies that M_x and M_y are consecutive maximal cliques, that is, $M_y = M_{x+1}$ and thus the algorithm would not have assigned both M_x and M_y to C_1, (C_0, respectively) a contradiction. □

Lemmas 3 and 4 lead to the following corollary.

Corollary 1. *The algorithm* 3COLOR(S) *computes a 3-sub-coloring of S.*

This corollary, along with Lemmas 1 and 2 leads to the following result.

Theorem 1. *There is a 3-approximation algorithm for the sub-coloring problem on interval graphs.*

We note that the sub-coloring obtained also yields a 6-approximation algorithm for partitioning an interval graph into the fewest number of *proper interval* graphs[4] Gardi [9] posed this as an open problem. The proof follows by noting that the sub-chromatic number of proper interval graphs is 2.

Theorem 2. *There is a 6-approximation algorithm for partitioning an interval graph into fewest number of proper interval graphs.*

3 Hypo-coloring of Interval Graphs

In this section, we show that the hypo-coloring problem is NP-complete, and give an $O(\log n)$-approximation algorithm via a reduction to the max-coloring problem.

3.1 NP-Completeness

The NP-completeness of hypo-coloring on interval graphs is shown by using a reduction from COLORING CIRCULAR-ARC GRAPHS [10]. Our proof is heavily influenced by the NP-completeness proof of minimum sum coloring on interval graphs by D. Marx [13] and the proof of NP-completeness of max-coloring on interval graphs by Pemmaraju et al. [20].

COLORING CIRCULAR-ARC GRAPHS
INPUT: A circular-arc graph $G = (V, E)$, and a number $k \in \mathbb{N}$.
QUESTION: Does G have a coloring of cost at most k?

We may assume that a circular arc representation of G is given to us since recognition and construction of a circular arc representation can be done in polynomial time [14]. Also as done in [13,20], we can assume that there exists a point on the circle contained in exactly k arcs. In the following proof, we view the hypo-coloring problem as a decision problem in which we are given an additional input, a positive integer W, and asked if the given instance has a hypo-coloring of cost at most W.

Theorem 3. *Hypo-coloring interval graphs is NP-Complete.*

Proof. Given a circular-arc graph G and parameter k, let r be a ray from the center of the circle that passes through k arcs of G. We construct an interval graph H from G by splitting the arcs intersecting r. More formally, let $I = \{I_1, \cdots, I_k\}$ be the arcs intersecting r. We replace each arc $I_i \in I$ by two arcs I'_i and I''_i, that start and end respectively at r. This gives us an interval graph and we can assume that the intervals $I' = \{I'_i \mid i = 1, 2, \ldots, k\}$ form the leftmost intervals, and the intervals $I'' = \{I''_i \mid i = 1, 2, \ldots, k\}$ form the rightmost intervals. We set the left end-points of the intervals I'_i such that $l(I'_i) < l(I'_j)$ whenever $i < j$ and we set $r(I''_i) < r(I''_j)$ whenever $i < j$. Here $l(I)$ and $r(I)$ respectively denote the left and right end-points of an interval I. Further, we add two sets of intervals $L = \{L_1, \cdots, L_k\}$ and $R = \{R_1, \cdots, R_k\}$, such that $r(L_i) = l(I'_i)$ and

---
[4] An interval graph is *proper* if there is an interval representation of G such that no interval properly contains another.

$l(R_i) = r(I_i'')$ for all $i = 1, \cdots, k$. The weights of the intervals are defined as follows. We let $w(L_i) = w(R_i) = 1 + i \cdot \epsilon$, for $\epsilon = \frac{1}{k+1}$), $w(I) = 1$, $\forall I \notin L \cup R$. This gives us the interval graph H. Note that scaling the weights by a factor of $k+1$ will guarantee that all vertex-weights in H are integral.

If to each interval in H, we assign a color c if it's corresponding inteval in G is assigned a color c, then I_i' and I_i'' are assigned the same color for each $i = 1, \cdots, k$, allowing us to assign the same colors to L_i and R_i as I_i' and I_i'', we obtain a hypo-coloring of cost $C = k + k(k+1)\epsilon/2$.

On the other hand, suppose there is a hypo-coloring of cost C or less. We first show that such a coloring must in fact be a proper coloring of H. To see this, consider just the subgraph of H induced by the intervals in $L \cup I'$. A hypo-coloring of this subgraph with cost at most C must itself be proper because any hypo-color class in this hypocoloring with a clique of size larger than 1 will force us to place one of the intervals in $L \cup I'$ in a new color class, incurring a cost of at least $C + 1$. Further, in such a proper coloring, if L_i is not placed in the same color class as R_i for each $i = 1, \cdots, k$, the coloring has cost at least $C + \epsilon$. Hence, a hypo-coloring of cost C or less must be proper, and place L_i and R_i in the same color class, for each i, which is only possible if I_i' and I_i'' are placed in the same color class for each $i = 1, \cdots, k$, and this yields a proper coloring of the circular-arc graph G. □

3.2 An $O(\log n)$-approximation for Hypo-coloring

In this section we show that an optimal solution to the max-coloring problem on any graph G is an $O(\log n)$-approximation to the hypo-coloring problem with input G. Since there is a 2-approximation algorithm for max-coloring on interval graphs [20] this implies that there is an $O(\log n)$-approximate solution for hypo-coloring interval graphs.

Theorem 4. *Given any graph G, an optimal max-coloring of G is an $O(\log n)$-approximation for hypo-coloring of G.*

Proof. Let OPT_H be an optimal hypo-coloring solution. We will prove the claim by showing that there is a feasible maxcoloring solution whose cost is $O(\log n)OPT_H$. Let $S_1, S_2, \ldots, S_k$ be the k color classes in OPT_H and let the cliques in S_i be given by $S_i^1, S_i^2, \ldots, S_i^{p_i}$. Let m_i be the maximum number of vertices in any clique in S_i. In other words, $m_i = \max_{1 \le j \le p_i} |S_i^j|$. Consider the max-coloring solution in which there are color classes $C_i^1, C_i^2, \ldots, C_i^{m_i}$ for each S_i in OPT_H. For $1 \le x \le m_i$, the color class C_i^x is formed by including the xth heaviest vertices from each of the cliques in S_i. Since the cliques are disjoint, so are the xth heaviest vertices from each of the cliques. Consider the heaviest vertex $v \in C_i^x$ and let v belong to clique S_i^y. Since v is the xth heaviest vertex in S_i^y there must be $x - 1$ vertices in S_i^y of weight at least $w(v)$ that are placed in $C_i^1, C_i^2, \ldots, C_i^{x-1}$. Hence, $w(C_i^x) \le \frac{W_i}{x}$, where W_i is the weight of the heaviest clique in S_i. Thus $C_i^1, C_i^2, \ldots, C_i^{m_i}$ is a feasible maxcoloring solution for input S_i with cost $\le W_i + \frac{W_i}{2} + \ldots + \frac{W_i}{m_i} = W_i H_{m_i} \le W_i(\ln m_i + 1)$. Thus the total cost of our maxcoloring solution for G becomes $\leq \sum_{i=1}^{k} W_i(\ln m_i + 1) \leq \sum_{i=1}^{k} W_i(\ln m + 1) = O(\log m) \sum_{i=1}^{k} W_i = O(\log m) OPT_H$ where m is the number of vertices in the largest clique among all batches $S_1, S_2, \ldots, S_k$. Since $m \leq n$, we have obtained a solution to maxcoloring of cost $O(\log n) OPT_H$. □

The above analysis can be shown to be tight. Owing to lack of space, the details are in the full version.

References

1. Achlioptas, D.: The complexity of g-free colorability. Discrete Math, 21–30 (1997)
2. Albertson, M.O., Jamison, R.E., Hedetnieme, S.T., Locke, S.C.: The subchromatic number of a graph. Discrete Math. 74, 33–49 (1989)
3. Awerbuch, B., Goldberg, A.V., Luby, M., Plotkin, S.A.: Network decomposition and locality in distributed computation. In: Proceedings of the 30th Annual Symposium on Foundations of Computer Science (FOCS), pp. 364–369 (1989)
4. Broere, I., Mynhardt, C.M.: Generalized colorings of outerplanar and planar graphs. In: Proceedings of the 10th International Workshop on Graph-Theoretic Concepts in Computer Science (WG), pp. 151–161 (1984)
5. Broersma, H., Fomin, F.V., Nesetril, J., Woeginger, G.J.: More about subcolorings. Computing 69, 187–203 (2002)
6. deWerra, D., Demange, M., Monnot, J., Paschos, V.T.: A hypocoloring model for batch scheduling. Discrete Applied Mathematics 146, 3–25 (2005)
7. Elbassioni, K.M., Raman, R., Ray, S., Sitters, R.: On the approximability of the maximum feasible subsystem problem with 0/1-coefficients. In: Proceedings of the 20th ACM-SIAM Symposium on Discrete algorithms (SODA), pp. 1210–1219 (2009)
8. Fiala, J., Jansen, K., Le, V.B., Seidel, E.: Graph subcoloring:complexity and algorithms. SIAM Journal on Discrete Mathematics 16, 635–650 (2003)
9. Gardi, F.: On partitioning interval and circular-arc graphs into proper interval subgraphs with applications. In: Latin American Theoretical Informatics Symposium (LATIN), pp. 129–140 (2004)
10. Garey, M.R., Johnson, D.S., Miller, G.L., Papadimitriou, C.H.: The complexity of coloring circular arcs and chords. SIAM J. Algebraic and Discrete Methods 1, 216–227 (1980)
11. Hartman, C.: Extremal Problems in Graph Theory (1997)
12. Fabian, K., Thomas, M., Roger, W.: On the locality of bounded growth. In: Proceedings of the 24th Annual ACM symposium on Principles of Distributed Computing (PODC), pp. 60–68. ACM, New York (2005)
13. Marx, D.: A short proof of the NP-completeness of minimum sum interval coloring. Operations Research Letters 33(4), 382–384 (2005)
14. McConnell, R.M.: Linear-time recognition of circular-arc graphs. Algorithmica 37(2), 93–147 (2003)
15. Mu'alem, A.W., Feitelson, D.G.: Utilization, predictability, workloads, and user runtime estimates in scheduling the ibm sp2 with backfilling. IEEE Trans. Parallel and Distributed Syst. 12(6), 529–543 (2001)
16. Mynhardt, C.M., Broere, I.: Generalized colorings of graphs. In: Proceedings of the 11th International Workshop on Graph-Theoretic Concepts in Computer Science (WG), pp. 583–594 (1985)

17. Panconesi, A., Srinivasan, A.: On the complexity of distributed network decomposition. J. Algorithms 20(2), 356–374 (1996)
18. Pemmaraju, S.V., Pirwani, I.A.: Good quality virtual realization of unit ball graphs. In: Arge, L., Hoffmann, M., Welzl, E. (eds.) ESA 2007. LNCS, vol. 4698, pp. 311–322. Springer, Heidelberg (2007)
19. Pemmaraju, S.V., Raman, R.: Approximation algorithms for the max-coloring problem. In: Caires, L., Italiano, G.F., Monteiro, L., Palamidessi, C., Yung, M. (eds.) ICALP 2005. LNCS, vol. 3580, pp. 1064–1075. Springer, Heidelberg (2005)
20. Pemmaraju, S.V., Raman, R., Varadarajan, K.: Buffer minimization using max-coloring. In: Proceedings of the 15th ACM-SIAM Symposium on Discrete Algorithms (SODA), pp. 562–571 (2004)
21. Shmueli, E., Feitelson, D.G.: Backfilling with look ahead to optimize the packing of parallel jobs. J. Parallel and Distributed Computing 65(9), 1090–1107 (1995)
22. Stacho, J.: Complexity of Generalized Colourings of Chordal Graphs (2008)

Parameterized Complexity of Generalized Domination Problems

Petr A. Golovach[1], Jan Kratochvíl[2], and Ondřej Suchý[2,*]

[1] Department of Informatics, University of Bergen,
5020 Bergen, Norway
petrg@ii.uib.no
[2] Department of Applied Mathematics
and
Institute for Theoretical Computer Science (ITI)[**],
Charles University, Prague, Czech Republic
{honza,suchy}@kam.mff.cuni.cz

Abstract. Given two sets σ, ρ of nonnegative integers, a set S of vertices of a graph G is (σ, ρ)-*dominating* if $|S \cap N(v)| \in \sigma$ for every vertex $v \in S$, and $|S \cap N(v)| \in \rho$ for every $v \notin S$. This concept, introduced by Telle in 1990's, generalizes and unifies several variants of graph domination studied separately before. We study the parameterized complexity of (σ, ρ)-domination in this general setting. Among other results we show that existence of a (σ, ρ)-dominating set of size k (and at most k) are W[1]-complete problems (when parameterized by k) for any pair of finite sets σ and ρ. We further present results on dual parametrization by $n-k$, and results on certain infinite sets (in particular for σ, ρ being the sets of even and odd integers).

1 Introduction

1.1 (σ, ρ)-Domination

Let σ, ρ be a pair of nonempty sets of nonnegative integers. A set S of vertices of a graph G is called (σ, ρ)-*dominating* if for every vertex $v \in S$, $|S \cap N(v)| \in \sigma$, and for every $v \notin S$, $|S \cap N(v)| \in \rho$. The concept of (σ, ρ)-domination was introduced by J.A. Telle [18,19] (and further elaborated on in [13,20]) as a unifying generalization of many previously studied variants of the notion of dominating sets. See Table 1 for some examples.

It is well known that the optimization problems such as MAXIMUM INDEPENDENT SET, MINIMUM DOMINATING SET, etc. are NP-hard. In many cases of the generalized domination already the existence of a (σ, ρ)-dominating set

---

[*] Work partially supported by the ERASMUS program, by the DFG, project NI 369/4 (PIAF), while visiting Friedrich-Schiller-Universität Jena, Germany (October 2008–March 2009) and by grant 201/05/H014 of the Czech Science Foundation.
[**] Supported by the Ministry of Education of the Czech Republic as project 1M0021620808.

Table 1. Overview of the special cases of (σ, ρ)-domination and their parameterized complexity (when parameterized by the size of the set). (Here $\mathbb{N}$ and $\mathbb{N}_0$ denote the sets of positive and nonnegative integers, respectively.)

σ	ρ	Problem name	Parameterized Complexity
$\mathbb{N}_0$	$\mathbb{N}$	Dominating Set	W[2]-complete
$\mathbb{N}$	$\mathbb{N}$	Total Dominating Set	W[2]-hard
$\mathbb{N}_0$	$\{1\}$	Efficient Dominating Set	W[1]-hard
$\{0\}$	$\mathbb{N}$	Indepependent Dominating Set	W[2]-complete
$\{0\}$	$\mathbb{N}_0$	Independent set	W[1]-complete
$\{0\}$	$\{1\}$	(1-)Perfect Code(Indep. Eff. Dom. Set)	W[1]-complete
$\{r\}$	$\mathbb{N}_0$	Induced r-Regular subgraph	W[1]-hard
$\{0\}$	$\{0,1\}$	Strong Stable Set	Unknown
$\{1\}$	$\{1\}$	Total Perfect Dominating Set	Unknown

becomes NP-hard (e.g., when both σ and ρ are finite and nonempty, and $0 \notin \rho$ [18]). Hence attention was paid to special graph classes, e.g. interval graphs ([15] shows polynomial-time solvability for any pair of finite σ, ρ), chordal graphs ([11] shows a P/NP-c dichotomy classification) or degenerate graphs [12].

Since the establishment of the Parameterized Complexity Theory by Downey and Fellows [7], domination-type problems have been among the first ones intensively studied in the framework of this theory. (We assume the reader is familiar with the concept of FPT and W[t] classes, otherwise we refer to [7,10] and [17] as excellent textbooks.) It is well known that INDEPENDENT SET is W[1]-complete [6] and DOMINATING SET is W[2]-complete [5,7] (when parameterized by the size of the set). A number of domination-type problems are considered in [2], where it is shown (among other results) that TOTAL DOMINATING SET is W[2]-hard and that EFFICIENT DOMINATING SET is W[1]-hard. INDEPENDENT DOMINATING SET is W[2]-complete [5], while EFFICIENT INDEPENDENT DOMINATING SET (also called PERFECT CODE) is W[1]-complete ([6] shows W[1]-hardness and [3] shows W[1]-membership). More results on parameterized complexity of problems from coding theory can be found in [9]. The complexity of finding an r-regular induced subgraph in a graph is studied in [16].

Parity constraints have been considered in [9]. A subset of a color class of a bipartite graph is called *odd* (*even*) if every vertex from the other class has an odd (even, respectively) number of neighbors in the set. Downey et al. show that deciding the existence of an odd set of size k, an odd set of size at most k, and an even set of size k are W[1]-hard problems; somewhat surprisingly, the complexity of EVEN SET OF SIZE AT MOST k remains open.

All these individual results concern special (σ, ρ)-dominating sets, and thus call for a unifying approach. Our paper attempts to be a starting one by giving general results for large classes of pairs σ, ρ. The second goal of our paper is to study (many of) the above problems from the dual parametrization point of view (looking for a set of size at least $n - k$, where k is the parameter), both for the domination-type and parity-type problems.

1.2 Notation and Overview of Our Results

We consider the following (σ, ρ)-domination problem

(σ,ρ)-DOMINATING SET OF SIZE AT MOST k
Input: A graph G.
Parameter: k.
Question: Is there a (σ,ρ)-dominating set in G of size at most k?

and its variants (σ,ρ)-DOMINATING SET OF SIZE k, (σ,ρ)-DOMINATING SET OF SIZE AT LEAST $n-k$, and (σ,ρ)-DOMINATING SET OF SIZE $n-k$, whose meaning should be clear. All these problems ar parameterized by k, and in the latter two, n denotes the number of vertices of the input graph. The first of our main results determines the parameterized complexity for finite sets σ and ρ.

Theorem 1. *Let σ and ρ be nonempty finite sets of nonnegative integers, $0 \notin \rho$. Then both (σ,ρ)-DOMINATING SET OF SIZE k and (σ,ρ)-DOMINATING SET OF SIZE AT MOST k are W[1]-complete problems.*

The hardness part is proved in Subsection 2.1, and the W[1]-membership is proved in Subsection 2.2 in a stronger form when σ is only required to be recursive but not necessarily finite.

We further study the dually parameterized problems and show in an even more general way that these problems become tractable. In Section 3 we prove the following theorem (here and throughout the paper, $\overline{X} = \mathbb{N}_0 \setminus X$ for a set X of integers).

Theorem 2. *Let σ and ρ be sets of nonnegative integers such that either σ or $\overline{\sigma}$ is finite, and similarly either ρ or $\overline{\rho}$ is finite. Then the (σ,ρ)-DOMINATING SET OF SIZE AT LEAST $n-k$ problem is in FPT.*

We show that a similar result cannot be expected for arbitrary recursive sets σ and ρ. Even for the parity case (when we denote **EVEN** $= \{0,2,4,6,\ldots\}$ and **ODD** $= \{1,3,5,\ldots\}$) we can prove W[1]-hardness.

Theorem 3. *Let $\sigma, \rho \in \{\textbf{EVEN}, \textbf{ODD}\}$. Then both (σ,ρ)-DOMINATING SET OF SIZE $n-k$ and (σ,ρ)-DOMINATING SET OF SIZE AT LEAST $n-k$ are W[1]-hard problems.*

As a tool for the previous result we consider the following parity problems on bipartite graphs. Suppose that G is a bipartite graph and R, B is a bipartition of its set of vertices (vertices of R are called *red* and vertices of B are *blue*). A nonempty set $S \subseteq R$ is called *even* if for every vertex $v \in B$, $|N(v) \cap S| \in \textbf{EVEN}$, and it is called *odd* if for every vertex $v \in B$, $|N(v) \cap S| \in \textbf{ODD}$. The following problem

EVEN SET OF SIZE AT LEAST $r - k$
Input: A bipartite graph $G = (R, B, E)$ and $r = |R|$.
Parameter: k.
Question: Is there an even set in R of size at least $r - k$?

and its variants EVEN SET OF SIZE $r - k$, ODD SET OF SIZE AT LEAST $r - k$, and ODD SET OF SIZE $r - k$ are the dually parameterized versions of bipartite parity problems studied in [9]. We prove in Section 4 that all four of them are W[1]-hard.

In the last section we present observations on FPT results for sparse graphs.

2 Complexity of the (σ, ρ)-DOMINATING SET OF SIZE AT MOST k Problems - Proof of Theorem 1

2.1 W[1]-Hardness

We are going to reduce a special variant of the SATISFIABILITY problem (the proof of W[1]-hardness of this problem is omitted here).

AT MOST α-SATISFIABILITY
Instance: A Boolean formula ϕ in conjunctive normal form, without negated variables.
Parameter: k.
Question: Does ϕ allow a satisfying truth assignment of weight at most k (i.e., at most k variables have value *true*) such that each clause of ϕ contains at most α variables which evaluate to *true*?

Suppose that σ and ρ are nonempty finite sets of nonnegative integers, $0 \notin \rho$. Let us denote $p_{min} = \min \sigma$, $p_{max} = \max \sigma$, $q_{min} = \min \rho$ and $q_{max} = \max \rho$. Further we set $t = \max\{i \in \mathbb{N}_0 : i \notin \rho, i + 1 \in \rho\}$ (since $0 \notin \rho$, t is correctly defined), and $\alpha = q_{max} - t \geq 1$. We are going to reduce AT MOST α-SATISFIABILITY. Due the space restrictions we give here only a sketch of the reduction. Complete description will appear in the journal version of the paper.

We first construct several auxiliary gadgets. These gadgets "enforce" on a given vertex the property of "not belonging to any (σ, ρ)-dominating set", and at the same time guarantee that this vertex has a given number of neighbors in any (σ, ρ)-dominating set in the gadget. To describe the properties formally, we will consider rooted graphs and introduce the following notion. Let G be a rooted graph with a set of root vertices X. We call a set $S \subseteq V(G)$ a (σ, ρ)-*dominating set for* G if $|N(v) \cap S| \in \sigma$ for every $v \in S \setminus X$, and $|N(v) \cap S| \in \rho$ for every $v \notin S$, $v \notin X$ (i.e., the conditions from the definition of (σ, ρ)-domination are required for all vertices except the roots).

The first gadget is a graph $F(s)$ (s is a positive integer) with s independent roots $x_1, \ldots, x_s$ of degree one, all adjacent to the same vertex, say a_1, which has the following property: Every (σ, ρ)-dominating set S for $F(s)$ contains a_1, contains none of the roots, and all such sets have the same size $f = f(\sigma, \rho)$.

The second gadget is a graph $F'(s)$ (s is a positive integer) with s independent roots $y_1, \ldots, y_s$ of degree one, all adjacent to the same vertex, say x. It has the following property: Every (σ, ρ)-dominating set S for $F'(s)$ contains none of the roots, neither it contains their common neighbor x, and all such sets have the same size $f' = f'(\sigma, \rho)$.

A selection gadget $R(l)$ (l is a positive integer) is a graph rooted in a clique X containing l vertices, and it satisfies the following property: Every (σ,ρ)-dominating set S for $R(l)$ contains exactly one root vertex, and all such sets have the same size $r = r(\sigma,\rho)$. Moreover, for every root vertex $x \in X$, there exists a (σ,ρ)-dominating set S in $R(l)$ which contains x (note that here we require that even the root vertices are dominated in a proper way).

Now we describe the reduction. Let ϕ be a formula as an input of the AT MOST α-SATISFIABILITY problem. Let $x_1, \ldots, x_n$ be its variables, and let $C_1, \ldots, C_m$ be the clauses.

We take k copies of the graph $R(n+1)$ denoted by $R_1, \ldots, R_k$, with the roots of R_i being denoted by $x_{i,j}$. For each clause C_s, a vertex C_s is added and joined by edges to all vertices $x_{i,j}$, $i = 1, \ldots, k$ such that the variable x_j occurs in the clause G_s. Now we distinguish two cases:

$t = 0$. In this case a copy of $F'(m)$ is introduced, and the m roots of this gadget are identified with vertices $C_1, \ldots, C_m$. In this case we set $k' = kr + f'$.

$t > 0$. We construct t copies of $F(m)$, and the roots of each copy are identified with $C_1, \ldots, C_m$. In this case we set $k' = kr + tf$.

The resulting graph is called G. The proof of W[1]-hardness is then concluded by the following lemma (whose proof is omitted).

Lemma 1. *The formula ϕ allows a satisfying truth assignment of weight at most k such that each clause of ϕ contains at most α variables with value* true *if and only if G has a (σ,ρ)-dominating set of size at most k'. Moreover, in such a case the size of any (σ,ρ)-dominating set is exactly k'.*

2.2 W[1]-Membership

Here we prove a slightly stronger claim.

Theorem 4. *Let σ be recursive, and suppose that ρ is finite. Then the (σ,ρ)-*DOMINATING SET OF SIZE AT MOST k *and* (σ,ρ)-DOMINATING SET OF SIZE k *problems are in* W[1].

To show the membership of the problems in W[1], we use the characterization of W[1] by Nondeterministic Random Access Machines as proposed in [10].

A nondeterministic random access machine (NRAM) model is based on the standard deterministic random access machine (RAM) model. A single nondeterministic instruction "GUESS" is added, whose semantics is: *Guess a natural number less than or equal to the number stored in the accumulator and store it in the accumulator.* Acceptance of an input by an NRAM is defined as usually for nondeterministic machines. The steps of computation of an NRAM that execute a GUESS instruction are called *nondeterministic steps*.

Definition 1. *An NRAM program $\mathbb{P}$ is* tail-nondeterministic k-restricted *if there are computable functions f and g and a polynomial p such that on every run with input $(x,k) \in \Sigma^* \times \mathbb{N}$ the program $\mathbb{P}$*

- performs at most $f(k) \cdot p(n)$ steps;
- uses at most the first $f(k) \cdot p(n)$ registers;
- contains numbers $\leq f(k) \cdot p(n)$ in any register at any time;

and all nondeterministic steps are among the last $g(k)$ steps of the computation. Here $n = |x|$.

The following characterization is crucial for our proof:

Theorem 5 ([10]). *A parameterized problem P is in $W[1]$ if and only if there is a tail-nondeterministic k-restricted NRAM program deciding P.*

Now we introduce our program SigmaRho that takes a graph G and a positive integer k as an input and there is an accepting computation of SigmaRho on G and k if and only if there is a (σ, ρ)-dominating set of size exactly k in G. We present it in a higher level language that can be easily translated to the NRAM instructions. It is straightforward to show that this program is tail-nondeterministic k-restricted, the formal proof will appear in journal version of the paper and we omit it here. Recall that $q_{max} = \max \rho$. Here $\binom{V}{r}$ denotes the set $\{R \subseteq V \mid |R| = r\}$.

Program SigmaRho$(G = (V, E), k)$

1   **for** $r := 1$ **to** $q_{max} + 1$ **do forall** $R \in \binom{V}{r}$ **do**

$$B(R) := |\bigcap_{u \in R} N_G(u)| = |\{v \mid v \in V, \forall u \in R : uv \in E\}|;$$

2   **Guess** k distinct vertices $v_1, \ldots, v_k$, denote $S = \{v_1, \ldots, v_k\}$;
3   **for** $i := 1$ **to** k **do if** $|\{v_j \mid v_i v_j \in E\}| \notin \sigma$ **then REJECT**;
4   **for** $r := q_{max} + 1$ **downto** 1 **do**

$$D(r) := \sum_{R \in \binom{S}{r}} (B(R) - |\bigcap_{u \in R} N_G(u) \cap S|) =$$

$$= \sum_{R \in \binom{S}{r}} |\{v \mid v \in V \setminus S, \forall u \in R : uv \in E\}|;$$

$$C(r) := D(r) - \sum_{t=r+1}^{q_{max}} \left(\binom{t}{r} \cdot C(t) \right);$$

   if $r \notin \rho$ **and** $C(r) \neq 0$ **then REJECT**;

5   **if** $0 \notin \rho$ **and** $\sum_{r \in \rho} C(r) \neq n - k$ **then REJECT; else ACCEPT;**

Lemma 2. *Let G be a graph and $k \in \mathbb{N}$. There is an accepting computation of SigmaRho on G and k if and only if there is a (σ, ρ)-dominating set of size (exactly) k in G.*

Proof. We will show that the program `SigmaRho` accepts the input if and only if the set S guessed in step 2 is a (σ, ρ)-dominating set of size k for the input graph G. It is easy to see that the members of the set S must satisfy the σ-condition due to step 3. Now observe that the number $D(r)$ computed in step 4 denotes the number of pairs (R, v) such that R is a subset of S of size r and v is a vertex not in S that has all vertices from R as neighbors (the first term counts all such vertices v in V and the second term subtracts such vertices v that are in S). Hence this $D(r)$ represents the number of vertices outside S which have at least r neighbors in S with multiplicities, in particular a vertex with t neighbors in S is counted $\binom{t}{r}$ times. Since in the first run of the cycle 4 with $r = q_{max} + 1$ we check that there is no vertex outside S with more than q_{max} neighbors in S, $C(r)$ represents the number of vertices outside S which have exactly r neighbors in S. It is now clear that if $r \notin \rho$ and there is a vertex outside S with r neighbors in S (i.e., $C(r) > 0$), then S cannot form a (σ, ρ)-dominating set. In the last step 5 we sum up the number of vertices outside S that satisfy the ρ-condition and thus S (which satisfies all the conditions checked by the previous steps) is (σ, ρ)-dominating if and only if this sum is equal to the total number of vertices outside S, i.e., $n - k$, or $0 \in \rho$.

Proof (Proof of Theorem 4). First observe that (σ, ρ)-DOMINATING SET OF SIZE AT MOST k can be easily reduced to (k calls of) (σ, ρ)-DOMINATING SET OF SIZE k. Hence it is enough to prove the membership for the second problem. But that is a direct consequence of Theorem 5 together with Lemma 2 and Program `SigmaRho` being tail-nondeterministic k-restricted.

3 Complexity of the (σ, ρ)-DOMINATING SET OF SIZE AT LEAST $n - k$ Problems

Theorem 2. *Let σ and ρ be sets of nonnegative integers such that either σ or $\overline{\sigma}$ is finite, and similarly either ρ or $\overline{\rho}$ is finite. Then the (σ, ρ)-DOMINATING SET OF SIZE AT LEAST $n - k$ problem is in FPT.*

Proof. We present an algorithm that is based on the bounded search tree technique. At the beginning the algorithm includes all vertices into the set S and then tries recursively excluding some of the vertices to make S (σ, ρ)-dominating. Once a vertex is excluded, it is never included in the set again (in the same branch of the algorithm). Obviously at most k vertices can be excluded from S to fulfill the size constraint.

We call a vertex v *satisfied* (with respect to the current set S) if it has the right number of neighbors in S (i.e., $v \in S$ and $|N(v) \cap S| \in \sigma$ or $v \notin S$ and $|N(v) \cap S| \in \rho$), otherwise we call it *unsatisfied*. Let $\widetilde{p}_{max}$ denote $\max \sigma$ if σ is finite and $\max \overline{\sigma}$ if $\overline{\sigma}$ is finite. Similarly let $\widetilde{q}_{max}$ denote $\max \rho$ or $\max \overline{\rho}$. (It is assumed here that $\max \emptyset = 0$.) Finally let b denote $\max\{\widetilde{p}_{max}, \widetilde{q}_{max}\}$. We call a vertex v *big* if $deg(v) > b + k$ and *small* otherwise.

The main idea of the algorithm is that there is at most one way to make an unsatisfied big vertex satisfied (to exclude it from S) and if this does not work, there is no (σ, ρ)-dominating set at all. On the other hand to satisfy a small vertex, we must either exclude it or one of its first b neighbors that were in S.

Procedure Exclude(S)
  if *there is no unsatisfied vertex* **then** **Return**(S);**Exit;**
  if $|S| = n - k$ **then Halt;**
  let v be an unsatisfied vertex;
  if v *is big* **then**
      if $v \in S$ *and* ρ *is infinite* **then** Exclude($S \setminus v$);
      else Halt;
  else
      if $v \in S$ **then** Exclude($S \setminus v$);
      let $\{u_1, \ldots, u_r\} = S \cap N(v)$ be the set of included neighbors of v;
      if $r = 0$ **then Halt;**
      for $i := 1$ **to** $\min\{b+1, r\}$ **do** Exclude($S \setminus \{u_i\}$).

The algorithm consists of a single call Exclude(V) and returns the set S returned by the procedure or NO if no set was returned.

4 Complexity for the Case $\sigma, \rho \in \{\text{EVEN}, \text{ODD}\}$

As a counterpart to the results of [9] we first show that all four parity problems for Red/Blue bipartite graphs are hard under the dual parametrization.

Theorem 6. *The* EVEN SET OF SIZE $r - k$, EVEN SET OF SIZE AT LEAST $r - k$, ODD SET OF SIZE $r - k$, *and* ODD SET OF SIZE AT LEAST $r - k$ *problems are all* W[1]-*hard.*

Proof. It was proved in [9] that
ODD SET OF SIZE AT MOST k
Input: A bipartite graph $G = (R, B, E)$.
Parameter: k.
Question: Is there an odd set in R of size at most k?
is W[1]-hard. It should be noted that W[1]-hardness was stated for the exact variant of the problem (i.e. for the question: Is there an odd set in R of size k?), but for our variant of the question, the proof of [9] works the same. We show that the problem remains W[1]-hard if all blue vertices have odd degrees (and also if all of them have even degrees). Then we deduce the claims by considering the set $R \setminus S$ for a would-be odd set $S \subset R$.

The main result of this section is the hardness of the (**EVEN**–**ODD**)-domination problems under the dual parametrization.

Theorem 3. *Let* $\sigma, \rho \in \{\textbf{EVEN}, \textbf{ODD}\}$. *Then the* (σ, ρ)-DOMINATING SET OF SIZE $n - k$ *and* (σ, ρ)-DOMINATING SET OF SIZE AT LEAST $n - k$ *problems are* W[1]-*hard.*

Proof. We prove this theorem for the (σ, ρ)-DOMINATING SET OF SIZE AT LEAST $n-k$ problem. The proof for the (σ, ρ)-DOMINATING SET OF SIZE $n-k$ is done by similar arguments. Also we give here only the proof for the case $\sigma = \rho = $ **EVEN**. The proofs for the other three cases use the similar ideas and are omitted here. We use the following lemma:

Lemma 3. *The* EVEN SET OF SIZE AT LEAST $r-k$ *problem remains* W[1]*-hard if all red vertices have even degrees.*

Proof. We reduce the EVEN SET OF SIZE AT LEAST $r-k$ problem by replacing each blue vertex by two vertices with the same neighborhoods. Trivially $S \subseteq R$ is an even set in the obtained graph if and only if it is an even set in the original graph.

If all red vertices have even degrees then $S \subseteq R$ is an even set if and only if $S \cup B$ is an (**EVEN, EVEN**)-dominating set. It follows immediately that G has an even set of size at least $r - k$ if and only if G has a (σ, ρ)-dominating set of size at least $n - k$ for $\sigma = \rho = $ **EVEN**.

5 Complexity of the (σ, ρ)-DOMINATING SET OF SIZE (AT MOST) k Problem for Sparse Graphs

It is well known that many problems which are difficult for general graphs can be solved efficiently for sparse graphs. Very general results of such kind were established in [4]. Let v be a vertex of a graph G. For a positive integer r, denote by $N_r[v]$ the *closed r-neighborhood* of v i.e. the set of vertices of G at distance at most r from v. Let $\mathcal{G}$ be a class of graphs. Suppose that there is a family of graphs $\{H_r\}$ such that for each graph $G \in \mathcal{G}$ and for any $v \in V(G)$,

$G[N_r(v)]$ excludes H_r as a minor for $r \geq 1$. It is said that the graph class $\mathcal{G}$ is *locally minor excluding*. It can be noted that e.g. planar graphs, graphs of bounded genus, H-minor-free graphs are locally minor excluding graph classes. It was proved in [4] that if $\mathcal{G}$ is a locally minor excluding class of graphs, then deciding first-order properties (i.e. properties which can be expressed in the first-order logic) is FPT on $\mathcal{G}$. The next claim follows immediately from this result.

Theorem 7. *Let σ and ρ be sets of nonnegative integers such that either σ or $\overline{\sigma}$ is finite, and similarly either ρ or $\overline{\rho}$ is finite. Then the (σ, ρ)-*DOMINATING SET OF SIZE (AT MOST) k *problem is* FPT *on locally minor excluding graph classes.*

It is known that some domination problems are FPT for a more general class of *degenerate* graphs (see e.g. [1,14]). These results can be easily generalized for (σ, ρ)-domination problems for some special sets σ and ρ. It is an interesting open problem whether the results of Theorem 7 can be extended to degenerate graphs.

References

1. Alon, N., Gutner, S.: Linear time algorithms for finding a dominating set of fixed size in degenerated graphs. In: Lin, G. (ed.) COCOON 2007. LNCS, vol. 4598, pp. 394–405. Springer, Heidelberg (2007)
2. Bodlaender, H.L., Kratsch, D.: A note on fixed parameter intractability of some domination-related problems. Private communication (1994)
3. Cesati, M.: Perfect code is $W[1]$-complete. Inf. Process. Lett. 81, 163–168 (2002)
4. Dawar, A., Grohe, M., Kreutzer, S.: Locally excluding a minor. In: LICS 2007, pp. 270–279. IEEE Computer Society Press, Los Alamitos (2007)
5. Downey, R.G., Fellows, M.R.: Fixed-parameter tractability and completeness. Congressus Numerantium, 161–178 (1992)
6. Downey, R.G., Fellows, M.R.: Fixed-parameter tractability and completeness. II. On completeness for $W[1]$. Theoret. Comput. Sci. 141, 109–131 (1995)
7. Downey, R.G., Fellows, M.R.: Parameterized Complexity. Springer, Heidelberg (1998)
8. Downey, R.G., Fellows, M.R.: Threshold dominating sets and an improved characterization of $W[2]$. Theoret. Comput. Sci. 209, 123–140 (1998)
9. Downey, R.G., Fellows, M.R., Vardy, A., Whittle, G.: The parametrized complexity of some fundamental problems in coding theory. SIAM J. Comput. 29 (electronic), 545–570 (1999)
10. Flum, J., Grohe, M.: Parameterized Complexity Theory. Springer, Heidelberg (2006)
11. Golovach, P.A., Kratochvíl, J.: Computational complexity of generalized domination: A complete dichotomy for chordal graphs. In: Brandstädt, A., Kratsch, D., Müller, H. (eds.) WG 2007. LNCS, vol. 4769, pp. 1–11. Springer, Heidelberg (2007)
12. Golovach, P.A., Kratochvíl, J.: Generalized domination in degenerate graphs: a complete dichotomy of computational complexity. In: Agrawal, M., Du, D.-Z., Duan, Z., Li, A. (eds.) TAMC 2008. LNCS, vol. 4978, pp. 182–191. Springer, Heidelberg (2008)
13. Heggernes, P., Telle, J.A.: Partitioning graphs into generalized dominating sets. Nordic J. Comput. 5, 128–142 (1998)
14. Kloks, T., Cai, L.: Parameterized tractability of some (efficient) Y - domination variants for planar graphs and t-degenerate graphs. In: Proceedings of the International Computer Symposium (ICS 2000), Taiwan (2000)
15. Kratochvíl, J., Manuel, P.D., Miller, M.: Generalized domination in chordal graphs. Nordic J. Comput. 2, 41–50 (1995)
16. Moser, H., Thilikos, D.M.: Parameterized complexity of finding regular induced subgraphs. In: Broersma, H., Dantchev, S.S., Johnson, M., Szeider, S. (eds.) AСiD 2006. Texts in Algorithmics, vol. 7, pp. 107–118. King's College, London (2006)
17. Niedermeier, R.: Invitation to Fixed Parameter Algorithms. Oxford University Press, USA (2006)
18. Telle, J.A.: Complexity of domination-type problems in graphs. Nordic J. Comput. 1, 157–171 (1994)
19. Telle, J.A.: Vertex partitioning problems: characterization, complexity and algorithms on partial k-trees. PhD thesis. Department of Computer Science, University of Oregon, Eugene (1994)
20. Telle, J.A., Proskurowski, A.: Algorithms for vertex partitioning problems on partial k-trees. SIAM J. Discrete Math. 10, 529–550 (1997)

Connected Feedback Vertex Set in Planar Graphs

Alexander Grigoriev[1] and René Sitters[2]

[1] Department of Quantitative Economics, Maastricht University, The Netherlands
a.grigoriev@ke.unimaas.nl
[2] Department of Econometrics and Operations Research, VU University, Amsterdam, The Netherlands
rsitters@feweb.vu.nl

Abstract. We study the problem of finding a minimum tree spanning the faces of a given planar graph. We show that a constant factor approximation follows from the unconnected version if the minimum degree is 3. Moreover, we present a polynomial time approximation scheme for both the connected and unconnected version.

1 Introduction

Given a planar graph, what is the smallest subgraph connecting all the faces? The simplicity and naturalness of this question is the main motivation for the study in this paper. Bodlaender et al. [5] call this the *face cover tree problem* and to the best of our knowledge they were the first to study it. They show that the problem can be solved efficiently for edge-weighted graphs of bounded treewidth. In this paper we consider unweighted planar graphs with the minimum degree at least three. This is a natural restriction since allowing vertices of degree two makes its complexity polynomially equivalent to the problem with polynomially bounded edge weights.

Interestingly, the problem does not depend on the embedding since any tree hitting all faces will, in fact, hit all cycles of the graph.

Lemma 1. *Let G be a connected planar graph and $T \subseteq G$ be a tree such that, for a given embedding of G, every face has at least one vertex in T. Then, every cycle of G has a vertex in T.*

Proof. Every cycle separates the embedded graph in an inner and outer part. Each of the two parts contains at least one face. Therefore, the cycle and the tree must have at least one vertex in common. □

The problem of finding the smallest set of vertices hitting all cycles is well-studied and known as the *feedback vertex set problem*. A natural variant for planar graphs is the problem of hitting all faces with a minimum number of vertices. By Lemma 1, the connected versions of these two problems are equivalent and independent of the embedding. This is the problem we study here and call it the *connected feedback vertex set problem*.

Planar Feedback Vertex Set (Planar FVS): Given an unweighted planar graph, find the smallest set S of vertices such that every *cycle* of the graph has at least one vertex in S.

Face Hitting Set (FHS): Given an unweighted planar graph with an embedding, find the smallest set S of vertices such that every *face* of the graph has at least one vertex in S.

Connected Planar Feedback Vertex Set (Connected Planar FVS): Given an unweighted planar graph, find the smallest tree T such that every *cycle* (or equivalently, every face in an embedding) of the graph has a vertex in T.

1.1 Related Results

The feedback vertex set problem is extensively studied. It is APX-hard in general graphs and can be approximated efficiently within a factor 2; see Becker and Geiger [4] and Bafna et al [1]. For planar graphs the problem is NP-hard [12] and a PTAS was given by Demaine and Hajiaghayi [8]. Goemans and Williamson apply the primal-dual method to obtain a (9/4)-approximation [13], which was later reduced to 2 by Chudak et al [7].

Regarding the connectivity constraint, two obvious related problems, are the problem of spanning all vertices and the problem of spanning all the edges of the graph. The latter is known as *connected vertex cover* and was introduced in 1977 by Garey and Johnson [11], who showed it to be NP-hard even when restricted to planar graphs with maximum degree 4. The 2-approximation algorithm for vertex cover in general graphs by Savage [14] transfers directly to the connected problem. Recently, Escoffier et al. [10] have shown that connected vertex cover admits a PTAS for planar graphs. A PTAS for connected dominating set in planar graphs was given in [8] as well.

1.2 Our Results

We give an overview on structural properties, complexity and approximability results for the connected feedback vertex set problem in planar graphs. We show that if the minimum vertex degree is three, then the ratio between the connected and unconnected problem is bounded by a constant. This provides a polynomial time constant approximation algorithm for connected planar FVS. Another interesting consequence of this structural result is that the diameter of a 3-polytope is in the order of the smallest set of vertices hitting all facets. Further, we show that FHS and connected planar FVS are strongly NP-hard and give polynomial time approximation schemes for both problems. Along the text we pose several interesting open questions.

2 Structural Results

2.1 Insightful Observations

We start with some simple lemmas to get an insight in the relation between FVS and FHS and the dependence on the embedding. Then we give the main result of this section on the relation between connected and unconnected FVS in planar graphs. We end with a small discussion on the application to diameters of polytopes.

Lemma 2. *For any planar graph G and embedding Γ_G with at least two faces we have $\mathrm{FHS}(\Gamma_G) \leq \mathrm{FVS}(G)$. Otherwise, $\mathrm{FHS}(\Gamma_G) = 1$ and $\mathrm{FVS}(G) = 0$.*

Lemma 3. *For any planar graph G with faces F we have $\mathrm{FVS}(G) \leq |F| - 1$ and this bound is tight.*

The proofs of Lemmas 2 and 3 are not complicated and we leave those for a reader.

Lemma 4. *For any planar graph G and embedding Γ_G we have $\mathrm{FVS}(G) \leq 2\mathrm{FHS}(\Gamma_G) - 1$ and this bound is tight for $0, 1$ or 2-connected graphs. If G is 3-connected then $\mathrm{FVS}(G) \leq 2\mathrm{FHS}(\Gamma_G) - 2$.*

Proof. Let S_1 be a minimum FHS in Γ_G. Now consider the graph H containing all uncovered cycles and let F_H be its faces. Each face f in H must contain a point from S_1 in its interior. Hence, $|S_1| \geq |F_H|$. Let S_2 be a minimum FVS in H. Then by Lemma 3 $|S_2| \leq |F_H| - 1 \leq |S_1| - 1$. Note that $S_1 \cup S_2$ is a FVS in G. Hence, $\mathrm{FVS}(G) \leq |S_1| + |S_2| \leq 2|S_1| - 1 = 2\mathrm{FHS}(\Gamma_G) - 1$. □

For 3-connected graphs the embedding is unique and so is the minimum value of FHS. In general the optimal value differs by at most a factor two for different embeddings and this bound is tight; see Figure 1(A).

Lemma 5. *Let Γ_1, Γ_2 be two embeddings of planar graph G. Then $\mathrm{FHS}(\Gamma_1) \leq 2\mathrm{FHS}(\Gamma_2) - 1$.*

Proof. If G contains only one face (i.e., it is a forest), then $\mathrm{FHS}(\Gamma_1) = \mathrm{FHS}(\Gamma_2) = 1$. In the other case Lemma 2 says $\mathrm{FHS}(\Gamma_1) \leq \mathrm{FVS}(G)$. By Lemma 4 we have $\mathrm{FVS}(G) \leq 2\mathrm{FHS}(\Gamma_2) - 1$. Combining these inequalities the lemma follows. □

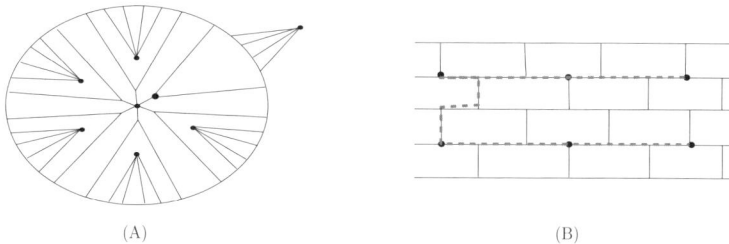

Fig. 1. (A) Tight example for the dependence on the embedding. If all k triangles are directed inwards, the optimal FHS has size $k + 1$. If all are directed outwards the optimal value is $2k + 1$. (B) In an infinite honeycomb graph the ratio between FHS and the connected FVS is 3.

2.2 FVS and FHS in Planar Graphs with Minimum Degree 3

The minimum FHS may be arbitrarily much smaller than the connected FVS if we allow vertices of degree 2. However, restricting to a minimum degree of 3 ensures a constantly bounded ratio between the connected and unconnected problem.

Theorem 1. *Let G be a connected planar graph with minimum degree 3 and let OPT^C be the optimal value for connected planar FVS. Further, let OPT be the optimal value for FHS, for some given embedding of G. Then, $\text{OPT}^C = 0$ if $\text{OPT} = 1$ and $\text{OPT}^C \leq 11\text{OPT} - 14$ otherwise.*

Proof. The case $\text{OPT} = 1$ is trivial. Now assume $\text{OPT} \geq 2$. The outline of the proof is as follows. Given a planar graph $G = (V, E)$ plus embedding, let S be a face hitting set and T be a minimum Steiner tree on S. We will construct curves in the embedding that go from edges in T to vertices in S such that no two curves intersect. On one hand, the number of curves will be $\Omega(|T|)$. On the other hand, we will see that non-intersection of curves implies that their number is $O(|S|)$. Combined we get $|T| = O(|S|)$.

To simplify the construction of curves we define a graph G' that follows from G after contractions of edges. We can partition T in a collection $\mathcal{P}$ of at most $2|S| - 2$ paths such that any two paths may only have an endpoint in common. By minimality of T, any path is a shortest path between its endpoints. We leave any path $P_i \in \mathcal{P}$ of length 1, 2 or 3 unchanged, where the length is the number of edges. If the length P_i is two, we denote its inner vertex by q_i. If the length is three we denote one of its inner vertices by q_i. Any path P_i of length at least four is reduced to length exactly four by contracting all points that are at distance at least two from the two endpoints of P_i in a single point q_i. Let the resulting (multi) graph be $G' = (V', E')$. The following properties are easy to verify. A short justification is given below.

(i) All points $q_1, q_2, \ldots,$ are different.
(ii) If $\text{length}(P_i) \leq 3$ then $\text{degree}(q_i) \geq \text{length}(P_i)$.
(iii) If $\text{length}(P_i) \geq 4$ then $\text{degree}(q_i) \geq \text{length}(P_i) - 1$.
(iv) Solution S is a FVS for G' as well.
(v) For any $v \in V'$, $\text{degree}(v) \geq 3$.

Explanation: (i) Paths only share endpoints and these are not contracted, i.e., no edge was contracted to any of those points. (ii) Obvious. (iii) Every point on P_i has at least one edge not in P_i. (iv) A contraction creates no extra faces. Vertices in S are not contracted. (v) The contraction of two adjacent points with degrees $d_1, d_2 \geq 3$ results in a point with degree $d_1 + d_2 - 2 \geq 4$.

Claim. There are no multiple edges in G'.

Proof. Since G is a simple graph, a multiple edge can only appear if at least one of the two endpoints is a contracted point. More precisely, either (i) there are edges (u_1, v) and (u_2, v) in E such that u_1 and u_2 are contracted in a single points q_i, or (ii) there are edges (u_1, v_1) and (u_2, v_2) in E such that u_1 and u_2 are contracted in q_i and v_1 and v_2 are contracted in q_j.

In case (i), the point v cannot be on the part of P_i between u_1 and u_2 since then all three points would be contracted in q_i. Therefore, the edges (u_1, v) and (u_2, v) plus the part of P_i between u_1 and u_2 form a simple cycle C in G. By Lemma 1, C must have a vertex from T and this can only be v. But then, we can strictly reduce the length of the T as follows. Remove from T the path from u_1 to u_2. Assume, w.l.o.g., that u_1 and v are in the same component. Now add edge (u_2, v) and remove the remaining redundant path to u_1. The argument for (ii) is similar. The edges edges $(u_1, v_1), (u_2, v_2)$ plus the part of P_i between u_1 and u_2 and the part of P_j between v_1 and v_2 form a simple cycle in G and must therefore contain a point from T. But there is no such point on these parts by definition. □

Note that a face in G' may have more than one vertex from S. For each face f we fix one vertex $s(f) \in S$. Now, consider a point q_i and let N_i be the set of neighboring edges whose endpoints are not in S. For each i with $|N_i| \geq 2$ we do the following. Consider two edges from N_i that are consecutive in the embedding, i.e., they appear consecutively among edges from N_i when we walk around q_i. Let f be a face that touches q_i between these edges. We draw a curve inside f from q_i to $s(f)$. Call this a *face-curve*. We do this for all $|N_i|$ pairs of consecutive edges of q_i. If, on the other hand, $|N_i| = 0$ or $|N_i| = 1$ we add a curve from q_i to each neighbor that is an element from S. Call these *edge curves*. Finally, for each path P_i of length one in G we define the point in the plane on the middle of edge P_i as r_i and draw one curve from r_i to $s(f)$, where f is a face adjacent to r_i. These curves are also called face-curves. Note that for each path $P_i \in \mathcal{P}$, we either defined a vertex q_i in G' or defined a point r_i in the embedding of G'.

We define the bipartite (multi-)graph $H = (\Pi \cup \Sigma, \mathcal{A})$ as follows. Let $\Pi = \{\pi_1, \ldots, \pi_{|\mathcal{P}|}\}$ and $\Sigma = \{\sigma_1, \ldots, \sigma_{|S|}\}$. For each curve defined in the process above there is an edge in H, i.e., for each curve from q_i or r_i to s_j there is an edge (π_i, σ_j).

Claim. The graph H is planar and degree$(\pi_i) \geq$ length$(P_i) - 1$ for each $P_i \in \mathcal{P}$.

Proof. The first follows directly from the following observations. All points q_i and r_i are different and none coincides with points from S. Each curve either lies inside a single face or corresponds to a single edge. All curves inside a face have a common endpoint.

If length$(P_i) = 1$ then degree$(\pi_i) = 1$. If length$(P_i) \in \{2, 3\}$ then degree$(q_i) \geq 3$. Now, either $|N_i| \geq 2$ in which case we added $|N_i|$ face-curves, or $|N_i| \leq 1$ in which case we added degree$(q_i) - |N_i| \geq 3 - 1 = 2$ edge-curves. Now assume length$(P_i) \geq 4$. Note that in that case no neighbor of q_i can be a vertex $s \in S$, since in that case we can reduce the length of $T \subseteq G$ by adding edge (q_i, s) and removing the path in T from q_i to one of the two endpoints of P_i. Hence, $|N_i| \geq$ degree$(q_i) \geq$ length$(P_i) - 1$. □

We will show that H has few edges. Consider the embedding of H defined naturally by the curves in the embedding of G'. In general, H may have faces of length two and may be disconnected. To facilitate the analysis we add edges to H until it is connected. Note that we can always do this without creating new faces of length two. Denote the new graph by H'. We prove through Claim 2.2 that H' does not have many edges by showing that it has no faces of length two. In the proof we use the next general statement on planar graphs.

Claim. Let $G = (V, E)$ be a simple planar graph with $|V| \geq 3$ and $s, w \in V$ such that $\mathrm{degree}(v) \geq 3$ for all $v \in V \setminus \{s, w\}$ and $\mathrm{degree}(w) \geq 1$. Then, there is at least one face that does not contain s on its boundary.

Proof. Remove s from G and consider a component C containing some $v \in V \setminus \{s, w\}$. Since $|V| \geq 3$ this component exists. If $w \in C$ then its degree is at least one. Any other vertex has degree at least two. The sum of the degrees in C is then at least $2n' - 1$, with n' the number of vertices in C. But then C is not a forest and must therefore have a cycle. Since s is not on the cycle there is a face not connected to s. □

Claim. There are no faces of length two in H'.

Proof. Suppose there is a face f of length two in H'. Since H' is connected, there cannot be a point $\sigma_j \in \Sigma$ inside f since it has to be connected to at least one of the two points of f, in which case f has length larger than two. Given that f has no points from Σ in its interior, the two curves in G' that correspond to the two edges of f do not enclose a point from S. We will show that this leads to a contradiction.

For each r_i we defined exactly one curve. So the two curves do not start from a point r_i. For each q_i we either defined edge-curves (at most one to each neighbor) or defined face-curves. Therefore the two curves must both be face-curves. Assume they go from q_i to s_j. Let $J \subseteq G'$ be the graph induced by all vertices enclosed by the two curves and including q_i and s_j. We know that s_j is the only vertex from S in J and, by construction, there is at least one edge $(q_i, w) \in N_i$ in J. Since $w \notin S$ we have $w \neq s_j$ and J has at least three vertices: q_i, w and s_j. Further, any vertex $v \notin \{q_i, s_j\}$ in J has degree at least 3 in J. By Claim 2.2, graph $J \subseteq G'$ has no multiple edges. Now, it follows from Claim 2.2 that there must be a face of J that is not connected to s_j. A contradiction. □

The proof of Theorem 1 now easily follows from an upper and lower bound on the number of edges $|\mathcal{A}|$ in H. By Claim 2.2 we have

$$|\mathcal{A}| = \sum_{\pi_i \in \Pi} \mathrm{degree}(\pi_i) \geq \sum_{P_i \in \mathcal{P}} (\mathrm{length}(P_i) - 1) = |T| - |\mathcal{P}|. \qquad (1)$$

Since we assumed $|S| \geq 2$, the number of vertices in H is $|\Pi| + |\Sigma| \geq 1 + 2 = 3$. Let n, m, f be, respectively, the number of vertices, the number of edges, and the number of faces in H'. Each face in H' is bounded by at least three edges so

$2m \geq 3f$. Since H' is connected we know from Euler's formula that $n+f = m+2$. Hence, $2m \geq 3f = 3m + 6 - 3n$ implying $m \leq 3n - 6$. We obtain,

$$|A| \leq m \leq 3n - 6 = 3(|\Pi| + |\Sigma|) - 6.$$

Combined with (1) we get that

$$|T| - |\mathcal{P}| \leq |A| \leq 3(|\Pi| + |\Sigma|) - 6 = 3(|\mathcal{P}| + |S|) - 6. \qquad (2)$$

In the definition of $\mathcal{P}$ we remarked that $|\mathcal{P}| \leq 2|S| - 2$. This combined with (2) gives

$$|T| \leq 4|\mathcal{P}| + 3|S| - 6 \leq 4(2|S| - 2) + 3|S| - 6 = 11|S| - 14.$$

If S is an *optimal* solution for the unconnected problem, then

$$\text{OPT}^C \leq |T| \leq 11|S| - 14 = 11\text{OPT} - 14. \qquad \square$$

Question 1. What is the right ratio for Theorem 1? We conjecture it is 3. See Figure 1(B).

2.3 Diameter of Polytopes

The 1-skeleton of a 3d-polytope is a 3-connected planar graph and vice versa. We proved that the smallest tree spanning all facets is in the order of the number of points hitting all facets. An easy corollary is that the diameter is not much larger. The famous Hirsch conjecture states that the diameter of any d-polytope is at most $n - d$, with n the number of facets. It is known to be true for $d = 3$. Note that the face hitting set may be much smaller than the number of faces. We believe the next easy corollary is of its own interest.

Corollary 1. *The diameter of a 3-dimensional polytope $\mathcal{P}$ is $O(\text{FHS}(\mathcal{P}))$.*

Proof. Let s_1, s_2 be vertices of the polytope $\mathcal{P}$ and S a smallest set covering all the facets. If $s_i \notin S$ we add a hyperplane that just cuts off s_i. The new polytope $\mathcal{P}'$ has at most two extra facets and we can cover all facets by at most $|S| + 2$ vertices. These vertices are spanned by tree of size at most $11(|S| + 2) - 14 = 11|S| + 4$. Clearly, the tree in $\mathcal{P}'$ induces a tree in $\mathcal{P}$ of at most the same size and which connects s_1 and s_2. $\qquad \square$

A similar statement for higher dimensions should depend on the dimension d. For example, the facets of a d-dimensional cube are covered by two opposite vertices while the diameter is d. Hence, the diameter of a d-cube $\mathcal{P}$ is $d/2 \cdot \text{FHS}(\mathcal{P})$, where FHS is the minimum *facet* hitting set. Barnette [3] proved that the diameter of a d-polytope is $O(2^d(n-d))$. Can we replace the n by the minimum FHS?

Question 2. Is there a function $f(d)$ such that for any d-dimensional polytope $\mathcal{P}$, the diameter is bounded by $f(d)\text{FHS}(\mathcal{P})$?

3 Complexity and Approximation

3.1 NP-Hardness of FHS and Connected FVS

Bodlaender et al. [5] show that connected FVS with maximum degree 4 is NP-hard even if every edge has either unit length or an input dependent length K. They reduce from the connected vertex cover problem in planar graphs. Garey and Johnson[11] show that the latter problem is already NP-hard if the maximum degree is 4. To prove NP-hardness for unit lengths we modify the original proof from [11]. NP-hardness of FHS follows easily from the reduction we use for the connected version.

Theorem 2. FHS *is NP-hard in planar graphs with maximum degree 6 and connected* FVS *is NP-hard in planar graphs with maximum degree 9.*

Proof. We concentrate on the proof for the connected FVS problem and we leave the proof for FHS to the extended journal version of this paper. We reduce from the vertex cover problem in planar graphs with maximum degree 3, which is known to be NP-hard [11]. Given a planar graph $G = (V, E)$ with maximum degree 3, we fix some embedding. Let F be the set of faces. We replace each edge by a graph on 10 vertices as in Figure 2. Call this a *bridge*. Let the size of a face be the length of a closed walk along the edges of the face. In each face f of size k we add two rings: an outside ring on $5k$ vertices and an inside ring on $15k$ vertices. Connections between the rings and bridges are illustrated in Figure 2. To enhance the counting we do this for the outer face as well. (Not shown in the figure.) The newly constructed graph G' has maximum vertex degree 9. We claim that G has a vertex cover of size s if and only if G' has a connected feedback vertex set of size $s + 12|E| + |F|$. We omit this technical proof and present it in the full length journal version of this paper. □

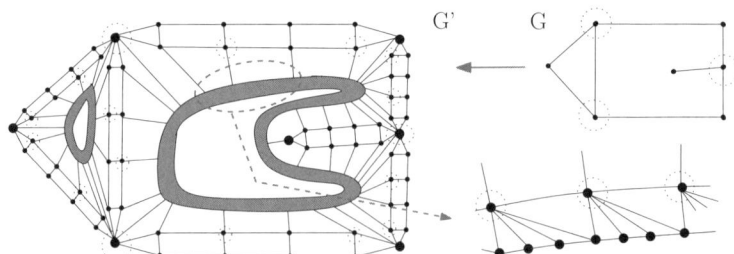

Fig. 2. The reduction. Construction for the outer face is not shown. The encircled vertices indicate a vertex cover in G and a connected feedback vertex set in G'.

3.2 Approximation Schemes for FHS and Connected FVS

First, we consider the connected FVS. We assume that the minimum vertex degree in the graph is at least 3. The polynomial time approximation scheme is a Baker's type algorithm; see Baker [2]. First, we define levels of the planar embedding following the recursive procedure: define level 1 of the embedding as the set of vertices incident to the exterior face of the embedding; assume we constructed level j, then level $j + 1$ is defined as the set of vertices incident to the exterior face of the embedding after removal of the first j levels.

Given a desired approximation precision $\varepsilon > 0$, let $k = 2\lceil(\log n)/\varepsilon\rceil$. Let T be a minimum tree hitting all faces in G and let $OPT = |V(T)| - 1$. For any $0 \leq j \leq k-1$ define set V_j as the union of levels i such that $i \equiv j \mod k$. Since sets $V_j, 0 \leq j \leq k-1$, define a partition of $V(G)$ into k subsets, there is a subset V_ℓ containing at most OPT/k vertices from T. Denote $q = |V_\ell \cap V(T)|$. Notice that we do not know in advance values ℓ and q. So, the algorithm enumerates all possibilities for those values and chooses the values providing the shortest tree hitting all faces. As $1 \leq \ell \leq k$ and $1 \leq q \leq OPT/k$, this enumeration adds a factor $O(n)$ in the running time. From now, we assume that the algorithm picked correct values ℓ and q.

Consider $k + 1$ consecutive levels with the first and the last levels from V_ℓ. We call a subgraph induced by such set of levels a *slice*. Clearly, any slice is a $(k + 1)$-outerplanar graph. By Bodlaender et al [5], the minimum FHS in k-outerplanar graphs can be found in time $O(n^3 + 2^{9.5539k})$. Thus, by definition of k, we can solve the problem on any slice in polynomial time. Using the same algorithm as in [5], we can solve in polynomial time even more general problem: given a slice and an arbitrary number $1 \leq r \leq n$, we have to find a minimum forest of at most r components that hits all faces of the slice. We omit the proof of this simple adjustment.

Notice that T can be seen as a collection of at most $q+1$ trees such that each of these trees is located in exactly one slice. Given the minimum forests for each slice and each number of components, by straightforward dynamic program we find the minimum forest hitting all faces of G with at most $q + 1$ components, each located in exactly one slice. Let T' be such forest. Notice, $E(T') \leq OPT$.

Now, we have a forest T' of at most $q+1$ components that hits all faces in G. Moreover, this forest is shorter than tree T. The only question remains: how to connect the components of T' at small cost? For any two components S and S' of T', let distance $d(S, S')$ be defined as the length of the shortest path connecting S and S'. On this metric, take a minimum spanning tree M. If $(S, S') \in E(M)$, we connect S and S' with the corresponding shortest path. In this way we obtain a connected graph that hits all faces. Hence, we can find a tree of length at most $OPT + \sum_{(S,S') \in E(M)} d(S, S')$.

Lemma 6. $\sum_{(S,S') \in E(M)} d(S, S') \leq \varepsilon OPT$.

Proof. As T' has at most $q+1$ components, M contains at most q edges. Assume there is an edge (S, S') in M of length greater than $\varepsilon OPT/q$. Since $OPT \geq qk \geq 2\log n/\varepsilon$, we have that $d(S, S') > 2\log n$. Consider the corresponding shortest path between S and S'. Take a vertex v in the middle of this path. Let the set of faces of G which are not incident to v be referred as $\mathcal{F}$. Consider the distance from v to $f \in \mathcal{F}$ by mean the length of the shortest path from v to the furthest vertex incident to f. By choice of v and the assumption that $d(S, S') > 2 \log n$, the distance from v to any face from $\mathcal{F}$ is greater than $\log n$. Therefore, the subgraph of G induced by all vertices on distance at most $1 + \log n$ from v is a tree. Since minimum degree in G is at least 3, the number of vertices in such tree is more than n. A contradiction.

Now, we summarize the main results of this section in the following theorem and corollary.

Theorem 3. *Given a planar graph G of minimum degree 3 and $\varepsilon > 0$, the algorithm above constructs in polynomial time a tree hitting all faces of G with length at most $(1 + \varepsilon)OPT$.*

Applying literally the same modifications to the Baker's algorithm as in Eppstein [9] and Bodlaender and Grigoriev [6] we derive the following corollary.

Corollary 2. *The connected feedback vertex set face hitting set problem on graphs embeddable on a surface of bounded genus and having minimum vertex degree 3 admits a polynomial time approximation scheme.*

Without the connectivity constraint the problem becomes much easier. A PTAS for FHS follows directly from the discussion above and we leave the proof to the reader.

Theorem 4. *The face hitting set problem on graphs embeddable on a surface of bounded genus admits a polynomial time approximation scheme.*

References

1. Bafna, V., Berman, P., Fujito, T.: A 2-Approximation Algorithm for the Undirected Feedback Vertex Set Problem. SIAM J. Discrete Math. 12, 289–297 (1999)
2. Baker, B.: Approximation algorithms for NP-complete problems on planar graphs. JACM 41, 153–180 (1994)
3. Barnette, D.: An upper bound for the diameter of a polytope. Discr. Math. 10, 9–13 (1974)
4. Becker, A., Geiger, D.: Optimization of Pearl's method of conditioning and greedy-like approximation algorithms for the vertex feedback set problem. Artificial Intelligence 83, 167–188 (1996)
5. Bodlaender, H.L., Feremans, C., Grigoriev, A., Penninkx, E., Sitters, R., Wolle, T.: On the minimum corridor connection and other generalized geometric problems. In: Erlebach, T., Kaklamanis, C. (eds.) WAOA 2006. LNCS, vol. 4368, pp. 69–82. Springer, Heidelberg (2007)
6. Bodlaender, H.L., Grigoriev, A.: Algorithms for graphs embeddable with few crossings per edge. Algorithmica 49, 1–11 (2007)
7. Chudak, F.A., Goemans, M.X., Hochbaum, D.S., Williamson, D.P.: A primal-dual interpretation of two 2-approximation algorithms for the feedback vertex set problem in undirected graphs. Oper. Res. Lett. 22, 111–118 (1998)
8. Demaine, E.D., Hajiaghayi, M.: The bidimensionality theory and its algorithmic applications. The Computer Journal 51, 292–302 (2008)

9. Eppstein, D.: Diameter and treewidth in minor-closed graph families. Algorithmica 27, 275–291 (2000)
10. Escoffier, B., Gourvès, L., Monnot, J.: Complexity and approximation results for the connected vertex cover problem. In: Brandstädt, A., Kratsch, D., Müller, H. (eds.) WG 2007. LNCS, vol. 4769, pp. 202–213. Springer, Heidelberg (2007)
11. Garey, M.R., Johnson, D.S.: The rectilinear Steiner tree problem is NP-complete. SIAM J. Appl. Math. 32, 826–834 (1977)
12. Garey, M.R., Johnson, D.S.: Computers and Intractability: A Guide to the Theory of NP-Completeness. Freeman, New York (1979)
13. Goemans, M.X., Williamson, D.P.: Primal-Dual Approximation Algorithms for Feedback Problems in Planar Graphs. Combinatorica 17, 1–23 (1997)
14. Savage, C.D.: Depth-first search and the vertex cover problem. Inf. Process. Lett. 14, 233–235 (1982)

Logical Locality Entails Frugal Distributed Computation over Graphs
(Extended Abstract)

Stéphane Grumbach and Zhilin Wu

INRIA – LIAMA,
Chinese Academy of Sciences, PO Box 2728, Beijing 100080, PR China
Stephane.Grumbach@inria.fr, zlwu@liama.ia.ac.cn

Abstract. First-order logic is known to have limited expressive power over finite structures. It enjoys in particular the locality property, which states that first-order formulae cannot have a global view of a structure. This limitation ensures their low sequential computational complexity. We show that the locality impacts as well on their distributed computational complexity. We use first-order formulae to describe the properties of finite connected graphs, which are the topology of communication networks, on which the first-order formulae are also evaluated. We show that over bounded degree networks and planar networks, first-order properties can be frugally evaluated, that is, with only a bounded number of messages, of size logarithmic in the number of nodes, sent over each link. Moreover, we show that the result carries over for the extension of first-order logic with unary counting.

1 Introduction

Logical formalisms have been widely used in many areas of computer science to provide high levels of abstraction, thus offering user-friendliness while increasing the ability to verify properties. In the field of databases, first-order logic constitutes the basis of relational query languages, which allow to write queries in a declarative manner, independently of the physical implementation. In this paper, we propose to use logical formalisms to express properties of the topology of communication networks, that can be verified in a distributed fashion over the networks themselves.

We focus on first-order logic over graphs. First-order logic has been shown to have limited expressive power over finite structures. In particular, it enjoys the locality property, which states that first-order formulae are local [Gai82], in the sense that local areas of the graphs are sufficient to evaluate them.

First-order properties have been shown to be computable with very low complexity in both sequential and parallel models of computation. It was shown that first-order properties can be evaluated in linear time over classes of bounded degree graphs [See95] and over classes of locally tree-decomposable graphs[1] [FG01].

---
[1] Locally tree-decomposable graphs generalize bounded degree graphs, planar graphs, and graphs of bounded genus.

These results follow from the locality of the logic. It was also shown that they can be evaluated in constant time over Boolean circuits with unbounded fan-in (AC^0) [Imm89]. These bounds lead us to be optimistic on the complexity of the distributed evaluation of first-order properties.

We consider communication networks based on the message passing model [AW04], where nodes exchange messages with their neighbors. The properties to be evaluated concern the graph which forms the topology of the network, and whose knowledge is distributed over the nodes, which are only aware of their 1-hop neighbors. We thus focus on connected graphs.

In distributed computing, the ability to solve problems locally has attracted a strong interest since the seminal paper of Linial [Lin92]. The ability to solve global problems in distributed systems, while performing as much as possible local computations, is of great interest to ensure scalability. Moreover relying as much as possible on local information improves fault-tolerance. Finally, restricting the computation to local areas allows to optimize time and communication complexity.

Naor and Stockmeyer [NS95] showed that there were non-trivial locally checkable labelings that are locally computable, while on the other hand some lower-bounds have been exhibited, thus resulting in non-local computability results [KMW04, KMW06].

Different notions of local computation have been considered. The most widely accepted restricts the time of the computation to be constant, that is independent of the size of the network [NS95], while allowing messages of size $O(\log n)$, where n is the size of the network. This condition is rather stringent. Naor and Stockmeyer [NS95] show their result for a restricted class of graphs (eg bounded odd degree). Godard et al. used graph relabeling systems as the distributed computational model, defined local computations as graph relabeling systems with locally-generated local relabeling rules, and characterized the classes of graphs that are locally computable [GMM04].

Our initial motivation is to understand the impact of the logical locality on the distributed computation, and its relationship with local distributed computation. It is easy to verify though that there are simple properties (expressible in first-order logic) that cannot be computed locally. Consider for instance the property "There exist at least two distinct triangles", which requires non-local communication to check the distinctness of the two triangles which may be far away from each other. Nevertheless, first-order properties do admit simple distributed computations.

We thus introduce frugal distributed computations. A distributed algorithm is *frugal* if during its computation only a bounded number of messages of size $O(\log n)$ are sent over each link. If we restrict our attention to bounded degree networks, this implies that each node is only receiving a bounded number of messages. Frugal computations resemble local computations over bounded degree networks, since the nodes are receiving only a bounded number of messages, although these messages can come from remote nodes through multi-hop paths.

We prove that first-order properties can be frugally evaluated over bounded degree networks and planar networks (Theorem 2 and Theorem 4). The proofs are obtained by transforming the centralized linear time evaluation algorithms [See95, FG01] into distributed ones satisfying the restriction that only a bounded number of messages are sent over each link. Moreover, we show that the results carry over to the extension of first-order logic with unary counting. While the transformation of the centralized linear time algorithm is simple for first-order properties over bounded degree networks, it is quite intricate for first-order properties over planar networks. The most intricate part is the distributed construction of an ordered tree decomposition for some subgraphs of the planar network, inspired by the distributed algorithm to construct an ordered tree decomposition for planar networks with bounded diameter in [GW09].

Intuitively, since in the centralized linear time computation each object is involved only a bounded number of times, in the distributed computation, a bounded number of messages sent over each link could be sufficient to evaluate first-order properties. So it might seem trivial to design frugal distributed algorithms for first-order properties over bounded degree networks and planar networks. Nevertheless, this is not the case, because in the centralized computation, after visiting one object, any other object can be visited, but in the distributed computation, only the *adjacent* objects (nodes, links) can be visited.

The paper is organized as follows. In the next section, we recall classical graph theory concepts, as well as Gaifman's locality theorem. In Section 3, we consider the distributed evaluation of first-order properties over respectively bounded degree and planar networks. Finally, in Section 4, we consider the distributed evaluation of first-order logic with unary counting.

2   Graphs, First-Order Logic and Locality

In this paper, our interest is focused to a restricted class of structures, namely finite graphs. Let $G = (V, E)$, be a finite graph. We use the following notations. If $v \in V$, then $deg(v)$ denotes the *degree* of v. For two nodes $u, v \in V$, the *distance* between u and v, denoted $dist_G(u, v)$, is the length of the shortest path between u and v. For $k \in \mathbb{N}$, the *k-neighborhood* of a node v, denoted $N_k(v)$, is defined as $\{w \in V | dist_G(v, w) \leq k\}$. If $\bar{v} = v_1...v_p$ is a collection of nodes in V, then the k-neighborhood of $\bar{v}$, denoted $N_k(\bar{v})$, is defined by $\bigcup_{1 \leq i \leq p} N_k(v_i)$. For $X \subseteq V$, let $\langle X \rangle^G$ denote the subgraph induced by X.

Let $G = (V, E)$ be a connected graph, a *tree decomposition* of G is a rooted labeled tree $\mathcal{T} = (T, F, r, B)$, where T is the set of vertices of the tree, $F \subseteq T \times T$ is the child-parent relation of the tree, $r \in T$ is the root of the tree, and B is a labeling function from T to 2^V, mapping vertices t of T to sets $B(t) \subseteq V$, called *bags*, such that

1. For each edge $(v, w) \in E$, there is a $t \in T$, such that $\{v, w\} \subseteq B(t)$.
2. For each $v \in V$, $B^{-1}(v) = \{t \in T | v \in B(t)\}$ is connected in T.

The *width* of $\mathcal{T}$, $width(\mathcal{T})$, is defined as $\max\{|B(t)|-1 | t \in T\}$. The tree-width of G, denoted $tw(G)$, is the minimum width over all tree decompositions of G. An *ordered tree decomposition* of width k of a graph G is a rooted labeled tree $\mathcal{T} = (T, F, r, L)$ such that:

- (T, F, r) is defined as above,
- L assigns each vertex $t \in T$ to a $(k+1)$-tuple $\overline{b^t} = (b_1^t, \cdots, b_{k+1}^t)$ of vertices of G (note that in the tuple $\overline{b^t}$, vertices of G may occur repeatedly),
- If $L'(t) := \{b_j^t | L(t) = (b_1^t, \cdots, b_{k+1}^t), 1 \leq j \leq k+1\}$, then (T, F, r, L') is a tree decomposition.

The *rank* of an (ordered) tree decomposition is the rank of the rooted tree, i.e. the maximal number of children of its vertices.

We consider first-order logic (FO) over the signature E, where E is a binary relation symbol. The syntax and semantics of first-order formulae are defined as usual [EF99]. The *quantifier rank* of a formula φ is the maximal number of nestings of existential and universal quantifiers in φ.

A *graph property* is a class of graphs closed under isomorphisms. Let φ be a first-order sentence, the graph property defined by φ, denoted $\mathcal{P}_\varphi$, is the class of graphs satisfying φ.

The distance between nodes can be defined by first-order formulae $dist(x, y) \leq k$ stating that the distance between x and y is no larger than k, and $dist(x, y) > k$ is an abbreviation of $\neg dist(x, y) \leq k$. In addition, let $\bar{x} = x_1...x_p$ be a list of variables, then $dist(\bar{x}, y) \leq k$ is used to denote $\bigvee_{1 \leq i \leq p} dist(x_i, y) \leq k$.

Let $\varphi(\bar{y})$ be a first-order formula with free variables $\bar{y}$, $k \in \mathbb{N}$, and $\bar{x}$ be a list of variables not occurring in $\varphi(\bar{y})$, then the formula bounding the quantifiers of $\varphi(\bar{y})$ to the k-neighborhood of $\bar{x}$, denoted $(\varphi(\bar{y}))^{(k)}(\bar{x})$, can be defined easily in first-order logic by using formulae $dist(\bar{x}, y) \leq k$. For instance, if $\varphi(\bar{y}) := \exists z \psi(\bar{y}, z)$, then
$$(\varphi(\bar{y}))^{(k)}(\bar{x}) := \exists z \left(dist(\bar{x}, z) \leq k \wedge (\psi(\bar{y}, z))^{(k)}(\bar{x}) \right).$$

We can now recall the notion of logical locality introduced by Gaifman [Gai82, EF99].

Theorem 1. *[Gai82] Let φ be a first-order formula with free variables $u_1, ..., u_p$, then φ can be written in Gaifman Normal Form, that is into a Boolean combination of (i) sentences of the form:*

$$\exists x_1 ... \exists x_s \left(\bigwedge_{1 \leq i < j \leq s} d(x_i, x_j) > 2r \wedge \bigwedge_i \psi^{(r)}(x_i) \right) \tag{1}$$

and (ii) formulae of the form $\psi^{(t)}(\bar{y})$, where $\bar{y} = y_1...y_q$ such that $y_i \in \{u_1, ..., u_p\}$ for all $1 \leq i \leq q$, $r \leq 7^{k-1}$, $s \leq p + k$, $t \leq (7^k - 1)/2$ (k is the quantifier rank of φ)[2].

---

[2] The bound on r has been improved to $4^k - 1$ in [KL04].

Moreover, if φ is a sentence, then the Boolean combination contains only sentences of the form (1).

The locality of first-order logic is a powerful tool to demonstrate non-definability results [Lib97]. It can be used in particular to prove that counting properties, such as the parity of the number of vertices, or recursive properties, such as the connectivity of a graph, are not first-order.

3 Distributed Evaluation of FO

We consider a message passing model of distributed computation [AW04], based on a communication network whose topology is given by a graph $G = (V, E)$ of diameter Δ, where E denotes the set of bidirectional *communication links* between nodes. From now on, we restrict our attention to *finite connected graphs*.

We assume that the distributed system is asynchronous and has no failure. The nodes have a unique *identifier* taken from $1, 2, \cdots, n$, where n is the number of nodes. Each node has distinct local ports for distinct links incident to it. The nodes have *states*, including final accepting or rejecting states.

For simplicity, we assume that there is only one query fired in the network by a *requesting node*. We also assume that a *breadth-first-search (BFS) tree* rooted on the requesting node has been pre-computed in the network[3], such that each node stores locally the identifier of its parent in the BFS-tree, and the states of the ports with respect to the BFS-tree, which are either "parent" or "child", denoting the ports corresponding to the tree edges, or "horizon", "upward", "downward", denoting the ports corresponding to the non-tree edges to some node with the same, smaller, or larger depth in the BFS-tree. The computation terminates, when the requesting node reaches a final state.

Let $\mathcal{C}$ be a class of graphs. A distributed algorithm is said to be *frugal* over $\mathcal{C}$ if there is a $k \in \mathbb{N}$ such that for any network $G \in \mathcal{C}$ of n nodes and any requesting node in G, the distributed computation terminates, with only at most k messages of size $O(\log n)$ sent over each link. If we restrict our attention to bounded degree networks, frugal distributed algorithms imply that each node only receives a bounded number of messages. Frugal computations resemble local computations over bounded degree networks, since the nodes receive only a bounded number of messages, although these messages can come from remote nodes through multi-hop paths.

Let $\mathcal{C}$ be a class of graphs, and φ an FO sentence, we say that φ can be distributively evaluated over $\mathcal{C}$ if there exists a distributed algorithm such that for any network $G \in \mathcal{C}$ and any requesting node in G, the computation of the distributed algorithm on G terminates, with the requesting node in the accepting state if and only if $G \models \varphi$. Moreover, if there is a frugal distributed algorithm to do this, then we say that φ can be frugally evaluated over $\mathcal{C}$.

---

[3] The pre-computation of the BFS tree can be done in $O(\Delta)$ distributed time and with $O(\Delta)$ messages sent over each link [BDLP08].

For centralized computations, it has been shown that Gaifman's locality of FO entails linear time evaluation of FO properties over classes of bounded degree graphs and classes of locally tree-decomposable graphs [See95, FG01]. In the following, we show that it is possible to design frugal distributed evaluation algorithms for FO properties over bounded degree and planar networks, by carefully transforming the centralized linear time evaluation algorithms into distributed ones with computations on each node well balanced.

3.1 Bounded Degree Networks

We first consider the evaluation of FO properties over bounded degree networks. We assume that each node stores the degree bound k locally.

Theorem 2. *FO properties can be frugally evaluated over bounded degree networks.*

Theorem 2 can be shown by using Hanf's technique [FSV95], in a way similar to the proof of Seese's seminal result [See95].

Let $r \in \mathbb{N}$, $G = (V, E)$, and $v \in V$, then the *r-type* of v in G is the isomorphism type of $(\langle N_r(v) \rangle^G, v)$. Let $r, m \in \mathbb{N}$, G_1 and G_2 be two graphs, then G_1 and G_2 are said to be (r,m)-*equivalent* if and only if for every r-type τ, either G_1 and G_2 have the same number of vertices with r-type τ or else both have at least m vertices with r-type τ. G_1 and G_2 are said to be k-*equivalent*, denoted $G_1 \equiv_k G_2$, if G_1 and G_2 satisfy the same FO sentences of quantifier rank at most k. It has been shown that:

Theorem 3. *[FSV95] Let $k, d \in \mathbb{N}$. There exist $r, m \in \mathbb{N}$ such that r depends only on k, m depends on both k and d, and for any graphs G_1 and G_2 with maximal degree no more than d, if G_1 and G_2 are (r, m)-equivalent, then $G_1 \equiv_k G_2$.*

Let us now sketch the proof of Theorem 2, which relies on a distributed algorithm consisting of three phases. Suppose the requesting node requests the evaluation of some FO sentence with quantifier rank k. Let r, m be the natural numbers depending on k, d specified in Theorem 3.

Phase I. The requesting node broadcasts messages along the BFS-tree to ask each node to collect the topology information in its r-neighborhood;
Phase II. Each node collects the topology information in its r-neighborhood;
Phase III. The r-types of the nodes in the network are aggregated through the BFS-tree to the requesting node up to the threshold m for each r-type. Finally the requesting node decides whether the network satisfies the FO sentence or not by using the information about the r-types.

It is easy to see that only a bounded number of messages are sent over each link in Phase I and II. Since the total number of distinct r-types with degree bound d depends only upon r and d and each r-type is only counted up to a threshold m, it turns out that over each link, only a bounded number of messages are sent in Phase III as well. So the above distributed evaluation algorithm is frugal over bounded degree networks.

3.2 Planar Networks

We now consider the distributed evaluation of FO properties over planar networks.

A *combinatorial embedding* of a planar graph $G = (V, E)$ is an assignment of a cyclic ordering of the set of incident edges to each vertex v such that two edges (u, v) and (v, w) are in the same face iff (v, w) is immediately before (v, u) in the cyclic ordering of v. Combinatorial embeddings, which encode the information about boundaries of the faces in usual embeddings of planar graphs into the planes, are useful for computing on planar graphs. Given a combinatorial embedding, the boundaries of all the faces can be discovered by traversing the edges according to the above condition.

We assume in this subsection that a combinatorial embedding of the planar network is distributively stored in the network, i.e. a cyclic ordering of the set of the incident links is stored in each node of the network.

Theorem 4. *FO properties can be frugally evaluated over planar networks.*

For the proof of Theorem 4, we first recall the centralized linear time algorithm to evaluate FO properties over planar graphs in [FG01][4].

Let $G = (V, E)$ be a planar graph and φ be an FO sentence. From Theorem 1, we know that φ can be written into Boolean combinations of sentences of the form (1),

$$\exists x_1 ... \exists x_s \left(\bigwedge_{1 \leq i < j \leq s} d(x_i, x_j) > 2r \wedge \bigwedge_i \psi^{(r)}(x_i) \right).$$

It is sufficient to show that sentences of the form (1) are linear-time computable over G. The centralized algorithm to evaluate FO sentences of the form (1) over planar graphs consists of the following four phases:

1. Select some $v_0 \in V$, let $\mathcal{H} = \{G[i, i+2r] | i \geq 0\}$, where $G[i, j] = \{v \in V | i \leq dist_G(v_0, v) \leq j\}$;
2. For each $H \in \mathcal{H}$, compute $K_r(H)$, where $K_r(H) := \{v \in H | N_r(v) \subseteq H\}$;
3. For each $H \in \mathcal{H}$, compute $P_H := \{v \in K_r(H) | \langle H \rangle^G \models \psi^{(r)}(v)\}$;
4. Let $P := \cup_H P_H$, determine whether there are s distinct nodes in P such that their pairwise distance is greater than $2r$.

In the computation of the 3rd and 4th phase above, an automata-theoretical technique to evaluate Monadic-Second-Order (MSO) formulae in linear time over classes of graphs with bounded tree-width [Cou90, FG06, FFG02] is used. In the following, we recall this centralized evaluation algorithm.

MSO is obtained by adding set variables and set quantifiers into FO, such as $\exists X \varphi(X)$ (where X is a set variable). MSO has been widely studied in the

---

[4] In fact, in [FG01], it was shown that FO is linear-time computable over classes of locally tree-decomposable graphs.

context of graphs for its expressive power. For instance, 3-colorability, transitive closure or connectivity can be defined in MSO [Cou08].

The centralized linear time evaluation of MSO formulae over classes of bounded tree-width graphs goes as follows. First an ordered tree decomposition $\mathcal{T}$ of the given graph is constructed. Then from the given MSO formula, a tree automaton $\mathcal{A}$ is obtained. Afterwards, $\mathcal{T}$ is transformed into a labeled tree $\mathcal{T}'$, finally $\mathcal{A}$ is run over $\mathcal{T}'$ (maybe several times for formulae containing free variables) to get the evaluation result.

In the rest of this section, we design a frugal distributed algorithm to evaluate FO sentences over planar networks by adapting the above centralized algorithm. The main difficulty is to distribute the computation among the nodes such that only a bounded number of messages are sent over each link during the computation.

Phase I. The requesting node broadcasts the FO sentence of the form (1) to all the nodes in the network through the BFS tree;
Phase II. For each $v \in V$, compute $C(v) := \{i \geq 0 | v \in G[i, i+2r]\}$;
Phase III. For each $v \in V$, compute $D(v) := \{i \geq 0 | N_r(v) \subseteq G[i, i+2r]\}$;
Phase IV. For each $i \geq 0$, compute $P_i := \{v \in V | i \in D(v), \langle G[i, i+2r] \rangle^G \models \psi^{(r)}(v)\}$;
Phase V. Let $P := \bigcup_i P_i$, determine whether there are s distinct nodes labeled by P such that their pairwise distance is greater than $2r$.

Phase I is trivial. Phase II is easy. In the following, we illustrate the computation of Phase III, IV, and V one by one.

We first introduce a lemma for the computation of Phase III.

For $W \subseteq V$, let $K_i(W) := \{v \in W | N_i(v) \subseteq W\}$. Let $D_i(v) := \{j \geq 0 | v \in K_i(G[j, j+2r])\}$.

Lemma 1. *For each $v \in V$ and $i > 0$, $D_i(v) = C(v) \cap \bigcap_{w:(v,w)\in E} D_{i-1}(w)$.*

With Lemma 1, $D(v) = D_r(v)$ can be computed in an inductive way to finish Phase III: Each node v obtains the information $D_{i-1}(w)$ from all its neighbors w, and performs the in-node computation to compute $D_i(v)$.

Now we consider Phase IV.

Because $\psi^{(r)}(x)$ is a local formula, $\psi^{(r)}(x)$ can be evaluated separately over each connected component of $G[i, i+2r]$ and the results are stored distributively.

Let C_i be a connected component of $G[i, i+2r]$, and $w_1^i, \cdots, w_l^i$ be all the nodes contained in C_i with distance i from the requesting node. Now we consider the evaluation of $\psi^{(r)}(x)$ over C_i.

Let C_i' be the graph obtained from C_i by including all ancestors of $w_1^i, \cdots, w_l^i$ in the BFS-tree, and C_i^* be the graph obtained from C_i' by contracting all the ancestors of $w_1^i, \cdots, w_l^i$ into one vertex, i.e. C_i^* has one more vertex, called the virtual vertex, than C_i, and this vertex is connected to $w_1^i, \cdots, w_l^i$. It is easy to see that C_i^* is a planar graph with a BFS-tree rooted on v^* and of depth at most $2r + 1$. So C_i^* is a planar graph with bounded diameter.

An ordered tree decomposition for planar networks with bounded diameter can be distributively constructed with only a bounded number of messages sent over each link as follows [GW09]:

- Do a depth-first-search to decompose the network into blocks, i.e. biconnected components;
- Construct an ordered tree decomposition for each nontrivial block: Traverse every face of the block according to the cyclic ordering at each node, triangulate all those faces, and connect the triangles into a tree decomposition by utilizing the pre-computed BFS tree;
- Finally the tree decompositions for the blocks are connected together into a complete tree decomposition for the whole network.

By using the distributed algorithm for the tree decomposition of planar networks with bounded diameter, we can construct distributively an ordered tree decomposition for C_i^*, while having the virtual vertex in our mind, and get an ordered tree decomposition for C_i.

With the ordered tree decomposition for C_i, we can evaluate $\psi^{(r)}(x)$ over C_i by using the automata-theoretical technique, and store the result distributively in the network (each node stores a Boolean value indicating whether it belongs to the result or not).

A distributed post-order traversal over the BFS tree can be done to find out the connected components of all $G[i, i+2r]$'s and construct the tree decompositions for these connected components one by one.

Finally we consider Phase V.

Label nodes in $\bigcup_i P_i$ with P.

Then consider the evaluation of FO sentence φ' over the vocabulary $\{E, P\}$,

$$\exists x_1 ... \exists x_s \left(\bigwedge_{1 \leq i < j \leq s} d(x_i, x_j) > 2r \wedge \bigwedge_i P(x_i) \right).$$

Starting from some node w_1 with label P, mark the vertices in $N_{2r}(w_1)$ as Q, then select some node w_2 outside Q, and mark those nodes in $N_{2r}(w_2)$ by Q again, continue like this, until w_l such that either $l = s$ or all the nodes with label P have already been labeled by Q.

If $l < s$, then label the nodes in $\bigcup_{1 \leq i \leq l} N_{4r}(v_i)$ as I. Each connected component of $\langle I \rangle^G$ has diameter no more than $4lr < 4sr$. We can construct distributively a tree decomposition for each connected component of $\langle I \rangle^G$, and connect these tree decompositions together to get a complete tree-decomposition of $\langle I \rangle^G$, then evaluate the sentence φ' by using this complete tree decomposition.

4 Beyond FO Properties

We have shown that FO properties can be frugally evaluated over respectively bounded degree and planar networks. In this section, we extend these results to FO unary queries and some counting extension of FO.

From Theorem 1, an FO formula $\varphi(x)$ containing exactly one free variable x can be written into a Boolean combinations of sentences of the form (1) and local formulae $\psi^{(t)}(x)$. Then it is not hard to prove the following result.

Theorem 5. *FO formulae $\varphi(x)$ with exactly one free variable x can be frugally evaluated over respectively bounded degree and planar networks, with the results distributively stored on the nodes of the network.*

Counting is one of the ability that is lacking to first-order logic, and has been added in commercial relational query languages (e.g. SQL). Its expressive power has been widely studied [GO92, GT95, Ott96] in the literature. Libkin [Lib97] proved that first-order logic with counting still enjoys Gaifman locality property. We prove that Theorem 2 and Theorem 4 carry over as well for first-order logic with unary counting.

Let FO(#) be the extension of first-order logic with unary counting. FO(#) is a two-sorted logic, the first sort ranges over the set of nodes V, while the second sort ranges over the natural numbers $\mathbb{N}$. The terms of the second sort are defined by: $t := \#x.\varphi(x) \mid t_1 + t_2 \mid t_1 \times t_2$, where φ is a formula over the first sort with one free variable x. Second sort terms of the form $\#x.\varphi(x)$ are called *basic* second sort terms.

The atoms of FO(#) extend standard FO atoms with the following two unary counting atoms: $t_1 = t_2 \mid t_1 < t_2$, where t_1, t_2 are second sort terms. Let t be a second sort term of FO(#), $G = (V, E)$ be a graph, then the interpretation of t in G, denoted t^G, is defined as follows:

- $(\#x.\varphi(x))^G$ is the cardinality of $\{v \in V | G \models \varphi(v)\}$;
- $(t_1 + t_2)^G$ is the sum of t_1^G and t_2^G;
- $(t_1 \times t_2)^G$ is the product of t_1^G and t_2^G.

The interpretation of FO(#) formulae is defined in a standard way.

Theorem 6. *FO(#) properties can be frugally evaluated over respectively bounded degree and planar networks.*

The proof of the theorem relies on a normal form of FO(#) formulae.

5 Conclusion

We have shown that logical formulae used to express properties of graphs, which constitute the topology of communication networks, can be evaluated very efficiently over these networks. Their distributed computation, although not local, can be done *frugally*, that is with a bounded number of messages of logarithmic size exchanged over each link, over respectively bounded degree and planar networks. The frugal computation, introduced in this paper, generalizes local computation and offers a large spectrum of applications. Moreover the results carry over to the extension of first-order logic with unary counting. The distributed time used in the frugal evaluation of FO properties over bounded degree networks is $O(\Delta)$, while that over planar networks is $O(n)$.

We assumed that a BFS tree is pre-computed and stored distributively in the network. Evidently the BFS-tree varies when the requesting node is chosen differently. Since a BFS-tree is a tree 2-spanner [CC95] of the network, we can actually assume that a tree 2-spanner, independent of the choice of the requesting node, is distributively pre-computed and stored in the network, and we still guarantee the frugality of the computation by adapting slightly the distributed evaluation algorithms in Section 3.

Beyond its interest for logical properties, the frugality of distributed algorithms, which ensures an extremely good scalability of their computation, raises fundamental questions, such as deciding what can be frugally computed. We leave as an open problem the question of deciding whether for instance a Hamiltonian path can be computed frugally.

References

[AW04] Attiya, H., Welch, J.: Distributed Computing: Fundamentals, Simulations and Advanced Topics. Wiley Interscience, Hoboken (2004)

[BDLP08] Boulinier, C., Datta, A.K., Larmore, L.L., Petit, F.: Space efficient and time optimal distributed BFS tree construction. Inf. Process. Lett. 108(5), 273–278 (2008)

[Bod93] Bodlaender, H.L.: A linear time algorithm for finding tree-decompositions of small treewidth. In: ACM STOC (1993)

[CC95] Cai, L., Corneil, D.G.: Tree Spanners. SIAM J. Discret. Math. 8(3), 359–387 (1995)

[Cou90] Courcelle, B.: Graph rewriting: An algebraic and logic approach. In: Handbook of Theoretical Computer Science, Volume B: Formal Models and Sematics (B), pp. 193–242. Elsevier and MIT Press (1990)

[Cou08] Courcelle, B.: Graph algebras and monadic second-order logic. To be published by Cambridge University Press (2008) (in preparation)

[EF99] Ebbinghaus, H.D., Flum, J.: Finite model theory. Springer, Heidelberg (1999)

[FFG02] Flum, J., Frick, M., Grohe, M.: Query evaluation via tree-decompositions. J. ACM 49(6), 716–752 (2002)

[FG01] Frick, M., Grohe, M.: Deciding first-order properties of locally tree-decomposable structures. J. ACM 48(6), 1184–1206 (2001)

[FG06] Flum, J., Grohe, M.: Parameterized Complexity Theory. Springer, Heidelberg (2006)

[FSV95] Fagin, R., Stockmeyer, L.J., Vardi, M.Y.: On a monadic NP vs monadic co-NP. Inf. Comput. 120(1), 78–92 (1995)

[Gai82] Gaifman, H.: On local and non-local properties. In: Proceedings of the Herbrand Symposium, Logic Colloquium 1981. North-Holland, Amsterdam (1982)

[GMM04] Godard, E., Métivier, Y., Muscholl, A.: Characterizations of classes of graphs recognizable by local computations. Theory Comput. Syst. 37(2), 249–293 (2004)

[GO92] Grädel, E., Otto, M.: Inductive definability with counting on finite structures. In: Computer Science Logic, CSL, pp. 231–247 (1992)

[GT95] Grumbach, S., Tollu, C.: On the expressive power of counting. Theor. Comput. Sci. 149(1), 67–99 (1995)

[GW09] Grumbach, S., Wu, Z.: On the distributed evaluation of MSO on graphs (2009) (manuscript)
[Imm89] Immerman, N.: Expressibility and parallel complexity. SIAM J. Comput. 18(3), 625–638 (1989)
[KL04] Keisler, H.J., Lotfallah, W.B.: Shrinking games and local formulas. Ann. Pure Appl. Logic 128(1-3), 215–225 (2004)
[KMW04] Kuhn, F., Moscibroda, T., Wattenhofer, R.: What cannot be computed locally. In: ACM PODC (2004)
[KMW06] Kuhn, F., Moscibroda, T., Wattenhofer, R.: The price of being nearsighted. In: Seventeenth ACM-SIAM SODA (2006)
[Lib97] Libkin, L.: On the forms of locality over finite models. In: LICS, pp. 204–215 (1997)
[Lin92] Linial, N.: Locality in distributed graph algorithms. SIAM J. Comput. 21(1), 193–201 (1992)
[NS95] Naor, M., Stockmeyer, L.J.: What can be computed locally? SIAM J. Comput. 24(6), 1259–1277 (1995)
[Ott96] Otto, M.: The expressive power of fixed-point logic with counting. J. Symb. Log. 61(1), 147–176 (1996)
[Pel00] Peleg, D.: Distributed computing: a locality-sensitive approach. Society for Industrial and Applied Mathematics, Philadelphia (2000)
[See95] Seese, D.: Linear time computable problems and logical descriptions. Electr. Notes Theor. Comput. Sci. 2 (1995)
[TW68] Thatcher, J.W., Wright, J.B.: Generalized finite automata theory with an application to a decision problem of second-order logic. Math. Systems Theory 2(1), 57–81 (1968)

On Module-Composed Graphs

Frank Gurski* and Egon Wanke

Heinrich-Heine-University Düsseldorf
Institute of Computer Science
D-40225 Düsseldorf, Germany
{gurski-wg09,wanke-wg09}@acs.uni-duesseldorf.de

Abstract. In this paper we introduce module-composed graphs, i.e. graphs which can be defined by a sequence of one-vertex insertions $v_1, \ldots, v_n$, such that the neighbourhood of vertex v_i, $2 \le i \le n$, forms a module of the graph defined by vertices $v_1, \ldots, v_{i-1}$.

We show that module-composed graphs are HHDS-free and thus homogeneously orderable, weakly chordal, and perfect. Every bipartite distance hereditary graph and every trivially perfect graph is module-composed. We give an $\mathcal{O}(|V| \cdot (|V| + |E|))$ time algorithm to decide whether a given graph $G = (V, E)$ is module-composed and construct a corresponding module-sequence.

For the case of bipartite graphs, we show that the set of module-composed graphs is equivalent to the well known class of distance hereditary graphs, which implies linear time algorithms for their recognition and construction of a corresponding module-sequence using BFS and Lex-BFS.

Keywords: special graph classes, homogeneous sets, HHDS-free graphs, distance hereditary graphs, bipartite graphs.

1 Introduction

In this paper we analyze special graphs $G = (V, E)$ which are defined by a bijective mapping $\varphi : V \to \{1, \ldots, |V|\}$, such that the neighbourhood of vertex $\varphi^{-1}(i)$ is characterized by special operations on the previously defined graph $G[\{\varphi^{-1}(1), \ldots, \varphi^{-1}(i-1)\}]$. For example a chordal graph is a graph, such that the neighbourhood of vertex $\varphi^{-1}(i)$ is a clique in graph $G[\{\varphi^{-1}(1), \ldots, \varphi^{-1}(i-1)\}]$. Further well known examples of such defined graph classes are trees, co-graphs, distance hereditary graphs, and k-trees, see [19] for a survey. The existence of such vertex orderings φ often has algorithmic applications, see [1,11].

We introduce the closely related new concept of module-composed graphs. We allow to insert vertex $\varphi^{-1}(i)$ into some defined graph $G[\{\varphi^{-1}(1), \ldots, \varphi^{-1}(i-1)\}]$ if the neighbourhood of vertex $\varphi^{-1}(i)$ is a module in graph $G[\{\varphi^{-1}(1), \ldots, \varphi^{-1}(i-1)\}]$.

---

* The work of the first author was supported by the German Research Association (DFG) grant RO 1202/11-1.

Our results are as follows. We classify module-composed graphs in general and for the case of bipartite graphs within the hierarchy of well known graph classes. We show that bipartite module-composed graphs are exactly bipartite distance hereditary graphs which implies a new characterization for bipartite distance hereditary graphs and linear time algorithms for the recognition and construction of a corresponding module-sequence using fundamental search strategies such as Lexicographic Breadth First Search (Lex-BFS) and Breadth First Search (BFS). For general module-composed graphs $G = (V, E)$ we give an $\mathcal{O}(|V| \cdot (|V| + |E|))$ time recognition algorithm based on a modular decomposition of the given graph.

2 Preliminaries

For some positive integer k, let $[k] = \{1, \ldots, k\}$ be the set of all positive integers between 1 and k.

Let $G = (V, E)$ be a graph. For $U \subseteq V$ we define by $G[U]$ the subgraph of G induced by the vertices of U. By $G - v$ we denote the subgraph $G[V - \{v\}]$ of G. The edge complement graph of G is denoted by co-G. Further by $2G$, we denote the disjoint union of two copies of graph G. For a set of graphs $\mathcal{F}$, we denote by $\mathcal{F}$-free graphs the set of all graphs that do not contain a graph of $\mathcal{F}$ as an induced subgraph. In Table 1 we show some special graphs to which we refer during the paper. A *hole* is a chordless cycle with at least five vertices. A *k-sun* is a chordal graph $G = (V, E)$ on $2k$ vertices for some $k \geq 3$ whose vertex set can be partitioned into $V = U \cup W$ such that $U = \{u_0, \ldots, u_{k-1}\}$ and $W = \{w_0, \ldots, w_{k-1}\}$ is an independent set. Additionally vertex u_i is adjacent to vertex w_j if and only if $i = j$ or $i = j+1 \mod k$. G is called a *sun* if it is a k-sun for some $k \geq 3$. If graph $G[U]$ is a clique, then G is called a *complete k-sun*.

For some vertex $v \in V$ we denote the *neighbourhood* of v by $N(v) = \{w \in V \mid \{v, w\} \in E\}$. Vertex set $M \subseteq V$ is called a *module (homogeneous set)* of G, if for all $(v_1, v_2) \in M^2$: $N(v_1) - M = N(v_2) - M$, i.e. v_1 and v_2 have identical

Table 1. Special graphs

neighbourhoods outside M. Module $M \subseteq V$ is called a *trivial module*, if $|M| = 0$, $|M| = 1$, or $M = V$.

3 Module-Composed Graphs

Definition 1. *A graph $G = (V, E)$ is module-composed, if there exists a linear ordering $\varphi : V \to [|V|]$, such that for every $2 \leq i \leq |V|$ the neighbourhood of vertex $\varphi^{-1}(i)$ in graph $G[\{\varphi^{-1}(1), \ldots, \varphi^{-1}(i-1)\}]$ forms a module. For some module-composed graph G ordering φ is called a* module-sequence *for G.*

In order to classify module-composed graphs within the hierarchy of well known graph classes and to prove the correctness of our recognition algorithm for module-composed graphs we start with some basic but important properties.

Lemma 1. 1. *Given a module-composed graph G, every induced subgraph of G is module-composed.*
2. *Given two module-composed graphs G_1, G_2, the disjoint union $G_1 \cup G_2$ is module-composed.*
3. *Given a module-composed graph G, the addition of a dominating vertex (vertex of maximum degree) leads a module-composed graph.*
4. *Given a module-composed graph G, the addition of a pendant vertex (vertex of degree one) leads a module-composed graph.*
5. *Graph $G = (V, E)$ is a module-composed graph, if and only if there exists at least one $v \in V$ such that $N(v)$ is a module in graph $G - v$ and for every such vertex v graph $G - v$ is a module-composed graph.*
6. *Every module-composed graph does not contain one of the following graphs as an induced subgraph (see Table 1): C_n, $n \geq 5$ (i.e. holes), co-C_n, $n \geq 5$ (i.e. anti-holes), house, domino, co-$(K_{3,3} - e)$, 3-sun, co-$2C_4$.*

Proof. 1. Let $G = (V, E)$ be a module-composed graph. If $M \subseteq V$ is a module of graph G then for every $u \in M$, $M - \{u\}$ is a module of graph $G - u$. Thus we can remove an arbitrary subset $V' \subseteq V$ and obtain a module-composed graph $G[V - V']$.
2. Given two module-composed graphs $G_1 = (V_1, E_1)$, $G_2 = (V_2, E_2)$ and two module-sequences φ_1, φ_2 for G_1 and G_2, respectively. It is easy to verify that sequence

$$\varphi(v) = \begin{cases} \varphi_1(v) & \text{if } v \in V_1 \\ \varphi_2(v) + |V_1| & \text{if } v \in V_2 \end{cases}$$

is a module-sequence for the disjoint union of these two graphs.
3. Given a module-composed graph $G = (V, E)$, a module-sequence φ for G and some vertex $v \notin V$, we can extend φ by $\varphi(v) = |V| + 1$ and obtain a module-sequence for G with dominating vertex v.
4. Similar to (3.)
5. Let v be the last vertex in a module-sequence for G, then by definition $N(v)$ is a module in graph $G - v$. By (1.) for every $v \in V$ induced subgraph $G - v$ is a module-composed graph.

Since $G - v$ is a module-composed graph there is some module-sequence for graph $G - v$ and since $N(v)$ is a module in graph $G - v$, we can extend this sequence by v for a module-sequence for G.

6. It is easy to verify that none of the given graphs G contains a vertex v such that $N(v)$ is a module in graph $G - v$. Thus by part (5.) the given graphs are not module-composed. □

The example of graph co-$2C_4$ shows that not every co-graph is module-composed. But the subclass of trivially perfect graphs is module-composed.

Lemma 2. *Trivially perfect graphs are module-composed.*

Proof. Every trivially perfect graph can be defined by an expression X using the disjoint union of two trivially perfect graphs and the addition of a dominating vertex. This implies by Lemma 1 (2.) and (3.) that we can construct a module-sequence for graph defined by X. □

Next we conclude results on super classes of module-composed graphs.

Lemma 3. *Module-composed graphs are {house,hole,domino,sun}-free (HHDS-free).*

Proof. By Lemma 1 (6.) the house, every hole, the domino, and the complete 3-sun are not module-composed. By a result shown in [10] each sun contains a complete sun as induced subgraph, which is obviously not module-composed. By Lemma 1 (1.) the result follows. □

Thus we conclude that module-composed graphs are perfect, homogeneously orderable, and weakly chordal. The example C_4 shows that module-composed graphs are not chordal in general.

4 Algorithms for Module-Composed Graphs

4.1 How to Find Module-Sequences

Next we give a polynomial time algorithm to recognize module-composed graphs. Within our algorithm we will use a modular decomposition of the input graph.

A graph $G = (V, E)$ is called *prime* if every module M of G is trivial, i.e. if $|M| = 0$, $|M| = 1$, or $M = V$. The smallest non-trivial prime graph is the P_4. A module M is *maximal* if there is no non-trivial module N such that $M \subseteq N$. A module is called *strong* if it does not overlap with any other module.

While the set of all modules of a graph G can be exponentially large, the set of strong modules is linear in the number of vertices. The inclusion order of the set of all strong modules defines a tree-structure which is denoted as *modular decomposition* T_G, see [17]. The root of T_G represents the graph G and the leaves of T_G correspond to the vertices of G. Every inner node, i.e. non-leaf node, w of T_G corresponds to an induced subgraph of G consisting of the leaves of T_G in subtree with root w, which is called the *representative graph* of w and is denoted by $G(w) = (V_{G(w)}, E_{G(w)})$. Vertex set $V_{G(w)}$ is a strong module of G. For some inner node v of T_G, the *quotient graph* $G[v]$ is obtained by substituting in $G(v)$ every strong module, represented by some child of v in T_G, by a single vertex. For some inner node v of T_G, quotient graph $G[v]$ is either a complete graph (v is denoted as *join node*), the edge complement of a complete graph (v is denoted as *co-join node*), or a prime graph (v is denoted as *prime node*).

Theorem 1. *Given a graph $G = (V, E)$, one can decide in time $\mathcal{O}(|V| \cdot (|V| + |E|))$ whether G is module-composed and in the case of a positive answer construct a module-sequence for G.*

Proof. We show that graph $G = (V, E)$ is module-composed, if and only if the algorithm of Table 2 returns a module-sequence for G.

Table 2. Recognition algorithm for module-composed graphs

Input: Graph $G = (V, E)$
Output: Module-sequence $\varphi : V \to [|V|]$ or the answer NO

(1) mod-com(G)
(2) $i = |V|$;
(3)   if (G disconnected){
(4)   for every connected component H of G: mod-com(H);
(5)   combine the obtained sequences to one sequence (see Lemma 1 (2.)) }
(6)   else {
(7)   construct modular decomposition T_G with root r;
(8)   if (r is join node) {
(9)   if ($\exists$ child v_l of r which is a leaf in T_G) {
(10)   for every such child v_l of r $\{\varphi(v_l) = i - -; G = G - v_l;\}$
(11)   mod-com(G);}
(12)   else if ($\exists$ child r_1 of r labeled by co-join and a child v_l of r_1 which
(13)   is a leaf in T_G) {
(14)   for every such vertex v_l $\{\varphi(v_l) = i - -; G = G - v_l;\}$
(15)   mod-com(G); }
(16)   }
(17)   else if (r is prime node) {
(18)   if ($\exists$ child v_l of r which is a leaf in T_G and corresponds to a vertex
(19)   of degree 1 in quotient graph $G[r]$) {
(20)   for every such child v_l of r $\{\varphi(v_l) = i - -; G = G - v_l;\}$
(21)   mod-com(G);}
(22)   else if ($\exists$ child r_1 of r labeled by co-join and corresponds to a vertex
(23)   of degree 1 in quotient graph $G[r]$ and a child v_l of r_1 which is a
(24)   leaf in T_G) {
(25)   for every such vertex v_l $\{\varphi(v_l) = i - -; G = G - v_l;\}$
(26)   mod-com(G); }
(27)   }
(28)   else
(29)   return NO;
(30)   }

Let $G = (V, E)$ be some module-composed graph and T_G be a modular decomposition for G. We can assume that G is connected, otherwise we consider every connected component of G, which is module-composed by Lemma 1 (1.). By Lemma 1 (5.) graph G contains at least one vertex v such that $N(v)$ is a module in $G - v$. It remains to show that our algorithm finds such a vertex within the set of leaves of tree T_G. Since G is connected the root r of T_G is a join node or a prime node.

If r is a join node then every son of r which is a leaf of T_G is a possible vertex v since its neighbourhood is $G - v$ and thus a (trivial) module in $G - v$. Further if there is a child r_1 of r labeled by co-join, then every child v_l of r_1 which is a leaf of T_G is a possible vertex v since its neighbourhood in $G - v$ is a module.

If r is a prime node then every son of r which is a leaf of T_G and corresponds to a vertex of degree 1 in quotient graph $G[r]$ is a possible vertex v since its neighbourhood in $G - v$ is a single vertex and thus a (trivial) module in $G - v$. Further if there is a child r_1 of r labeled by co-join and corresponds to a vertex of degree 1 in quotient graph $G[r]$, then every child v_l of r_1 which is a leaf of T_G is a possible vertex v since its neighbourhood in $G - v$ is a single vertex and thus a (trivial) module in $G - v$.

A case distinction shows that no further situations for vertex v are possible which guarantees that our algorithm finds some vertex v such that $N(v)$ is a module in $G - v$. Since by Lemma 1 (5.) graph $G - v$ remains module-composed the same argumentation holds in every iteration until our algorithm returns a vertex ordering φ for the input graph. Ordering φ is a module-sequence for G, since we either remove vertices v whose neighbourhood is a module in $G - v$ or combine module-sequences by Lemma 1 (2.) for connected components within our algorithm.

The reverse direction holds by definition.

The construction of the modular decomposition T_G in Line (7) of our algorithm in Table 2 can be realized in time $\mathcal{O}(|V| + |E|)$ by [8,16] which implies that the total running time of the given algorithm is $\mathcal{O}(|V| \cdot (|V| + |E|))$. □

It remains open whether there exists a linear time algorithm to recognize module-composed graphs. Within our given algorithm we compute in every iteration a new modular decomposition. Thus if we could find for some given modular decomposition for some graph G, a modular decomposition for graph $G - v$ in time less time than $\mathcal{O}(|V|+|E|)$, e.g. in time $\mathcal{O}(|V|)$, we could reduce the running time of our algorithm.

4.2 Easy Problems on Module-Composed Graphs

Since module-composed graphs are HHD-free, we conclude by the results shown in [14] the following theorem.

Theorem 2. *For every module-composed graph which is given together with a module-sequence the size of a largest independent set, the size of a largest clique,* the chromatic number, and the minimum number of cliques covering the graph can be computed in linear time.

A very usual approach when looking for solvable problems on special graph classes is to consider their tree-width [3] or clique-width [7]. In the case of module-composed graphs both parameters are unbounded which is easy to show.

Remark 1. First, since every complete graph is module-composed, the set of all module-composed graphs can not have bounded tree-width. Further, every graph which can be constructed from a single vertex by a sequence of one vertex extentions by a dominating vertex or a pendant vertex is module-composed by Lemma 1 (3.) and (4.). But the set of all such defined graphs has unbounded clique-width [19] and thus the set of all module-composed graphs can not have bounded clique-width.

5 Independent Module-Composed Graphs

Next we want to analyze module-composed graphs for a restricted case.

Definition 2. *A graph $G = (V, E)$ is* independent module-composed, *if there exists a linear ordering $\varphi : V \to [|V|]$, such that for every $2 \leq i \leq |V|$ the neighbourhood of vertex $\varphi^{-1}(i)$ in graph $G[\{\varphi^{-1}(1), \ldots, \varphi^{-1}(i-1)\}]$ forms a module which is an independent set. For some independent module-composed graph G ordering φ is called an* independent module-sequence *for G.*

5.1 Characterizations for Independent Module-Composed Graphs

It is easy to see that independent module-composed graphs do not contain any of the graphs of Table 1 as induced subgraph.

Lemma 4. *Independent module-composed graphs are {house,hole,domino,gem}-free (HHDG-free).*

HHDG-free are also known as distance hereditary graphs [13,2], which can be defined by a so-called *pruning sequence*. A pruning sequence for a graph $G = (V, E)$ with n vertices is a sequence of pairs $(v_1, l_1), \ldots, (v_n, l_n)$ where l_i is either a single vertex graph, denoted by •, or from $\{v_1, \ldots, v_{i-1}\} \times \{\text{leaf}, \text{true-twin}, \text{false-twin}\}$. A graph is defined by a pruning sequence as follows. Let G_i, $1 \leq i \leq n$, be the graph defined by pruning sequence $(v_1, l_1), \ldots, (v_i, l_i)$. Then graph G_i is obtained from G_{i-1} by inserting vertex v_i and some edges defined as follows. If $l_i = \bullet$ then no edges are inserted. If $l_i = (v_j, \text{leaf})$ then vertex v_i will be connected with vertex v_j. If $l_i = (v_j, \text{false-twin})$ then vertex v_i will be connected with all neighbors of vertex v_j. If $l_i = (v_j, \text{true-twin})$ then vertex v_i will be connected with vertex v_j and all neighbors of vertex v_j.

Theorem 3 ([13,2]). *Let G be some graph. The following conditions are equivalent.*

1. G is distance hereditary.
2. G is HHDG-free.
3. G can be generated by a pruning sequence.

Our next theorem shows that for the case of bipartite graphs, i.e. odd-cycle-free graphs, the notion module-composed even is equivalent to the notion of distance hereditary.

Theorem 4. *Let G be some graph. The following conditions are equivalent.*

1. G is bipartite module-composed.
2. G is independent module-composed.
3. G is bipartite distance hereditary.
4. G is domino, hole, and odd-cycle-free.
5. G can be generated by a pruning sequence without true twins.

Proof. (3.) $\Leftrightarrow$ (4.) $\Leftrightarrow$ (5.) Well known results from [2].
(1.) $\Rightarrow$ (2.) In the case of bipartite graphs any neighbourhood is an independent set and thus every module-sequence even is an independent module-sequence.
(2.) $\Rightarrow$ (4.) By Lemma 4 we know that independent module-composed are domino and hole-free, and since C_3 is not independent module-composed these graphs are even odd-cycle-free.
(5.) $\Rightarrow$ (1.) By our first equivalence we know that G is bipartite. So we have to show that G is module-composed. Let G be a graph defined by a pruning sequence $(v_1, l_1), \ldots, (v_n, l_n)$ without true twins. We now define a module-sequence $\varphi : V \to [|V|]$ for G. We start with $\varphi(v_1) = 1$. Assume we have defined a module-sequence $\varphi : \{v_1, \ldots, v_i\} \to [i]$ for the first i vertices. A module-sequence for the first $i+1$ vertices is obtained as follows. We distinguish the following three cases.
 - If $l_{i+1} = \bullet$, then φ is extended by $\varphi(v_{i+1}) = i+1$.
 - If $l_{i+1} = (v_j, \text{leaf})$, then φ is extended by $\varphi(v_{i+1}) = i+1$.
 - If $l_{i+1} = (v_j, \text{false-twin})$, then let $x = \varphi^{-1}(v_j)$. For every vertex $v \in \{v_1, \ldots, v_i\}$ with $\varphi(v) > x$ we increase $\varphi(v)$ by one and define $\varphi(v_{i+1}) = x+1$. That is vertex v_{i+1} ins inserted in our module-sequence immediately after its false twin v_j.

Notice that because of the movement of the false twins it may happen that some leaf vertices of G_i are no longer of degree one in graph G_{i+1}. But since in G_{i+1} the neighbourhood of such leaf vertices of G_i is a set of twin vertices of G_{i+1}, our transformation leads a feasible module-sequence. □

In general, the class of module-composed graphs is incomparable with the class of distance hereditary graphs, since there are module-composed graphs which are not distance hereditary, e.g. the gem, and there are distance hereditary graph which are not module-composed, e.g. the co-$(K_{3,3} - e)$.

By Theorem 4 module-composed graphs can be regarded as a generalization of bipartite distance hereditary graphs. For a generalization of arbitrary distance hereditary graphs by homogeneously orderable graphs we refer to the paper [4].

5.2 How to Find Independent Module-Sequences

The problem to decide whether a given graph is bipartite distance hereditary and to construct a corresponding pruning sequence can be done in linear time by the well known characterization for bipartite graphs as 2-colorable graphs and existing linear time recognition algorithms for distance hereditary graphs shown in [13,2,15]. By Theorem 4, this immediately implies a linear time algorithm for recognizing independent module-composed graphs. A corresponding independent module-sequence can be constructed in linear time from a pruning sequence as shown in the proof of Theorem 4.

Next we discuss how to use well known fundamental search strategies Lexicographic Breadth First Search (Lex-BFS) and Breadth First Search (BFS) for recognizing independent module-composed graphs, see [5] for a survey on Lex-BFS.

First we show how to produce an independent module-sequence for some given independent module-composed graph.

Theorem 5. *Given an independent module-composed graph $G = (V, E)$, every reverse Lex-BFS ordering constructs in time $\mathcal{O}(|V|+|E|)$ an independent module-sequence for G.*

Proof. By Lemma 3 we know that independent module-composed graphs are HHDS-free, which allows us to apply the results on Lex-BFS orderings shown for HHD-free graphs shown in [14]. If $(\{v_1, v_2, v_3, v_4\}, \{\{v_1, v_2\}, \{v_2, v_3\}, \{v_3, v_4\}\})$ induces a P_4 in some graph $G = (V, E)$ we denote v_2 and v_3 as *midpoints* of the P_4. A vertex v in graph G is *semi-simplicial*, if v is no midpoint of any induced P_4 in G. A vertex ordering $\varphi : V \to [|V|]$ is denoted as *semi perfect elimination ordering* if for every $1 \leq i \leq |V|$ vertex $\varphi^{-1}(i)$ is semi-simplicial in graph $G[\{\varphi^{-1}(i), \ldots, \varphi^{-1}(n)\}]$. That is, for every semi perfect elimination ordering φ of some graph G, we know that every independent set $N(\varphi^{-1}(i))$ in $G[\{\varphi^{-1}(i+1), \ldots, \varphi^{-1}(n)\}]$ is even a module in $G[\{\varphi^{-1}(i+1), \ldots, \varphi^{-1}(n)\}]$.

In [14] it is shown that for every HHD-free graph $G = (V, E)$ every Lex-BFS ordering $\varphi : V \to [|V|]$ is a semi perfect elimination ordering. Since for the case of bipartite graphs every neigbourhood of some vertex is an independent set, φ obviously is even a reverse independent module-sequence. Thus for every independent module-composed graph G every reverse Lex-BFS ordering is an independent module-sequence for G, which can be found in time $\mathcal{O}(|V| + |E|)$ [14]. □

Next we show how to decide whether a given graph is independent module-composed and if so, how to produce an independent module-sequence.

Theorem 6. *Given a graph $G = (V, E)$, one can decide using BFS in time $\mathcal{O}(|V| + |E|)$ whether G is independent module-composed and in the case of a positive answer construct an independent module-sequence.*

Proof. To decide whether a given graph $G = (V, E)$ is bipartite distance hereditary can be done by Corollary 5 shown in [2] using the fundamental search strategy of BFS which produces a classification of the vertices into levels, with respect to a start vertex u. Level i is the set of vertices with distance i to vertex u and is denoted by $N_i(u)$.

Theorem 7 (Corollary 5 of [2]). *Let G be a connected graph and let u be a vertex of G. Then G is bipartite distance hereditary if and only if all levels $N_k(u)$ are edgeless, and for every vertices v,w in $N_k(u)$ and neighbours x and y of v in $N_{k-1}(u)$, we have $N(x) \cap N_{k-2}(u) = N(y) \cap N_{k-2}(u)$, and further $N(v) \cap N_{k-1}(u)$ and $N(w) \cap N_{k-1}(u)$ are either disjoint or one is contained in the other.*

A BFS starting at a vertex u can compute the level sets $N_k(u)$ in time $\mathcal{O}(|V|+|E|)$ and using these levels, the conditions of Corollary 5 of [2] can be verified in the same time.

A BFS numbering $\varphi : V \to [|V|]$ of the vertices with respect to some vertex u can be used to obtain an independent module-sequence φ_1 as follows. We start with $\varphi_1(v) = \varphi(v), \forall v \in V$. For the first $|N_0(u)| + |N_1(u)|$ vertices we obviously can choose $\varphi_1(v) = \varphi(v)$. For the vertices of $w \in N_k(u)$, $k \geq 2$, we know that their neighbours in set $N_{k-1}(u)$ are modules which can be ordered by a series of inclusions $N^1 \subseteq N^2 \subseteq \ldots \subseteq N^j$. We rearrange the order of the vertices in $N_k(u)$ with respect to φ_1 such that for every such series of inclusions $\varphi_1(w_1) < \varphi_1(w_2)$ if and only if $N_{k-1}(u) \cap N(w_1) \supseteq N_{k-1}(u) \cap N(w_2)$. This leads an independent module-sequence for graph G if G is bipartite distance hereditary. □

5.3 Easy Problems on Independent Module-Composed Graphs

In contrast to general module-composed graphs, independent module-composed graphs have bounded clique-width, since distance hereditary graphs have

Table 3. Inclusions of special graph classes

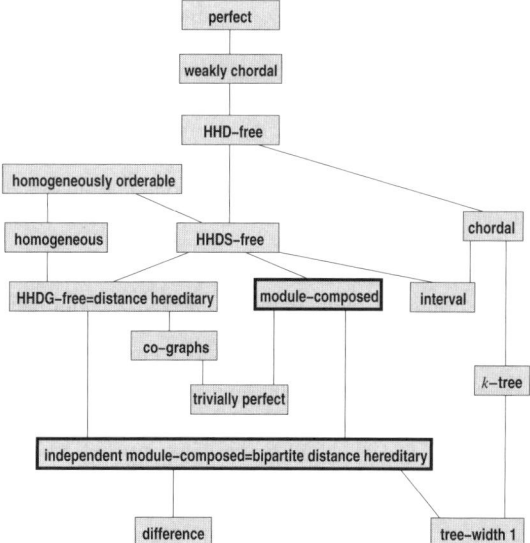

clique-width at most 3 [12]. This implies that a lot of hard problems can be solved in polynomial time on independent module-composed graphs [6,9]. Further algorithms for hard problems on bipartite distance hereditary graphs can be found in [20,18].

6 Graph class Inclusions

In Table 3 we summarize the relation of module-composed graphs and related graph classes.

7 Conclusions

In this paper we introduced the new concept of module-composed graphs. We have classified module-composed graphs in general and for the case of bipartite graphs within the hierarchy of well known graph classes. Independent module-composed graphs turn out to be a well known graph class and can be recognized in linear time. For general module-composed graphs $G = (V, E)$ we showed an $\mathcal{O}(|V| \cdot (|V| + |E|))$ time recognition algorithm.

It remains open whether there are equivalent characterizations of module-composed graphs and a linear time recognition algorithm. In any case BFS and Lex-BFS approaches failed to improve our given algorithm.

References

1. Arnborg, S., Proskurowski, A.: Linear time algorithms for NP-hard problems restricted to partial k-trees. Discrete Applied Mathematics 23, 11–24 (1989)
2. Bandelt, H.-J., Mulder, H.M.: Distance-hereditary graphs. Journal of Combinatorial Theory, Series B 41, 182–208 (1986)
3. Bodlaender, H.L.: A partial k-arboretum of graphs with bounded treewidth. Theoretical Computer Science 209, 1–45 (1998)
4. Brandstädt, A., Dragan, F.F., Nicolai, F.: Homogeneously orderable graphs. Theoretical Computer Science 172, 209–232 (1997)
5. Corneil, D.G.: Lexicographic breadth first search - a survey. In: Hromkovič, J., Nagl, M., Westfechtel, B. (eds.) WG 2004. LNCS, vol. 3353, pp. 1–19. Springer, Heidelberg (2004)
6. Courcelle, B., Makowsky, J.A., Rotics, U.: Linear time solvable optimization problems on graphs of bounded clique-width. Theory of Computing Systems 33(2), 125–150 (2000)
7. Courcelle, B., Olariu, S.: Upper bounds to the clique width of graphs. Discrete Applied Mathematics 101, 77–114 (2000)
8. Cournier, A., Habib, M.: A new linear time algorithm for modular decomposition. In: Tison, S. (ed.) CAAP 1994. LNCS, vol. 787, pp. 68–84. Springer, Heidelberg (1994)
9. Espelage, W., Gurski, F., Wanke, E.: How to solve NP-hard graph problems on clique-width bounded graphs in polynomial time. In: Brandstädt, A., Le, V.B. (eds.) WG 2001. LNCS, vol. 2204, pp. 117–128. Springer, Heidelberg (2001)

10. Farber, M.: Characterizations of strongly chordal graphs. Discrete Mathematics 43, 173–189 (1983)
11. Gavril, F.: Algorithms for minimum coloring, maximum clique, minimum covering by cliques, and maximum independent set of a chordal graph. SIAM Journal on Computing 1(2), 180–187 (1972)
12. Golumbic, M.C., Rotics, U.: On the clique-width of some perfect graph classes. International Journal of Foundations of Computer Science 11(3), 423–443 (2000)
13. Hammer, P.L., Maffray, F.: Completely separable graphs. Discrete Applied Mathematics 27, 85–99 (1990)
14. Jamison, B., Olariu, S.: On the semi-perfect elimination. Advances in applied mathematics 9, 364–376 (1988)
15. Lanlignel, J.-M., Raynaud, O., Thierry, É.: Pruning graphs with digital search trees. Application to distance hereditary graphs. In: Reichel, H., Tison, S. (eds.) STACS 2000. LNCS, vol. 1770, pp. 529–541. Springer, Heidelberg (2000)
16. McConnell, R.M., Spinrad, J.: Modular decomposition and transitive orientation. Discrete Mathematics 201(1-3), 189–241 (1999)
17. Möhring, R.H., Radermacher, F.J.: Substitution decomposition for discrete structures and connections with combinatorial optimization. Annals of Discrete Mathematics 19, 257–365 (1984)
18. Müller, H., Nicolai, F.: Polynomial time algorithms for hamiltonian problems on bipartite distance-hereditary graphs. Information Processing Letters 46(5), 225–230 (1993)
19. Rao, M.: Clique-width of graphs defined by one-vertex extensions. Discrete Mathematics 308(24), 6157–6165 (2008)
20. Yeh, H.-G., Chang, G.J.: The path-partition problem in bipartite distance-hereditary graphs. Taiwanese Journal of Mathematics 2(3), 353–360 (1998)

An Even Simpler Linear-Time Algorithm for Verifying Minimum Spanning Trees

Torben Hagerup

Institut für Informatik, Universität Augsburg, 86135 Augsburg, Germany
hagerup@informatik.uni-augsburg.de

Abstract. A new linear-time algorithm is presented for the *tree-path-maxima* problem of, given a tree T with real edge weights and a list of pairs of distinct nodes in T, computing for each pair (u, v) on the list a maximum-weight edge on the path in T between u and v. Linear-time algorithms for the tree-path-maxima problem were known previously, but the new algorithm may be considered significantly simpler than the earlier solutions. A linear-time algorithm for the tree-path-maxima problem implies a linear-time algorithm for the *MST-verification* problem of determining whether a given spanning tree of a given undirected graph G with real edge weights is a minimum-weight spanning tree of G.

1 Introduction

A spanning tree of an undirected graph G with real edge weights is a *minimum spanning tree* (*MST*) if its weight, i.e., the total weight of its edges, is minimal among all weights of spanning trees of G. The *MST-construction* problem of computing an MST of a given connected undirected graph with real edge weights is a fundamental problem with numerous applications [1,10].

In a comparison-based model that allows constant-time binary comparisons of edge weights but no other operations on weights, the best currently known upper bound on the complexity of the MST-construction problem was established by Chazelle [5], who gave an algorithm that runs in $O(m\alpha(m, n))$ time on input graphs with n vertices and m edges, where α is an inverse of the Ackermann function. Pettie and Ramachandran [17] described a provably optimal deterministic algorithm based on table-lookup for small subproblems, but nothing better than $O(m\alpha(m, n))$ is known about its running time. A randomized algorithm that runs in $O(n + m)$ time with high probability was found by Karger, Klein and Tarjan [11]; another such algorithm was subsequently indicated by Pettie and Ramachandran [18]. Fredman and Willard [9] devised a deterministic algorithm with a running time of $O(n+m)$ for the word-RAM model, which assumes the weights to be integers on which a range of operations can be executed in constant time. Extensive surveys of material related to the MST problem were compiled by Graham and Hell [10] and by Mareš [15].

A number of researchers have considered the following problem related to MST construction.

MST-verification problem: Given an undirected graph G with real edge weights and a spanning tree T of G, decide whether T is an MST of G.

The MST-verification problem for a graph with n vertices and m edges obviously reduces in $O(n + m)$ time to the MST-construction problem for the same graph, but it might be easier (except, of course, in models of computation for which even the MST-construction problem is known to be solvable in linear time). Komlós [14] described how to solve the MST-verification problem for a graph with n vertices and m edges with $O(n + m)$ binary comparisons of edge weights, but did not indicate an algorithm with an overall running time of $O(n + m)$. Dixon, Rauch and Tarjan [8] gave the first linear-time algorithm for MST verification. It is based on decomposing an input tree into small microtrees, many of which share the same tree shape, computing for each possible microtree shape the decision tree implied by Komlós' result, using the decision trees to process the actual microtrees, and solving the remaining slightly smaller overall problem using an efficient but not linear-time algorithm of Tarjan [19]. Subsequently King [13] proposed an implementation of Komlós' algorithm that reduces the overhead spent outside of weight comparisons to $O(n + m)$. Abstractly speaking, a linear time bound is achieved by packing several elements of a set into a single computer word and operating on all of them simultaneously at unit cost. Edge weights are only subjected to binary comparisons, however, as required by the comparison-based model. A linear-time algorithm that works on a pointer machine was proposed by Buchsbaum et al. [3].

Most of the algorithms for MST verification discussed above actually solve the following problem.

Tree-path-maxima problem: Given a tree T with real edge weights and a list of pairs of distinct nodes in T, determine for each pair (u, v) on the list a heaviest edge on the path in T between u and v.

A spanning tree T of an undirected graph G with real edge weights is an MST of G if and only if the weight of every nontree edge $\{u, v\}$ in G is at least as large as the maximum weight of an edge on the path in T between u and v. (To see this, observe that T can be constructed by Kruskal's algorithm if the condition is satisfied and that a cheaper spanning tree can be obtained via an edge swap if not.) Therefore the MST-verification problem for a graph with n vertices and m edges reduces, within $O(n + m)$ time, to the tree-path-maxima problem for an n-node tree and a list of at most m node pairs. The tree-path-maxima problem has additional applications, however, notably to the randomized construction of minimum spanning trees [7,11,12]. It can be viewed in a more abstract setting, where the computation of the maximum is replaced by a general semigroup operation. In this setting, the complexity of the problem is known to be $\Theta(m\alpha(m, n))$ [4,6,20]. In an online setting in which the query corresponding to each pair (u, v) must be answered before the next pair becomes known, the same bound applies even to the original tree-path-maxima problem [16].

King [13] demonstrated that the tree-path-maxima problem reduces, again within linear time, to the special case of itself characterized below. A *full branching tree* is a rooted tree in which all leaves are on the same level and all inner nodes have at least 2 children.

Special tree-path-maxima problem: Given a full branching tree T with real edge weights and a list of pairs (u,v) of nodes in T such that u is a proper ancestor of v, determine for each pair (u,v) on the list a heaviest edge on the path in T between u and v.

King's reduction replaces the original input tree T by a tree with at most twice as many nodes that reflects the structure of an execution of Borůvka's MST-construction algorithm [2], applied to T, and it uses a data structure for determining the lowest common ancestor (LCA) of two nodes to replace each original node pair (u,v) by the two pairs (z,u) and (z,v), where z is the LCA of u and v (if z coincides with u or v, the corresponding pair is omitted).

This paper describes a new linear-time algorithm in the comparison-based model for the special tree-path-maxima problem. Used with King's reduction, it yields linear-time algorithms for the general tree-path-maxima problem and the MST-verification problem. Similarly as King's algorithm, the new algorithm is an implementation of Komlós' algorithm for full branching trees, and its running-time analysis depends crucially on a lemma provided by Komlós.

The algorithm of Dixon, Rauch and Tarjan [8] appears moderately complicated and not very suited for practical use. King's algorithm [13] probably fares better on both accounts. However, it employs a nontrivial scheme for encoding edges, represents objects of several logically distinct types in single computer words, and distinguishes between big and small nodes, which causes the processing at a node to take one of at least four different forms. In contrast, the new algorithm uses no encoding of edges, represents only one very simple type of object with a richer logical structure than an integer in single computer words, and processes all nodes in exactly the same uniform manner. As an additional demonstration of the simplicity of the new algorithm, Section 5 offers a complete working implementation of it in the programming language D (http://www.digitalmars.com/d). The code fits on three pages. It seems unlikely that any of the earlier algorithms has a comparably succinct description.

2 The ↓ Operation on Sets

The new algorithm is most conveniently formulated and proved correct with the aid of a set operation ↓, written infix and defined as follows: For all sets A and B of integers,

$$A \downarrow B = \{b \in B \mid \exists a \in A : a < b \text{ and there is no } b' \in B \text{ with } a < b' < b\}.$$

Informally, $A \downarrow B$ can be described as follows: Each $a \in A$ picks the next larger $b \in B$, if any; $A \downarrow B$ is the set of bs picked. Several as may pick the same b, which is nonetheless included only once. Some as may pick no b at all. A number of elementary properties of ↓ follow more or less directly from the definition. Let us agree to take $\inf \emptyset = \infty$ and $\sup \emptyset = -\infty$ and employ the natural ordering conventions for the set of integers extended by ∞ and $-\infty$.

Lemma 1. *For all finite sets A, B and C of integers, the following relations hold:*

(a) $A \downarrow B \subseteq B$.
(b) $|A \downarrow B| \leq |A|$.
(c) $(A \cup B) \downarrow C = (A \downarrow C) \cup (B \downarrow C)$.
(d) $A \downarrow (B \cup C) \subseteq (A \downarrow B) \cup (A \downarrow C)$.
(e) *If* $A \downarrow B = \emptyset$, *then* $A \downarrow C = A \downarrow (C \setminus B)$.
(f) *If* $A \downarrow B \subseteq C \subseteq B$, *then* $A \downarrow B = A \downarrow C$.
(g) *If* $\sup(B \cap C) < \inf(B \setminus C)$, *then* $A \downarrow (B \cap C) = (A \downarrow B) \cap C$.
(h) *If* $A \subseteq B$, *then* $A \downarrow C = A \downarrow (B \downarrow C)$.

Proof. Given sets A and B of integers and a $b \in B$, let us call an $a \in A$ an (A,B)-*witness* of b if $a < b$ and there is no $b' \in B$ with $a < b' < b$. Thus $A \downarrow B$ is the set of those elements of B that have an (A,B)-witness.

(a) Trivial.

(b) No two distinct elements of B can have the same (A,B)-witness.

(c) An $(A \cup B, C)$-witness of some $c \in C$ is clearly an (A,C)-witness of c if it belongs to A and a (B,C)-witness of c if it belongs to B. Thus $(A \cup B) \downarrow C \subseteq (A \downarrow C) \cup (B \downarrow C)$. On the other hand, every (A,C)- or (B,C)-witness of c is an $(A \cup B, C)$-witness of c, so $(A \downarrow C) \cup (B \downarrow C) \subseteq (A \cup B) \downarrow C$.

(d) An $(A, B \cup C)$-witness of some $x \in B \cup C$ is an (A,B)-witness of x if $x \in B$ and an (A,C)-witness of x if $x \in C$.

(e) Assume that $A \downarrow B = \emptyset$. Then $A \downarrow (B \cap C) = \emptyset$ and, by part (d), $A \downarrow C \subseteq A \downarrow (C \setminus B)$. Conversely, let $c \in A \downarrow (C \setminus B)$ and let a be an $(A, C \setminus B)$-witness of c. By definition, there is no $c' \in C \setminus B$ with $a < c' < c$. If there were a $c' \in C \cap B$ with $a < c' < c$, the smallest element of $\{b \in B \mid a < b\}$ would belong to $A \downarrow B$, a contradiction. Therefore $c \in A \downarrow C$.

(f) Assume that $A \downarrow B \subseteq C \subseteq B$. Let $b \in A \downarrow B$ and let a be an (A,B)-witness of b. Then the relation $a < b' < b$ does not hold for any $b' \in B$. In particular, it does not hold for any $b' \in C$. Since $b \in C$, it follows that $b \in A \downarrow C$.

Conversely, let $c \in A \downarrow C$ and let a be an (A,C)-witness of c. Assume that $c \notin A \downarrow B$. Then, since $c \in C$ and hence $c \in B$, there is a $b \in B$ with $a < b < c$. If b is chosen smallest with this property, then $b \in A \downarrow B$ and $b \notin C$, a contradiction.

(g) Assume that $\sup(B \cap C) < \inf(B \setminus C)$. Let $x \in A \downarrow (B \cap C)$ and let a be an $(A, B \cap C)$-witness of x. Then there is no $x' \in B \cap C$ with $a < x' < x$. Since $x \in B \cap C$, there is no $x' \in B \setminus C$ with $x' < x$ either. Thus $x \in (A \downarrow B) \cap C$.

Conversely, let $x \in (A \downarrow B) \cap C$ and let a be an (A,B)-witness of x. Then $a < x' < x$ holds for no $x' \in B$ and, in particular, for no $x' \in B \cap C$. Since $x \in B \cap C$, $x \in A \downarrow (B \cap C)$.

(h) Assume that $A \subseteq B$. Let $c \in A \downarrow C$ and let a be an (A,C)-witness of c. Since $a \in B$, $c \in B \downarrow C$. Moreover, since $B \downarrow C \subseteq C$, there is no $c' \in B \downarrow C$ with $a < c' < c$, so $c \in A \downarrow (B \downarrow C)$.

Conversely, let $c \in A \downarrow (B \downarrow C)$ and let a be an $(A, B \downarrow C)$-witness of c, so that $a < c' < c$ holds for no $c' \in B \downarrow C$. Assume that $a < c' < c$ holds for some $c' \in C$ and choose c' smallest with this property. Then, since $a \in B$, $c' \in B \downarrow C$, a contradiction. Thus $c \in A \downarrow C$.

3 The High-Level Algorithm

Consider an input instance of the special tree-path-maxima problem consisting of an n-node tree $T = (V, E)$ and a list $(u_1, v_1), \ldots, (u_m, v_m)$ of m pairs of nodes in T such that for $i = 1, \ldots, m$, u_i is a proper ancestor of v_i. Let r and h be the root of T and its height, respectively. For all $u \in V$, let $d(u)$ be the depth of u in T and, if $u \neq r$, let $w(u)$ be the weight of the edge between u and its parent in T.

Stepping through the list $(u_1, v_1), \ldots, (u_m, v_m)$ once, the algorithm begins by storing the set $L_v = \{i \mid 1 \leq i \leq m \text{ and } v_i = v\}$ with each node $v \in V$. In traversals of T, it subsequently computes h and, for each node $u \in V$, its depth $d(u)$ and the set

$$D_u = \{d(u_i) \mid u_i \text{ is a proper ancestor of } u \text{ and } v_i \text{ is a descendant of } u\},$$

which may be thought of, informally, as the set of depths of endpoints above u of query paths that contain u. According to the recursive formula

$$D_u = \{d(u_i) \mid v_i = u\} \cup \bigcup_{v \text{ is a child of } u} (D_v \setminus \{d(u)\}),$$

the computation of D_u for all $u \in V$ can easily be carried out in a bottom-up fashion, as indicated by King [13] (who writes $LCA(u)$ for what we call D_u).

For each $v \in V$, denote by $P_v(j)$, for $j = 0, \ldots, d(v)$, the ancestor of v in T of depth j. Moreover, let

$$M_v = \{j \mid 1 \leq j \leq d(v) \text{ and } w(P_v(j)) > w(P_v(k)) \text{ for } k = j + 1, \ldots, d(v)\}.$$

Informally, M_v is the set of depths of suffix maxima on the path in T from r to v, where a suffix maximum is a nonroot node u such that $w(u)$ is strictly larger than every weight of an edge that follows u on the path. Obviously, M_v is strictly decreasing in the sense that if $i, j \in M_v$ and $i < j$, then $w(P_v(i)) > w(P_v(j))$.

It is easy to see that the correct answer to the query represented by a node pair (u, v) is $P_v(j)$, where j is the single element of the set $\{d(u)\} \downarrow M_v$. Therefore the problem at hand essentially reduces to computing M_v for all $v \in V$. Since it is not known how to do this in linear time, instead we find $S_v = D_v \downarrow M_v$ for all $v \in V$. By part (h) of Lemma 1 and the fact that $d(u) \in D_v$, $\{d(u)\} \downarrow S_v = \{d(u)\} \downarrow (D_v \downarrow M_v) = \{d(u)\} \downarrow M_v$, so that S_v serves just as well as M_v. The remainder of the section explains how to determine S_v for all $v \in V$.

Let v be a nonroot node in T and let u be its parent. Then M_v can be obtained from M_u by removing zero or more largest elements of M_u and adding the single element $d(v)$. As a consequence, $\sup(M_u \cap M_v) < \inf(M_u \setminus M_v)$. Moreover, by part (c) of Lemma 1, $(D_v \setminus D_u) \downarrow M_u \subseteq \{d(u)\} \downarrow M_u = \emptyset$. Using these relations as well as parts of Lemma 1 (indicated above the relevant relation symbols), we find

$$S_v = D_v \downarrow M_v$$
$$\stackrel{(c)}{=} ((D_v \cap D_u) \downarrow M_v) \cup ((D_v \setminus D_u) \downarrow M_v)$$
$$\stackrel{(c),(e)}{\subseteq} (D_u \downarrow M_v) \cup ((D_v \setminus D_u) \downarrow (M_v \setminus M_u))$$
$$\stackrel{(d)}{\subseteq} (D_u \downarrow (M_u \cap M_v)) \cup (D_u \downarrow (M_v \setminus M_u)) \cup ((D_v \setminus D_u) \downarrow (M_v \setminus M_u))$$
$$\stackrel{(g),(a)}{\subseteq} ((D_u \downarrow M_u) \cap M_v) \cup (M_v \setminus M_u)$$
$$= (S_u \cap M_v) \cup \{d(v)\}$$
$$\stackrel{(a)}{=} \{j \in S_u \mid w(P_v(j)) > w(v)\} \cup \{d(v)\}$$
$$= \{j \in S_u \mid j \leq \sup\{j' \in S_u \mid w(P_v(j')) > w(v)\}\} \cup \{d(v)\}.$$

Let
$$R = \{j \in S_u \mid j \leq \sup\{j' \in S_u \mid w(P_v(j')) > w(v)\}\} \quad \text{and}$$
$$R' = \{j \in S_u \mid j \leq \sup\{j' \in D_v \downarrow S_u \mid w(P_v(j')) > w(v)\}\}.$$

By what we proved above, $D_v \downarrow M_v \subseteq R \cup \{d(v)\} \subseteq M_v$. Therefore, by part (f) of Lemma 1, $D_v \downarrow M_v = D_v \downarrow (R \cup \{d(v)\})$. Moreover, we have $\sup(S_u \cap R) = \sup R < \inf(S_u \setminus R)$, so, by parts (g) and (a) of Lemma 1, $D_v \downarrow R \subseteq D_v \downarrow S_u$ and $R' \subseteq R$. Let $j \in D_v \downarrow R$. By part (a) of Lemma 1, if $j \notin R'$, we have $j > \sup\{j' \in D_v \downarrow S_u \mid w(P_v(j')) > w(v)\} \geq \sup\{j' \in D_v \downarrow R \mid w(P_v(j')) > w(v)\} = \sup(D_v \downarrow R)$, a contradiction. We may conclude that $D_v \downarrow R \subseteq R'$ and, by another application of part (f) of Lemma 1, that $D_v \downarrow R = D_v \downarrow R'$. Since $d(v) > \sup(D_v \downarrow R)$, this is easily seen to imply that $S_v = D_v \downarrow (R \cup \{d(v)\}) = D_v \downarrow (R' \cup \{d(v)\})$. The algorithm therefore proceeds to compute S_v for all $v \in V$ in a top-down traversal of T according to the recursive formula

$$S_v = \begin{cases} \emptyset & \text{if } v = r; \\ D_v \downarrow (\{j \in S_u \mid j \leq \sup\{j' \in D_v \downarrow S_u \mid w(P_v(j')) > w(v)\}\} \cup \{d(v)\}), & \\ & \text{if } v \text{ has parent } u. \end{cases}$$

During the traversal, the ancestors of the node currently visited are kept in an array sorted by depth, so that it is easy to access $P_v(j)$ for arbitrary $j \in \{0, \ldots, d(v)\}$ during the visit of v.

The value $\sup\{j' \in D_v \downarrow S_u \mid w(P_v(j')) > w(v)\}$ is found with binary search by initializing a set S to $D_v \downarrow S_u$ and proceeding as follows: If $S = \emptyset$, return $-\infty$ (or 0, which has the same effect). Otherwise, as long as $|S| > 1$, compute the median k of S, defined here to be the element in S of rank $\lfloor |S|/2 + 1 \rfloor$ (i.e., ties are broken in favor of the larger element), and replace S by $\{j \in S \mid j \geq k\}$ if $w(P_v(k)) > w(v)$ and by $\{j \in S \mid j < k\}$ otherwise. Finally, if the single element k of S satisfies $w(P_v(k)) > w(v)$, return k—it is the largest element with that property; otherwise return $-\infty$.

During the visit of a node v in the traversal, the algorithm also answers the queries whose indices are stored in L_v: As discussed above, for each $i \in L_v$, it computes $\{d(u_i)\} \downarrow S_v$ and outputs $P_v(j)$, where j is the single element of the resulting set.

Outside of the binary searches, the algorithm carries out $O(1)$ operations per node and per query. Some of the operations manipulate entire sets. However, as will be shown in the next section, each set operation required by the algorithm, including the evaluation of $A \downarrow B$, can be executed in constant time. Since $|S_v| \leq |D_v|$ for all $v \in V$ according to part (b) of Lemma 1, this yields a running time of $O(n + m + \sum_{v \in V} \log(|D_v| + 1))$. Komlós [14] has shown that when T is a full branching tree, as assumed here, then $\sum_{v \in V} \log(|D_v| + 1) = O(n + m)$. Therefore the running time of the algorithm is $O(n + m)$.

4 Realizing the Set Operations

The sets L_v, with $v \in V$, must support only the operations of initialization to the empty set, insertion of an element and iteration over the entire set. Therefore they can obviously be realized via, e.g., linked lists. Every other set manipulated by the algorithm is a set of depths in T, i.e., it is a subset of $\{0, \ldots, h\}$. Let us call such a set an *h-set*. Since T is a full branching tree, $h \leq \log_2 n$. It is therefore reasonable to assume that integers in the range $0, \ldots, 2^{h+1} - 1$ can be stored in single computer words in the usual 2's-complement representation and operated on in constant time by standard computer instructions, and we will make this assumption. This allows us to represent an h-set A via the integer $\sum_{i \in A} 2^i$, which makes the implementation of most of the relevant set operations a routine matter, as indicated in the table below. Let us number the bit positions of nonnegative integers $0, 1, \ldots$, starting at the least significant bit position.

Set expression	C equivalent	Comment
$\emptyset$	0	all bits cleared
$A \cup B$	A|B	bitwise OR
$A \setminus B$	A&~B	bitwise AND and complement
$\{j\}$	1<<j	2^j, obtained by left shift of 1
$\{i \in A \mid i < j\}$	A&(1<<j)-1	mask with bits $0, \ldots, j-1$ set
$\{i \in A \mid i \geq j\}$	A&~((1<<j)-1)	mask with bits $0, \ldots, j-1$ cleared
$A \downarrow B$	B&(~(A|B)^(A+(A|~B)))	see below

To understand the last line of the table, let A and B be h-sets, consider the (arithmetic) addition of the integers representing the sets A and $A \cup \overline{B}$, where $\overline{B}$ is the complement of B with respect to the universe $\{0, \ldots, h\}$, and assume that the addition is carried out with the usual school method, i.e., from bit position 0 to bit position h and with the use of carries. Let $i \in \{0, \ldots, h\}$. If $i \in A$, then both operands of the addition have a 1 in bit position i, which turns position i into a *carry-generate* position that sends a carry into position $i+1$ (if such a bit position exists), independently of all other bits of the operands. Similarly, if i

belongs to neither A nor B, position i is a *carry-propagate* position that passes a carry on to position $i+1$ (if it exists) if and only if it receives a carry from position $i-1$. The final case, i belongs to B but not to A, yields a *carry-absorb* position that never sends a carry to position $i+1$. In summary, each element i of A "throws" a carry that runs through positions $i+1, i+2, \ldots$ until it is "caught" (if ever) by an element $j > i$ of B (if $j \in A$, a new carry is simultaneously thrown). It is now easy to see that $A \downarrow B$ is the set of those elements of B whose corresponding bit positions receive a carry. It is possible to test whether this is the case for position i by forming the exclusive OR (C representation: ^) of the three bits in position i of the two operands of the addition and of their sum. In the light of the simplifying relation $A \oplus (A \cup \overline{B}) = \overline{A \cup B}$, where $\oplus$ denotes the symmetric set difference, this leads to the C expression given in the last line of the table.

The necessary C operations can be reduced to a core set consisting of bitwise complement, bitwise AND or OR and left shift of 1 by $j \in \{0, \ldots, h\}$ bit positions, since the remaining operations reduce to these with a constant-factor overhead via de Morgan's laws, etc. If even the core operations are not available at unit cost, they can be replaced by lookup in tables that can be precomputed in $O(n)$ time.

The only operation that still needs to be explained is the computation of the median of an h-set, for which we have to resort to the table-lookup method.

For all nonnegative integers n and k, denote by $\left\{{n \atop k}\right\}$ the set of all subsets of size k of $\{0, \ldots, n-1\}$. Moreover, for arbitrary sets of sets A and B, let $A \talloblong B = \{a \cup b \mid a \in A \text{ and } b \in B\}$. In complete analogy with a well-known recursive formula for binomial coefficients, the formula

$$\left\{{n \atop k}\right\} = \begin{cases} \emptyset, & \text{if } n < k; \\ \{\emptyset\}, & \text{if } n \geq k = 0; \\ \left(\{\{n-1\}\} \talloblong \left\{{n-1 \atop k-1}\right\}\right) \cup \left\{{n-1 \atop k}\right\}, & \text{if } n \geq k \geq 1 \end{cases}$$

holds for all nonnegative integers n and k. It is easily translated into a recursive function *subsets* that, for $n, k \in \{0, \ldots, h\}$, can compute $\left\{{n \atop k}\right\}$ (as a sequence of integers, each of which represents an h-set) in $O(|\left\{{n \atop k}\right\}|)$ time.

Given a set A of sets of integers and an integer k, let $A + k$ denote the set of those sets obtained from the sets in A by adding k to each of their elements. Then, for $s \in \{0, \ldots, h\}$, the set $median^{-1}(s)$ of those h-sets whose median is s is

$$\bigcup_{k=0}^{s} \left[\left(\left\{{h-s \atop k}\right\} + s + 1\right) \talloblong \{\{s\}\} \talloblong \left(\left\{{s \atop k}\right\} \cup \left\{{s \atop k+1}\right\}\right)\right],$$

since an h-set belongs to $median^{-1}(s)$ if and only if it consists of s and, for some $k \in \{0, \ldots, s\}$, k elements larger than s and either k or $k+1$ elements smaller than s. With the aid of the function *subsets* discussed above, this formula is easily translated to a function that, for each $s \in \{0, \ldots, h\}$, can compute $median^{-1}(s)$ in

186   T. Hagerup

$O(|median^{-1}(s)|)$ time. And with this function, finally, it is a trivial matter to fill
in a table of size 2^{h+1} that maps each h-set to its median in $O(2^h) = O(n)$ time.

5   An Implementation in D

```
int[] tree_path_maxima(int root,int[] child,int[] sibling,
   double[] weight,int[] upper,int[] lower) {
// Returns an array of answers to the queries of an instance
// (T,((u_1,v_1),...,(u_m,v_m))) of the special tree-path-maxima
// problem. For some natural number n, T is a weighted full
// branching tree on the node set {0,...,n-1}, represented
// via the variables root, which indicates the root of T,
// child[n], which maps each node u of T to its leftmost child
// (in some arbitrary left-to-right order of T), or to -1 if u is
// a leaf, sibling[n], which maps each node v of T to its right
// sibling, or to -1 if v has no right sibling, and weight[n],
// which maps each node v of T, except for the root of T, to the
// weight of the edge between v and its parent. For i=1,...,m,
// u_i and v_i are the i'th element of upper[m] and lower[m],
// respectively, and the i'th element of the array returned is
// a vertex v of T such that the edge between v and its parent
// in T is a heaviest edge on the path in T between u_i and v_i.
// The running time is O(n+m).

 int height=0,n=child.length,m=upper.length;
 int[] depth=new int[n],D=new int[n],L=new int[n],
   Lnext=new int[m],answer=new int[m],median,P;

 void init(int u,int d) { // d = depth of u
  depth[u]=d;
  if (d>height) height=d; // height of T = maximum depth
  for (int i=L[u];i>=0;i=Lnext[i]) D[u]|=1<<depth[upper[i]];
  for (int v=child[u];v>=0;v=sibling[v]) {
   init(v,d+1);
   D[u]|=D[v]&~(1<<d);
  }
 }

 int[] median_table(int h) {
 // Returns a table of size 2^(h+1) whose entry in position i, for
 // i=0,...,2^(h-1)-1, is the median of the set represented by i.
  int[] T=new int[(1<<h)+1],median=new int[1<<h+1];
```

```
int subsets(int n,int k,int p) {
// Stores the subsets of size k of {0,...,n-1} in T,
// starting in position p, and returns p plus their number.
 if (n<k) return p;
 if (k==0) { T[p]=0; return p+1; }
 int q=subsets(n-1,k-1,p);
 for (int i=p;i<q;i++) T[i]|=1<<(n-1);
 return subsets(n-1,k,q);
}

 for (int s=0;s<=h;s++)
  for (int k=0;k<=s;k++) {
   int p=subsets(h-s,k,0);
   int q=subsets(s,k,p);
   q=subsets(s,k+1,q);
   for (int i=0;i<p;i++) {
    int b=(1<<s+1)*T[i]+(1<<s); // fixed high bits
    for (int j=p;j<q;j++) median[b+T[j]]=s; // variable low bits
   }
  }
 return median;
} // end median_table

int down(int A,int B) {
// Returns A "downarrow" B
 return B&(~(A|B)^(A+(A|~B)));
}

void visit(int v,int S) { // S = S of parent

 int binary_search(double w,int S) {
 // Returns max({j in S | weight[P[j]]>w} union {0})
  if (S==0) return 0;
  int j=median[S];
  while (S!=1<<j) { // while |S|>1
   S&=(weight[P[j]]>w)?~((1<<j)-1):(1<<j)-1;
   j=median[S];
  }
  return (weight[P[j]]>w)?j:0;
 }

 P[depth[v]]=v; // push current node on stack
 int k=binary_search(weight[v],down(D[v],S));
```

```
  S=down(D[v],S&(1<<(k+1)-1)|(1<<depth[v]));
  for (int i=L[v];i>=0;i=Lnext[i])
    answer[i]=P[median[down(1<<depth[upper[i]],S)]];
  for (int z=child[v];z>=0;z=sibling[z]) visit(z,S);
} // end visit

L[]=-1; Lnext[]=-1; // initialize all array elements to -1
for (int i=0;i<m;i++) { // distribute queries to lower nodes
  Lnext[i]=L[lower[i]];
  L[lower[i]]=i;
}
D[]=0; // initialize all array elements to 0
init(root,0);
P=new int[height+1];
median=median_table(height);
visit(root,0);
return answer;
} // end tree_path_maxima
```

References

1. Ahuja, R.K., Magnanti, T.L., Orlin, J.R.: Network Flows: Theory, Algorithms, and Applications. Prentice-Hall, Upper Saddle River (1993)
2. Borůvka, O.: O jistém problému minimálním. Práce Mor. Přírodověd. Spol. v Brně 3, 37–58 (1926)
3. Buchsbaum, A.L., Georgiadis, L., Kaplan, H., Rogers, A., Tarjan, R.E., Westbrook, J.R.: Linear-time algorithms for dominators and other path-evaluation problems. SIAM J. Comput. 38, 1533–1573 (2008)
4. Chazelle, B.: Computing on a free tree via complexity-preserving mappings. Algorithmica 2, 337–361 (1987)
5. Chazelle, B.: A minimum spanning tree algorithm with inverse-Ackermann type complexity. J. Assoc. Comput. Mach. 47, 1028–1047 (2000)
6. Chazelle, B., Rosenberg, B.: The complexity of computing partial sums off-line. Internat. J. Comput. Geometry Appl. 1, 33–45 (1991)
7. Chiang, Y., Goodrich, M.T., Grove, E.F., Tamassia, R., Vengroff, D.E., Vitter, J.S.: External-memory graph algorithms. In: Proc. 6th Annual ACM-SIAM Symposium on Discrete Algorithms (SODA), pp. 139–149 (1995)
8. Dixon, B., Rauch, M., Tarjan, R.E.: Verification and sensitivity analysis of minimum spanning trees in linear time. SIAM J. Comput. 21, 1184–1192 (1992)
9. Fredman, M.L., Willard, D.E.: Trans-dichotomous algorithms for minimum spanning trees and shortest paths. J. Comput. System Sci. 48, 533–551 (1994)
10. Graham, R.L., Hell, P.: On the history of the minimum spanning tree problem. Annals Hist. Comput. 7, 43–57 (1985)
11. Karger, D.R., Klein, P.N., Tarjan, R.E.: A randomized linear-time algorithm to find minimum spanning trees. J. Assoc. Comput. Mach. 42, 321–328 (1995)

12. Katriel, I., Sanders, P., Träff, J.L.: A practical minimum spanning tree algorithm using the cycle property. In: Di Battista, G., Zwick, U. (eds.) ESA 2003. LNCS, vol. 2832, pp. 679–690. Springer, Heidelberg (2003)
13. King, V.: A simpler minimum spanning tree verification algorithm. Algorithmica 18, 263–270 (1997)
14. Komlós, J.: Linear verification for spanning trees. Combinatorica 5, 57–65 (1985)
15. Mareš, M.: The saga of minimum spanning trees. Comput. Sci. Rev. 2, 165–221 (2008)
16. Pettie, S.: An inverse-Ackermann type lower bound for online minimum spanning tree verification. Combinatorica 26, 207–230 (2006)
17. Pettie, S., Ramachandran, V.: An optimal minimum spanning tree algorithm. J. Assoc. Comput. Mach. 49, 16–34 (2002)
18. Pettie, S., Ramachandran, V.: Randomized minimum spanning tree algorithms using exponentially fewer random bits. ACM Trans. Algorithms 4(1), 5:1–5:27 (2008)
19. Tarjan, R.E.: Applications of path compression on balanced trees. J. Assoc. Comput. Mach. 26, 690–715 (1979)
20. Yao, A.C.: Space-time tradeoff for answering range queries. In: Proc. 14th Annual ACM Symposium on Theory of Computing (STOC), pp. 128–136 (1982)

The k-Disjoint Paths Problem on Chordal Graphs

Frank Kammer and Torsten Tholey

Institut für Informatik, Universität Augsburg, D-86135 Augsburg, Germany
{kammer,tholey}@informatik.uni-augsburg.de

Abstract. Algorithms based on a bottom-up traversal of a tree decomposition are used in literature to develop very efficient algorithms for graphs of bounded treewidth. However, such algorithms can also be used to efficiently solve problems on chordal graphs, which in general do not have a bounded treewidth. By combining this approach with a sparsification technique we obtain the first linear-time algorithm for chordal graphs that solves the k-disjoint paths problem. In this problem k pairs of vertices are to be connected by pairwise vertex-disjoint paths. We also present the first polynomial-time algorithm for chordal graphs capable of finding disjoint paths solving the k-disjoint paths problem with minimal total length. Finally, we prove that the version of the disjoint paths problem, where k is part of the input, is $\mathcal{NP}$-hard on chordal graphs.

1 Introduction

In the k-disjoint paths problem (k-DPP), k pairs of vertices are to be connected by pairwise vertex-disjoint paths. This appears to be a hard problem since, for many classes of graphs, efficient algorithms are unknown or do not exist. Indeed, Fortune, Hopcroft, and Wyllie [3] have shown that the problem is $\mathcal{NP}$-hard on directed graphs, even if k is restricted to 2. As shown by Lynch [9] and by Knuth (see the paper of Karp [7]) the same is true on undirected graphs for the *disjoint paths problem* (DPP), where k, in contrary to the k-DPP, is part of the input. It is a common approach in combinatorial optimization to construct for $\mathcal{NP}$-hard problems so-called *fixed parameter algorithms* that solve the original problem in polynomial time if one or more of the input parameters are fixed. We present a linear time algorithm for the k-DPP, which then can be considered also as a fixed parameter algorithm for the DPP.

As usual in graph theory, we let n and m denote the number of vertices and edges, respectively, of the graph under consideration. For every fixed k, Robertson and Seymour, in their series of papers, developed a polynomial algorithm for the k-DPP on undirected graphs. Perković and Reed [13] presented an algorithm with an improved running time. Unfortunately, the constants hidden in the O-notation of the running time of the algorithms above are extremely large and make these algorithms unfeasible in practice. Algorithms with better practical running times are known for several classes of graphs such as undirected graphs of bounded treewidth [16] and directed acyclic graphs [3]. However, for many classes of graphs, e.g., for general, for planar, or for chordal undirected graphs, algorithms more efficient than the algorithm of Perković and Reed are known only for the special case $k = 2$. The first polynomial-time algorithms for the case $k = 2$ on general undirected graphs are given, e.g., in [11,17,18,19]. Perl and Shiloach [14] presented the first polynomial-time algorithm for the 2-DPP on undirected chordal graphs and on undirected planar graphs, namely with a linear running time. A simpler algorithm for chordal graphs can be found in [8].

The importance of chordal graphs is due to several facts. On the one hand, chordal graphs have nice properties that can be used to design efficient algorithms for many problems. For example, as shown by Fulkerson and Gross [4], chordal graphs are exactly the set of graphs with a perfect elimination order, and this order can be used to compute a maximal independent set, a maximum clique or an optimal coloring on chordal graphs in linear time [5]. On the other hand, concerning the practical relevance of chordal graphs, Gavril [6] has shown that the set of chordal graphs is equal to the set of *subtree graphs*, where a subtree graph is the intersection graph of a family of subtrees of a tree. Let us call a tuple $(G_1, \ldots, G_k, G)$ of graphs to be an intersection model for the intersection graph of $G_1, \ldots, G_k$ if the latter are subgraphs of G. Many practical problems in different areas such as computer science and genetics can be modeled by an intersection model and solved by a transformation to problems on the corresponding intersection graph; e.g., see [15]. In general, it seems that translating a problem on an intersection model into a problem on the corresponding intersection graph makes the problem easier to solve. However, in this paper we study the reverse direction. We translate the k-DPP on a chordal graph into a problem on the corresponding intersection model $(T_1, \ldots, T_k, T)$ or, more precisely, on a tree decomposition defined by this model, and we derive a simple approach to solve the k-DPP on chordal graphs. From another point of view our paper shows that algorithms based on a bottom-up traversal of a tree decomposition are useful not only for graphs of bounded treewidth, but can also be used for efficiently solving different problems on chordal graphs, even on those of unbounded treewidth. We only use the fact that we can choose the so-called bags of a tree decomposition as cliques. Following a similar approach, Okamato, Uno, and Uehara [12] have recently shown that the number of independent sets in a chordal graph can be counted in linear time.

In Section 2 we present an algorithm for solving the k-DPP on a chordal graph with a running time of $O(n^{2k+2} + m)$. As shown in Section 3, this algorithm can be modified to connect given pairs of vertices by pairwise disjoint paths such that the number of edges used by the paths is minimized among all such solutions. Note that so far no polynomial-time algorithm was known for solving this latter problem for every fixed k on chordal graphs.

In Section 4, as the main result of our paper, we show that the tree decomposition based algorithm of Section 2 can be combined with a sparsification technique in order to reduce the running time for solving the k-DPP on chordal graphs to $O(m + (2k)^{4k+2} n)$. This means that we obtain a linear fixed parameter algorithm for the DPP. Moreover, the additional constants hidden in the O-notation are of moderate size. For every fixed k, we obtain a running time which improves the running time of the previous best known algorithm for solving the k-DPP on chordal graphs, namely the algorithm of Perković and Reed mentioned earlier in this introduction for solving the problem on general graphs. Moreover, our algorithm is easy to implement and—for small values of k—it is practical. It is not surprising that the running time increases exponentially in k since—as a further new result—we can proof that the DPP (with k being non-fixed) is $\mathcal{NP}$-hard even for chordal graphs. Details can be found in Section 5.

2 The k-Disjoint-Paths Problem

Many graph problems can be solved in polynomial time by traversing a so-called tree decomposition bottom-up if the so-called treewidth is taken to be a constant.

Definition 1. *A* tree decomposition *for a graph $G = (V, E)$ is a pair (T, B), where $T = (V_T, E_T)$ is a tree and B is a mapping that maps each node w of V_T to a subset $B(w)$ of V—called the* bag *of w—such that (1) $\cup_{w \in V_T} B(w) = V$, (2) for each edge $e \in E$, there exists a node $w \in V_T$ with $e \subseteq B(w)$, (3) $B(x) \cap B(y) \subseteq B(w)$ for all $w, x, y \in V_T$ with w being a node on the path from x to y in T. The* treewidth *of T is the maximal cardinality of a bag minus one and the* size *of a tree decomposition is the sum of the cardinalities of its bags.*

The *treewidth* of a graph G, denoted by $tw(G)$, is the smallest width of a tree decomposition for G. One of the problems that can be solved efficiently on graphs with constant treewidth is the k-disjoint paths problem [16]. Unfortunately, the treewidth of chordal graphs is not bounded by a constant but we can find a very special tree decomposition that helps us to solve the k-DPP even on chordal graphs. For a set $U \subseteq V$ of a graph $G = (V, E)$, we define $G[U]$ to be the subgraph of G induced by the vertices in U.

Definition 2. *A* clique tree *for a graph $G = (V, E)$ is a tree decomposition (T, B) with the additional property that (4) B is a bijection between the nodes of T and $\{U \subseteq V \mid G[U]$ is a maximal clique in $G\}$.*

It is well known that chordal graphs are exactly the graphs that have a clique tree [2,6,20] and that a clique tree of linear size can be constructed in linear time [1]. As one can show by using property (4) a clique tree has $O(|V|)$ nodes. As input of our algorithm we will take a *weak clique tree* that is defined as a clique tree if we replace (4) by the following property: **(4')** the vertices of each bag induce a clique in G. More precisely, our algorithm starts with constructing, for the graph $G = (V, E)$ on which we search for k disjoint paths, in $O(|V| + |E|)$ time a weak clique tree (T, B) of size $O(|V| + |E|)$ for G with T being a rooted tree having $O(|V|)$ nodes. For example we can take a standard clique tree. We call the vertices to be connected by disjoint paths the *terminals* of G. For a node w of T, we let $G(w)$ be the subgraph of G induced by all vertices contained in at least one set $B(w')$ for a descendant w' of w in T, where w is also a descendant of itself. In order to obtain a simpler description of our algorithm, we describe our problem as a coloring problem. A *coloring* of a graph G' or a vertex set U is a mapping that assigns a color to some of the vertices in G' and in U, respectively. For an instance of the k-DPP, we always define an *initial coloring* that, for each pair of terminals to be connected, colors both terminals with the same color different from the colors of the other terminals. In general, we call a coloring C_2 of a vertex set V_2 an *extension* of a coloring C_1 of a vertex set $V_1 \subseteq V_2$ if all colored vertices of V_1 are also colored by C_2 with the same color. Let us call a coloring of a graph G *legal* if it is an extension of the initial coloring of the terminals and if, for each pair t_1 and t_2 of terminals sharing a color c, there is a path from t_1 to t_2 in the subgraph of G induced by the vertices of color c. Moreover, we call a legal coloring *good* with respect to a (weak) clique tree (T, B) for G if each bag of a node in T contains at most two vertices of color c.

For a node w with a father $p(w)$, let us call the set of vertices in $B(w) \cap B(p(w))$ the *transition set* of w denoted by $A(w)$. On the one hand, our algorithm will need to know the colors of all vertices in a bag for finally obtaining a complete coloring of G. On the other hand, our algorithm will extend stepwise colorings of $G(w)$ for a node w to colorings of $G(p(w))$ and for determining the colors for the new vertices of this extension (namely the vertices in $G(p(w)) - G(w)$) it only needs to know the colors of the vertices in $A(w)$. Thus, we define a *full* and a *reduced characteristic* for a node w of T as a coloring of the nodes in $B(w)$ and $A(w)$, respectively, such that for each color c at most two vertices are colored with c. Let Z be a full or a reduced characteristic of a node w. Then Z is *valid* if and only if

1. there exists a good coloring C of $G(w)$ extending Z and the initial coloring of the terminals in $G(w)$.
2. for each color c the following is true: if there is exactly one terminal in $G(w)$ of color c, a vertex in $A(w)$ is colored with c by Z.

A coloring C with properties 1 and 2 is then called *conform* to Z. We also call two characteristics *compatible* if one characteristic is an extension of the other.

There is a connection between the k-DPP and good colorings: If the k-DPP has a solution, take a solution with minimal total length. Then a coloring that colors the vertices of each path of the solution with the color of its endpoints is a good coloring because the following is true: if one of the disjoint paths visits three vertices v_1, v_2, and v_3 of one bag in this order, we obtain a shorter path by replacing the subpath from v_1 to v_3 by edge $\{v_1, v_3\}$. In the reverse direction assume that we are given a good coloring conform to a valid full characteristic of the root of T. Then disjoint paths connecting the terminals of the same color can be obtained by a depth-first search on each subgraph induced by the vertices of one color. Hence, we can solve the k-DPP by computing a good coloring conform to a valid full characteristic of the root of T.

For all nodes of T, we want to determine bottom-up all valid full and all valid reduced characteristics. If we restrict a coloring of $G(w)$ conform to a valid full characteristic of a node w to the graph $G(w')$ for a descendant w' of w, this restricted coloring is conform to a valid full as well as to a valid reduced characteristic of w'. In the reverse direction, a full characteristic Z of a node w is valid if and only if

- the terminals in $B(w)$ are colored by Z with their original color,
- for each child w' of w, the reduced characteristic of w' compatible to Z is valid,
- each color is used by Z to color at most two vertices, and
- for each color c, the following is true: if there is exactly one terminal in $G(w)$ of color c, a vertex in $A(w)$ is colored with c by Z.

Because of the above conditions there exists a good coloring of $G(w)$ extending Z; in particular, using the last condition we can conclude by induction that, for each pair of terminals t_1 and t_2 sharing a color c, there is a path from t_1 to t_2 in the subgraph of G induced by the vertices of color c. By iterating over all valid full characteristics of a node w in T we can easily compute a lookup-table storing 1 for each valid reduced characteristic of w and 0 for each non-valid reduced characteristic of w. The time for updating the whole table for w is of size $O(\ell \cdot tw(G) \deg(w))$, where ℓ is the number of full characteristics of w.

Hence, by a bottom-up traversal of T we can find a good coloring for G if such a coloring exists. We next analyze the running time of our algorithm. For each node, we can test whether a certain full characteristic is valid in at most $O((tw(G) + k) \deg(w))$ time by testing the four properties listed above. For testing the last condition, note the following: in $O(k \deg(w))$ time, we can update the set of colors c for which there is exactly one terminal of color c in $G(w)$ if we are given the corresponding sets for the children of w. There are at most $(tw(G) + 1)^{2k}$ full characteristics for a node w since for each color we can choose twice either no vertex or one of the $\leq tw(G) + 1$ vertices in $B(w)$. Since the time needed to initialize the lookup-table for the reduced characteristics of a node w is bounded linear in $O(|A(w)|)$ times the number of full characteristics, we obtain the next lemma.

Lemma 3. *The k-DPP can be solved in $O((tw(G)+1)^{2k}(tw(G)+k)|V|+|E|) = O(|V|^{2k+2} + |E|)$ time on a chordal graph $G = (V, E)$.*

3 Shortest k-Disjoint Paths

Define the weight of a coloring as the number of edges whose endpoints are both colored and have the same color. The cost of a characteristic Z of a node w is the minimal possible weight of a coloring of $G(w)$ conform to Z. In order to output disjoint paths using a minimal number of edges, we also have to compute the costs of the characteristics. Given, for a full characteristic Z of a node w, the costs $W(Z_1), \ldots, W(Z_\ell)$ of the reduced characteristics $Z_1, \ldots, Z_\ell$ of the children of w compatible to Z, one can compute the cost $W(Z)$ of Z as follows: Initialize $W(Z)$ with $W(Z_1)+W(Z_2)+\ldots+W(Z_\ell)$. Subsequently, for each color c used by Z to color two vertices $v_1, v_2 \in B(w)$, add one minus the number of children of w with their bags containing both, v_1 and v_2. This update takes $O(\min\{k, |B(w)|\} \cdot \deg(w))$ time and does not increase the asymptotic running time.

Theorem 4. *On a chordal graph $G = (V, E)$ one can find in $O(|V|^{2k+2} + |E|)$ time paths solving an instance of the k-DPP using a minimal number of edges among all sets of paths solving the instance.*

4 A Speedup

In this section we present a speed-up of the algorithm in Section 2. Once again, we first construct in $O(|V|+|E|)$ time a weak clique tree (T, B) of size $O(|V|+|E|)$ for our input graph $G = (V, E)$ with T having $O(|V|)$ nodes. We assume that there is no edge $\{t_1, t_2\}$ in G for a pair $\{t_1, t_2\}$ of terminals that are to be connected in G. Otherwise, our problem would be reduced to the $(k-1)$-DPP on $G[V - \{t_1, t_2\}]$. For each pair (t_1, t_2) of terminals t_1 and t_2 that should be connected by a path, let us choose $\Gamma(t_1)$ and $\Gamma(t_2)$ as the unique pair of nodes in T with their bags containing t_1 and t_2, respectively, such that the distance between the nodes is minimal. We choose for an arbitrary terminal t the node $\Gamma(t)$ as the root of T. Let f be a fixed bijection from V to $\{1, \ldots, |V|\}$ assigning the highest numbers to the terminals of G. For nodes w_1 and w_2 of T and for a vertex v of G, we define the $(w_1, w_2)_B$-*count* of v as a tuple $(a, f(v))$, where a is the number of nodes w' on the path from w_1 to w_2 in T whose bags $B(w')$ contain v. We say that a vertex v_1 with $(w_1, w_2)_B$-count (a_1, b_1) has a larger $(w_1, w_2)_B$-count than a vertex v_2 with $(w_1, w_2)_B$-count (a_2, b_2) if and only if either $a_1 = a_2$ and $b_1 > b_2$ or $a_1 > a_2$ holds. For a node $w \in T$, we let $I(w) = \{t \mid t \text{ is terminal with } \Gamma(t) \text{ contained in the subtree of } T \text{ rooted in } w\}$.

In order to improve the efficiency of the algorithm presented in Section 2, we replace (T, B) by a new tuple (T^*, B^*), where T^* will be a subtree of T and where B^* will be a function that maps each node w of T^* to a subset of $B(w)$ of size $\leq 4k^2$. In order to describe (T^*, B^*) more precisely, we need some further definitions. A bag $B(w)$ of a node w is called *small* if $|B(w)| \leq 2k$ and *big* otherwise. For each node w of T and for each terminal t, we define $D(w, t)$ as the set of the $\min\{2k, |B(w)|\}$ vertices of $B(w)$ with the largest $(w, \Gamma(t))_B$-count. We also let $D(w)$ be the union of $D(w, t)$ over all $t \in I(w)$ and of the set of all terminals in $B(w) \setminus I(w)$.

We now obtain T^* from T by deleting all nodes w with $I(w) = \emptyset$. We choose the same root for T^* as for T and, for each node w, we insert the vertices of $D(w)$ into $B^*(w)$. Moreover, for each child w' of w, we insert an arbitrary subset of $D(w) \cap B(w')$ of size $\min\{2k, |D(w) \cap B(w')|\}$ into $B^*(w')$. Let $t \in I(w')$. Keep in mind that, if $|B(w) \cap B(w')| \geq 2k$, then $D(w, t)$—and consequently also $B^*(w)$—contains $2k$ vertices of $B(w')$ since these vertices have the largest $(w, \Gamma(t))_B$-count. Thus, if $|B(w) \cap B(w')| \geq 2k$, the rules for node w add $2k$ vertices of $B^*(w)$ to $B^*(w')$, i.e., $|B^*(w) \cap B^*(w')| \geq 2k$. Note that by our definition $B(w) = B^*(w)$ holds for each small bag $B(w)$. We also can conclude:

Lemma 5. *Let v be a vertex of G, w' be a node of T^* with $v \in B^*(w')$, and w'' be the node of lowest depth with $v \in B(w'')$. Then $v \in B^*(w)$ holds for each node w on the path from w' to w'' in T^*.*

Proof. Since $B(w)$ and $B^*(w)$ share the same terminals, Lemma 5 holds if v is a terminal. If it is not, we merely need to show that, for each node w on the path from w' to w'' in T with $v \in B(w)$, the following holds: if w has a father x with $v \in B(x)$, we also have $v \in B^*(x)$. Since $B^*(x) = B(x)$ if $B(x)$ is small, we only need to consider the case, where $B(x)$ is big. Let us consider the case, where $B(w)$ is small. Because of $v \in B^*(w)$ there is a $t \in I(w)$ for which $v \in D(w,t)$ or $v \in D(x,t)$ holds. Since $|B(w)| \leq 2k$, $v \in D(w,t) \cap B(x)$ also implies $v \in D(x,t)$. Consequently, $v \in B^*(x)$. Let us finally consider the case where both, $B(w)$ and $B(x)$, are big. If the insertion rule for x inserts v into $B^*(w)$, we have $v \in B^*(x)$. Otherwise, the only reason for v being contained in $B^*(w)$ is that $v \in D(w,t)$ for a terminal $t \in I(w)$. Then v must also be one of the vertices with the $2k$ largest $(x, \Gamma(t))_B$-counts and therefore is also contained in $B^*(x)$. □

Corollary 6. *For each vertex v of G, the subtree of T^* induced by the nodes w with $v \in B^*(w)$ is connected.*

Let G^* be the graph obtained from G by removing all vertices v and all edges $\{v_1, v_2\}$ from G for which there is no longer any node w with $v \in B^*(w)$ and $\{v_1, v_2\} \subseteq B^*(w)$, respectively.

Lemma 7. (T^*, B^*) *is a weak clique tree for G^* of width $4k^2 - 1$.*

Proof. By our construction and Corollary 6 all properties of a weak clique tree hold for (T^*, B^*). Concerning the treewidth, for the root r of T, we have $|B^*(r)| \leq |D(r)| \leq 4k^2$ since $|I(r)| = 2k$. By our choice of r there is a terminal t_1 with $\Gamma(t_1) = r$. We have $|D(w)| \leq 4k^2 - 2k$ for all nodes $w \neq r$ in T since the subtree of T rooted in w does not contain $\Gamma(t_1)$. Consequently, $|B^*(w)| \leq 4k^2$. □

Lemma 8. *The k-DPP has a solution on G if and only if this is true for G^*.*

Proof. Clearly, an instance of the k-DPP is solvable on G if this true for G^*. For the reverse direction we merely need to show that a solution of the k-DPP on G allows us to construct a good coloring of G^* with respect to (T^*, B^*). Moreover, for a legal coloring C and a pair of terminals t_1 and t_2 colored with c by C, let us call a pair of incident nodes w_1 and w_2 on the unique path from $\Gamma(t_1)$ to $\Gamma(t_2)$ a *color break* with respect to c (and C) if no vertex in $B^*(w_1) \cap B^*(w_2)$ is colored with c. Let $\mathcal{C}$ be the set of all legal colorings of G that color at most two vertices of each bag in (T^*, B^*) with the same color. The solvability of the k-DPP implies $\mathcal{C} \neq \emptyset$ since there exists—as shown in Section 2—at least one good coloring with respect to (T, B) and since each good coloring is contained in $\mathcal{C}$. In the reverse direction, our Lemma holds if there is a $C \in \mathcal{C}$ without any color break since C then is a good coloring of G^* with respect to (T^*, B^*).

Assume now that we can find no coloring in $\mathcal{C}$ without color breaks. Let us choose a fixed numbering with $1, \ldots, k$ for the colors assigned to the terminals and a coloring $C \in \mathcal{C}$ such that the lowest number among the colors with a color break is as large as possible. Moreover, if c is the color with the lowest number for which there is a color break and if t_1 and t_2 are the terminals of color c, we choose the coloring $C \in \mathcal{C}$ with the above properties such that there is a maximal distance between $\Gamma(t_1)$ and the node w_{σ_0} of the first color break $(w_{\sigma_0}, w_{\sigma_0+1})$ on the path $p = (w_1, w_2, w_3, \ldots)$ from $\Gamma(t_1)$ to $\Gamma(t_2)$ in T. Let v be a vertex of color c with $v \in B(w_{\sigma_0}) \cap B(w_{\sigma_0+1})$ and $v \notin B^*(w_{\sigma_0}) \cap B^*(w_{\sigma_0+1})$.

Assume $|B^*(w_{\sigma_0}) \cap B^*(w_{\sigma_0+1})| < 2k$. Then $|B(w_{\sigma_0}) \cap B(w_{\sigma_0+1})| < 2k$. Let $w \in \{w_{\sigma_0}, w_{\sigma_0+1}\}$ be the father of the other node $w' \in \{w_{\sigma_0}, w_{\sigma_0+1}\}$ and $t \in \{t_1, t_2\} \cap I(w')$. We can conclude $v \in D(w, t)$ and consequently $v \in B^*(w)$. Since $|B(w_{\sigma_0}) \cap B(w_{\sigma_0+1})| < 2k$, the rule for w adds all vertices of $D(w, t) \cap B(w')$ including v into $B^*(w')$, a contradiction to $v \notin B^*(w_{\sigma_0}) \cap B^*(w_{\sigma_0+1})$.

Hence $|B^*(w_{\sigma_0}) \cap B^*(w_{\sigma_0+1})| \geq 2k$. Since no vertex in $B^*(w_{\sigma_0}) \cap B^*(w_{\sigma_0+1})$ is colored with c and since C is a coloring which uses each color at most twice in a bag of (T^*, B^*) and thus in $B^*(w_{\sigma_0}) \cap B^*(w_{\sigma_0+1})$, it follows that there must be an uncolored vertex in $B^*(w_{\sigma_0}) \cap B^*(w_{\sigma_0+1})$. Let us define u to be the uncolored vertex in $B^*(w_{\sigma_0}) \cap B^*(w_{\sigma_0+1})$ that among all uncolored vertices has the largest $(w_{\sigma_0+1}, \Gamma(t_2))_B$-count. We next show that we can construct a coloring $C^* \in \mathcal{C}$ without any new color breaks for the colors different from c for which—if it has a color break with respect to c—the first such color break occurs after the pair $\{w_{\sigma_0}, w_{\sigma_0+1}\}$ on p. This leads to a contradiction to our choice of C and proofs our lemma.

After having initially set $C^* = C$ we modify C^* as follows. First of all, we color u with c. We then define $w_{\sigma_{-1}}$ and w_{σ_1} as the first and the last node on p, respectively, such that $u \in B^*(w_{\sigma_{-1}}) \cap B^*(w_{\sigma_1})$. Let S be the set consisting of u and all vertices colored with c by C contained in a bag of $\{B(w_{\sigma_{-1}}), \ldots, B(w_{\sigma_1})\}$.

Second, modify C^* as follows: For the set X of all nodes reachable from one of the nodes in $\{w_{\sigma_{-1}}, \ldots, w_{\sigma_1}\}$ without visiting $w_{\sigma_{-1}-1}$ or w_{σ_1+1}, uncolor all c-colored vertices in $B^*(X)$ apart from u, the vertex $v' \in S \cap B^*(w_{\sigma_{-1}})$ with the largest $(w_{\sigma_{-1}}, \Gamma(t_1))_{B^*}$-count, and the vertex $v'' \in S$ with the largest $(w_{\sigma_1}, \Gamma(t_2))_B$-count. Note that $v'' \in B(w_{\sigma_1+1})$ or $w_{\sigma_1} = \Gamma(t_2)$.

Third, if $v'' \in B(w_{\sigma_0+1}) \setminus D(w_{\sigma_0+1}, t_2)$, no vertex is colored with c by C in $D(w_{\sigma_0+1}, t_2)$. Then, let $\tilde{u}$ be the vertex of highest $(w_{\sigma_0+1}, \Gamma(t_2))_B$-count in $D(w_{\sigma_0+1}, t_2)$ not colored by C, let w_{σ_2} be the last node on p with $\tilde{u} \in B(w_{\sigma_2})$, and let v''' be the vertex in $B(w_{\sigma_2})$ with $C(v''') = c$ that among all such vertices has the largest $(w_{\sigma_2}, \Gamma(t_2))_B$-count. In particular, $v''' \in B(w_{\sigma_2+1})$ or we have $w_{\sigma_2} = \Gamma(t_2)$ and $v''' = t_2$. If $v'' \in B(w_{\sigma_0+1}) \setminus D(w_{\sigma_0+1}, t_2)$ and $v'' \neq u \neq \tilde{u}$, so-called *extra modifications* of C^* are required: For the set Y of all nodes reachable from one of the nodes in $\{w_{\sigma_1}, \ldots, w_{\sigma_2}\}$ without visiting w_{σ_1-1} or w_{σ_2+1}, change C^* by uncoloring v'' and all c-colored vertices in $B^*(Y)$. See Fig. 1. (Some ranges are explained later in more detail.) Coloring v' with c guarantees that there is no new color break between $\Gamma(t_1)$ and w_{σ_0} on p.

Let us first consider the case, where no extra modifications are applied. By coloring v'' with c we can guarantee that C^* is a legal coloring. Hence, if C^* does not belong to $\mathcal{C}$, there is a bag $B^*(w)$ containing u, v', v'' and $|\{u, v', v''\}| = 3$. By Corollary 6 we can choose w w.l.o.g. as a node on p. Due to our choice of v' we know that $v' \in B^*(w_{\sigma_{-1}})$. There is no node w' on the subpath of p from w_{σ_0+1} to $\Gamma(t_2)$ with $v' \in B^*(w')$ since otherwise $v' \in B^*(w_{\sigma_0}) \cap B^*(w_{\sigma_0+1})$

by Corollary 6. Thus, w is on the subpath of p from $\Gamma(t_1)$ to w_{σ_0}. Note that $v'' \in B^*(w)$ implies $v'' \in B(w)$. We consider two subcases:

- w_{σ_0+1} is the father of w_{σ_0}. Since $v'' \in B(w_{\sigma_1})$, we also have $v'' \in B(w_{\sigma_0}) \cap B(w_{\sigma_0+1})$. Therefore, $v'' \in B^*(w_{\sigma_0}) \cap B^*(w_{\sigma_0+1})$ according to Lemma 5 and we obtain a contradiction since $(w_{\sigma_0}, w_{\sigma_0+1})$ is a color break.
- w_{σ_0} is the father of w_{σ_0+1}. Since no extra modifications are applied and since $u \neq v''$, we have $v'' \notin B(w_{\sigma_0+1})$ or $v'' \in D(w_{\sigma_0+1}, t_2) \subseteq B^*(w_{\sigma_0+1})$ or $u = \tilde{u}$. In the first case, v'' has a smaller $(w_{\sigma_1}, \Gamma(t_2))_B$-count than u. In the second case, we have $v'' \in B^*(w_{\sigma_0}) \cap B^*(w_{\sigma_0+1})$ by Corollary 6. If only the third case holds, because of $u \in D(w_{\sigma_0+1}, t_2)$ and $v'' \notin D(w_{\sigma_0+1}, t_2)$, vertex u has a larger $(w_{\sigma_1}, \Gamma(t_2))_B$-count than v''. Consequently, a contradiction occurs in each case.

Let us finally assume that the extra modifications are being applied. Then $u \neq \tilde{u}$ and hence $\tilde{u} \notin B^*(w_{\sigma_0})$. Like in the previous case without extra modifications, v' can not be contained in one of the bags $B^*(w_{\sigma_0+1}), \ldots, B^*(\Gamma(t_2))$. Thus, no bag of (T^*, B^*) contains v' and $\tilde{u}$. Since $D(w_{\sigma_0+1}, t_2) \cap \{v'', \tilde{u}\} = \{\tilde{u}\}$, the $(w_{\sigma_0+1}, \Gamma(t_2))_B$-count of v'' is smaller than that of $\tilde{u}$. Therefore and because of $v'' \in B(w_{\sigma_0+1})$, the $(w_{\sigma_1}, \Gamma(t_2))_B$-count of v'' is smaller than that of $\tilde{u}$. Note that $v''' \notin B(w_{\sigma_1})$ since otherwise $v''' \in S$ and the fact that $v''' \in B(w_{\sigma_2+1})$ or $v''' = t_2$ would imply that v''' has a larger $(w_{\sigma_1}, \Gamma(t_2))_B$-count than that of $\tilde{u}$ and that of v''. Thus, no bag of (T, B) contains u and v'''. Consequently, C^* colors at most two vertices in each bag of (T^*, B^*). Our choice of v''' guarantees that C^* is a legal coloring.

We have shown that $C^* \in \mathcal{C}$ and that the distance between $\Gamma(t_1)$ and the first node of a color break on p with respect to c and C^*—if indeed there is a color break—is larger than the corresponding distance for C. This is a contradiction to our choice of C. □

We can therefore solve the k-DPP on a chordal graph G as follows: we first determine for each node w of T the set $I(w)$ and subsequently all sets $D(w, s)$ and $D(w)$ ($s \in I(w)$). This can easily be done in a bottom-up traversal of T in at

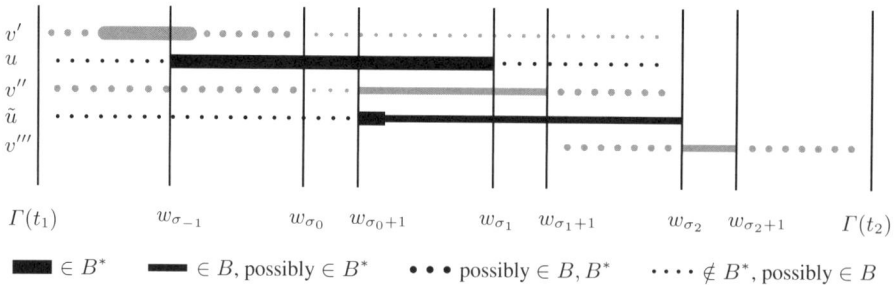

Fig. 1. The ranges of the variables in the case where the extra modifications are applied. Black lines represent vertices uncolored by C.

most $O(m+k^3n)$ time. Hence, we can replace G and (T,B) by G^* and (T^*,B^*) in the same running time. We then apply the algorithm of the Section 2 on G^*.

Theorem 9. *The k-DPP on chordal graphs with n vertices and m edges can be solved in $O(m+(4k^2)^{2k+1}n) = O(m+(2k)^{4k+2}n)$ time.*

5 Hardness of the Disjoint-Paths Problem

Theorem 10. *The disjoint-paths problem on chordal graphs is NP-hard.*

Proof. We can prove the theorem by a reduction from a restricted case of *1-in-3 SAT*. In 1-in-3 SAT we are given a formula in conjunctive normal form with 3 variables per clause and we have to find an assignment of the variables such that exactly one of the three literals is true in every clause. A formula in conjunctive normal form is *monotone* if every literal is positive and it is *cubic* if every variable occurs exactly three times. In [10] it is shown that 1-in-3 SAT is NP-complete even on monotone and cubic formulas. We now reduce an instance of 1-in-3 SAT consisting of a monotone and cubic formula F to an instance of the DPP on a chordal graph G. Fig. 2 should represent a clique tree of G. Each subgraph induced by the vertices of a bag should be a clique whose edges are colored gray in Fig. 2—however, not all existing edges are shown in the figure. Black lines represent paths of length 0. Therefore, the endpoints of black lines represent the same vertex even if they appear in different shapes.

In detail, we construct G as follows: For each variable x and each clause C in F, we introduce a *variable* and a *clause gadget*, respectively, as shown in Fig. 2. A variable gadget has six terminals $a_1, a_2, a_3, b_1, b_2, b_3$ and a clause gadget six terminals $y_1, y_2, y_3, z_1, z_2, z_3$. Each gadget is connected to one big clique Γ—see the rightmost bag Fig. 2. Γ contains 6ℓ vertices where ℓ is the number of clauses. We next divide the terminals into pairs such that the resulting instance of the DPP has a solution if and only if F has a satisfying assignment. If a clause C contains a variable x as the i-th variable and if it is the j-th occurrence of variable x in F that is part of C, the pairs (a_j, y_i) and (b_j, z_i) are added to our instance of the DPP, where the four terminals a_j, b_j, y_i, and z_i belong to the gadgets for x and C. Moreover, we identify one triangular and one square vertex in the gadget of x with one triangular and one square vertex, respectively, in the gadget of C different from the triangular and square vertices chosen for other variables or clauses. For a simpler notation, the terminals a_1, a_2 and a_3 shown in Fig. 2 are called *A-terminals* and the remaining terminals *B-, Y-* and *Z-terminals*, respectively.

Let us consider a satisfying assignment of F. For a variable x, we construct in the gadget of x six paths from terminals to the triangular and square vertices such that, if x is set to true, the paths starting in the A-terminals are routed exclusively to the triangular vertices; otherwise, they are routed exclusively to the square vertices. Since each clause C has exactly one true variable, for each clause gadget, exactly one path from an A-terminal to the gadget of C arrives at a triangular vertex, whereas the other two paths from an A-terminal to the

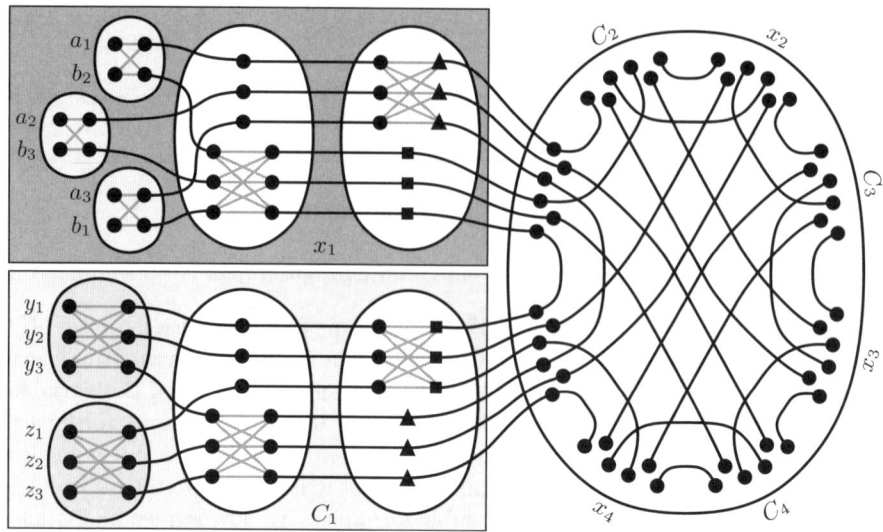

Fig. 2. Reduction from $(x_1 \vee x_2 \vee x_4) \wedge (x_1 \vee x_2 \vee x_3) \wedge (x_2 \vee x_3 \vee x_4) \wedge (x_1 \vee x_3 \vee x_4)$

gadget of C arrive at a square vertex. Thus, we can forward the three paths to the Y-terminals of C. Similarly, we can forward the paths from the B-terminals to the Z-terminals. Hence, we have found a solution to our instance of the DPP.

Let us now consider a solution of our DPP. It remains to show that F can be satisfied. In our construction, the number of vertices in the big clique Γ is equal to the number of pairs that have to be connected in our instance. Moreover, each path has to use at least one and therefore exactly one vertex of Γ. Note that for each clause C and each variable x of C, Γ contains exactly two vertices common with the gadgets of C and x, respectively: one triangular and one square vertex. Thus, these two vertices must be the two vertices visited by the two paths connecting two pairs of terminals in the gadgets of C and x. As a consequence, for any fixed variable gadget the paths starting from the A-terminals must pass either exclusively through the triangular vertices or exclusively through the square vertices of this gadget—see the variable gadget in Fig. 2. We define a variable x of F to be true if the paths of the A-terminals from the gadget of x pass through the triangular vertices. Since the A-terminals are connected to the Y-terminals, exactly one path from an A-terminal uses a triangular vertex of each clause gadget, i.e., exactly one variable of each clause is set to true. □

References

1. Blair, J.R.S., England, R.E., Thomason, M.G.: Cliques and their separators in triangulated graphs, Tech. Report UT-CS-88-73, Dept. Comput. Sci., University of Tennessee, Knoxville (1988)
2. Buneman, P.: A characterisation of rigid circuit graphs. Discrete Math. 9, 205–212 (1974)

3. Fortune, S., Hopcroft, J., Wyllie, J.: The directed subgraph homeomorphism problem. Theoret. Comput. Sci. 10, 111–121 (1980)
4. Fulkerson, D.R., Gross, O.A.: Incidence matrices and interval graphs. Pacific J. Math. 15, 835–855 (1965)
5. Gavril, F.: Algorithms for minimum coloring, maximum clique, minimum covering by cliques, and maximum independent set of a chordal graph. SIAM J. Comput. 1, 180–187 (1972)
6. Gavril, F.: The intersection graphs of subtrees in trees are exactly the chordal graphs. J. Combin. Theory Ser. B 16, 47–56 (1974)
7. Karp, R.M.: On the computational complexity of combinatorial problems. Networks 5, 45–68 (1975)
8. Krishnan, S.V., Pandu Rangan, C., Seshadri, S.: A new linear algorithm for the two path problem on chordal graphs. In: Kumar, S., Nori, K.V. (eds.) FSTTCS 1988. LNCS, vol. 338, pp. 49–66. Springer, Heidelberg (1988)
9. Lynch, J.F.: The equivalence of theorem proving and the interconnection problem (ACM) SIGDA Newsletter 5, 31–36 (1975)
10. Moore, C., Robson, J.M.: Hard tiling problems with simple tiles. Discrete and Comput. Geom. 26, 573–590 (2001)
11. Ohtsuki, T.: The two disjoint path problem and wire routing design. In: Saito, N., Nishizeki, T. (eds.) Graph Theory and Algorithms. LNCS, vol. 108, pp. 207–216. Springer, Heidelberg (1981)
12. Okamoto, Y., Uno, T., Uehara, R.: Linear-time counting algorithms for independent sets in chordal graphs. In: Kratsch, D. (ed.) WG 2005. LNCS, vol. 3787, pp. 433–444. Springer, Heidelberg (2005)
13. Perković, L., Reed, B.: An improved algorithm for finding tree decompositions of small width. International Journal of Foundations of Computer Science (IJFCS) 11, 365–372 (2000)
14. Perl, Y., Shiloach, Y.: Finding two disjoint paths between two pairs of vertices in a graph. J. ACM 25, 1–9 (1978)
15. Roberts, F.S.: Discrete Mathematical Models with Applications to Social, Biological, and Environmental Problems. Prentice Hall, Englewood Cliffs (1976)
16. Scheffler, P.: A practical linear time algorithm for disjoint paths in graphs with bounded tree-width, Report No. 396/1994, TU Berlin, FB Mathematik (1994)
17. Seymour, P.D.: Disjoint paths in graphs. Discrete Math. 29, 293–309 (1980)
18. Shiloach, Y.: A polynomial solution to the undirected two paths problem. J. ACM 27, 445–456 (1980)
19. Thomassen, C.: 2-linked graphs. European J. Combin. 1, 371–378 (1980)
20. Walter, J.R.: Representations of chordal graphs as subtrees of a tree. J. Graph Theory 2, 265–267 (1978)

Local Algorithms for Edge Colorings in UDGs

Iyad A. Kanj[1], Andreas Wiese[2], and Fenghui Zhang[3]

[1] School of Computing, DePaul University, 243 S. Wabash Avenue, Chicago, IL 60604, USA
ikanj@cs.depaul.edu
[2] Institute für Mathematik, Technische Universität Berlin, Germany
wiese@math.tu-berlin.de
[3] Department of Computer Science, Texas A&M University, College Station, TX 77843-3112, USA
fhzhang@cs.tamu.edu

Abstract. In this paper we consider two problems: the EDGE COLORING and the STRONG EDGE COLORING problems on unit disk graphs (UDGs). Both problems have important applications in wireless sensor networks as they can be used to model link scheduling problems in such networks. It is well known that both problems are NP-complete, and approximation algorithms for them have been extensively studied under the centralized model of computation. Centralized algorithms, however, are not suitable for ad-hoc wireless sensor networks whose devices typically have limited resources, and lack the centralized coordination.

We develop local distributed approximation algorithms for the EDGE COLORING and the STRONG EDGE COLORING problems on unit disk graphs. For the EDGE COLORING problem, our local distributed algorithm has approximation ratio 2 and locality 50. We show that the locality upper bound can be improved to 28 while keeping the same approximation ratio, at the expense of increasing the computation time at each node. For the STRONG EDGE COLORING problem on UDGs, we present two local distributed algorithms with different tradeoffs between their approximation ratio and locality. The first algorithm has ratio 128 and locality 22, whereas the second algorithm has ratio 10 and locality 180.

1 Introduction

The EDGE COLORING problem is to color the edges of a given graph G using the minimum number of colors so that no two edges of the same color are adjacent. The STRONG EDGE COLORING problem is to color the edges of a given graph G with the minimum number of colors so that no two edges with the same color are of distance less than 2. The EDGE COLORING and the STRONG EDGE COLORING problems are known to be NP-complete even on restricted classes of graphs [4,9]. Since both problems have numerous applications in networks where they model channel assignments/scheduling problems (see [1,2,3,7,11,12], among others), it is natural to seek approximation algorithms for them.

For the EDGE COLORING problem, Vizing's theorem [13] showed that any graph with maximum degree Δ has an edge coloring that uses at most $\Delta + 1$ colors;

however, his result was nonconstructive. Misra and Gries [10] gave a polynomial-time constructive proof of Vizing's theorem, thus showing that the problem can be approximated to within an additive constant of 1. Ramanathan [11] gave a very simple centralized greedy algorithm for the problem of ratio 2. Under the distributive model of computation, Gandam et al. [3] gave a distributed approximation algorithm based on Misra and Gries' [10] constructive proof of Vizing's theorem that approximates the problem to within an additive constant of 1. Kodialam and Nandagopal [7] gave a simple distributive algorithm of ratio 2, which was based on the centralized greedy algorithm of Ramanathan [11].

For the STRONG EDGE COLORING problem on planar graphs, Barrett el al. [1] gave a centralized algorithm that approximates the STRONG EDGE COLORING problem to ratio 17. This ratio has recently been improved to 2 by Ito et al. [5].

The assumed underlying graph model and the assumed computational model in the above results, however, do not seem appropriate for ad-hoc wireless sensor networks. In wireless sensor networks, devices can in principal communicate if they are in each other's transmission range. Therefore, a general graph model, or even a plane (embedded planar) graph model, is too flexible in the sense that it does not reflect the restrictions on the connectivity of such networks. Moreover, the topology of such networks undergoes constant change, and the devices in those ad-hoc networks have limited energy/power. Therefore, any assumed computational model should take into account the decentralized nature of such networks, and should be sensitive to issues such as scalability, robustness, and fault tolerance. In terms of the underlying graph model, when studying wireless sensor networks, it is natural to embed them in a Euclidean metric space. A common simple embedding assumes that the space is two dimensional, and that the transmission range of all devices is the same. In that case, the network is modeled as a *Unit Disk Graph*, abbreviated UDG henceforth, in the Euclidean plane: the nodes of the UDG correspond to the mobile wireless devices, and its edges connect pairs of nodes whose corresponding devices are in each other's transmission range equal to one unit. While this model is idealized, it has the advantage of being easier to work with. Meaningful theoretical and practical results can be derived under this model that, hopefully, will carry (at least partially) to more general models. Moreover, there are real examples where such models make sense: boats on water surfaces, vehicles in a relatively flat desert, etc.... In terms of the computational model, most of the above issues (scalability, robustness, fault tolerance) can be dealt with under the *local* distributed computational model, as defined by Linial [8]. A distributed algorithm is said to be k-*local* (where $k \geq 0$ is an integer) if the computation at each node of the graph depends solely on the initial state (in our case the ID and coordinates) of the nodes at distance (number of edges) at most k from the node (i.e., within k hops from the node). An algorithm is called *local* if it is k-local for some integer constant k. Efficient local distributed algorithms are naturally fault-tolerant and robust because faults and changes can be handled locally by such algorithms. These algorithms are also scalable because the computation performed by a device is not affected by the total size of the network.

Local distributed algorithms for the EDGE COLORING problems on UDGs have been considered in [2]. However, the results in [2] only deal with a restricted subclass of UDGs called the "Yao-Like" subgraphs, and give an approximation algorithm for the EDGE COLORING problem within an additive constant of 1 from the optimal solution. For the STRONG EDGE COLORING problem on UDGs, we are only aware of the distributed approximation algorithm given by Barrett [1] which achieves an O(1) ratio; however, this algorithm is distributed but *not local*.

In this paper we develop local distributed approximation algorithms for the the EDGE COLORING and the STRONG EDGE COLORING problems on UDGs. For the EDGE COLORING problem, we present a local distributed algorithm of approximation ratio 2 and locality 50; this algorithm works for a generalization of UDGs, called *quasi-UDGs*. We show that the locality upper bound can be improved to 28, while keeping the same approximation ratio, at the expense of increasing the computation time at each node. For the STRONG EDGE COLORING problem on UDGs, we present two local distributed algorithms with different tradeoffs between their approximation ratio and locality. The first algorithm has approximation ratio 128 and locality 22, whereas the second algorithm has ratio 10 and locality 180.

2 Definitions and Notations

We assume familiarity with the basic graph-theoretic notations and terminologies.

Given a set of nodes S in the Euclidean plane, the Euclidean graph $\mathcal{E}$ on S is the complete graph whose node-set is S. The *unit disk graph*, shortly *UDG*, G on S is the subgraph of $\mathcal{E}$ with the same node-set as $\mathcal{E}$, and such that (u,v) is an edge of G if and only if $|(u,v)| \leq 1$, where $|(u,v)|$ is the Euclidean length of edge (u,v).

Let $0 < r \leq 1$ be a constant. The *quasi-UDG* on S with parameter r, is the subgraph G of $\mathcal{E}$ with the same node-set as $\mathcal{E}$, and such that for any two nodes u and v in G: if $|(u,v)| \leq r$ then $|(u,v)|$ is an edge of G, if $r < |(u,v)| \leq 1$ then (u,v) may or may not be an edge of G, and if $|(u,v)| > 1$ then (u,v) is not an edge of G. Clearly, a UDG is a quasi-UDG with $r = 1$.

Let H be a graph. We denote by $V(H)$ and $E(H)$ the set of nodes and the set of edges of H, respectively. The *length* of a path P in H, denoted $|P|$, is the number of edges in P. A *shortest path* between two nodes u and v in H is a path between u and v with the minimum length. A node v is said to be an *i-hop neighbor* of u in H, if the length of a shortest path between u and v in H is at most i. If u is an i-hop neighbor of v in H, we will say that the *hop distance* between u and v in H is at most i. For a node $u \in H$, and a natural number i, define $N_i[u]$ to be the set of i-hop neighbors of u in H.

For two edges e and e' in H, the *distance* between e and e' is the minimum length of a path, among all paths in H connecting an endpoint of e to an endpoint of e'. Two distinct edges are *adjacent* if their distance is 0, or equivalently, if they share an endpoint. An *edge coloring* of H is an assignment of colors[1] to the edges in $E(H)$ such that no two adjacent edges in H are assigned the same color. A *strong edge coloring* of a graph H is an assignment of colors to the edges in $E(H)$ such that no two edges of distance at most 1 are assigned the same color. A *minimum edge coloring* of H is an edge coloring of H that uses the minimum number of colors. Similarly, a *minimum strong edge coloring* of H is a strong edge coloring of H that uses the minimum number of colors.

An *approximation algorithm* for a minimization problem $\mathcal{Q}$ is an algorithm that for each instance of $\mathcal{Q}$ computes a solution to the instance. The *ratio* of an approximation algorithm for a minimization problem is the maximum value, over all instances of the problem, of the size of the solution to the instance returned by the algorithm over the minimum-size solution to the instance.

The algorithms designed in this paper are k-local distributed algorithms. Each node in these algorithms starts by computing its k-hop neighbors, and performs only local computations afterwards. For a fixed k, it was shown in [6] that the k-hop neighborhoods of the nodes in a UDG (or a quasi-UDG) can be computed by a local distributed algorithm in which the total number of messages sent by all the nodes in the UDG is $O(n)$, where n is the number of nodes in the UDG. Therefore, the message complexity of each of the presented local distributed algorithms is $O(n)$.

3 Preliminaries

Let $\alpha > 2$ be a constant. Fix an infinite square tiling (i.e., a grid) $\mathcal{T}$ of the plane of tile dimensions $\alpha \times \alpha$.

Let T_1 be the translation with vector $(0,0)$ (the identity translation), T_2 the translation of vector $(\alpha/2, 0)$ (horizontal translation), T_3 the translation of vector $(0, \alpha/2)$ (vertical translation), and T_4 the translation of vector $(\alpha/2, \alpha/2)$ (diagonal translation). We have the following simple lemma whose proof can be easily verified by the reader (note that $\alpha > 2$).

Fact 3.1. *Let G be a quasi-UDG, and let (u,v) be any edge in G. There exists a translation T in $\{T_1, T_2, T_3, T_4\}$ such that the translations of the nodes u and v under T, i.e., $T(u)$ and $T(v)$, reside in the interior of the same tile of $\mathcal{T}$.*

The following lemma uses a folklore packing argument to bound the length of a path between two nodes in a UDG that reside within a region of bounded area of the plane (see for example [14]).

Lemma 3.1. *Let G be a quasi-UDG of parameter $0 < r \leq 1$. Let H be a connected induced subgraph of G residing in a region R of the plane. Let R' be a region of area a' that contains R such that for any node p in R the disk centered at p and of radius $r/2$ is contained in R'. Then for any two nodes u and v of H, there exists a path in H between u and v of length at most $\lfloor 8a'/(\pi r^2) \rfloor$.*

---

[1] Without loss of generality, we shall assume that the colors are natural numbers.

4 Edge Coloring

In this section we present a local distributed algorithm that approximates the EDGE COLORING problem on quasi-UDGs which are a super class of UDGs. The idea behind the algorithm is to tile the plane as discussed in Section 3, and then to have the nodes residing in the same tile color the edges interior to their tile using the greedy algorithm given in [7,11]. This is a proper coloring since two edges contained in two distinct tiles are not adjacent. However, not every edge in the graph is interior to a tile because an edge may cross the horizontal or vertical (or both) boundary of a tile. To deal with this issue, we affect an appropriate set of translations to the nodes so that, for any edge in the graph, its translation under at least one of the translations is contained in some tile. This ensures that every edge of the graph will eventually be colored appropriately. Implementing this algorithm under a centralized model of computation is straightforward. However, implementing this algorithm under a localized distributed model poses some potential issues since the effect of the color of an edge over other edges needs to be limited, and some consensus problems need to be resolved.

We use the tiling $\mathcal{T}$ described in Section 3. Let G be a quasi-UDG with parameter r, where $0 < r \leq 1$. Each node $p \in G$ executes the algorithm **EdgeColoring-APX** given in Figure 1.

Lemma 4.1. *The algorithm* **EdgeColoring-APX** *is a k-local distributed algorithm, where $k = \lfloor (22\alpha^2 + 8r^2 + 32\alpha r)/(\pi r^2) \rfloor$.*

1: p collects the coordinates of the nodes in $N_k[p]$ in G, where $k = \lfloor (22\alpha^2 + 8r^2 + 32\alpha r)/(\pi r^2) \rfloor$
2: **for** round $i = 1, 2, 3, 4$ **do**
3:     let $G_i(p)$ be a copy of the subgraph of G consisting of the set $E_i(p)$ of uncolored edges whose endpoints are in $N_k[p]$, and such that, for any edge $(u, v) \in E_i(p)$, $T_i(u)$ and $T_i(v)$ are in the same tile of $\mathcal{T}$
4:     let $C_i^1(p), \ldots, C_i^\ell(p)$, where $\ell \geq 1$, be the connected components of $G_i(p)$
5:     **for** $j = 1, \ldots, \ell$ **do**
6:        p orders all the edges in $C_i^j(p)$ using the lexicographic order into the sequence of edges $E_i^j(p)$
7:        **for** each edge e in $E_i^j(p)$ **do**
8:           color e in $G_i(p)$ with the smallest available color, i.e., the smallest color that has not been used in the previous rounds to color any of the edges adjacent to e
9:        **end for**
10:    **end for**
11:    **for** each edge $e \in G_i(p)$ incident on p **do**
12:        p colors e in G with the same color in $G_i(p)$
13:    **end for**
14: **end for**

Fig. 1. The algorithm **EdgeColoring-APX**

Proof. It is clear that the computation at each node depends solely on the coordinates of its k-hop neighbors, where $k = \lfloor (22\alpha^2 + 8r^2 + 32\alpha r)/(\pi r^2) \rfloor$. □

For each $i \in \{1, 2, 3, 4\}$, let G_i be the subgraph of G consisting of the edges $(u, v) \in G$ such that $T_i(u)$ and $T_i(v)$ are in the same tile of $\mathcal{T}$; we call each connected component C in G_i an *i-cluster*, and we say that i is the *label* of C. Note that, by definition, any two distinct i-clusters are disjoint. A *cluster* is an i-cluster for some $i \in \{1, 2, 3, 4\}$. A sequence of clusters is said to be a *potential affecting sequence*, if the labels of the clusters on this sequence are strictly increasing, and each two consecutive clusters in the sequence are adjacent, i.e., share at least one node in G. Note that a potential affecting sequence of clusters has length at most 4. The notion of a potential affecting sequence will be used to confine the "effect" of the color of an edge on the color of another edge, as shown by the following lemma whose proof is omitted for lack of space:

Lemma 4.2. *Let $S = (C_1, C_2, C_3, C_4)$ be a potential affecting sequence of clusters (we allow C_i, $i \in \{1, 2, 3, 4\}$, to be empty). Then for any two nodes u and v in S, u is a k-hop neighbor of v in G, where $k = \lfloor (22\alpha^2 + 8r^2 + 32\alpha r)/(\pi r^2) \rfloor$ and r is the parameter of the quasi-UDG G.*

Lemma 4.3. *The algorithm* **EdgeColoring-APX** *is an approximation algorithm of ratio 2 for the* EDGE COLORING *problem on quasi-UDGs.*

Proof. We first show that the algorithm computes an edge coloring of a given quasi-UDG G.

Let u be a node in G. By Fact 3.1, every edge incident on u belongs to one of the subgraphs $G_i(u)$, $i \in \{1, 2, 3, 4\}$, defined in line 3 of algorithm. Since u applies the greedy algorithm to the edges of $G_i(u)$ coloring an edge in $G_i(u)$ with a color that has not been used so far by an edge incident on it, node u will color its incident edges properly. Therefore, it suffices to show that for any edge (u, v), both u and v assign the *same* color to edge (u, v) to conclude that the coloring of G by the algorithm is consistent, and hence is an edge coloring of G.

For an edge $e \in G$, define $label(e)$ to be the minimum $i \in \{1, 2, 3, 4\}$ such that e is contained in an i-cluster. We say that an edge e *directly affects* another edge e' if e and e' are adjacent and either $label(e) < label(e')$ or $label(e) = label(e')$ and e comes before e' in the lexicographic order. We say that an edge e *affects* an edge e' if there exists an *affecting sequence* of edges $(e = e_0, e_1, \ldots, e_j = e')$ such that for $\ell = 0, \ldots, j-1$, e_ℓ directly affects $e_{\ell+1}$. Observe that the labels of the edges in any affecting sequence must be non-decreasing. Therefore, all the edges with the same label i in an affecting sequence form a connected subgraph of G, and hence are contained within a single i-cluster. It follows that, for any edge $e \in G$, any affecting sequence of edges containing e must be contained in some potential affecting sequence of clusters that contains e.

By looking at how the algorithm works, if the color of an edge e "influences" the color of an edge e', then edge e affects e'. For a potential affecting sequence S and an edge (u, v) in some cluster in S, both u and v in the algorithm collect the coordinates of all their k-hop neighbors, where $k = \lfloor (22\alpha^2 + 8r^2 + 32\alpha r)/(\pi r^2) \rfloor$.

Therefore, by Lemma 4.2, both u and v have collected the coordinates of every node in S. It follows that both u and v must assign edge (u, v) the same color because both u and v have the coordinates of the endpoints of all edges affecting (u, v) and will color these edges in the same order using the same algorithm.

This shows that the algorithm computes a proper edge coloring of G.

To prove that the algorithm has approximation ratio 2, let apx_G be the number of colors used by the algorithm to color the edges of G, and let opt_G be the number of colors in a minimum edge coloring of G. Note that $opt_G \geq \Delta$, where Δ is the maximum degree of G. Let $e = (u, v)$ be the edge with the highest color number, i.e., $color(e) = \max_{e' \in E(G)} color(e')$. Let Δ_u and Δ_v be the degrees of nodes u and v. Since $color(e)$ is the smallest color number that is not used by any edge incident on u or v, it follows that $color(e) \leq (\Delta_u - 1) + (\Delta_v - 1) + 1$. Since e has the highest color number among all edges in G, we have $apx_G \leq (\Delta_u - 1) + (\Delta_v - 1) + 1 \leq 2 \cdot \Delta - 1 \leq 2 \cdot opt_G - 1$. □

Theorem 4.1. *The algorithm* **EdgeColoring-APX** *is a k-local distributed approximation algorithm for the* EDGE COLORING *problem on quasi-UDGs, where $k = \lfloor (22\alpha^2 + 8r^2 + 32\alpha r)/(\pi r^2) \rfloor$, $0 < r \leq 1$ is the quasi-UDG parameter, and $\alpha > 2$ is a constant. For a UDG ($r = 1$), and by choosing α to be slightly larger than 2, the algorithm* **EdgeColoring-APX** *is a 50-local distributed approximation algorithm for* EDGE COLORING *of ratio 2.*

The above upper bound on the locality of the algorithm (i.e., k) can be improved by using smaller dimensions for the tiles; this will reduce the size of the region containing any affecting sequence, and hence decrease the upper bound on k. However, if we decrease the dimensions of the tiles, the above set of translations will no longer be sufficient to color all the edges in G (some edges may no longer reside in the interior of a tile under any of the above translations). To overcome this problem, we will need to use a family of translations, rather than a single translation, along each of the horizontal, vertical, and diagonal, directions. By fixing the dimensions of the tiles to be $(1 + \epsilon) \times (1 + \epsilon)$, where $\epsilon > 0$ is a constant, and picking an appropriate family of translations, we can prove that, in the worst case, any affecting sequence will be contained in a region whose area is at most $r^2 + (3\epsilon + 5)r + \epsilon^2 + 5\epsilon + 5$. This will give an upper bound of $8(r^2 + (3\epsilon + 5)r + \epsilon^2 + 5\epsilon + 5)/(\pi r^2)$ on k. In Table 1 we show the values of k corresponding to the values $\epsilon = 0.1, \ldots, 0.9$, and the asymptotic value of k when $\epsilon \to 0$. We note that, as ϵ decreases, the number of translations needed increases, and hence, the local computation time at the nodes increases.

Theorem 4.2. *For any constant $\epsilon > 0$, there exists a k-local distributed approximation algorithm of ratio 2 for the* EDGE COLORING *problem on quasi-UDGs,*

Table 1. Locality for different tile sizes

ϵ	0.9	0.8	0.7	0.6	0.5	0.4	0.3	0.2	0.1	$\to 0$
k	48	45	43	41	38	36	34	32	30	28

where $k = \lfloor 8(r^2 + (3\epsilon+5)r + \epsilon^2 + 5\epsilon + 5)/(\pi r^2) \rfloor$, and $0 < r \leq 1$ is the quasi-UDG parameter.

5 Strong Edge Coloring

In this section we present local distributed algorithms that approximate the STRONG EDGE COLORING problem on UDGs. Although the same approach used for the EDGE COLORING problem—in the previous section—works for the STRONG EDGE COLORING problem, this approach does not lead to good bounds on the locality of the algorithm. Therefore, we will adopt a different approach here. We note that the techniques in this section can be extended to quasi-UDGs; however, for simplicity, we restrict our attention to UDGs.

The local distributed algorithms we present use a centralized algorithm as a building block. We start by presenting this centralized algorithms.

5.1 The Centralized Algorithm

Barrett et al. [1] proposed a centralized greedy algorithm for approximating the STRONG EDGE COLORING problem on UDGs that works as follows. The nodes are first ordered using a lexicographic order. This lexicographic order on the nodes is used to induce a certain order on the edges (a bottom-up order). The edges are then considered with respect to this order, and an edge e is colored with the smallest color that has not been used to color any edge of distance at most 1 from e. If opt_G is the number of colors in a minimum strong edge coloring of G, then it was proved in [1] that the greedy algorithm computes a strong edge coloring of G that uses at most $8opt_G + 1$ colors. We will refer to the algorithm in [1] as the **Centralized-StrongEdgeColoring** algorithm.

We can show that, irrespective of the ordering in which the edges in G are considered, the algorithm **Centralized-StrongEdgeColoring** produces a strong edge coloring of G that uses at most $10opt_G$ colors. This property will be essential to bounding the approximation ratio of the algorithm we present in Subsection 5.3. The proof of this upper bound on the ratio is very similar to the proof given in [1] that the algorithm **Centralized-StrongEdgeColoring** has ratio $8opt_G + 1$ when the specific bottom-up ordering is used.

Theorem 5.1. *For any ordering $\mathcal{O}$ of the edges in G, the algorithm* **Centralized-StrongEdgeColoring***, when it considers the edges in G with respect to the ordering $\mathcal{O}$, has approximation ratio 10.*

5.2 The Local Distributed Algorithm

In this subsection we present a local distributed algorithm that approximates the STRONG EDGE COLORING problem on UDGs. The approach is similar in flavor to the one used in Section 4. Using a different approach, we shall improve on the approximation ratio significantly at the expense of worsening the locality in Subsection 5.3.

Consider the same rectilinear tiling $\mathcal{T}$ of the plane discussed in Section 4 whose tiles are $\alpha \times \alpha$ squares, where $\alpha > 2$. We can label the tiles in $\mathcal{T}$ with the labels 1, 2, 3, 4, so that any two tiles with the same label are separated by at least one tile. We denote by $label(t)$ the label of a tile $t \in \mathcal{T}$.

Fact 5.1. *Let G be a UDG, and let e and e' be two edges in G such that the endpoints of e reside in the interior of a tile t, and the endpoints of e' reside in the interior of a tile t', where $t \neq t'$, and such that $label(t) = label(t')$. Then the distance between e and e' is at least 2.*

Proof. (Sketch) The statement follows from the facts that: (1) any two different tiles with the same label are separated by at least one tile, and (2) the dimension of a tile is greater than 1. □

Let T_1, T_2, T_3, and T_4, be the translations described in Section 4, and note that since $\alpha > 2$, Lemma 3.1 still holds true. Let C_i^1, C_i^2, C_i^3, and C_i^4, for $i = 1, 2, 3, 4$, be 16 mutually disjoint color classes. We assume that each of the color classes contains enough colors to color the edges of G, and that the colors in each class are ordered from smallest to largest.

Suppose that $\mathcal{A}$ is a centralized approximation algorithm of ratio $\rho_{\mathcal{A}}$ for the STRONG EDGE COLORING problem on UDGs. Intuitively, the algorithm can be summarized as follows. The algorithm runs in 4 rounds, each round corresponds to one of the above translations. Different color classes are used in different rounds to ensure that edges that are colored in different rounds do not conflict. In a given round i, translation T_i is applied to all the edges, and only the edges whose translations are interior to the tiles in $\mathcal{T}$ are colored as follows: the edges whose translations are in the same connected component of a tile of label j are colored with colors from class C_i^j, using the centralized algorithm $\mathcal{A}$. This ensures that edges whose translations end up in tiles of different labels are colored differently. Since different tiles of the same label are far enough from each other, and the centralized algorithm $\mathcal{A}$ is used to color the edges within the same tile, edges that are colored in the same round are colored properly.

More formally, each node p in G applies the algorithm **Strong-Edge-Coloring-APX** given in Figure 2.

Lemma 5.1. *The algorithm **Strong-Edge-Coloring-APX** is a k-local distributed algorithm, where $k = \lfloor 8(\alpha + 1)^2 / \pi \rfloor$.*

Lemma 5.2. *The algorithm **Strong-Edge-Coloring-APX** computes a valid strong edge coloring of G.*

Lemma 5.3. *The algorithm **Strong-Edge-Coloring-APX** approximates the STRONG EDGE COLORING problem on UDGs to a ratio $16 \cdot \rho_{\mathcal{A}}$, where $\rho_{\mathcal{A}}$ is the approximation ratio of $\mathcal{A}$.*

Proof. Let j be the round among the 4 rounds of the algorithm in which the maximum number of colors, apx_j, is used. It follows from the choice of j that the total number of colors used by the algorithm, call it apx_G, is at most $4 \cdot apx_j$.

1: p collects the coordinates of the nodes in $N_k[p]$ in G, where $k = \lfloor 8(\alpha+1)^2/\pi \rfloor$
2: **for** round $i = 1, 2, 3, 4$ **do**
3:    p applies translation T_i and computes its virtual coordinates under T_i
4:    if $T_i(p)$ is interior to some tile t_0 with label $\ell_0 \in \mathcal{T}$, where $\ell_0 \in \{1,2,3,4\}$, p determines the set $S_i(p)$ of all the nodes in $N_k[p]$ whose translations under T_i reside in the same connected component as $T_i(p)$ in the interior of tile t_0; Let $H_i(p)$ be the subgraph of G induced by $S_i(p)$
5:    p applies the algorithm $\mathcal{A}$ to the subgraph $H_j(p)$ to compute a strong edge coloring of $H_j(p)$, using only colors from the color class $C_{\ell_0}^j$, and starting with the smallest color in $C_{\ell_0}^j$; if an edge $e \in H_j(p)$ has already been colored in a previous round, p overwrites the previous color of e
6: **end for**

Fig. 2. The algorithm **Strong-Edge-Coloring-APX**

Let ℓ_j be the label of the color class from which the maximum number of colors, $apx_{\ell_j}^j$ is used in round j. Since there are 4 labels, it follows that $apx_{\ell_j}^j \leq 4 \cdot apx_j$, and hence, $apx_G \leq 16 \cdot apx_{\ell_j}^j$. Let opt_G be the number of colors in a minimum strong edge coloring of G.

From the way the algorithm works, in round j, every set of nodes S in G whose translations are in the same connected component in the interior of some tile with label ℓ_j, apply the algorithm $\mathcal{A}$ to compute a strong edge coloring of the edges of the subgraph of G induced by S, using the same set of colors $C_{\ell_j}^j$, and in the same order (all starting with the smallest color in $C_{\ell_j}^j$). Therefore, there exists a set of nodes S_j in G, whose translations reside in the same connected component in the interior of some tile, such that algorithm $\mathcal{A}$ uses $apx_{\ell_j}^j$ colors to properly color the edges of the subgraph H_j induced by S_j. Since $\mathcal{A}$ has approximation ratio $\rho_{\mathcal{A}}$, a minimum strong edge coloring of H_j requires at least $apx_{\ell_j}^j/\rho_{\mathcal{A}}$ colors. Since H_j is an induced subgraph of G, a minimum strong edge coloring of G requires at least $apx_{\ell_j}^j/\rho_{\mathcal{A}}$ colors. It follows that $opt_G \geq apx_{\ell_j}^j/\rho_{\mathcal{A}}$, and $16 \cdot apx_j \leq 16 \cdot \rho_{\mathcal{A}} \cdot opt_G$. This shows that the algorithm properly colors the edges of G using no more than $16 \cdot \rho_{\mathcal{A}} \cdot opt_G$ colors, and hence has ratio $16 \cdot \rho_{\mathcal{A}}$. □

Theorem 5.2. *There exists a 22-local distributed algorithm that, given a UDG G, computes a strong edge coloring of G using at most $128 \cdot opt_G + 16$ colors, where opt_G is the number of colors in a minimum strong edge coloring of G.*

Proof. Since a node p in the algorithm **Strong-Edge-Coloring-APX** can consider the edges in H_p in any order, p can order these edges according to the bottom-up ordering used in [1]. Under this specific ordering, as was mentioned before, the algorithm **Centralized-StrongEdgeColoring** computes a strong edge coloring of H_p using at most $8 \cdot opt_{H_p} + 1$ colors, where opt_{H_p} is the number of colors in a minimum strong edge coloring of H_p. Using the algorithm **Centralized-StrongEdgeColoring** as the subroutine $\mathcal{A}$ in the algorithm

Strong-Edge-Coloring-APX, and setting α to a value slightly larger than 2, the statement follows from Lemma 5.1, Lemma 5.2, and Lemma 5.3. □

5.3 The Improved Algorithm

In this subsection we present a local distributed algorithm for the STRONG EDGE COLORING problem on UDGs with a smaller approximation ratio, but larger locality, than the algorithm presented in Subsection 5.2. The algorithm uses the same tiling $\mathcal{T}$, but we require that $\alpha > 3$. The tiles are labeled with the labels $1, 2, 3, 4$ as in Subsection 5.2.

Each node is assigned to the tile which contains it. Ambiguities caused by nodes on the boundaries of tiles are resolved by assigning them to the tile with the smallest label which contains them (any other resolving method works as well). We observe that two tiles of the same label have a Euclidean distance more than 3. Therefore, if we place a bounding square box of dimensions $(\alpha + 1) \times (\alpha + 1)$ centered at each tile, two bounding boxes of two tiles with the same label have a Euclidean distance larger than 1. Consequently, two edges contained in different bounding boxes of two tiles with the same label have distance at least 2, and can be colored in the same round. The improved algorithm is given in Figure 3.

1: p collects the coordinates of the nodes in $N_k[p]$ in G, where $k = \lfloor (32\alpha^2 + 80\alpha + 40)/\pi \rfloor$
2: **for** round $i = 1, 2, 3, 4$ **do**
3:    let $G_i(p)$ be a copy of the subgraph of G consisting of the set $E_i(p)$ of uncolored edges whose endpoints are in $N_k[p]$, and such that, for any edge $(u, v) \in E_i(p)$, u and v are in the bounding box of some tile of label i
4:    p colors all the uncolored edges in $G_i(p)$ using the algorithm **Centralized-StrongEdgeColoring**
5:    **for** each edge $e \in G_i(p)$ incident on p **do**
6:       p colors e in G with the same color in $G_i(p)$
7:    **end for**
8: **end for**

Fig. 3. The algorithm **Improved-StrongEdgeColoring-APX**

Lemma 5.4. *The algorithm is a k-local distributed algorithm, where* $k = \lfloor (32\alpha^2 + 80\alpha + 40)/\pi \rfloor$, *that computes a strong edge coloring of a given UDG.*

Lemma 5.5. *The algorithm is an approximation algorithm of ratio 10 for the* STRONG EDGE COLORING *problem on UDGs.*

Proof. By Lemma 5.4, the algorithm **Improved-StrongEdgeColoring-APX** is an approximation algorithm for the STRONG EDGE COLORING problem on UDGs. To prove that the algorithm has ratio 10, note that the algorithm **Improved-StrongEdgeColoring-APX** is equivalent to the algorithm **Centralized-StrongEdgeColoring** applied to the edges of G in the order they were colored by the algorithm **Improved-StrongEdgeColoring-APX**. It follows from Theorem 5.1 that the algorithm has ratio 10. □

Theorem 5.3. *Given a UDG G and a constant $\alpha > 3$, the algorithm* **Improved-StrongEdgeColoring-APX** *is a a k-local distributed algorithm, where $k = \lfloor (32\alpha^2 + 80\alpha + 40)/\pi \rfloor$, that computes a strong edge coloring of G using at most $10 opt_G$ colors, where opt_G is the number of colors in a minimum strong edge coloring of G. By choosing α to be slightly larger than 3, the algorithm* **Improved-StrongEdgeColoring-APX** *is a 180-local distributed algorithm of ratio 10.*

References

1. Barrett, C., Kumar, V., Marathe, M., Thite, S., Istrate, G.: Strong edge coloring for channel assignment in wireless radio networks. In: PERCOMW 2006, pp. 106–110 (2006)
2. Czyzowicz, J., Dobrev, S., Kranakis, E., Opatrny, J., Urrutia, J.: Local edge colouring of yao-like subgraphs of unit disk graphs. In: Prencipe, G., Zaks, S. (eds.) SIROCCO 2007. LNCS, vol. 4474, pp. 195–207. Springer, Heidelberg (2007)
3. Gandham, S., Dawande, M., Prakash, R.: Link scheduling in sensor networks: distributed edge coloring revisited. In: INFOCOM, pp. 2492–2501 (2005)
4. Holyer, I.: The NP-completeness of edge-coloring. SIAM J. Comput. 10(4), 718–720 (1981)
5. Ito, T., Kato, A., Zhou, X., Nishizeki, T.: Algorithms for finding distance-edge-colorings of graphs. J. Discrete Algorithms 5(2), 304–322 (2007)
6. Kanj, I., Wiese, A., Zhang, F.: Computing the k-hop neighborhoods locally. Technical report # 08-007 at:
   http://www.cdm.depaul.edu/research/Pages/TechnicalReports.aspx
7. Kodialam, M., Nandagopal, T.: Characterizing achievable rates in multi-hop wireless mesh networks with orthogonal channels. IEEE/ACM Trans. Netw. 13(4), 868–880 (2005)
8. Linial, N.: Locality in distributed graph algorithms. SIAM J. Comput. 21(1), 193–201 (1992)
9. Mahdian, M.: On the computational complexity of strong edge coloring. Discrete Applied Mathematics 118(3), 239–248 (2002)
10. Misra, J., Gries, D.: A constructive proof of vizing's theorem. IPL 41 (1992)
11. Ramanathan, S.: A unified framework and algorithm for channel assignment in wireless networks. Wirel. Netw. 5(2), 81–94 (1999)
12. Ramanathan, S., Lloyd, E.L.: Scheduling algorithms for multihop radio networks. IEEE/ACM Trans. Netw. 1(2), 166–177 (1993)
13. Vizing, V.: On the estimate of the chromatic class of p-graphs. Diskret. Analiz 3, 25–30 (1964)
14. Wiese, A., Kranakis, E.: Local construction and coloring of spanners of location aware unit disk graphs. Technical report # 07-18 at,
    http://www.scs.carleton.ca/~kranakis/Papers/TR-07-18.pdf

Directed Rank-Width and Displit Decomposition*

Mamadou Moustapha Kanté[1] and Michaël Rao[2]

[1] LIMOS - Université Blaise Pascal, France
[2] LABRI - Université Bordeaux 1, France
mamadou.kante@isima.fr, rao@labri.fr

Abstract. *Rank-width* is a graph complexity measure that has many structural properties. It is known that the rank-width of an undirected graph is the maximum over all induced prime graphs with respect to *split decomposition* and an undirected graph has rank-width at most 1 if and only if it is a distance-hereditary graph. We are interested in an extension of these results to directed graphs. We give several characterizations of directed graphs of rank-width 1 and we prove that the rank-width of a directed graph is the maximum over all induced prime graphs with respect to *displit decomposition*, a new decomposition on directed graphs.

1 Introduction

Rank-width [18,19] is a graph complexity measure introduced by Oum and Seymour in their investigations on recognition algorithms for undirected graphs of *clique-width* [4] at most k, for fixed k. It is known that a class of graphs has bounded rank-width if and only if it has bounded clique-width [19]. However, rank-width has better algorithmic properties: undirected graphs of rank-width at most k can be recognized by a cubic-time algorithm [13] and are characterized by a finite list of undirected graphs to exclude as *vertex-minors* [18].

Another interesting fact is that rank-width is related to *split decomposition*. The split decomposition, introduced by Cunningham [5], is a generalisation of the well known *modular decomposition* [10,16]. It was defined on graphs (directed or not), but only the undirected case has been widely studied in literature. Split decomposition of undirected graphs can be computed in linear time [7], and can be used in several problems such as: circle graph recognition [9,21], parity graph recognition [3,7], and solving some optimization problems [5,3,11,20]. The rank-width of an undirected graph is the maximum over the rank-width of its induced prime graphs with respect to split decomposition. Moreover, undirected graphs of rank-width at most 1 are exactly *distance hereditary* graphs [18], which are graphs that are *completely decomposable* by the split decomposition.

Despite all these positive results of rank-width on clique-width, clique-width has an undeniable advantage on rank-width: it is defined for undirected as well as directed graphs and its definition can be extended to relational structures. In

---

* Research supported by the French ANR-project "Graph decompositions and algorithms" (GRAAL).

his investigations for an extension of rank-width to relational structures, Kanté defined in [15] a notion of rank-width for directed graphs, called GF(4)-*rank-width*, and that generalized the rank-width of undirected graphs. He, moreover, generalized two results on undirected graphs: directed graphs of GF(4)-rank-width k can be recognized by a cubic-time algorithm and are also characterized by a finite list of directed graphs to exclude as vertex-minors. It is thus natural to ask whether we can generalize all the results known for rank-width of undirected graphs.

In this paper, we are interested in a characterization of directed graphs of GF(4)-rank-width 1, similar to the one for undirected graphs. In the literature, there exist several characterizations of undirected graphs of rank-width 1 that we recall in the following.

Theorem 1 ([1,12,18]). *Let G be a connected undirected graph. Then the following conditions are equivalent:*

1. *G is completely decomposable by the split decomposition (i.e., every node in the split decomposition tree is degenerated).*
2. *G can be obtained from a single vertex by creating twins or adding pendant vertices.*
3. *G has rank-width 1.*
4. *For every $W \subseteq V_G$ with $|W| \geq 4$, $G[W]$ has a non trivial split.*
5. *G is (house, hole, domino, gem)-free.*
6. *G is distance hereditary (i.e., for every $x, y \in V_G$, every chordless path between x and y has the same length).*

The main result of this paper is the extension of Theorem 1 to directed graphs (Theorem 6). We will show in particular that directed graphs of GF(4)-rank-width 1 are obtained by orienting in a certain way distance hereditary graphs and are exactly directed graphs completely decomposable by the *displit decomposition*, a new decomposition that generalizes split decomposition. As a consequence we get that the GF(4)-rank-width of a directed graph is the maximum over the GF(4)-rank-width of its induced prime graphs with respect to displit decomposition.

The paper is organized as follows. We give some notations in Section 2 and recall the notion of GF(4)-rank-width in Section 3. In Section 4 we define the notion of displit decomposition and derive some basic properties. In Section 5 we prove our main result. We conclude by a comparison between the split decomposition of directed graphs introduced by Cunningham [5] and the displit decomposition.

2 Preliminaries

When the context is clear we will write u to denote the set $\{u\}$. We denote by 2^V the power-set of a set V and we let $\mathbb{N}$ be the set of natural integers. A function $f : 2^V \to \mathbb{N}$ is said *symmetric* if for any $X \subseteq V$, $f(X) = f(V \backslash X)$; it is said *sub-modular* if for any $X, Y \subseteq V, f(X \cup Y) + f(X \cap Y) \leq f(X) + f(Y)$.

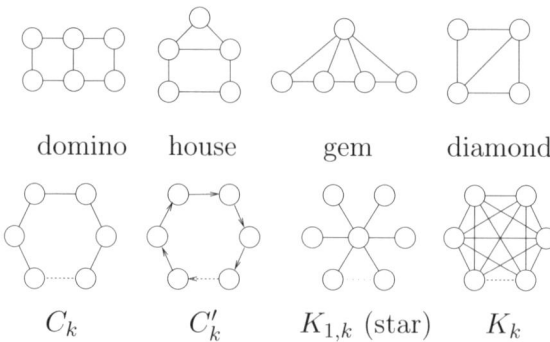

domino   house   gem   diamond

C_k C'_k $K_{1,k}$ (star) K_k

For sets R and C, an (R,C)-*matrix* is a matrix where the rows are indexed by elements in R and columns indexed by elements in C. For an (R,C)-matrix M, if $X \subseteq R$ and $Y \subseteq C$, we let $M[X,Y]$ be the sub-matrix of M where the rows and the columns are indexed by X and Y respectively. If M is an (X,Y)-matrix, M^t denotes the transposed (Y,X)-matrix. A *Y-vector* is an (X,Y)-matrix where $|X|=1$. The matrix rank function is denoted by rk.

A *directed graph* (or *digraph*) G is a couple (V_G, E_G) where V_G is the set of *vertices* and E_G, the set of *edges*, is a set of ordered pairs (x,y) with $x,y \in V_G$ and $x \neq y$. We consider *undirected graphs* as special cases of directed graphs where $(x,y) \in E_G \Leftrightarrow (y,x) \in E_G$ (edges are denoted xy in this case). Unless otherwise specified, a graph is considered as directed. If G is a digraph and x a vertex of G, we denote by $N_G^+(x)$ the set $\{y \mid (x,y) \in E_G\}$, by $N_G^-(x)$ the set $\{y \mid (y,x) \in E_G\}$ and by $N_G(x)$ the set $N_G^+(x) \cup N_G^-(x)$. The *degree* of x is $|N_G(x)|$.

For a graph G, we denote by $G[X]$ the sub-graph of G induced by $X \subseteq V_G$ and we let $G - X$ be the sub-graph $G[V_G \backslash X]$. If G is a digraph, let $u(G)$ be the undirected graph obtained from G by forgetting the directions of edges, *i.e.*, $u(G) = (V_G, E_G \cup \{(y,x) \mid (x,y) \in E_G\})$. A digraph G is said *strongly connected* if for every pair $x,y \in V_G$, there is a sequence $x_0 = x, x_1, \ldots x_k = y$ such that $(x_i, x_{i+1}) \in E_G$ for every $i \in \{0, \ldots k-1\}$, and it is said *connected* if $u(G)$ is connected.

An undirected graph is *acyclic* if it does not contain simple cycles of length at least 3. A *tree* is an acyclic connected undirected graph. In order to avoid confusions, the vertices of trees will be called *nodes*. The nodes of degree at most 1 in trees are called *leaves* and denoted by L_T. A *sub-cubic* tree is a tree such that the degree of each node is at most 3.

A *layout* of a set V is a pair $(T, \mathcal{L})$ of an undirected tree T and a bijective function $\mathcal{L}: V \to L_T$. For each edge (u,v) of T, we let X_{uv} be the set of leaves reachable from u by a path going through v. Each edge (u,v) of T induces a bipartition $\{X_{uv}, L_T \backslash X_{uv}\}$ of L_T, and thus a bipartition $\{X^{uv}, V \backslash X^{uv}\} = \{\mathcal{L}^{-1}(X_{uv}), \mathcal{L}^{-1}(L_T \backslash X_{uv})\}$ of V.

3 Rank-Width of Digraphs

In [15] Kanté defined a notion of rank-width for digraphs named GF(4)-*rank-width*. This notion is based on a function, called *cut-rank function*, that measures how some bipartitions of sets of vertices are connected. The cut-rank function is based on a representation of digraphs by matrices over the field GF(4). We recall that GF(4) has four elements $\{0, 1, \mathfrak{o}, \mathfrak{o}^2\}$ with the property that $1 + \mathfrak{o} + \mathfrak{o}^2 = 0$ and $\mathfrak{o}^3 = 1$ and is of characteristic 2.

For a digraph G, we denote by M_G the (V_G, V_G)-matrix over GF(4) where:

$$M_G[x,y] = \begin{cases} 0 & \text{if } (x,y) \notin E_G \text{ and } (y,x) \notin E_G \\ \mathfrak{o} & \text{if } (x,y) \in E_G \text{ and } (y,x) \notin E_G \\ \mathfrak{o}^2 & \text{if } (y,x) \in E_G \text{ and } (x,y) \notin E_G \\ 1 & \text{if } (x,y) \in E_G \text{ and } (y,x) \in E_G. \end{cases}$$

For every subset X of V_G, we let $\mathrm{cutrk}_G^{(4)}(X)$, called *cut-rank function*, be $\mathrm{rk}\left(M_G[X, V_G \setminus X]\right)$.

Lemma 1 ([15]). *For every digraph G, the function $\mathrm{cutrk}_G^{(4)}$ is symmetric and sub-modular.*

Definition 1 (GF(4)-Rank-Width). *A sub-cubic layout of a digraph G is a layout $(T, \mathcal{L})$ of V_G where T is sub-cubic. Let $(T, \mathcal{L})$ be a sub-cubic layout of a digraph G. The GF(4)-rank-width of an edge (u, v) of T is $\mathrm{cutrk}_G^{(4)}(X^{uv})$. The GF(4)-rank-width of a sub-cubic layout $(T, \mathcal{L})$ is the maximum GF(4)-rank-width over all edges of T. The GF(4)-rank-width of G, denoted by $\mathrm{rwd}^{(4)}(G)$, is the minimum GF(4)-rank-width over all sub-cubic layouts of G.*

Observation 1. *Since GF(4) is an extension of GF(2), for every undirected graph G, we have $\mathrm{rwd}^{(4)}(G) = \mathrm{rwd}(G)$, where $\mathrm{rwd}(G)$ denotes the rank-width of G.*

4 Displit Decomposition

4.1 Bi-Partitive Families

Two bipartitions $\{X_1, X_2\}$ and $\{Y_1, Y_2\}$ of a set V *overlap* if $X_i \cap Y_j \neq \emptyset$ for every $i, j \in \{1, 2\}$.

Definition 2 (Bi-Partitive Family). *Let V be a finite set and let $\mathcal{F}$ be a family of bipartitions of V. Then $\mathcal{F}$ is* bi-partitive *if:*

- $\{\emptyset, V\} \notin \mathcal{F}$,
- *for all $v \in V$, $\{\{v\}, V \setminus \{v\}\} \in \mathcal{F}$ and*
- *for all $\{X_1, X_2\} \in \mathcal{F}$ and $\{Y_1, Y_2\} \in \mathcal{F}$ such that $\{X_1, X_2\}$ and $\{Y_1, Y_2\}$ overlap, then $\{X_i \cap Y_j, V \setminus (X_i \cap Y_j)\} \in \mathcal{F}$, for every $i, j \in \{1, 2\}$.*

A member $\{X_1, X_2\}$ of a bi-partitive family $\mathcal{F}$ is trivial *if $|X_1| \leq 1$ or $|X_2| \leq 1$, and is* strong *if there is no $\{Y_1, Y_2\} \in \mathcal{F}$ such that $\{X_1, X_2\}$ and $\{Y_1, Y_2\}$ overlap.*

Bi-partitive families have been studied in [6]. They are very close to partitive families [2,16] introduced in order to generalize properties of modular decomposition. An example of a bi-partitive family is the family of splits[1] in a strongly connected digraph [5]. The following proposition gives another example of a bi-partitive family.

Proposition 1 (Folklore). *Let $f : 2^V \to \mathbb{N}$ be a symmetric and sub-modular function and let $m = \min_{\emptyset \subsetneq X \subsetneq V} f(X)$. Then the family $\mathcal{F} = \{\{X, V\backslash X\} \mid f(X) = m\}$ is bi-partitive.*

Proof. Let $\{X, V\backslash X\}$ and $\{Y, V\backslash Y\}$ be in $\mathcal{F}$ such that $\{X, V\backslash X\}$ and $\{Y, V\backslash Y\}$ overlap. Thus $f(X \cap Y) + f(X \cup Y) \leq 2m$. Since $X \cap Y$ and $X \cup Y$ are non-empty, $f(X \cap Y) \geq m$ and $f(X \cup Y) \geq m$. Thus $f(X \cap Y) = f(X \cup Y) = m$ and $\{X \cap Y, V\backslash(X \cap Y)\}$ and $\{X \cup Y, V\backslash(X \cup Y)\}$ are in $\mathcal{F}$. □

A major result on bi-partitive families, that we recall in the following theorem, is that every bi-partitive family can be represented by a unique labeled tree.

Theorem 2. *Let $\mathcal{F}$ be a bi-partitive family on a finite set V. Then there is a unique layout $(T, \mathcal{L})$ of V, called the* representative layout, *such that each internal node of T has at least 3 neighbors, is marked* `degenerate`, `linear` *or* `prime` *and:*

- *For every $(u, v) \in E_T$, the bipartition $\{X^{uv}, V\backslash X^{uv}\}$ is a strong bipartition in $\mathcal{F}$ and there is no other strong bipartition in $\mathcal{F}$.*
- *For every internal node u of T:*
  - *If u is* `degenerated`, *then for every $\emptyset \subsetneq W \subsetneq N_T(u)$, the bipartition $\{\cup_{v \in W} X^{uv}, V\backslash \cup_{v \in W} X^{uv}\}$ is in $\mathcal{F}$.*
  - *If u is* `linear`, *there is an ordering $v_1, \ldots, v_k$ of $N_T(u)$ such that for every $1 \leq i \leq j < k$, the bipartition $\{\cup_{\ell \in \{i,\ldots,j\}} X^{uv_\ell}, V\backslash \cup_{\ell \in \{i,\ldots,j\}} X^{uv_\ell}\}$ is in $\mathcal{F}$.*
- *There is no other bipartition in $\mathcal{F}$.*

(By convention, an internal node of degree 3 is always degenerated.)

Remark 1. Theorem 2 is proved in [6] using a different formalism. It follows also directly from results on partitive families [2,16] using the simple bijection $f(\mathcal{F}) = \{X \subseteq V\backslash\{v\} \mid \{X, V\backslash X\} \in \mathcal{F}\}$ between bi-partitive families on V and partitive families on $V\backslash\{v\}$, where $v \in V$ is fixed.

Remark 2. If $\mathcal{F}$ is a bi-partitive family with the additional property:

- for all $\{X_1, X_2\} \in \mathcal{F}$ and $\{Y_1, Y_2\} \in \mathcal{F}$ such that $\{X_1, X_2\}$ and $\{Y_1, Y_2\}$ overlap, $\{X_1 \Delta Y_1, X_1 \Delta Y_2\} \in \mathcal{F}$ [2],

---

[1] A *split* in a digraph G is a bipartition $\{X, V_G\backslash X\}$ of V_G, where $\emptyset \subsetneq X \subsetneq V_G$, such that for every $u, v \in X$, $(N_G^+(u) \setminus X \neq \emptyset) \wedge (N_G^+(v) \setminus X \neq \emptyset) \Rightarrow (N_G^+(u) \setminus X = N_G^+(v) \setminus X)$, and $(N_G^-(u) \setminus X \neq \emptyset) \wedge (N_G^-(v) \setminus X \neq \emptyset) \Rightarrow (N_G^-(u) \setminus X = N_G^-(v) \setminus X)$.
[2] For two sets X and Y, we let $X \Delta Y$ be the set $X\backslash Y \cup Y\backslash X$.

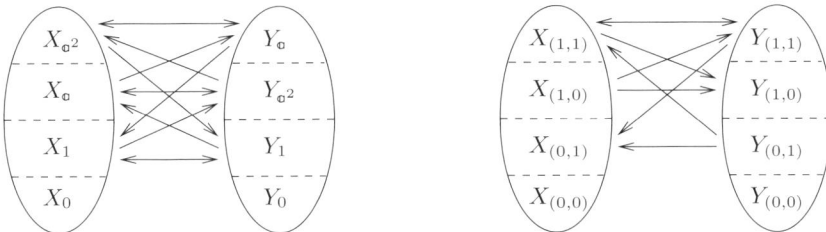

Fig. 1. Schematic view of a displit (left) and a Cunningham's split (right)

then $\mathcal{F}$ is said to be *strongly bi-partitive*. The representative layout of a strongly bi-partitive family has no `linear` node. Cunningham showed that the family of splits in a connected undirected graph is strongly bi-partitive [5]. Another example is the family of bi-joins in an undirected graph [17].

4.2 Displits

Definition 3 (Displit). *Let G be a digraph. A bipartition $\{X_1, X_2\}$ of V_G is a displit if $X_1 \neq \emptyset$, $X_2 \neq \emptyset$ and $\text{cutrk}_G^{(4)}(X_1) \leq 1$.*

Figure 1 shows a comparison between displits and splits on digraphs. A digraph G is *degenerated* (for the displit decomposition) if every bipartition of V_G is a displit, and G is *prime* if every displit in G is trivial. Finally G is *linear* if there is an ordering $x_1, \ldots, x_n$ of its vertices such that the family of displits in G is $\{\{\{x_i, \ldots, x_j\}, V_G \setminus \{x_i, \ldots, x_j\}\} \mid 1 \leq i \leq j < n\}$. By convention, a graph with at most 3 vertices is only degenerated.

By Proposition 1, the family of displits in a connected digraph is bi-partitive. By Theorem 2, this family can be represented by a unique labeled layout, that we call *displit decomposition*.

Observation 2. *If $\{X_1, X_2\}$ is a displit in G, then $\{X_1, X_2\}$ is a split in $u(G)$. The converse is not necessarily true.*

4.3 Quotient Graphs

Let $(T, \mathcal{L})$ be a displit decomposition of a connected digraph G and let u be an internal node of T. We recall that for every node v in $N_T(u)$, X_{uv} is the set of leaves reachable from u by a path going through v. The set $\{X^{uv} = \mathcal{L}^{-1}(X_{uv}) \mid v \in N_T(u)\}$ is a proper partition of V_G, and for every $v \in N_T(u)$, $\{X^{uv}, V_G \setminus X^{uv}\}$ is a displit.

For every $v \in N_T(u)$, we choose a vertex x_v in X^{uv} such that x_v is adjacent to a vertex in $V_G \setminus X^{uv}$. Such a x_v always exists since G is connected. Let $C(u)$ be the graph of vertex set $N_T(u)$ and of edge set $\{(v, w) \mid (x_v, x_w) \in E_G\}$. It is worth noticing that $C(u)$ is isomorphic to $G[\{x_v \mid v \in N_T(u)\}]$, and that $C(u)$ is not unique for a node u. Then we will consider $C(u)$ as an induced sub-graph of G. We now prove or state some technical lemmas.

Lemma 2. *Let $\{X,Y\}$ be a displit in G, and let $x \in X$ and $y \in Y$ such that x is adjacent to y. Let $\{X',Y'\}$ be a bipartition of V_G with $Y' \subseteq Y$. Then $\operatorname{cutrk}_G^{(4)}(Y') = \operatorname{cutrk}_{G'}^{(4)}(Y')$, where $G' = G[Y \cup \{x\}]$.*

Proof. Obviously $\operatorname{cutrk}_{G'}^{(4)}(Y') \leq \operatorname{cutrk}_G^{(4)}(Y')$. By definition of displits, there is an X-vector A and a Y-vector B such that $M_G[X,Y] = A^t \cdot B$. Since x is adjacent to a vertex in Y, $A[x] \neq 0$. Thus $M_G[X,Y'] = A[x]^{-1} \cdot A^t \cdot M_G[\{x\},Y']$. Therefore, $\operatorname{rk}(M_G[X'\backslash(X\backslash\{x\}),Y']) = \operatorname{rk}(M_G[X',Y'])$ since all rows in $M_G[X,Y']$ are generated by the row $M_G[\{x\},Y']$. □

Lemma 3. *Let $(T,\mathcal{L})$ be a displit decomposition of a digraph G and let u be a node of T. If u is prime (resp. degenerated, linear), then $C(u)$ is prime (resp. degenerated, linear).*

Proof. Let $\{X,Y\}$ be a bi-partition of $V_{C(u)}$, let $X' = \bigcup_{v \in X} X^{uv}$ and let $Y' = V_G \backslash X'$. We show that $\{X,Y\}$ is a displit in $C(u)$ if and only if $\{X',Y'\}$ is a displit in G. Trivially, if $\{X',Y'\}$ is a displit in G, then $\{X,Y\}$ is a displit in $C(u)$.

Now suppose that $\{X,Y\}$ is a displit in $C(u)$. $\{X',Y'\}$ does not overlap $\{X^{uv},V_G\backslash X^{uv}\}$ for every $v \in N_T(u)$. We apply $|N_T(u)|$ times Lemma 2, for all $\{X^{uv},V_G\backslash X^{uv}\}$. Thus $\{X',Y'\}$ is a displit if and only if $\{X,Y\}$ is a displit. □

The following lemmas give characterization of degenerated and linear digraphs. (Proofs are omitted.)

Lemma 4. *If G is degenerated with at least 4 vertices, then either $u(G)$ is a star, or G is C'_3 where each of the 3 vertices is substituted by a complete graph (maybe with 0 vertex).*

Lemma 5. *If G is linear and has at least 4 vertices, then there is an ordering $(x_1,\ldots,x_n)$ of vertices of V_G, and a function $f : V_G \to \{0,1,2\}$ such that for all $j > i$:*

- *$(x_i,x_j) \in E_G$ if $f(x_i) \equiv f(x_j) \pmod{3}$ or $f(x_i) \equiv f(x_j)+1 \pmod{3}$,*
- *$(x_j,x_i) \in E_G$ if $f(x_i) \equiv f(x_j)-1 \pmod{3}$ or $f(x_i) \equiv f(x_j)+1 \pmod{3}$,*
- *there are no other edges in the graph.*

Theorem 3. *Let G be a connected digraph with at least 3 vertices, and let $(T,\mathcal{L})$ be its displit decomposition. Then $\operatorname{rwd}^{(4)}(G) = \max\{\operatorname{rwd}^{(4)}(C(u)) \mid u \in V_T\backslash L_T\}$.*

Proof. Let $m = \max\{\operatorname{rwd}^{(4)}(C(u)) \mid u \in V_T\backslash L_T\}$. Obviously $m \leq \operatorname{rwd}^{(4)}(G)$ (since $C(u)$ is an induced sub-graph of G). For every $u \in V_T\backslash L_T$, let $(T_u,\mathcal{L}_u)$ be a sub-cubic layout of $C(u)$ of GF(4)-rank-width at most m. We suppose w.l.o.g. that the T_u are pairwise disjoint. We construct a sub-cubic layout $(T',\mathcal{L}')$ of G of GF(4)-rank-width at most m. Let T' be the union of all T_u (for $u \in V_T\backslash L_T$), after the identification of the vertices u in T_v and v in T_u for every $(u,v) \in E_{T-L_T}$, and after contraction of every vertex of degree 2. For all $x \in V_G$, let $\mathcal{L}'(x) = \mathcal{L}_u(\mathcal{L}(x))$ where $\{u\} = N_T(\mathcal{L}(x))$.

It is not hard to see that $(T',\mathcal{L}')$ is a sub-cubic layout of G. Moreover, by Lemma 2, in T' every edge has GF(4)-rank-width at most m. □

4.4 Decomposition Algorithm

It is known that the split decomposition of an undirected graph can be computed in linear time [7], and the split decomposition of a digraph in time $O(m \log(n))$ [14]. We present here a simple $O(nm)$ algorithm to compute the displit decomposition of a digraph. This algorithm is a simple adaptation of [9]. Due to space limitation, we present only the main lines, stated in the following two lemmas without proofs.

Lemma 6. *Let x and y be two vertices of a connected digraph G. We can compute in time $O(n+m)$ a non trivial displit $\{X,Y\}$ such that $x \in X$ and $y \in Y$ (if it exists).*

Lemma 7. *Given a digraph G, we can compute in time $O(nm)$ a family $\mathcal{F}$ of non overlapping displits such that for every displit $\{X,Y\}$ in G, either $\{X,Y\} \in \mathcal{F}$, or there is a bipartition $\{X',Y'\} \in \mathcal{F}$ such that $\{X,Y\}$ and $\{X',Y'\}$ overlap.*

The family constructed in the previous lemma contains obviously all strong displits in G. A final $O(nm)$ procedure finds every non-strong displits in $\mathcal{F}$. This leads to the following theorem.

Theorem 4. *The displit decomposition of every digraph can be computed in time $O(nm)$.*

5 Digraphs of GF(4)-Rank-Width 1

In [15] Kanté defined a notion of *vertex-minor* for digraphs that extended the one for undirected graphs. He also characterized the class of digraphs of GF(4)-rank-width at most k in the following.

Theorem 5 ([15]). *For each k, there is a finite list $\mathcal{C}_k$ of digraphs having at most $(6^{k+1} - 1)/5$ vertices such that a digraph G has GF(4)-rank-width at most k if and only if no digraph in $\mathcal{C}_k$ is isomorphic to a vertex-minor of G.*

When $k = 1$, the digraphs to exclude as vertex-minors have at most 7 vertices. However, we do not know any polynomial-time algorithm that checks whether a given graph is a vertex-minor of another. We will give in this section several characterizations of digraphs of GF(4)-rank-width 1. As a consequence we get an algorithm for recognizing digraphs of GF(4)-rank-width 1.

A vertex x of a digraph G is a *pendant vertex* of another vertex y if y is the only neighbor of x in G. Two vertices x and y of a digraph G are called *dtwins* if x and y verify one of the following exclusive conditions ($A = N^+_{G-y}(x)$, $B = N^-_{G-y}(x)$):

1. $N^+_{G-x}(y) = A$, $N^-_{G-x}(y) = B$ or,
2. $N^+_{G-x}(y) = B$, $N^-_{G-x}(y) = (B \backslash A) \cup (A \backslash B)$ or,
3. $N^+_{G-x}(y) = (A \backslash B) \cup (B \backslash A)$, $N^-_{G-x}(y) = A$.

We say that a digraph is *completely decomposable by the displit decomposition* if every node in the displit decomposition is degenerate or linear. The main result of this paper is the following theorem, analogous to Theorem 1.

Theorem 6. *Let G be a connected digraph with at least 2 vertices. Then the following conditions are equivalent:*

1. *G is completely decomposable by the displit decomposition.*
2. *G can be obtained from a single vertex by creating dtwins or adding pendant vertices.*
3. *G has GF(4)-rank-width 1.*
4. *For every $W \subseteq V$ with $|W| \geq 4$, $G[W]$ has a non-trivial displit.*
5. *$u(G)$ is distance-hereditary and for every $W \subseteq V$ with $|W| \leq 5$, we have $\mathrm{rwd}^{(4)}(G[W]) \leq 1$.*

Condition 5 gives a characterization of digraphs of GF(4)-rank-width 1 by forbidden induced sub-graphs: a digraph has GF(4)-rank-width 1 if and only if it is $(\mathcal{H}, \mathcal{C})$-free, where $\mathcal{H}$ is the set of digraphs G such that $u(G)$ is a house, a gem, a domino or a hole ($C_k, k \geq 5$), and $\mathcal{C}$ is the set of connected digraphs G with at most 5 vertices such that $\mathrm{rwd}^{(4)}(G) > 1$ and for every $x \in V_G$, $\mathrm{rwd}^{(4)}(G-x) \leq 1$.

Before proving Theorem 6, let us state and prove two technical propositions. The following is immediate from the definitions.

Proposition 2. *Let x and y be two vertices of a digraph G. Then $\{x, y\}$ is a displit if and only if x and y are dtwins or x is a pendant vertex of y or y is a pendant vertex of x.*

The following proposition is a straightforward adaptation of [18, Proposition 7.1].

Proposition 3. *Let x and y be dtwins of a digraph G such that $G - x$ has at least one edge. Then $\mathrm{rwd}^{(4)}(G - x) = \mathrm{rwd}^{(4)}(G)$.*

Proof. By definition of GF(4)-rank-width we have $\mathrm{rwd}^{(4)}(G - x) \leq \mathrm{rwd}^{(4)}(G)$. We will prove that $\mathrm{rwd}^{(4)}(G-x) \geq \mathrm{rwd}^{(4)}(G)$. Let $(T, \mathcal{L})$ be a sub-cubic layout of GF(4)-rank-width $k = \mathrm{rwd}^{(4)}(G - x)$ of $G - x$. By definition, there is a bijection $\mathcal{L}$ between V_{G-x} and L_T. Let $v = \mathcal{L}(y)$ and let $u \in V_T$ such that $uv \in E_T$. Let T' be obtained from T as follows: $V_{T'}$ is the set $V_T \cup \{u', w\}$ (where u' and w are two new nodes) and $E_{T'}$ the set $(E_T \backslash \{uv\}) \cup \{uu', u'v, u'w\}$. We let $\mathcal{L}' : V_G \to L_{T'}$ be such that $\mathcal{L}'(x) = w$ and for every $z \in V_G \backslash x$, $\mathcal{L}'(z) = \mathcal{L}(z)$.

It is clear that $(T', \mathcal{L}')$ is a sub-cubic layout of G. We claim that the GF(4)-rank-width of $(T', \mathcal{L}')$ is equal to the GF(4)-rank-width of $(T, \mathcal{L})$.

It is clear that the GF(4)-rank-width of the edges $u'v$ and $u'w$ are at most 1. Since x and y are dtwins, the GF(4)-rank-width of the edge uu' is at most 1 (Proposition 2). Moreover, the other edges of T' are in T, then their GF(4)-rank-width in $(T', \mathcal{L}')$ is equal to their GF(4)-rank-width in $(T, \mathcal{L})$ (Lemma 2). Since $G - x$ has at least one edge we have $\mathrm{rwd}^{(4)}(G - x) \geq 1$. Therefore $\mathrm{rwd}^{(4)}(G-x) \geq \mathrm{rwd}^{(4)}(G)$. □

We can now begin the proof of Theorem 6.

Proof (Proof of Theorem 6). $1 \to 2$). By induction on $|V_G|$. It is trivial if $|V_G| \leq 2$. Otherwise, let $(T, \mathcal{L})$ be the displit decomposition of G, and let u be a leaf in $T - L_T$. If u is degenerated, let $\{v, w\} \subseteq N_T(u) \cap L_T$. Otherwise, u is linear and has at least 4 neighbors. Let $v_1, \ldots v_k$ be its ordering. If $N_T(u) \setminus L_T \subseteq \{v_2, \ldots, v_{k-1}\}$, take $v = v_1$ and $w = v_k$. Otherwise, take $v = v_2$ and $w = v_3$. In all cases, $\{\mathcal{L}^{-1}(\{v, w\}), V_G \setminus \mathcal{L}^{-1}(\{v, w\})\}$ is a displit. By Proposition 2, either $x = \mathcal{L}^{-1}(v)$ and $y = \mathcal{L}^{-1}(w)$ are dtwins, or one is a pendant vertex of the other. If x and y are dtwins or x is a pendant vertex of y, we let $G' = G - x$, otherwise $G' = G - y$. By induction G' is obtained from a single vertex by creating dtwins or adding pendant vertices.

$2 \to 3$). By induction on $|V_G|$. It is trivial if $|V_G| \leq 2$. Otherwise, let $x \in V_G$ be the last added vertex. If x is a pendant vertex, let $\{y\} = N_G(x)$, otherwise let y be the dtwin of x. By induction, $\mathrm{rwd}^{(4)}(G - x) = 1$. Using Proposition 3, $\mathrm{rwd}^{(4)}(G) = 1$.

$3 \to 4$). If $\mathrm{rwd}^{(4)}(G) \leq 1$, then for every $W \subseteq V_G$, $\mathrm{rwd}^{(4)}(G[W]) \leq 1$. When $|W| \geq 4$, a sub-cubic layout of $G[W]$ has an edge (u, v) such that $\{X^{uv}, V \setminus X^{uv}\}$ is non-trivial, and thus $G[W]$ has a non-trivial displit.

$4 \to 1$). Suppose that G is not completely decomposable. Then the displit decomposition of G has a prime node u. By definition of a representative layout, the degree of u is at least 4. By Lemma 3, the quotient graph $C(u)$ is prime and is an induced sub-graph of G with at least 4 vertices.

$3 \to 5$). By Observation 2, $\mathrm{rwd}(u(G)) = 1$ since the layout of GF(4)-rank-width 1 for G is a layout of rank-width 1 for $u(G)$. Thus by Theorem 1, $u(G)$ is distance hereditary. Moreover, for every $W \subseteq V$, we have $\mathrm{rwd}^{(4)}(G[W]) \leq 1$.

$5 \to 3$). Due to space limitation we will give only a sketch of the proof. Suppose that G is a digraph such that $\mathrm{rwd}^{(4)}(G) > 1$ and such that $u(G)$ is distance hereditary. Let W be a minimal subset of V_G such that $\mathrm{rwd}^{(4)}(G[W]) > 1$. Working on the split decomposition of $u(G[W])$, one can show successively that:

- $u(G[W])$ has no pendant vertex,
- if $u(G[W])$ has a false twin, then $G[W]$ has at most 4 vertices,
- if $u(G[W])$ has no false twin and no pendant vertex, then $u(G)$ is complete,
- and if $u(G[W])$ is complete, then $G[W]$ has at most 5 vertices.

Thus there is a $W \subseteq V_G$ of size at most 5 such that $\mathrm{rwd}^{(4)}(G[W]) > 1$. □

As a corollary of Theorems 4 and 6, we get an algorithm for recognizing digraphs of GF(4)-rank-width 1.

Corollary 1. *Digraphs of* GF(4)*-rank-width* 1 *can be recognized in time* $O(nm)$.

6 Concluding Remarks

Differences with Cunningham's split decomposition of digraphs. Cunningham showed that the family of splits in a strongly connected digraph is bi-partitive.

He also gave a characterization of degenerated and linear digraphs for the split decomposition: a digraph is degenerated for the split decomposition if and only if it is complete or is a star, and is linear if and only if it is a *circle of transitive tournaments* (CTT) [5].

The displit decomposition and the split decomposition of digraphs are both generalization of the split decomposition of undirected graphs. A first difference is that for the displit decomposition the graph has only to be connected.

The quotient graphs of the displit decomposition are induced sub-graphs of the original graph; this is not necessarily true for the split decomposition of digraphs.

Finally, the split decomposition and the displit decomposition are mutually exclusive. For all $k \geq 3$, the graph C'_k is linear for the split decomposition (and thus completely decomposable) since it is a CTT, but it is prime for the displit decomposition since $u(C'_k)$ is prime for the split decomposition. In the other hand, we can construct an infinite family of graphs linear for the displit decomposition and prime for the split decomposition.

Links between bi-rank-width and Cunningham's split decomposition. Kanté defined another digraph parameter called *bi-rank-with*, and showed relations between GF(4)-rank-width and bi-rank-width [15]. A strongly connected digraph is completely decomposable by Cunningham's split decomposition if and only if it has bi-rank-width 2. It is open to find another characterization for digraphs of bi-rank-width 2.

Generalization to 2-structures. A *2-structure* is a complete digraph with labels on edges. We mention that GF(4)-rank-width and displit decomposition can be generalized to 2-structures over finite fields. For a field $\mathbb{F}$, we obtain a decomposition for 2-structures over $\mathbb{F}$ with a characterization theorem similar to Theorem 6. An interesting case is GF(3), which gives a decomposition theory for oriented graphs (*i.e.*, directed anti-symmetric graph).

References

1. Bandelt, H.J., Mulder, H.M.: Distance-Hereditary Graphs. Journal of Combinatorial Theory, Series B 41(2), 182–208 (1986)
2. Chein, M., Habib, M., Maurer, M.C.: Partitive Hypergraphs. Discrete Mathematics 37(1), 35–50 (1981)
3. Cicerone, S., Di Stefano, D.: On the Extension of Bipartite Graphs to Parity Graphs. Discrete Applied Mathematics 95(1-3), 181–195 (1999)
4. Courcelle, B., Olariu, S.: Upper Bounds to the Clique-Width of Graphs. Discrete Applied Mathematics 101, 77–114 (2000)
5. Cunningham, W.H.: Decomposition of Directed Graphs. SIAM Journal on Algebraic and Discrete Methods 3(2), 214–228 (1982)
6. Cunningham, W.H., Edmonds, J.: A Combinatorial Decomposition Theory. Canadian Journal of Mathematics 32, 734–765 (1980)

7. Dahlhaus, E.: Parallel Algorithms for Hierarchical Clustering and Applications to Split Decomposition and Parity Graph Recognition. Journal of Algorithms 36(2), 205–240 (2000)
8. Ehrenfeucht, A., Harju, T., Rozenberg, G.: The Theory of 2-Structures - A Framework for Decomposition and Transformation of Graphs. World Scientific, Singapore (1999)
9. Gabor, C.P., Hsu, W.L., Supowit, K.J.: Recognizing Circle Graphs in Polynomial-Time. Journal of the ACM 36(3), 435–473 (1989)
10. Gallai, T.: Transitiv orientierbare Graphen. Acta Mathematica Academiae Scientiarum Hungaricae 18(1-2), 25–66 (1967)
11. Gavoille, C., Paul, C.: Distance Labeling Scheme and Split Decomposition. Discrete Mathematics 273(1), 115–130 (2003)
12. Hammer, P., Maffray, F.: Completely Separable Graphs. Discrete Applied Mathematics 27(1-2), 85–99 (1990)
13. Hliněný, P., Oum, S.: Finding Branch-Decompositions and Rank-Decompositions. SIAM Journal on Computing 38(3), 1012–1032 (2008)
14. Joeris, B.L., Lundberg, S., McConnell, R.M.: O(mlogn) Split Decomposition of Strongly-Connected Graphs. In: Proceedings of Graph Theory, Computational Intelligence and Thought. LNCS. Springer, Heidelberg (2008)
15. Kanté, M.M.: The Rank-Width of Directed Graphs (2009) (in revision)
16. Möhring, R.H., Radermacher, F.J.: Substitution Decomposition for Discrete Structures and Connections with Combinatorial Optimization. Annals of Discrete Mathematics 19, 257–356 (1984)
17. de Montgolfier, F., Rao, M.: The Bi-Join Decomposition. Electronic Notes in Discrete Mathematics 22, 173–177 (2005)
18. Oum, S.: Rank-Width and Vertex-Minors. Journal of Combinatorial Theory, Series B 95(1), 79–100 (2005)
19. Oum, S., Seymour, P.D.: Approximating Clique-Width and Branch-Width. Journal of Combinatorial Theory, Series B 96(4), 514–528 (2006)
20. Rao, M.: Solving some NP-complete problems using split decomposition. Discrete Applied Mathematics 156(14), 2768–2780 (2008)
21. Spinrad, J.: Recognition of Circle Graphs. Journal of Algorithms 16, 264–282 (1994)

An Algorithmic Study of Switch Graphs

Bastian Katz[1], Ignaz Rutter[1], and Gerhard Woeginger[2]

[1] Faculty of Informatics, Universität Karlsruhe (TH), KIT
{katz,rutter}@iti.uka.de
[2] Department of Mathematics and Computer Science, TU Eindhoven
gwoegi@win.tue.nl

Abstract. We derive a variety of results on the algorithmics of switch graphs. On the negative side we prove hardness of the following problems: Given a switch graph, does it possess a bipartite / planar / triangle-free / Eulerian configuration? On the positive side we design fast algorithms for several connectivity problems in undirected switch graphs, and for recognizing acyclic configurations in directed switch graphs.

1 Introduction

What is a switch graph? A *switch* s on an underlying vertex set V is a pair (p_s, T_s) where $p_s \in V$ is the *pivot* vertex and where $T_s \subseteq V$ is a non-empty set of *target* vertices. The vertex set V and some set S of switches on V together form a *switch graph* $G = (V, S)$. A *configuration* of a switch graph is a mapping $c : S \to V$ such that $c(s) \in T_s$ for all $s \in S$. The configuration selects exactly one edge $e_c(s) := \{p_s, c(s)\}$ for every switch $s \in S$, and thus yields a corresponding multi-set $E_c = \{e_c(s) : s \in S\}$ of edges. The corresponding multi-graph is denoted $G_c = (V, E_c)$; see Fig. 1 for an illustration. Biologically speaking, a switch graph represents the genotype of an entire population of graphs, and every configuration specifies the phenotype of one concrete member in this population.

A brief history of switch graphs. Over the last 30 years a huge number of fairly unrelated combinatorial structures has been introduced under the name *switch graph* or *switching graph*; see the introduction of [5] for some pointers to the literature. The switching graph model of Meinel [6] comes very close to the model

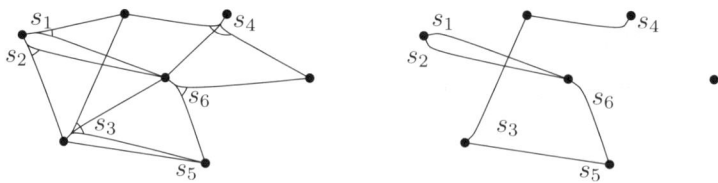

Fig. 1. To the left: A switch graph G with six switches, where s_5 has only a single target. To the right: A configuration yielding a multigraph G_c.

that is investigated in this paper. Another somewhat restricted type of switch graph has been introduced by Cook [1] who studied cyclic configurations as an abstraction of certain features in Conway's game of life. In Cook's model the vertices are not allowed to have degrees higher than three, and every switch has an obligatory incident edge that must show up in every configuration. Reinhardt [8] essentially studies Cook's model, but drops the constant degree constraint. Reinhardt constructs a polynomial-time $O(|V|^4)$ algorithm that decides whether there exists a configuration that contains a simple path between two prespecified vertices. He also links switch graphs to certain matching problems in computational biology [9]. We note that Cook's and Reinhardt's switch graph models can both easily be emulated by our switch graph model.

Groote and Ploeger [5] concentrate on switch graphs with *binary* switches (where every target set contains two elements). Their work is motivated by certain questions around the modal μ-calculus, and among other results they study the complexity of certain graph properties on switch graphs. For instance they show that in directed binary switch graphs, one can decide in polynomial time whether there is a configuration that connects (respectively disconnects) two prespecified vertices. Our current paper was heavily inspired by the conclusions section of [5]; our results in Theorems 3, 5, and 7 answer open questions that have been posed in [5].

Results of this paper. Every graph property $\mathcal{P}$ naturally leads to a corresponding algorithmic problem on switch graphs: Given a switch graph, does there exist a configuration with property $\mathcal{P}$? We will derive a collection of positive and negative results for various graph properties.

- It is NP-hard to decide whether a given switch graph has a configuration that is (a) bipartite, (b) planar, or (c) triangle-free. The three hardness proofs are presented in Section 3.
- We establish a number of matroid properties for switch graphs that possess a connected configuration. This yields a simple $O(|S| + |V|^2)$ time greedy algorithm for finding a configuration that minimizes the number of connected components (and of course also settles the question whether there is a connected configuration); see Section 4.
- We provide a fast algorithm to detect a configuration that connects two given vertices in an undirected switch graph. This substantially improves the time complexity of Reinhardt's result [8]; see Section 5.
- Finding a configuration in which all vertex degrees are even is easy, but finding a configuration with an Eulerian cycle is NP-hard for forward directed switch graphs as well as for undirected switch graphs. Moreover, it is NP-hard to find a configuration that is biconnected (for undirected switch graphs) or strongly connected (for forward directed switch graphs); see Section 6.
- Deciding whether a forward directed switch graph allows an acyclic configuration can be done in linear time. In contrast to this, finding a configuration that minimizes the number of directed cycles is NP-hard; see Section 7.

We stress that our negative results hold in the most restricted binary switch model, whereas our positive results apply to the general model.

2 Basic Definitions

Let $G = (V, S)$ be a switch graph. For a subset $S' \subseteq S$ of switches and a configuration c, we denote $E_c(S') = \{e_c(s) : s \in S'\}$ and $G_c(S') = (V, E_c(S'))$. We denote $V(s) := T_s \cup \{p_s\}$ and $V(S') := \bigcup_{s \in S'} V(s)$. For $S' \subseteq S$ and $V' \subseteq V$, we denote by $S'(V') := \{s \in S' \mid V(s) \subseteq V'\}$ the set of *inner switches* of V'. Observe that $V(S'(V')) \subseteq V'$. A switch graph has *fan-out* k if $|T_s| \leq k$ for all $s \in S$. It is called *binary* if $|T_s| \leq 2$ holds for all $s \in S$. Throughout we will use $n := |V|$ and $m := |S|$.

Although the paper mainly deals with undirected graphs, all definitions easily carry over to directed switches and directed multi-graphs. In a *forward* switch $s = (p_s, T_s)$, arcs must be directed from pivot to target. In a *reverse* switch $s = (T_s, p_s)$, arcs must be directed from target to pivot. A *directed switch graph* may contain both, forward and reverse switches. A *forward directed switch graph* contains only forward switches.

Note that all problems we consider in this paper ask for configurations of a given switch graph with properties that can be tested in polynomial time. Since there are at most n^m configurations, all NP-hard problems presented in this paper are also NP-complete.

3 Bipartite, Planar, Triangle-Free Graphs

In this section, we show hardness of finding configurations that are bipartite, triangle-free or planar.

Theorem 1. *For binary undirected switch graphs, it is NP-hard to decide if there is a bipartite configuration* (SWITCHBIPARTITE).

Proof. We sketch a reduction from SETSPLITTING: Given a ground set $X = \{x_1, \ldots, x_n\}$ and a set T of 3-element subsets of X, it is NP-hard to decide whether there is a partition of X into two sets X_1, X_2, such that every $t \in T$ has non-empty intersection with both, X_1 and X_2. For a given instance of SETSPLITTING, we construct a switch graph $G = (V, S)$, containing vertices $x_1, \ldots, x_n$ for the elements of X. For each triplet $t_i \in T$ we introduce a switch $s_i = (x_j, t_i - \{x_j\})$ for an arbitrary $x_j \in t_i$.

Every solution X_1, X_2 to SETSPLITTING yields a bipartite configuration: Color the vertices x_i according to X_1, X_2. Then every triplet t contains both colors, which allows to set the corresponding switch to connect two vertices of distinct colors. Conversely, every bipartition of some configuration G_c induces a bipartition of the x_i. For any triplet in T, the switches prevent the corresponding three vertices from receiving all the same color, and thus the induced partition yields a solution to SETSPLITTING.

Theorem 2. *For binary undirected switch graphs, it is NP-hard to decide if there is a triangle-free configuration* (SWITCHTRIANGLEFREE).

$(x_1 \vee \overline{x}_2 \vee x_3) \wedge$
$(x_2 \vee \overline{x}_3 \vee \overline{x}_4)$

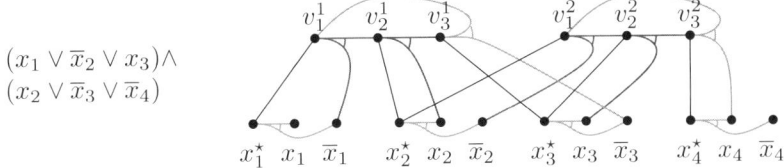

Fig. 2. Reduction of 3SAT to SWITCHTRIANGLEFREE

Proof. The proof is by reduction from 3SAT. Let ϕ be an instance of 3SAT. Without loss of generality we assume that each clause contains three different variables. For each variable x_i we create three vertices $x_i^\star$, x_i and $\overline{x}_i$, and a variable switch $s_i = (x_i^\star, \{x_i, \overline{x}_i\})$. For every clause C_j, we add three new vertices v_1^j, v_2^j, v_3^j. If the kth literal in clause C_j corresponds to variable x_i, we introduce an edge $(v_k^j, \{x_i^\star\})$. If it is x_i, we introduce a switch $(v_k^j, \{v_{k+1}^j, \overline{x}_i\})$. If it is $\overline{x}_i$, we introduce a switch $(v_k^j, \{v_{k+1}^j, x_i\})$, defining $v_4^j := v_1^j$. See Fig. 2 for an example.

The variable switch intuitively picks the true literal. A clause switch can only connect outside the clause, if its corresponding literal is satisfied. Consequently, in a satisfying truth assignment we can connect at least one switch of every clause to the outside, thus avoiding all triangle. Conversely a triangle-free configuration specifies a truth assignment for the variables such that every clause contains at least one satisfied literal. Otherwise, the corresponding clause switch would induce a triangle.

Theorem 3. *For binary undirected switch graphs, it is NP-hard to decide if there is a planar configuration* (SWITCHPLANAR).

Proof. The proof is by reduction from monotone planar 3SAT. Planar 3SAT is a well-known NP-hard restriction of 3SAT where additionally the variable-clause graph is assumed to be planar. Monotone planar 3SAT is even more restricted: the literals of each clause must be either all positive or all negative. Moreover the variable clause graph can be drawn in the plane without crossings such that all the variables are on the x-axis, the clauses with positive literals are above the x-axis and the clauses with negative literals are below the x-axis. Monotone planar 3SAT is NP-hard [3].

Let ϕ be an instance of planar monotone 3SAT. For every variable x, we introduce a variable gadget as depicted in Fig. 3 (a) with one variable switch s_x and switches with pivots ℓ_x^i and $\ell_{\overline{x}}^i$ for every occurrence of a literal x or $\overline{x}$ in a clause. For every clause C, we introduce a clause gadget as depicted in Fig. 3 (b) which basically is a K_5 of which three edges can be disabled by setting a switch appropriately. We identify the pivots of these switches (and the pivots themselves) with the vertices ℓ_x^i or $\ell_{\overline{x}}^i$ for the respective literal x or $\overline{x}$ and an i induced by the drawing of ϕ. An example is given in Fig. 3 (c).

A solution to ϕ induces a planar drawing by switching s_x for a true x to x and the switches with pivots ℓ_x^i to the inner of the variable gadget, the switches with pivots $\ell_{\overline{x}}^i$ to the inner of the clause gadget. For a false x, we set switches

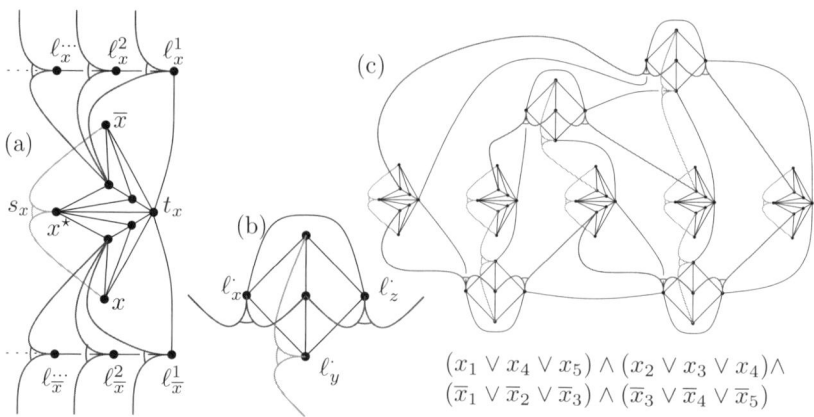

Fig. 3. Reduction of monotone planar 3SAT to SWITCHPLANAR

accordingly. Conversely, in a planar drawing at most two switches of a clause gadget may be switched to the inner of the clause, which otherwise would be a K_5. If, in turn, some switch with pivot ℓ_x^i is switched to the inner of the variable gadget, the switch s_x must be switched to x, since otherwise contraction of the path $t_x \ell_x^1 \ldots \ell_x^i$ would make the upper half of the variable gadget a K_5. Analogously, any switch with pivot $\ell_{\overline{x}}^i$ being switched to the inner of the variable clause forces s_x to be switched to $\overline{x}$. □

4 Global Connectivity

In this section we discuss the question whether a given switch graph $G = (V, S)$ has a connected configuration. It turns out that this question has many ties to matroid theory, which allows us to invoke some powerful machinery from mathematical programming.

First, we consider two matroids $(E, \mathcal{I}_1)$ and $(E, \mathcal{I}_2)$ that both have the same ground set E, which is the set of all the multi-edges over V that can possibly result from some switch in S. The set system $\mathcal{I}_1$ consists of all cycle-free subsets of E. The set system $\mathcal{I}_2$ consists of all subsets of E that contain at most one multi-edge from each switch. Then $(E, \mathcal{I}_1)$ forms a *graphic matroid* and $(E, \mathcal{I}_2)$ forms a *partition matroid*. Obviously, the switch graph $G = (V, S)$ has a connected configuration, if and only if there exists a set $E' \subseteq E$ of cardinality $n - 1$ that belongs to both $\mathcal{I}_1$ and $\mathcal{I}_2$. This is a standard matroid intersection problem, which can be solved in polynomial time [4].

It is not hard to see that the intersection $(E, \mathcal{I}_1 \cap \mathcal{I}_2)$ itself does not form a matroid. In the following, we will model the problem in terms of a single matroid, which yields simpler and faster algorithms. Our approach is based on a third structure $(S, \mathcal{I}_3)$ that is defined over the ground set S of switches. A subset $S' \subseteq S$ lies in $\mathcal{I}_3$, if there exists a configuration c such that $E_c(S')$ is cycle-free (or in other words, such that $E_c(S')$ belongs to $\mathcal{I}_1$). The sets in $\mathcal{I}_3$ are called *independent* sets.

Theorem 4. *The structure $(S, \mathcal{I}_3)$ forms a matroid.*

Proof. Clearly the set system $\mathcal{I}_3$ contains the empty set and is closed under taking subsets. It remains to show that for two independent sets $A, B \subseteq S$ with $|A| < |B|$, there is an $s \in B - A$ such that also $A \cup \{s\}$ is independent.

Since A and B are independent, there exist configurations a and b for which the corresponding edge sets $E_a(A)$ and $E_b(B)$ are cycle-free. Among all such configurations a and b, we consider a pair that maximizes the number of switches that are in $A \cap B$ and that configure into the same edge both in configuration $E_a(A)$ and in configuration $E_b(B)$; such switches are called *good* switches. Since $E_a(A)$ and $E_b(B)$ are cycle-free, they belong to $\mathcal{I}_1$ in the underlying graphic matroid. Since $|E_a(A)| = |A| < |B| = |E_b(B)|$, there exists an edge $e \in E_b(B) - E_a(A)$ such that $E_a(A) \cup \{e\}$ is cycle-free.

Let $s_e \in B$ denote the switch that in configuration b generates edge e. We claim that this switch s_e cannot be in A: Otherwise configuration a would configure this switch s_e into an edge f. Then we can modify configuration a into a new configuration c by switching s_e into e instead of f. The resulting edge set $E_c(A)$ is still cycle-free, whereas the number of good switches has increased. That is a contradiction. Hence $s_e \notin A$, and $A \cup \{s_e\}$ is an independent set of switches. □

Our next goal is to get a better understanding of independence in $(S, \mathcal{I}_3)$.

Lemma 1. *A set $S' \subseteq S$ is independent if and only if $|T| < |V(T)|$ holds for all $T \subseteq S'$.*

Proof. One direction of the proof is easy: If there is a $T \subseteq S'$ with $|T| \geq |V(T)|$, then every configuration c induces a cycle on T since $|E_c(T)| = |T| \geq |V(T)|$ holds.

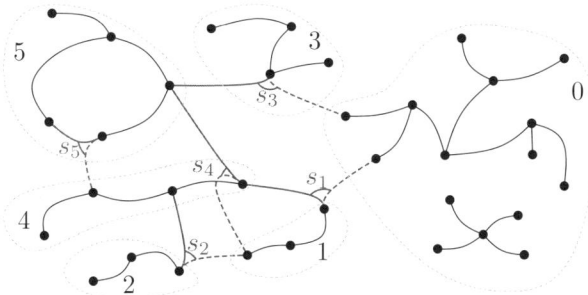

Fig. 4. A proper switch sequence $s_1, \ldots, s_5$ that eventually breaks the cycle. Switches are reinserted changing s_5, s_4, s_1.

For the other direction of the proof, we consider a set $S' \subseteq S$ that contains some switch $s \in S'$ for which $S'' = S' - \{s\}$ is independent. We will show that then either S' itself is independent, or that it contains an appropriate subset T with $|T| \geq |V(T)|$. Indeed, since S'' is independent there exists a configuration c for which $E_c(S'')$ is cycle-free. Adding an arbitrary edge e from switch s to $E_c(S'')$ produces a configuration c for S' whose edge set $E := E_c(S'') \cup \{e\}$ contains a single cycle. Let C_0 denote the connected component that contains the cycle, and let V_0 denote the set of all vertices in $V(S')$ that are not in C_0.

We work through a number of removal phases. In the ith phase ($i \geq 1$) we select a switch $s_i = (p_i, T_i)$ that contributes an edge e_i to component C_{i-1}, and whose target set T_i contains a target t_i^* in some set $V_{j(i)}$ with $0 \leq j(i) \leq i-1$. We remove the edge e_i from the edge set, and thus split component C_{i-1} into two connected parts. The part containing the cycle becomes the new component C_i, and the vertices in the other (cycle-free) part form the set V_i. Then the $(i+1)$th removal phase starts.

There are two possibilities how this process can terminate: Either (i) there is no appropriate switch with a target in $V_0, \ldots, V_{i-1}$, or (ii) removing the edge destroys the cycle in component C_{i-1}. In case (i), we choose T as the set of switches in component C_{i-1}; then $|T| \geq |V(T)|$, and we are done. In case (ii) we will show how to reinsert and how to reconfigure the removed edges and switches step by step in reverse order $s_k, s_{k-1}, \ldots, s_1$ so that the resulting edge set is cycle-free (here k denotes the number of the last phase).

Throughout we will maintain the following invariant: Just after the reconfiguration of switch s_i ($1 \leq i \leq k$), there exists an index $\ell(i)$ with $0 \leq \ell(i) < i$, such that the vertex set $V_{\ell(i)} \cup \bigcup_{h \geq i} V_h$ forms a cycle-free connected component with respect to the current edge set. This component is called the *crucial* component; intuitively speaking we will make it grow until it covers all of $V(S')$. We start the growing process with switch s_k, which by definition has a target t_k^* in the set $V_{j(k)}$ with $j(k) < k$. By reinserting the edge $\{p_k, t_k^*\}$ for switch s_k and by setting $\ell(k) := j(k)$, we satisfy the invariant. In handling a switch s_i with $i < k$ we distinguish two cases: First, if $i \neq \ell(i+1)$ then we simply reinsert its old edge e_i and keep $\ell(i) := \ell(i+1)$. This merges the vertices in V_i into the crucial component while maintaining the invariant. In the second case $i = \ell(i+1)$. We insert the new edge $\{p_i, t_i^*\}$, and set $\ell(i) := j(i)$. This merges the vertices in $V_{j(i)}$ into the crucial component, and again maintains the invariant. This reconfiguration process eventually produces a cycle-free configuration for S', and thus completes the proof. □

The statement of Lemma 1 is combinatorial, but its proof is algorithmical and yields as a by-product a fast independence test for the matroid $(S, \mathcal{I}_3)$: Given an acyclic configuration of an independent set S'' we can check in $O(kn)$ time whether a given switch s can be added to S'' without destroying independence: The independence of S'' implies that $|S''+s| \leq n$, which also bounds the number of removal phases. To achieve selection of removable switches within a total of $O(kn)$ time, we direct all edges in $E_c(S')$ which are not part of the cycle to point away from it. Whenever a switch s_i is removed, we use this information to mark all vertices in V_i. Obviously, this only adds $O(n)$ time. A switch s is a candidate if it has a marked target and both p_s and $c(s)$ are unmarked. A set of candidates can be maintained in $O(kn)$ total time.

If the test is positive, we obtain a corresponding cycle-free configuration for $S'' \cup \{s\}$. If the test is negative, we get the final component C_{k-1} that contains the cycle. Let U denote the set of all switches in S'' that contribute an edge to C_{k-1}. Then $|U| = |V(C_{k-1})| - 1$, and none of the switches in U has a target outside of $V(C_{k-1})$. Hence in any cycle-free configuration of S'' the switches in U induce a connected graph on $V(C_{k-1})$; such a set U of switches is called a *tight* set. These ideas lead to the following theorem, which is the main result of this section. Note that it as a special case yields a polynomial-time algorithm for recognizing switch graphs with connected configurations.

Theorem 5. *For a given switch-graph with fan-out k, we can determine in $O(km + kn^2)$ time a configuration that minimizes the number of connected components.*

Proof. Any basis $\mathcal{B}$ of the matroid $(S, \mathcal{I}_3)$ yields a cycle-free configuration with the maximum number of edges, and hence a configuration with the minimum number of connected components; the switches not in $\mathcal{B}$ then can be set arbitrarily. Hence it is enough to determine a basis, and this is done by the standard greedy algorithm.

We start with the empty set, and test the switches one by one. If the test is positive, we add the switch and update the cycle-free configuration. If the test is negative, we forget the switch and contract all switches in the corresponding tight set U. These contractions can be done in overall $O(n\alpha(n))$ time by using a union-find data structure. Every test on a non-trivial graph costs $O(kn)$ time. Every positive test adds an edge to a cycle-free edge set; hence there are at most $n-1$ of these tests. Every negative test on a non-trivial graph contracts some vertices; hence there are at most $n-1$ of these tests. Every negative test on a trivial graph (that has been contracted to a single vertex) costs $O(k)$ time. All in all, this yields the claimed time complexity. □

5 Local Connectivity

In this section, we investigate configurations that connect two given vertices a and b by a path. In the following, we call a sequence of switches a *forward path* if every switch's pivot is a target of its predecessor. A *contraction* of a switch s in a switch graph is defined as the switch graph identifying all vertices in $T_s \cup \{p_s\}$.

Lemma 2. *Let $G = (V, S)$ be a switch graph and s be any switch such that in $(V, S - s)$, there is a (possibly trivial) forward path from p_s to b. Let G' be the result from contracting s. Then G can be switched to connect a and b if and only if this is possible for G'.*

Proof. First, by contracting a switch, it is not possible to lose connectivity. We will thus assume that it is possible to find a configuration c' that connects a and

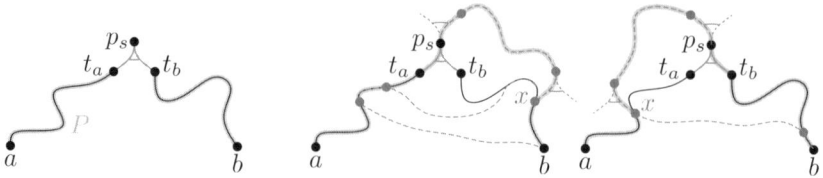

Fig. 5. A path in $G'_{c'}$ witnesses a path in G_c for some configuration c

b in G' and show that this witnesses such a configuration c for G. We denote the path in $G'_{c'}$ as sequence of switches P. If P forms a single path in $G_{c'}$, or if P connects either a or b to p_s, finding a connecting configuration is trivial. Otherwise, P forms two paths, connecting a to some $t_a \in T_s$ and b to some $t_b \in T_s$. In this situation, depicted in Fig. 5, we can make use of the fact that in $S - s$, there is a forward path from p_s to b. Its first switch is not part of P, and we simply follow the forward path until we hit some vertex x on P. Now, switching the forward path from p_s to x gives us a bypass on either $p_s - t_a$ or $p_s - t_b$ and switching s accordingly connects a and b. □

This lemma provides a simple test for a-b-connectivity: If there is a configuration c that connects a and b in G_c, there either is a forward path from b to a or there is a contractable switch, since there must be a first switch that is used "forward" on the path from a to b in G_c. The proof of Lemma 2 is constructive, naively implemented, it yields an $O(n^2 + knm)$ time algorithm to test the existence of and compute a connecting configuration by storing a forward path for each contraction. The time complexity can be further improved to almost linear (for the proof, we refer the reader to the full paper):

Theorem 6. *For a given switch-graph $G = (V, S)$ with fan-out k and two vertices $a, b \in V$, we can determine in $O(km + n\alpha(n))$ time, where α denotes the inverse Ackermann function, a configuration that connects a and b, if such a configuration exists.*

6   Even Degrees and Eulerian Graphs

Lemma 3. *For an undirected switch graph $G = (V, S)$, a configuration in which all vertex degrees are even can be detected in polynomial time.*

Proof. We use the results of Cornuéjols [2] on the general factor problem: Let (W, E) be an undirected graph, and for every $v \in W$ let $D(v)$ be a subset of $\{1, \ldots, |W|\}$. Does there exist a subset $F \subseteq E$, such that in the graph (W, F) every vertex has its degree in $D(v)$? Cornuéjols [2] shows that this problem can be decided in polynomial time, as long as the sets $D(v)$ do not contain any gap of length 2. (A set D of integers contains a gap of length 2, if it contains two elements d_1 and d_2, such that $d_2 \geq d_1 + 3$ and such that none of the numbers $d_1 + 1, \ldots, d_2 - 1$ is in D.)

For the proof of the lemma, construct a bipartite auxiliary graph between the set of switches and the set of vertices in the switch graph. Put an edge between any switch s and all targets in T_s. A vertex $v \in V$ is called odd (even), if it is the pivot of an odd (even) number of switches. For any switch $s \in S$ set $D(s) = \{1\}$. For any even vertex $v \in V$ set $D(v) = \{0, 2, 4, \ldots\}$, and for any odd vertex $v \in V$ set $D(v) = \{1, 3, 5, \ldots\}$. Note that none of these sets contains a gap of length 2. It can be seen that the auxiliary graph has a factor obeying the degree constraints if and only if the graph G has a configuration in which all vertex degrees are even. □

Theorem 7. *For binary undirected switch graphs it is NP-hard to decide if there is an Eulerian or a biconnected configuration. For forward directed switch graphs it is NP-hard to decide if there is an Eulerian or a strongly connected configuration.*

Proof. We reduce from DIRECTEDHAMILTONIANCYCLE which is known to be NP-hard for directed graphs with out-degree bounded by two [7].

Let $G = (V, E)$ be a directed graph with out-degrees 1, 2. We define a switch graph $H = (V, S)$ as follows. For each vertex $v \in V$ we add a switch $s_v = (v, N(v))$ where $N(v) = \{u \in V \mid (v, u) \in E\}$. Now since for every configuration c, H_c has n vertices and n edges, the following properties are equivalent:

(i) G has a directed Hamiltonian cycle
(ii) H has a directed Eulerian configuration as a directed switch graph
(iii) H has a strongly connected configuration as a directed switch graph
(iv) H has a biconnected configuration as an undirected switch graph
(v) H has a Eulerian cycle as an undirected switch graph

7 Acyclic and Almost Acyclic Graphs

This section mainly deals with forward directed switch graphs (as defined in Section 2): We check in polynomial time whether such a graph has a DAG configuration, and we show that finding a configuration with the minimum number of directed cycles is NP-hard.

Hence, let $G = (V, S)$ be a forward directed switch graph, and observe the following. First: The out-degree of every vertex in G_c is independent of the chosen configuration. Second: If all vertices in a digraph have out-degree at least 1, then the graph contains a directed cycle. Third: If G contains a sink v (that is, a vertex v with out-degree 0), then it is safe to configure all switches s with $v \in T_s$ towards this sink. These three observations suggest a simple procedure: As long as the graph contains a sink v, we first set $c(s) := v$ for all switches s with $v \in T_s$, and then remove v together with all these switches. The procedure either stops with an empty graph (and an acyclic configuration), or with a non-empty subgraph of G in which all vertices have out-degree at least 1 (in which case there is no acyclic configuration). The algorithm can easily be implemented to run in linear time. In contrast, finding an acyclic configuration in general directed switch graphs and minimizing the number of cycles in forward directed switch graphs is hard.

Theorem 8. *For a forward directed switch graph, it can be decided in $O(n+m)$ time if it has an acyclic configuration. If an acyclic configuration exists, it can be found within the same time complexity.*

Theorem 9. *For a directed switch graph, it is NP-hard to decide if it has an acyclic configuration* (SWITCHDIRECTEDACYCLIC).

Proof. The proof is by reduction from 3SAT. Let ϕ be an instance of 3SAT with variables $x_1, \ldots, x_n$ and clauses $C_1, \ldots, C_m$. We construct a switch graph G_ϕ as follows: We start with two vertices z and w and the arc (w, z). For each variable x_i we create two corresponding vertices $x_i, \overline{x}_i$ and a reverse switch $s_i = (\{x_i, \overline{x}_i\}, w)$. For each clause C_i we add a vertex v_i and the arc (z, v_i). Let x_u, x_v, x_w be the variables occurring in clause C_i. We set $\ell_u = x_u$ if x_u occurs negated in C_i and $\ell_u = \overline{x}_u$ otherwise. We define ℓ_v, ℓ_w analogously. We then add a *clause switch* $s_i^C = (v_i, \{\ell_u, \ell_v, \ell_w\})$. See Fig. 6 for an example.

A satisfying truth assignment for ϕ yields an acyclic configuration c of G_ϕ: For each variable x_i we set $c(s_i) = x_i$ if x_i is assigned the value true and $c(s_i) = \overline{x}_i$ otherwise. Since each clause of ϕ is satisfied in this configuration at least one target of every clause switch has out-degree 0. Hence every clause switch can easily be configured to avoid all cycles.

Furthermore, an acyclic configuration c of G_ϕ yields a satisfying truth assignment for ϕ: We set variable x_i to true if $c(s_i) = x_i$ and to false otherwise. As the configuration is acyclic every clause switch must have a sink as target, and this sink represents a satisfied literal in the corresponding clause.

Note that although the clause switches have fan-out 3, the result also holds for binary switch graphs, as we can replace each switch with fan-out 3 by two binary switches without affecting the number of cycles with respect to any configuration.

Theorem 10. *For a forward directed switch graph G and an integer $k > 0$, it is NP-hard to decide if there is a configuration with at most k cycles* (SWITCH-MINIMUMDIRECTEDCYCLES).

Proof. We show how to simulate binary reverse switches with usual binary forward switches at the cost of one cycle per reverse switch. Let G' be an instance

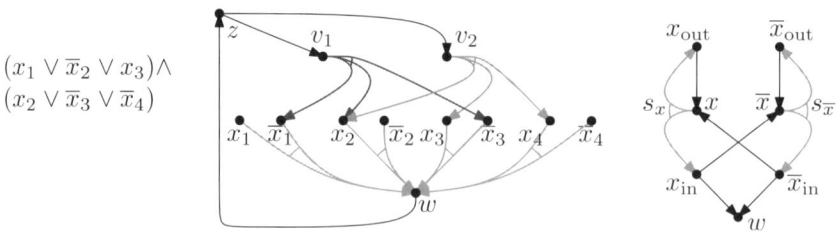

Fig. 6. Reduction of 3SAT to SWITCHDIRECTEDACYCLIC with reverse switches (left). Replacement of reverse switches for reduction of 3SAT to SWITCHMINIMUMDIRECTEDCYCLES (right).

of SWITCHDIRECTEDACYCLIC with k binary reverse switches. We construct a directed switch graph G by replacing each reverse switch $s = (\{x, \overline{x}\}, w)$ by the construction depicted in Fig. 6. As each replacement creates at least one cycle every configuration of G has at least k cycles. Each replacement has four distinct configurations. Two of them directly correspond to a configuration of the original reverse switch, namely the ones where one of the vertices $x, \overline{x}$ is connected to its in- and the other one to its out-vertex. We say that a configuration of G is *good* for the replacement in this case. There is a bijection between the acyclic configurations of G' and the configurations of G with k cycles that are good for each replacement.

Let c be a configuration of G with k cycles. We can modify c such that it is good for each replacement without increasing the number of cycles: The case $c(s_x) = x_{\text{out}}, c(s_{\overline{x}}) = \overline{x}_{\text{out}}$ can be excluded, as it would induce two cycles. In case $c(s_x) = x_{\text{in}}, c(s_{\overline{x}}) = \overline{x}_{\text{in}}$ we can change $c(s_x) := x_{\text{out}}$ without increasing the number of cycles thus making c good for the replacement.

Acknowledgments. We thank Alexander Wolff for pointing us to switch graph problems. We thank the anonymous referees for helpful comments.

References

1. Cook, M.: Still Life Theory. In: New Constructions in Cellular Automata, vol. 226, pp. 93–118. Oxford University Press, Oxford (2003)
2. Cornuéjols, G.: General factors of graphs. Journal of Combinatorial Theory, Series B 45, 185–198 (1988)
3. de Berg, M., Khosravi, A.: Optimal binary space partitions (Manuscript) (2008)
4. Edmonds, J.: Submodular functions, matroids, and certain polyhedra. In: Proceedings of the Calgary International Conference on Combinatorial Structures and Their Applications, Calgary, pp. 69–87 (1969)
5. Groote, J.F., Ploeger, B.: Switching graphs. In: Proceedings of the 2nd Workshop on Reachability Problems (RP 2008). ENTCS, pp. 119–135 (2008)
6. Meinel, C.: Switching graphs and their complexity. In: Kreczmar, A., Mirkowska, G. (eds.) MFCS 1989. LNCS, vol. 379, pp. 350–359. Springer, Heidelberg (1989)
7. Plesńik, J.: The NP-completeness of the Hamiltonian Cycle Problem in planar digraphs with degree bound two. Information Processing Letters 8(4), 199–201 (1979)
8. Reinhardt, K.: The simple reachability problem in switch graphs. In: Nielsen, M., Kucera, A., Miltersen, P.B., Palamidessi, C., Tuma, P., Valencia, F.D. (eds.) SOFSEM 2009. LNCS, vol. 5404, pp. 461–472. Springer, Heidelberg (2009)
9. Sharan, R., Gramm, J., Yakhini, Z., Ben-Dor, A.: Multiplexing schemes for generic SNP genotyping assays. Journal of Comp. Biology 15, 514–533 (2005)

Hardness Results and Efficient Algorithms for Graph Powers

Van Bang Le[1] and Ngoc Tuy Nguyen[1,2,⋆]

[1] Universität Rostock, Institut für Informatik, D-18051 Rostock, Germany
le@informatik.uni-rostock.de
[2] Hong Duc University, Thanh Hoa, Vietnam
nntuy@yahoo.com

Abstract. The k-th power H^k of a graph H is obtained from H by adding new edges between every two distinct vertices having distance at most k in H. Lau [Bipartite roots of graphs, *ACM Transactions on Algorithms* 2 (2006) 178–208] conjectured that recognizing k-th powers of some graph is NP-complete for all fixed $k \geq 2$ and recognizing k-th powers of a bipartite graph is NP-complete for all fixed $k \geq 3$. We prove that these conjectures are true. Lau and Corneil [Recognizing powers of proper interval, split and chordal graphs, *SIAM J. Discrete Math.* 18 (2004) 83–102] proved that recognizing squares of chordal graphs and squares of split graphs are NP-complete. We extend these results by showing that recognizing k-th powers of chordal graphs is NP-complete for all fixed $k \geq 2$ and providing a quadratic-time recognition algorithm for squares of strongly chordal split graphs. Finally, we give a polynomial-time recognition algorithm for cubes of graphs with girth at least ten. This result is related to a recent conjecture posed by Farzad et al. [Computing graph roots without short cycles, *Proceedings of STACS* 2009, pp. 397–408] saying that k-th powers of graphs with girth at least $3k - 1$ is polynomially recognizable.

1 Introduction and Results

In a graph $H = (V_H, E_H)$, the distance $d_H(x, y)$ between two vertices x and y in H is the number of edges of a shortest path in G connecting x and y. Given a positive integer k, the k-th power of H, written H^k, is the graph obtained from H by adding new edges between any pair of vertices at distance at most k in H; formally, $H^k = (V_H, \{xy \mid 1 \leq d_H(x, y) \leq k\})$. A graph G is the k-th power of a graph H if $G = H^k$, and in this case, H is a k-th root of G. For the cases of $k = 2$ and $k = 3$, we say that H^2 and H^3 is the *square*, respectively, is the *cube* of H and H is a *square root* of $G = H^2$, respectively, a *cube root* of $G = H^3$. Graph powers and roots are fundamental graph-theoretic concepts and have been extensively studied in the literature, both in theoretic and algorithmic

---

⋆ This author is supported by the Ministry of Education and Training, Vietnam, Grant No. 3766/QD-BGD & DT.

senses; see, e.g., [1,4,8,9,10,11] for recent results and the numerous references listed there.

Let $k \geq 2$ be a given integer and let $\mathcal{C}$ be a given graph class. We consider the following problems:

k-TH POWER OF GRAPH
Instance: A graph G.
Question: Does there exist a graph H such that $G = H^k$?

k-TH POWER OF $\mathcal{C}$ GRAPH
Instance: A graph G.
Question: Does there exist a graph H in $\mathcal{C}$ such that $G = H^k$?

In general, it is very unlikely that good characterizations of powers can exist as Motwani and Sudan [14] proved that SQUARE OF GRAPH is NP-complete. In [9], Lau proved that SQUARE OF BIPARTITE GRAPH is polynomially solvable but CUBE OF BIPARTITE GRAPH and CUBE OF GRAPH are NP-complete. He then strongly believes that the following conjectures should be true:

Conjecture 1 ([9]). *k-TH POWER OF GRAPH is NP-complete for all fixed $k \geq 2$.*

Conjecture 2 ([9]). *k-TH POWER OF BIPARTITE GRAPH is NP-complete for all fixed $k \geq 3$.*

Our first set of results (Theorems 4 and 5) consists of the proofs showing that both Conjectures 1 and 2 are indeed true. Moreover, our results in Section 5 will imply, on the positive side, that cubes of bipartite graphs without cycles of length at most eight can be recognized efficiently.

SQUARE OF $\mathcal{C}$ GRAPH remains NP-complete for various classes $\mathcal{C}$, such as chordal graphs and split graphs [10]. Our next set of results (Theorems 6, 7, and 8) consists of the proof that k-TH POWER OF CHORDAL GRAPH is NP-complete for all fixed $k \geq 2$ and, on the positive side, a good characterization of squares of strongly chordal split graphs that leads to a quadratic-time algorithm for solving SQUARE OF STRONGLY CHORDAL SPLIT GRAPH. Notice that the computational complexity of k-TH POWER OF CHORDAL GRAPH was unknown before and k-TH POWER OF SPLIT GRAPH is trivial for $k \geq 3$; k-th powers of split graphs, $k \geq 3$, are exactly the complete graphs.

Very recently, square roots with girth conditions have been considered in [4]; the girth of a graph is the smallest length of a cycle in the graph. It is shown in [4] that SQUARE OF GRAPH WITH GIRTH ≤ 4 is NP-complete, while SQUARE OF GRAPH WITH GIRTH ≥ 6 is polynomially solvable; the case of square roots with girth 5 still remains open. In [4], the following conjecture is proposed:

Conjecture 3 ([4]). *k-POWER OF GRAPH WITH GIRTH $\geq 3k-1$ is polynomially solvable.*

Our last set of results (Theorems 11 and 12) is related to Conjecture 3 and consists of a good characterization of cubes of graphs with girth at least ten and a polynomial-time algorithm for solving CUBE OF GRAPH WITH GIRTH ≥ 10. Moreover, our Theorem 11 will imply a good characterization of cubes of trees as well as of bipartite graphs without 'short' cycles.

In the last section we give our conclusion as well as discuss some open problems.

2 Preliminaries

All graphs considered are finite, undirected and simple. Since the power of a graph is the union of the powers of the connected components of that graph, we may assume that all graphs considered are connected. The *diameter* of a graph is the maximum distance in the graph.

Let $G = (V_G, E_G)$ be a graph. We often write $xy \in E_G$ for $\{x, y\} \in E_G$. Following [14,10], we sometimes also write $x \leftrightarrow y$ for the adjacency of x and y in the graph in question; this is particularly the case when we describe reductions in NP-completeness proofs. For disjoint sets of vertices X and Y, we write $X \leftrightarrow Y$, meaning each vertex in X is adjacent to each vertex in Y; if $X = \{x\}$, we simply write $x \leftrightarrow Y$.

The *neighborhood* $N_G(v)$ in G of a vertex v is the set all vertices in G adjacent to v and the *closed neighborhood* of v in G is $N_G[v] = N_G(v) \cup \{v\}$. For $U \subseteq V_G$ we write $N_G(U) = \bigcup_{u \in U} N_G(u)$ and $N_G[U] = N_G(U) \cup U$. The *k-th neighborhood* $N_G^k(v)$ of v is the set of vertices at distance k from v. Set $\deg_G(v) = |N_G(v)|$, the *degree* of v in G. We call vertices of degree one *end-vertices*. A *universal vertex* is one that is adjacent to all other vertices.

A *u, v-path* is a path P connecting two vertices u, v; u and v are the *end-vertices of P*. For $k \geq 1$, let P_k denote a chordless path with k vertices and $k - 1$ edges, and for $k \geq 3$, let C_k denote a chordless cycle with k vertices and k edges. A *complete* graph is one in which every two distinct vertices are adjacent. A graph is *chordal* if it does not contain any induced C_ℓ, $\ell \geq 4$.

A set of vertices $Q \subseteq V_G$ is called a *clique* in G if every two distinct vertices in Q are adjacent; a *maximal clique* is a clique that is not properly contained in another clique. A vertex is *simplicial* if its neighborhood is a clique. $\mathcal{C}(G)$ denotes the set of all maximal cliques of G. A *stable set* is a set of pairwise non-adjacent vertices. A graph is *bipartite* if its vertex set can be partitioned into two stable sets. A *split graph* is one whose vertex set can be partitioned into a clique and a stable set. Clearly, split graphs are chordal. Given a set of vertices $X \subseteq V_G$, the subgraph induced by X is written $G[X]$ and $G - X$ stands for $G[V \setminus X]$.

Due to space limitations, most of the proofs are omitted.

3 Hardness Results

In proving NP-completeness results we will consider the well-known NP-complete problem SET SPLITTING ([5, Problem SP4]), also known as HYPERGRAPH 2-COLORABILITY.

SET SPLITTING
Instance: Collection D of subsets of a finite set S.
Question: Is there a partition of S into two disjoint subsets S_1 and S_2 such that each subset in D intersects both S_1 and S_2?

Throughout this section, we will consider the following small instance of SET SPLITTING to illustrate our reductions:

Example. $S = \{u_1, u_2, u_3, u_4, u_5, u_6, u_7\}$ and $D = \{d_1, d_2, d_3\}$ with $d_1 = \{u_2, u_3, u_4\}$, $d_2 = \{u_1, u_5\}$ and $d_3 = \{u_3, u_4, u_6, u_7\}$. In this example, $S_1 = \{u_1, u_2, u_3\}$ and $S_2 = \{u_4, u_5, u_6, u_7\}$ is a possible solution.

We will also make use of the *tail structure*, described first in [14] and generalized later in [9]. The tail structure of a vertex v enables us to pin down exactly the neighborhood of v in any k-th root H of G.

Lemma 1 ([9]). *Let $G = (V_G, E_G)$ be a connected graph with $\{v_1, \ldots, v_{k+1}\} \subset V_G$ where $N_G(v_1) = \{v_2, \ldots, v_{k+1}\}$ and $N_G(v_i) \subset N_G[v_{i+1}]$ for all $1 \leq i \leq k$. Then in any k-th root H of G, (1) $N_H(v_1) = \{v_2\}$, (2) $N_H(v_i) = \{v_{i-1}, v_{i+1}\}$ for all $2 \leq i \leq k$, and (3) $N_H(v_{k+1}) - v_k = N_G(v_2) - \{v_1, \ldots, v_{k+1}\}$.*

The vertices $v_1, \ldots, v_k$ are 'tail vertices' of v_{k+1}; see Figure 1 for an illustration.

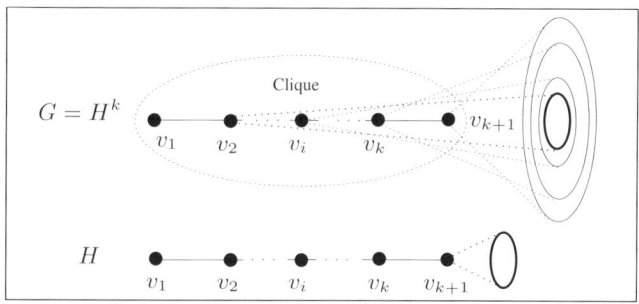

Fig. 1. Tail in G and in k-th root H of G

We remark that k-TH POWER OF $\mathcal{C}$ GRAPH is obvious in NP whenever recognizing $\mathcal{C}$ is polynomially because guessing a k-th root H, verifying if H is in $\mathcal{C}$ and if $G = H^k$ can be done in polynomial time. This is the case for all graph classes considered in this paper.

We first prove that, for fixed $k \geq 3$, k-TH POWER OF BIPARTITE GRAPH is NP-complete by reducing SET SPLITTING to it. Our reduction generalizes those in [9] for CUBE OF BIPARTITE GRAPH. Let $S = \{u_1, \ldots, u_n\}$, $D = \{d_1, \ldots, d_m\}$ where $d_j \subseteq S$, $1 \leq j \leq m$, be an instance of SET SPLITTING, and let $k \geq 3$ be a fixed integer. We construct an instance $G = G(D, S)$ for k-TH POWER OF BIPARTITE GRAPH as follows.

The vertex set of G consists of:

- U_i, for all $1 \leq i \leq n$. Each 'element vertex' U_i corresponds to the element u_i in S.
- D_j, for all $1 \leq j \leq m$. Each 'subset vertex' D_j corresponds to the subset d_j in D.

- $D_j^1, \ldots, D_j^k$, for all $1 \leq j \leq m$. k 'tail vertices' $D_j^1, \ldots, D_j^k$ of the subset vertex D_j.
- $P_1^1, \ldots, P_1^{k-2}$ and $P_2^1, \ldots, P_2^{k-2}$ are $k-2$ pairs of 'partition vertices'.
- Connection vertex: X.

The edge set of G consists of:

- Edges of tail vertices:
  (E_1) $D_j, D_j^1, \ldots, D_j^k$ form a clique;
  (E_2) For all $1 \leq t \leq k-1$: $D_j^t \leftrightarrow \{U_i \mid u_i \in d_j, 1 \leq i \leq n\}$;
  (E_3) For all $1 \leq t \leq k-2$: $D_j^t \leftrightarrow X$, $D_j^t \leftrightarrow \{D_{j'} \mid d_j \cap d_{j'} \neq \emptyset\}$,
  $D_j^t \leftrightarrow \{P_1^h \mid 1 \leq h \leq k-t-1\}$ and $D_j^t \leftrightarrow \{P_2^h \mid 1 \leq h \leq k-t-1\}$;
  (E_4) For all $1 \leq t \leq k-3$: $D_j^t \leftrightarrow \{U_i \mid 1 \leq i \leq n\}$,
  $D_j^t \leftrightarrow \{D_{j'}^h \mid 1 \leq h \leq k-t-2, d_j \cap d_{j'} \neq \emptyset\}$;
  (E_5) For all $1 \leq t \leq k-4$: $D_j^t \leftrightarrow \{D_{j'} \mid 1 \leq j' \leq m\}$;
  (E_6) For all $1 \leq t \leq k-5$: $D_j^t \leftrightarrow \{D_{j'}^h \mid 1 \leq h \leq k-t-4, 1 \leq j' \leq m\}$.
- Edges of subset vertices:
  (E_7) $D_j \leftrightarrow \{X, P_1^1, \ldots, P_1^{k-2}, P_2^1, \ldots, P_2^{k-2}\}$,
  $D_j \leftrightarrow \{U_i \mid 1 \leq i \leq n\}$, $D_j \leftrightarrow \{D_{j'} \mid d_j \cap d_{j'} \neq \emptyset\}$.
  (E_8) If $k \geq 4$: $D_j \leftrightarrow \{D_{j'} \mid 1 \leq j' \leq m\}$.
- Edges of element vertices:
  (E_9) $U_1, \ldots, U_n$ form a clique, and $U_i \leftrightarrow \{X, P_1^1, \ldots, P_1^{k-2}, P_2^1, \ldots, P_2^{k-2}\}$.
- Edges of partition vertices:
  (E_{10}) $P_1^1, \ldots, P_1^{k-2}, X, U_1, \ldots, U_n$ form a clique, $P_2^1, \ldots, P_2^{k-2}, X, U_1, \ldots, U_n$ form a clique.
  (E_{11}) For all $1 \leq t \leq k-3$, $P_1^t \leftrightarrow \{P_2^h \mid 1 \leq h \leq k-t-2\}$, $P_2^t \leftrightarrow \{P_1^h \mid 1 \leq h \leq k-t-2\}$.

Clearly, G can be constructed from D, S in polynomial time. For an illustration, in case $k = 4$, the example instance yields the graph G is depicted in Figure 2.

In this and other figures, each ellipse corresponds to a clique and we omit the clique edges to keep the figures simpler. The two dotted lines from a vertex to the cliques mean that the vertex is adjacent to all vertices in those cliques.

Lemma 2. *If there exists a partition of S into two disjoint subsets S_1 and S_2 such that each subset in D intersects both S_1 and S_2, then there exists a bipartite graph H such that $G = H^k$.*

Proof. Let H have the same vertex set as G. The edges of H are as follows; see also Figure 3.

- Edges of subset vertices and its tail vertices: For all $2 \leq t \leq k$, $D_j^t \leftrightarrow D_j^{t-1}$ and $D_j^1 \leftrightarrow D_j$, and $D_j \leftrightarrow \{U_i \mid u_i \in d_j, 1 \leq i \leq n\}$.
- Edges of partition vertices:
  $P_1^1 \leftrightarrow \{U_i \mid u_i \in S_1, 1 \leq i \leq n\}$ and $P_2^1 \leftrightarrow \{U_i \mid u_i \in S_2, 1 \leq i \leq n\}$, and for all $2 \leq t \leq k-2$, $P_1^t \leftrightarrow P_1^{t-1}$ and $P_2^t \leftrightarrow P_2^{t-1}$.
- Edges of connection vertex: $X \leftrightarrow \{U_1, \ldots, U_n\}$.

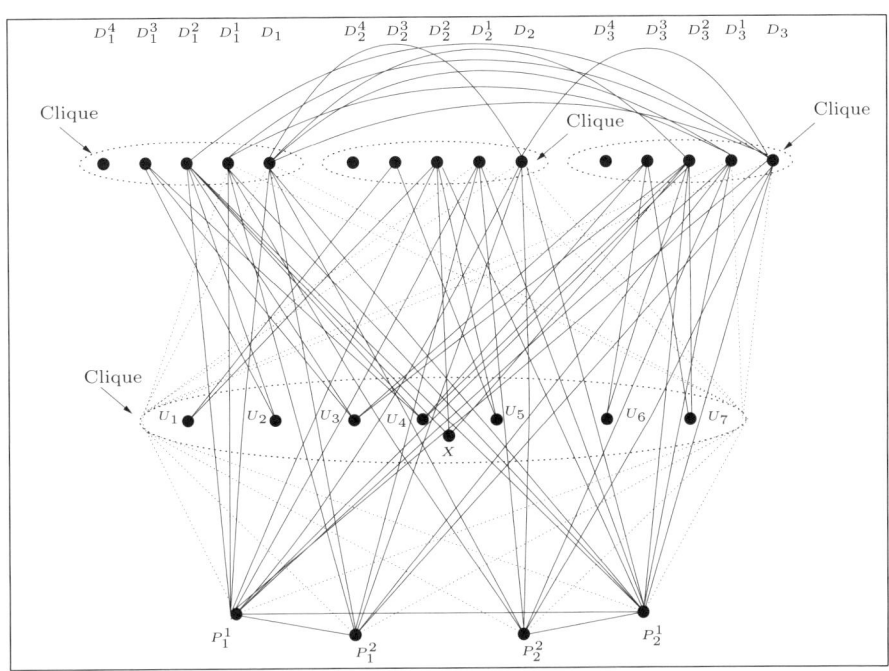

Fig. 2. The graph G for the example instance of SET SPLITTING and $k = 4$

For the example instance, the k-th root graph H corresponds to the solution S_1, S_2 is shown in Figure 3. □

Lemma 3. *If H is a k-th root of G, then there exists a partition of S into two disjoint subsets S_1 and S_2 such that each subset in D intersects both S_1 and S_2.*

Notice that in the Lemma 3, we did not use the property that H is a bipartite graph. In fact, any k-th root of G would tell us how to do SET SPLITTING. In particular, any bipartite k-th root H of G will do. Hence, by Lemmas 2 and 3, we conclude

Theorem 4. *k-TH POWER OF BIPARTITE GRAPH is NP-complete for all fixed $k \geq 3$.*

By the same reason, k-TH POWER OF $\mathcal{C}$ GRAPH is NP-complete for all fixed $k \geq 3$ whenever $\mathcal{C}$ contains all bipartite graphs (such as triangle-free graphs, parity graphs, perfect graphs, etc.). In particular, applied for the class of all graphs, this observation and the NP-completeness of SQUARE OF GRAPH [14] together give

Theorem 5. *k-TH POWER OF GRAPH is NP-complete for all fixed $k \geq 2$.*

Lau and Corneil [10] shown that SQUARE OF CHORDAL GRAPH is NP-complete. We are able to extend this result by showing that k-TH POWER OF CHORDAL

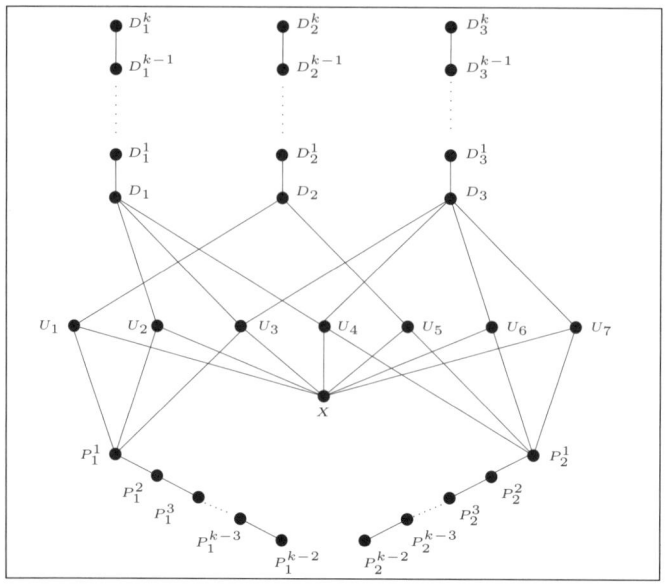

Fig. 3. A bipartite k-th root H in Lemma 2 to the example solution S_1, S_2

GRAPH is NP-complete for all fixed $k \geq 2$. The reduction is almost the same as the previous one and generalizes those in [10] for SQUARE OF CHORDAL GRAPH.

Theorem 6. k-TH POWER OF CHORDAL GRAPH *is NP-complete for all fixed* $k \geq 2$.

4 Squares of Strongly Chordal Split Graphs

Lau and Corneil [10] shown that SQUARE OF SPLIT GRAPH is NP-complete. In contrast, we will show in this section that there exists a good characterization of squares of strongly chordal split graphs that gives a recognition algorithm in time $O(\min\{n^2, m\log n\})$ for such squares.

A chordal graph is *strongly chordal* if it does not contain any ℓ-sun as an induced subgraph; here a ℓ-sun, $\ell \geq 3$, consists of a stable set $\{u_1, u_2, \ldots, u_\ell\}$ and a clique $\{v_1, v_2, \ldots, v_\ell\}$ such that for $i \in \{1, \ldots, \ell\}$, u_i is adjacent to exactly v_i and v_{i+1} (index arithmetic modulo ℓ). We will make use of the following well-known fact:

Lemma 4 ([2,12,16]). *Powers of strongly chordal graphs are strongly chordal.*

In a graph, a vertex is *maximal* if its closed neighborhood is maximal with respect to set-inclusion. For split graphs $H = (V_H, E_H)$ we write $H = (C \cup S, E_H)$, meaning $V_H = C \cup S$ is a partition of the vertex set of H into a clique C and a stable set S.

Lemma 5. *Let $H = (C \cup S, E_H)$ be a connected split graph without 3-sun. Then Q is a maximal clique in H^2 if and only if $Q = N_H[v]$ for some maximal vertex $v \in C$ of H.*

Theorem 7. *G is square of a strongly chordal split graph if and only if G is strongly chordal and $\left|\bigcap_{Q \in \mathcal{C}(G)} Q\right| \geq |\mathcal{C}(G)|$.*

Theorem 8. *Given an n-vertex and m-edge graph G, recognizing if G is the square of some strongly chordal split graph H can be done in time $O(\min\{n^2, m \log n\})$, and if so, such a square root H for G can be constructed in the same time.*

Proof. By the constructive proof of Theorem 7, the following Algorithm 1 correctly computes a strongly chordal split graph H that is a square root for G, if any.

Algorithm 1

Input: Connected graph $G = (V_G, E_G)$ with $n = |V_G|$ and $m = |E_G|$.
Output: A strongly chordal split graph H with $G = H^2$ if such H exists or 'NO' otherwise.

1. **if** G is strongly chordal **then**
2.   compute all maximal cliques $Q_1, \ldots, Q_q$ of G
3.   compute $C = \bigcap_{1 \leq i \leq q} Q_i$
4.   **if** $|C| \geq q$ **then**
5. $V_H := V_G$; $E_H := \{xy \mid x, y \in C\}$
6.   **for** $i := 1$ **to** q **do**
7.   choose a vertex $v_i \in C$ with $v_i \neq v_j$ for $i \neq j$
8.   **for** $i := 1$ **to** q **do**
9. $E_H := E_H \cup \{vv_i \mid v \in Q_i \setminus C\}$
10.   return H
11.   **else return** 'NO'
12. **else return** 'NO'

The time complexity of Algorithm 1 is dominated by the time consumed at lines 1 and 2. Testing if G is strongly chordal can be done in time $O(\min\{n^2, m \log n\})$ ([3,13,15,17]). Assuming G is strongly chordal, all maximal cliques $Q_1, \ldots Q_q$ of G can be listed in linear time (cf. [6,17]); note that $q \leq n$. So, the total time of the algorithm is bounded by $O(\min\{n^2, m \log n\})$. □

5 Powers versus Girth

The girth of a graph is the smallest length of a cycle in the graph. In other words, G has girth k if and only if G contains a cycle of length k but does not contain any (induced) cycle of length $\ell = 3, \ldots, k-1$. Note that the girth of an n-vertex and m-edge graph can be computed in $O(nm)$ [7].

In [4] square roots with girth conditions have been considered. It is shown there that SQUARE OF GRAPH WITH GIRTH ≤ 4 is NP-complete, while SQUARE OF GRAPH WITH GIRTH ≥ 6 is polynomially solvable, and it has been conjectured that k-POWER OF GRAPH WITH GIRTH $\ge 3k-1$ is polynomially solvable. In this section, we give a good characterization of graphs that are cubes of a graph having girth at least 10. Our characterization leads to an $O(nm^2)$-time recognition for such graphs.

The following fact is the key observation for further discussions.

Lemma 6. *Let $G = (V, E_G)$ be a connected, non-complete graph such that $G = H^3$ for some graph $H = (V, E_H)$ with girth at least 10. Then $Q \subseteq V$ is a maximal clique in G if and only if $Q = N_H[u, v]$ for some edge $uv \in E_H$ with $\deg_H(u) \ge 2$ and $\deg_H(v) \ge 2$.*

Corollary 1. *If $G = (V_G, E_G)$ is the cube of some graph with girth at least 10, then G has at most $|E_G|$ maximal cliques.*

Definition 1. *Let G be an arbitrary graph. An edge e of G is called* forced *if e is the intersection of two distinct maximal cliques in G.*

The meaning of forced edges is that if the graph considered is the cube of some graph with girth at least 10, then its forced edges must belong to the edge set of any cube root with such girth condition.

Observation 9 *Let $G = H^3$ for some graph H with girth at least 10. Then, an edge of G is forced if and only if it is the mid-edge of an P_6 in H.*

Definition 2. *A connected graph G is said to be* trivial *if it contains a non-empty clique C such that $G \setminus C$ is the disjoint union of at most $|C| - 1$ cliques and every vertex in C is adjacent to every vertex in $G \setminus C$.*

Observation 10 (i) *A graph is trivial if and only if it is the cube of some tree of diameter at most 4;*
(ii) *Trivial graphs can be recognized in linear time.*

By Observation 10, we need consider non-trivial graphs only. A *star* is a tree with at least two vertices and diameter at most two. For an edge e, let $\mathcal{C}_e$ denote the set of all maximal cliques containing e.

Proposition 1. *Let G be a connected, non-trivial graph such that $G = H^3$ for some graph H with girth at least 10, and let F be the subgraph of G consisting of all forced edges of G. Then*

(i) *F is a connected induced subgraph of H;*
(ii) *For each $e \in F$, there exists a unique maximal clique $Q_e \in \mathcal{C}_e$ such that*
 (a) *for every two disctint non-disjoint forced edges e and e', $e \cup e' \subseteq Q_e \cap Q_{e'}$,*
 (b) *for every $Q \in \mathcal{C}(G) \setminus \{Q_e \mid e \in F\}$, and for all forced edges e_1, e_2 in Q, $Q_{e_1} \cap Q = Q_{e_2} \cap Q$;*

(iii) For each $e \in F$, $\mathcal{C}_e \setminus \{Q_e\}$ can be partitioned into non-empty disjoint sets $\mathcal{A}_e$ and $\mathcal{B}_e$ with
  (a) $Q \cap Q' = e$ if and only if $Q \in \mathcal{A}_e$ and $Q' \in \mathcal{B}_e$ or vice versa,
  (b) setting $A_e = \bigcap_{Q \in \mathcal{A}_e} Q$, $B_e = \bigcap_{Q \in \mathcal{B}_e} Q$, all pairs of maximal cliques in $\mathcal{A}_e$ have the same intersection A_e, all pairs of maximal cliques in $\mathcal{B}_e$ have the same intersection B_e,
  (c) $Q_e = A_e \cup B_e$, and $|A_e| \geq |\mathcal{A}_e| + 2$, $|B_e| \geq |\mathcal{B}_e| + 2$,
  (d) $F[A_e \cap V_F]$ and $F[B_e \cap V_F]$ are stars with distinct universal vertices in e;
(iv) $\mathcal{C}(G) = \bigcup_{e \in F} \mathcal{C}_e$;
(v) $V_G \setminus \bigcup_{e \in F} Q_e$ consists of exactly the simplicial vertices of G.

Theorem 11. *Let G be a connected, non-trivial graph. Let F be the subgraph of G consisting of all forced edges in G. Then, G is the cube of a graph with girth at least 10 if and only if F is connected and has girth at least 10 and G satisfies the conditions* (ii) – (v) *listed in Proposition 1.*

Theorem 12. *Given an n-vertex m-edge graph G, recognizing if G is the cube of some graph H with girth at least 10 can be done in time $O(nm^2)$, and if so, such a cube root H for G can be constructed in the same time.*

Proof. Note that by Corollary 1, any cube of an m-edge graph with girth at least ten has at most m maximal cliques. Then, use the algorithm in [18] to list the maximal cliques of G in time $O(nm^2)$. If there are more than m maximal cliques, G is not the cube of any graph with girth at least ten. Otherwise, the (at most m) maximal cliques of G are available. Then computing the forced edges of G to form the subgraph F of G, as well as the lists $\mathcal{C}_e$ for each $e \in F$ can be done in time $O(m^2)$ in an obvious way.

Moreover, the partitions $\mathcal{C}_e = \mathcal{A}_e \cup \{Q_e\} \cup \mathcal{B}_e$ satisfying (iii) (if any) for all forced edges e can be found in time $O(nm^2)$. Given these partitions, conditions (ii) – (v) in Proposition 1 then can be tested within the same time bound, as well as the square root H, in case all conditions are satisfied, can be constructed by Algorithm 2 below; $\mathcal{K} := \{Q_e \mid e \in F\}$. □

Remark. In the proof of Theorem 11, if F is a tree or a (C_4, C_6, C_8)-free bipartite graph, then the root H for G is also a tree, respectively, a (C_4, C_6, C_8)-free bipartite graph. Thus, if we replace the condition on F in Theorem 11 by 'F is a (C_4, C_6, C_8)-free bipartite graph', we obtain a good characterization and an $O(nm^2)$-time recognition for cubes of bipartite roots of this kind, while CUBE OF BIPARTITE GRAPH is NP-complete in general [9].

6 Conclusion and Open Problems

Although it has been gererally expected that k-TH POWER OF GRAPH and k-TH POWER OF BIPARTITE GRAPH are NP-complete for all fixed $k \geq 2$, respectively, $k \geq 3$, this paper contains the first proofs that these problems are indeed

Algorithm 2

```
1.  H := F
2.  for each Q_e ∈ K do
3.     let e = xy where x is universal in F[A_e ∩ V_F] and
4.     y is universal in F[B_e ∩ V_F]     // cf. (iii)(d)
5.     put all edges ux, vy into H, u ∈ A_e \ V_F, v ∈ B_e \ V_F
6.     // H[Q_e] = N_H[x,y] for all e = xy ∈ F
7.  for each Q ∉ K do
8.     let Q ∈ C_e for some forced edge e
9.     choose a vertex c_Q ∈ (Q ∩ Q_e) \ V_F; c_Q ≠ c_{Q'} for Q ≠ Q' ∉ K
10.    // Note that Q ∩ Q_e = A_e or Q ∩ Q_e = B_e, hence the choices
11.    // of c_Q's are possible by (iii)(c); c_Q is independent of e by (ii)(b)
13.    put all edges vc_Q, v ∈ Q \ V_H, into H
14. return H
```

NP-complete. We also have proved that k-TH POWER OF CHORDAL GRAPH is NP-complete for all fixed $k \geq 2$. On the positive side, we have found efficient algorithms for recognizing squares of strongly chordal split graphs and of cubes of graph with girth at least ten.

Some interesting open questions are: What is the computational complexity of recognizing powers of strongly chordal graphs? of chordal bipartite graphs (bipartite graphs without cycles of length at least six)? and of graphs with 'large' girth (cf. Conjecture 3)? We note that strongly chordal split graphs and chordal bipartite graphs are closely related: a bipartite graph is chordal bipartite iff completing one of its bipartition part yielding a strongly chordal split graph. Hence, given our result on squares of strongly chordal split graphs, CUBE OF CHORDAL BIPARTITE GRAPH could be polynomially solvable. Note also that the open questions we posed here are of particular interest as they generalize tree powers, which have been widely investigated in the literature.

References

1. Chang, M.-S., Ko, M.-T., Lu, H.-I.: Linear-time algorithms for tree root problems. In: Arge, L., Freivalds, R. (eds.) SWAT 2006. LNCS, vol. 4059, pp. 411–422. Springer, Heidelberg (2006)
2. Dahlhaus, E., Duchet, P.: On strongly chordal graphs. Ars Combin. 24 B, 23–30 (1987)
3. Farber, M.: Characterizations of strongly chordal graphs. Discrete Math. 43, 173–189 (1983)
4. Farzad, B., Lau, L.C., Le, V.B., Nguyen, N.T.: Computing graph roots without short cycles. In: Proceedings of the 26th International Symposium on Theoretical Aspects of Computer Science (STACS 2009), pp. 397–408 (2009)
5. Garey, M.R., Johnson, D.S.: Computers and Intractability–A Guide to the Theory of NP-Completeness. Freeman, New York (1979); Twenty-third printing (2002)
6. Golumbic, M.C.: Algorithmic Graph Theory and Perfect Graphs. Academic Press, New York (1980)

7. Itai, A., Rodeh, M.: Finding a minimum circuit in a graph. SIAM J. Computing 7, 413–423 (1978)
8. Kearney, P.E., Corneil, D.G.: Tree powers. J. Algorithms 29, 111–131 (1998)
9. Lau, L.C.: Bipartite roots of graphs. ACM Transactions on Algorithms 2, 178–208 (2006); Proceedings of the 15th Annual ACM-SIAM Symposium on Discrete Algorithms (SODA 2004), pp. 952–961
10. Lau, L.C., Corneil, D.G.: Recognizing powers of proper interval, split and chordal graphs. SIAM J. Discrete Math. 18, 83–102 (2004)
11. Lin, Y.-L., Skiena, S.S.: Algorithms for square roots of graphs. SIAM J. Discrete Math. 8, 99–118 (1995)
12. Lubiw, A.: Γ-free matrices, Master Thesis, Dept. of Combinatorics and Optimization, University of Waterloo, Canada (1982)
13. Lubiw, A.: Doubly lexical orderings of matrices. SIAM J. Computing 16, 854–879 (1987)
14. Motwani, R., Sudan, M.: Computing roots of graphs is hard. Discrete Appl. Math. 54, 81–88 (1994)
15. Paige, R., Tarjan, R.E.: Three partition refinement algorithms. SIAM J. Computing 16, 973–989 (1987)
16. Raychaudhuri, A.: On powers of strongly chordal and circular arc graphs. Ars Combin. 34, 147–160 (1992)
17. Spinrad, J.P.: Efficient Graph Representations. Fields Institute Monographs, Toronto (2003)
18. Tsukiyama, S., Ide, M., Ariyoshi, H., Shirakawa, I.: A new algorithm for generating all the maximal independent sets. SIAM J. Computing 6, 505–517 (1977)

Graph Partitioning and Traffic Grooming with Bounded Degree Request Graph[*]

Zhentao Li[1] and Ignasi Sau[2,3]

[1] School of Computer Science - McGill University - Montreal, Canada
zhentao.li@mail.mcgill.ca
[2] Mascotte project - INRIA/CNRS/UNS - Sophia Antipolis, France
[3] Graph Theory and Combinatorics group of UPC - Barcelona, Spain
ignasi.sau@sophia.inria.fr

Abstract. We study a graph partitioning problem which arises from traffic grooming in optical networks. We wish to minimize the equipment cost in a SONET WDM ring network by minimizing the number of Add-Drop Multiplexers (ADMs) used. We consider the version introduced by Muñoz and Sau [12] where the ring is unidirectional with a grooming factor C, and we must design the network (namely, place the ADMs at the nodes) so that it can support *any* request graph with maximum degree at most Δ. This problem is essentially equivalent to finding the least integer $M(C, \Delta)$ such that the edges of any graph with maximum degree at most Δ can be partitioned into subgraphs with at most C edges and each vertex appears in at most $M(C, \Delta)$ subgraphs [12]. The cases where $\Delta = 2$ and $\Delta = 3, C \neq 4$ were solved by Muñoz and Sau [12]. In this article we establish the value of $M(C, \Delta)$ for many more cases, leaving open only the case where $\Delta \geq 5$ is odd, $\Delta \pmod{2C}$ is between 3 and $C - 1$, $C \geq 4$, and the request graph does not contain a perfect matching. In particular, we answer a conjecture of [12].

Keywords: optical networks, traffic grooming, ADM, graph decomposition, cubic graph.

1 Introduction

Traffic grooming is the generic term for packing low rate signals into higher speed streams in optical networks [4, 7, 11, 15]. By using traffic grooming, it is possible to bypass the electronics at the nodes which are not sources or destinations of traffic, and therefore reduce the cost of the network. Typically, in a Wavelength Division Multiplexing (WDM) network, instead of having one SONET Add Drop Multiplexer (ADM) on every wavelength at every node, it is possible to have ADMs only for the wavelengths used at that node; the other wavelengths being optically routed without electronic switching. The so called traffic grooming problem consists of minimizing the total number of ADMs to be used, in

---

[*] This work has been partially supported by: 1st author: NSERC; 2nd author: European project IST FET AEOLUS, PACA region of France, Ministerio de Ciencia e Innovación, European Regional Development Fund under project MTM2008-06620-C03-01/MTM, and Catalan Research Council under project 2005SGR00256.

order to reduce the overall cost of the network. The problem is easily seen to be NP-hard for an arbitrary set of requests in very simple topologies. In fact, hardness and approximation results exist for traffic grooming in ring, star, and tree networks [9,8,2]. Here we consider unidirectional SONET/WDM ring networks with symmetric requests. In this case, the routing is unique and to each request between two nodes, we assign a wavelength and some bandwidth on this wavelength. If the traffic is uniform and any given wavelength can carry at most C requests, we can assign at most $\frac{1}{C}$ of the bandwidth to each request. C is known as the *grooming factor*. Furthermore, if the traffic requirement is symmetric, we may assume that symmetric requests are assigned the same wavelength, as it is easy to show (by exchanging wavelengths) that there exists an optimal solution where all symmetric requests are given the same wavelength. Then each pair of symmetric requests uses $\frac{1}{C}$ of the bandwidth in the whole ring. If the two end-nodes are u and v, we need one ADM at node u and one at node v. The main point is that if two requests have a common end-node, they can share an ADM if they are assigned the same wavelength.

The traffic grooming problem for a unidirectional SONET ring with n nodes, grooming ratio C, and a symmetric request graph R has been modeled as a graph partition problem as follows (see [3,10]). Each edge of R corresponds to a pair of symmetric requests, and edges are colored by their assigned wavelength λ. All edges of color λ induce a connected subgraph B_λ of R, where each node corresponds to an ADM. The grooming constraint, i.e. the fact that a wavelength can carry at most C requests, translates to an upper bound C on the number of edges in each B_λ. The cost corresponds to the total number of vertices used in the subgraphs, and the objective is therefore to minimize $\sum_\lambda |V(B_\lambda)|$. While most of previous work has focused on the case where the requests are given as input [2,4,7,8,9,11,3,10], we consider the case where only the network topology is given, together with a bound Δ on the request graph. We would like to place, for each value of the grooming factor C, a minimum number of ADMs at each node in such a way that they could support *any* traffic pattern where each node is the end-node of at most Δ requests. This model was recently introduced in [12], and it is interesting because the network can support dynamic traffic without replacement of the ADMs. The problem can be formulated as a graph partition problem as follows.

Δ-DEGREE-BOUNDED TRAFFIC GROOMING IN UNIDIRECTIONAL RINGS
Input: Three integers n (size of the ring), C (grooming factor), and Δ (maximum degree).
Output: An assignment of $A(v)$ ADMs to each vertex v of the ring, in such a way that *for any request graph G* with maximum degree at most Δ, there exists a partition of $E(G)$ into subgraphs $\{B_\lambda\}_{1 \leq \lambda \leq \Lambda} = \mathcal{B}$, such that:
(i) $|E(B_\lambda)| \leq C$ for all λ; and
(ii) each vertex $v \in V(G)$ appears in at most $A(v)$ subgraphs.
Objective: Minimize $\sum_v A(v)$.

The optimum to the above problem for each n, C, Δ is denoted by $A(n, C, \Delta)$. Given a graph with maximum degree at most Δ, a partition of G into subgraphs with at most C edges is called a *C-edge-partition* of G.

Previous work and our contribution. The cases where $\Delta = 2$ and the cases $\Delta = 3$, $C \neq 4$ were solved in [12]. In this article we establish the value of $M(C, \Delta)$ for the following cases: when $\Delta = 3$ and $C = 4$ (answering a conjecture of [12], c.f. Section 3), when $\Delta \geq 4$ is even for any C (c.f. Section 4), and when $\Delta \geq 5$ is odd (c.f. Section 5) and either $C \in \{2,3\}$, $\Delta \pmod{2C} = 1$, $\Delta \pmod{2C} \geq C$, or the request graph contains a perfect matching. We first fix the notation below and give some preliminaries in Section 2.

Notation. The (multi)graphs considered in this paper are finite and without self-loops. Edges are denoted $\{u, v\}$. The *degree* of a vertex v is the number of edges containing v as an end-point. The *maximum degree* of a (multi)graph is the maximum degree over all its vertices. A Δ-*graph* is a (multi)graph with maximum degree at most Δ. $\mathcal{G}_\Delta$ denotes the class of all Δ-graphs. A Δ-*regular* (multi)graph is a graph in which all vertices have degree Δ. An *almost Δ-regular* (multi)graph is a (multi)graph in which all vertices have degree Δ except possibly one which has degree $\Delta - 1$. A *bridge* in a (multi)graph G is an edge whose removal disconnects G. A *matching* in a (multi)graph $G = (V, E)$ is a subset $M \subseteq E$ which contains each vertex at most once. A *perfect matching* is a matching containing all vertices. A *digon* is a cycle of length 2. A *trail* in a (multi)graph is a sequence $\{\{x_1, x_2\}, \{x_2, x_3\}, \ldots, \{x_{k-1}, x_k\}\}$ of distinct edges in which the second end of an edge is the first end of the next edge (the same pair of vertices may appear more than once if there is more than one edge between them). Vertices $x_2, x_3, \ldots, x_{k-1}$ of a trail are called *midpoints*. The *length* of a trail is the number of edges in it. Given a (multi)graph $G = (V, E)$ and a subset of vertices $V' \subseteq V$, we denote by $G - V'$ the (multi)graph obtained from G by removing the vertices in V', the edges incident with vertices in V', and isolated vertices (if any). Similarly, given a subset of edges $E' \subseteq E$, we denote by $G - E'$ the (multi)graph obtained from G by removing the edges in E' and isolated vertices (if any).

2 Reducing the Problem

We begin by applying some easy reductions to the problem and recalling some results from [12] that will be used throughout. Let $M(C, \Delta)$ be the smallest number M such that $A(n, C, \Delta) \leq Mn$ for all n. It is known that $M(C, \Delta)$ is an integer for all values of C, Δ [12]. If the request graph is further restricted to belong to a subclass of graphs $\mathcal{C} \subseteq \mathcal{G}_\Delta$, then the corresponding positive integer is denoted by $M(C, \Delta, \mathcal{C})$.

By the discussion above, $A(n, C, \Delta)$ is of the form $A(n, C, \Delta) = M(C, \Delta)n - \alpha(C, \Delta)$, where $M(C, \Delta)$ and $\alpha(C, \Delta)$ are integers depending only on C and Δ. Suppose that a Δ-graph H requires at least $M(C, \Delta) + 1$ ADMs at some vertex. Since any Δ-graph must be supported with the same ADMs, by relabeling the vertices of H we could force at least $M(C, \Delta) + 1$ ADMs in $\Omega(n)$ nodes of the network. This would contradict the definition of $M(C, \Delta)$. Therefore, each vertex can appear in at most $M(C, \Delta)$ subgraphs. So we may conclude the following.

Remark 1. *For each value of C and Δ, Δ-DEGREE-BOUNDED TRAFFIC GROOMING IN UNIDIRECTIONAL RINGS reduces to finding the least integer $M(C, \Delta)$ such that the edges of any Δ-graph can be partitioned into subgraphs with at most C edges and each vertex appears in at most $M(C, \Delta)$ subgraphs.*

This allows us to give an equivalent definition of $M(C, \Delta)$. Let $G \in \mathcal{G}_\Delta$ and let $\mathcal{P}_C(G)$ be the set of C-edge-partitions of G. For $P \in \mathcal{P}_C(G)$, let $\mathrm{occ}(P)$ be the maximum number of occurrences of a vertex in the partition, that is,

$$\mathrm{occ}(P) = \max_{v \in V(G)} |\{B_\lambda \in P : v \in B_\lambda\}|,$$

and then $M(C, \Delta) = \max_{G \in \mathcal{G}_\Delta} \left(\min_{P \in \mathcal{P}_C(G)} \mathrm{occ}(P) \right).$

In the remainder of this paper, we use Remark 1 and focus on determining $M(C, \Delta)$ for each value of C and Δ. Observe also that any Δ-graph H is a subgraph of some Δ-regular graph G (with possibly more vertices). Note also that if we restrict a partition of G to the vertices of H, the number of occurrences of the vertices cannot increase. Therefore,

Remark 2. $M(C, \Delta) = M(C, \Delta, \mathcal{C})$, *where $\mathcal{C}$ is the class of Δ-regular graphs.*

The following two results will be used throughout the article.

Lemma 1 (Muñoz and Sau [12]). *The following statements hold trivially:*

(i) $M(C, 1) = 1$ *for all $C \geq 1$.*
(ii) $M(1, \Delta) = \Delta$ *for all $\Delta \geq 1$.*
(iii) If $C' \geq C$, then $M(C', \Delta) \leq M(C, \Delta)$.
(iv) If $\Delta' \geq \Delta$, then $M(C, \Delta') \geq M(C, \Delta)$.
(v) $M(C, \Delta) \leq \Delta$ *for all $C, \Delta \geq 1$.*

Proposition 1 (Muñoz and Sau [12]). $M(C, \Delta) \geq \left\lceil \frac{C+1}{C} \frac{\Delta}{2} \right\rceil$ *for all $C, \Delta \geq 1$.*

In [12] it is proved that $M(C, 2) = 2$ for any $C \geq 1$, that $M(C, 3) = 3$ for $C \leq 3$, and that $M(C, 3) = 2$ for $C \geq 5$. The latter result was proved using a result of Thomassen [14], settling a conjecture of Bermond et al. [5], stating that the edges of a cubic graph can be 2-colored such that each monochromatic component is a path of length at most 5.

Let us now discuss how these ideas can be extended to other values of C, Δ. A *linear C-forest* in a graph is a forest consisting of paths of length at most C. The *linear C-arboricity* of a graph G is the minimum number of linear C-forests required to partition $E(G)$, and is denoted by $la_C(G)$ [5]. Let $la_C(\Delta) = \max_{G \in \mathcal{G}_\Delta} la_C(G)$. Clearly $M(C, \Delta) \leq la_C(\Delta)$ for all C, Δ, since the paths in a *linear C-forest* are graphs with at most C edges. Therefore, the following upper bound given by Alon et al. [1] also applies to $M(C, \Delta)$.

Theorem 3 (Alon et al. [1]). *There is an absolute constant $\beta > 0$ such that for $\sqrt{\Delta} > C \geq 2$,*

$$la_C(\Delta) \leq \frac{C+1}{C}\frac{\Delta}{2} + \beta\sqrt{C\Delta \log \Delta}. \tag{1}$$

The first term of the right-hand side of Equation (1) is equal to the lower bound of Proposition 1, so Theorem 3 provides an additive $\mathcal{O}(\sqrt{C\Delta \log \Delta})$-approximation of $M(C, \Delta)$ for $\sqrt{\Delta} > C \geq 2$. We improve this bound for $M(C, \Delta)$ in Sections 4 and 5 (without using the linear C-arboricity), providing an additive 1-approximation of $M(C, \Delta)$ for any value of C and Δ, which is optimal for any even Δ, and in many cases for odd Δ.

3 Case $\Delta = 3, C = 4$

Muñoz and Sau conjectured that $M(4, 3) = 2$ [12], which we now prove. We first need the following classical result and an easy generalization (although it is well known, we provide a short proof here for the sake of completeness).

Theorem 4 (Petersen [13]). *Any cubic bridgeless graph has a perfect matching.*

Corollary 1. *Any cubic bridgeless multigraph without self-loops has a perfect matching.*

Proof: Let G be a cubic multigraph without self-loops. We can assume that G has no triple edges, otherwise G has only 2 vertices and any of the 3 edges is a perfect matching. Consider the simple graph G' built from G as follows: for each digon $\{\{u, v\}, \{u, v\}\}$, add 2 new vertices s_{uv} and t_{uv}, and replace the digon with the edges $\{u, s_{uv}\}, \{u, t_{uv}\}, \{v, s_{uv}\}, \{v, t_{uv}\}$, and $\{s_{uv}, t_{uv}\}$. By Theorem 4, G' has a perfect matching M'. We now construct a perfect matching M of G from M'. For each edge $e \in M'$ such that e was also an edge of G, put e in M'. For each digon $\{\{u, v\}, \{u, v\}\}$ of G, if any of the pairs $\{\{u, s_{uv}\}, \{v, t_{uv}\}\}$ or $\{\{u, t_{uv}\}, \{v, s_{uv}\}\}$ is in M', put one of the copies of $\{u, v\}$ in M. Otherwise, $\{s_{uv}, t_{uv}\}$ belongs to M' and we do nothing. It is easy to check that M is a perfect matching of G. □

We are ready to prove the main result of this section.

Theorem 5. *The edges of every almost 3-regular multigraph G without self-loops can be partitioned into a set $\mathcal{W} = \{W_1, W_2, \ldots, W_k\}$ of trails of length at most 4 such that each vertex appears as the midpoint of a trail.*

Proof: Suppose the theorem is false and let G be a counterexample with the minimum number of vertices. G is connected as otherwise, we can take the union of the partitions of its connected components, which exist by minimality of G.

    Case 1: G contains a bridge $e = \{u, v\}$. Then $G - \{e\}$ has exactly two components: U containing u and V containing v. Without loss of generality, we

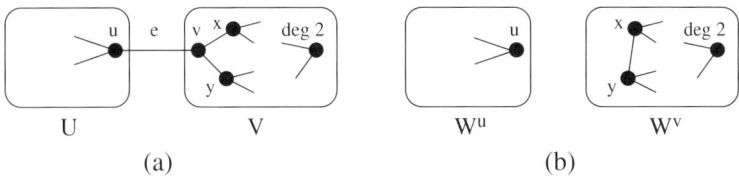

Fig. 1. (a) A bridge $e = \{u, v\}$ in an almost 3-regular graph G with components U and V of $G - \{e\}$. (b) Graphs smaller than G from which we obtain a partition into trails W^u and W^v.

may choose U to be the component with no degree 2 vertex in G and e is chosen so that U is maximal with this property. Thus this component U of $G - \{e\}$ is almost 3-regular (only u has degree 2). By minimality of G, U can be partitioned into a set $\mathcal{W}^u$ of trails as in the statement of the theorem.

If v has degree 2 in G then $V - \{v\}$ is almost 3-regular. By minimality of G, $V - \{v\}$ can be partitioned into a set $\mathcal{W}^v$ of trails as in the theorem. Now the only edges of G not in any trail in $\mathcal{W}^u \cup \mathcal{W}^v$ are those incident to v. Thus taking $\mathcal{W}^u \cup \mathcal{W}^v$ together with a trail consisting of the 2 edges incident to v (which has v as a midpoint) yields the required partition of the edges of G into trails. This contradicts the fact that G is a counterexample.

If v has degree 3 in G, let x, y be the neighbors of v in V (see Fig. 1(a)). We can assume $x \neq y$ (i.e., $\{v, x\}$ and $\{v, y\}$ are not parallel edges) since otherwise, the third edge incident to $x = y$ is a cut edge whose choice (instead of e) would increase the size of U. Let H be the graph obtained from $V - \{v\}$ by adding an edge $f = \{x, y\}$ (see Fig. 1(b)). By minimality of G, H can be partitioned into a set $\mathcal{W}^v$ of trails. We now attempt to transform $\mathcal{W}^u \cup \mathcal{W}^v$ into a partition of G into trails.

The edge f appears in some trail $\{W_1, \{x, y\}, W_2\}$ of $\mathcal{W}^v$, where W_1 is a (possibly empty) trail ending at x and W_2 is a (possibly empty) trail starting at y. At least one of the subtrails $\{W_1, \{x, y\}\}$ or $\{\{x, y\}, W_2\}$ has fewer than 3 edges. Without loss of generality, it is $\{W_1, \{x, y\}\}$. Replace this trail with $\{W_1, \{x, v\}, \{v, u\}\}$ which has length at most 4, and $\{\{v, y\}, W_2\}$ which has length less than or equal to $\{W_1, \{x, y\}, W_2\}$. Note that x and v are midpoints of the first trail and y is the midpoint of the second trail. Furthermore, any other vertex which was a midpoint in $\{W_1, \{x, y\}, W_2\}$ is still a midpoint (since W_1 and W_2 appear as subtrails). Thus the union of $\mathcal{W}^u$ and $\mathcal{W}^v$ with the above replacement yields a partition of G into trails of length at most 4 with the desired property, which is a contradiction.

Case 2: G does not contains a bridge. If G is 3-regular, let $G' = G$. Otherwise, let G' be the graph obtained from G by replacing the vertex of degree 2 with an edge between its endpoints. Note that G' is 3-regular and contains no bridges. Therefore, by Corollary 1, G' contains a perfect matching $M \subseteq E(G')$.

Since G' is 3-regular, $G' - M$ is 2-regular. Thus, $G' - M$ is a union of disjoint cycles. We can orient the cycles of $G' - M$ so that each vertex v has exactly one edge e_v pointing towards v. For each edge $\{u, v\} \in M$, $W_{uv} = \{e_u, \{u, v\}, e_v\}$

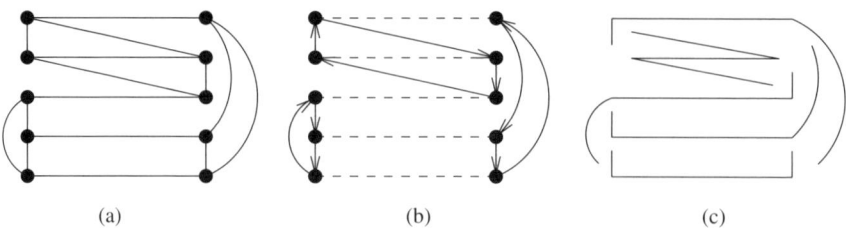

Fig. 2. (a) A 3-regular graph G' with no bridges. (b) A matching M of G' (shown in dashed lines) and an orientation of the cycles of $G' - M$. (c) A partition of the edges of G' into trails of length 3 using M and the orientation of the cycle of $G' - M$ in (b).

is a trail of length 3 (see Fig. 2). Note that $\mathcal{W} = \{W_{uv} \mid \{u,v\} \in M\}$ is a partition of the edges of G' into trails of length 3. Furthermore, every vertex u in the matching appears as the midpoint of the trail corresponding to the edge of the matching in which u appears. Since M is a perfect matching, every vertex appears as the midpoint of some trail in $\mathcal{W}$. Thus $G' \neq G$ as otherwise, we have constructed a partition as required by the theorem. So G has a vertex v of degree 2 which we replaced with an edge $e = \{x, y\}$ to obtain G'. Let $W = \{W_1, \{x,y\}, W_2\}$ be the trail in $\mathcal{W}$ containing e, and recall that W has length 3. Replacing W with $\{W_1, \{x,v\}, \{v,y\}, W_2\}$ in $\mathcal{W}$ yields a partition of $E(G)$ into trails of length at most 4, which is a contradiction. □

Note that the simple trees with some vertex of degree 3 and the digon with a pendant edge at each side are *not* allowed in the partition stated in Theorem 5, since these graphs cannot be thought of as trails. The following corollary answers the conjecture of [12].

Corollary 2. $M(4,3) = 2$.

Proof: By Remark 2, we may restrict ourselves to 3-regular graphs. Thus, a 3-regular graph G is almost 3-regular and we may apply Theorem 5 to obtain a partition $\mathcal{W}$. Let $\mathcal{B} = \{E(W)\}_{W \in \mathcal{W}}$. Each vertex of G appears in at most two elements of $\mathcal{B}$, as G is 3-regular and each vertex appears as the midpoint of some trail in $\mathcal{W}$. □

4 Case $\Delta \geq 4$ Even

In this section we establish the value of $M(C, \Delta)$ for $\Delta \geq 4$ even and any value of C.

Theorem 6. *Let $\Delta \geq 4$ be even. Then for any $C \geq 1$, $M(C, \Delta) = \left\lceil \frac{C+1}{C} \frac{\Delta}{2} \right\rceil$.*

Proof: The lower bound follows from Proposition 1. Let us give an explicit construction for any Δ-regular graph $G = (V, E)$. Orient the edges of G in an Eulerian tour, and assign to each vertex $v \in V$ its $\Delta/2$ out-edges, namely E_v^+. For each $v \in V$, partition E_v^+ into $\left\lceil \frac{\Delta}{2C} \right\rceil$ stars with C edges centered at v

(except, possibly, one star with fewer edges). Each vertex v appears as a leaf in stars centered at other vertices exactly $\Delta - \Delta/2 = \Delta/2$ times. Therefore, the number of occurrences of each vertex in this partition is

$$\left\lceil \frac{\Delta}{2C} \right\rceil + \frac{\Delta}{2} = \left\lceil \frac{\Delta}{2}\left(1 + \frac{1}{C}\right) \right\rceil = \left\lceil \frac{C+1}{C}\frac{\Delta}{2} \right\rceil.$$

□

5 Case $\Delta \geq 5$ Odd

The cases where Δ is odd turn out to be inherently much more complicated than the cases where Δ is even. In Section 5.1, we present a general construction which differs from the lower bound of Proposition 1 by at most 1, and we determine when this construction is optimal. In Section 5.2, we present an optimal construction for graphs with a perfect matching. Finally, in Section 5.3, we provide an improved lower bound when $\Delta \equiv C \pmod{2C}$, which meets our upper bound.

5.1 General Upper Bound

The following proposition provides a general upper bound, which differs from the lower bound of Proposition 1 by at most 1.

Proposition 2. *Let $\Delta \geq 5$ be odd. Then for any $C \geq 1$, $M(C, \Delta) \leq \left\lceil \frac{C+1}{C}\frac{\Delta}{2} + \frac{C-1}{2C} \right\rceil$.*

Proof: Let G be a Δ-regular graph. Since Δ is odd, $|V(G)|$ is even. Add a perfect matching M to G to obtain a $(\Delta+1)$-regular multigraph G'. Orient the edges of G' in an Eulerian tour, and assign to each vertex $v \in V(G')$ its $(\Delta+1)/2$ out-edges E_v^+. Remove the edges of M and, as in the case Δ even, partition E_v^+ into stars with at most C edges. To count the number of occurrences of each vertex, we distinguish two cases. If an edge of M is in E_v^+, then v appears as center in $\left\lceil \frac{\Delta-1}{2C} \right\rceil$ stars and as a leaf in $\Delta - \frac{\Delta-1}{2}$ stars. Summing both terms yields

$$\left\lceil \frac{\Delta-1}{2C} \right\rceil + \Delta - \frac{\Delta-1}{2} = \left\lceil \frac{C+1}{C}\frac{\Delta}{2} + \frac{C-1}{2C} \right\rceil.$$

Otherwise, if no edge of M is in E_v^+, the number of occurrences of v is

$$\left\lceil \frac{\Delta+1}{2C} \right\rceil + \Delta - \frac{\Delta+1}{2} = \left\lceil \frac{C+1}{C}\frac{\Delta}{2} + \frac{1-C}{2C} \right\rceil \leq \left\lceil \frac{C+1}{C}\frac{\Delta}{2} + \frac{C-1}{2C} \right\rceil.$$

□

The upper bound of Proposition 2 and the lower bound of Proposition 1 are equal for, roughly speaking, half of the pairs C, Δ, as shown in the following corollary.

Corollary 3. *Let $\Delta \geq 5$ be odd. If $\Delta \pmod{2C} = 1$ or $\Delta \pmod{2C} \geq C+1$, then $M(C, \Delta) = \left\lceil \frac{C+1}{C}\frac{\Delta}{2} \right\rceil$.*

Proof: Let $\Delta = \lambda \cdot 2C + h$, with h odd, $1 \leq h \leq 2C - 1$. Writing $k := \lambda(C+1) + \frac{h-1}{2}$, the lower bound of Proposition 1 equals $k + \lceil \frac{1}{2} + \frac{h}{2C} \rceil$, and the upper bound of Proposition 2 equals $k + \lceil 1 + \frac{h-1}{2C} \rceil$. If $h = 1$ both bounds equal $k + 1$, and if $h \geq C + 1$ both bounds equal $k + 2$. □

In particular, when $C = 2$ and Δ is odd, $\Delta \pmod{2C}$ is either 1 or 3, and then by Corollary 3 the lower bound is attained.

Corollary 4. *For any $\Delta \geq 5$ odd, $M(2, \Delta) = \lceil \frac{3\Delta}{4} \rceil$.*

We shall see in Theorem 7 that if $\Delta \equiv C \pmod{2C}$, then $M(C, \Delta) = \lceil \frac{C+1}{C} \frac{\Delta}{2} \rceil + 1$.

5.2 Optimal Construction for Graphs with a Perfect Matching

If the input graph has a perfect matching, we can prove that the lower bound is attained for all values of C.

Proposition 3. *Let $\Delta \geq 5$ be odd and let $\mathcal{C}$ be the class of Δ-regular graphs than contain a perfect matching. Then $M(C, \Delta, \mathcal{C}) = \lceil \frac{C+1}{C} \frac{\Delta}{2} \rceil$.*

Proof: First, it is easy to check that the proof of the lower bound of Proposition 1 in [12] still carries over when restricted to the class of graphs with a perfect matching. To prove the upper bound, if G is Δ-regular with a perfect matching M, we orient the edges of $G - M$ in an Eulerian tour, and assign to each vertex $v \in V(G)$ its $\frac{\Delta-1}{2}$ out-edges E_v^+. We distinguish three cases.

(1) $\Delta < C$. For each edge $\{u, v\} \in M$, build the tree with Δ edges consisting of $\{u, v\}$, $\frac{\Delta-1}{2}$ edges from E_u^+, and $\frac{\Delta-1}{2}$ edges from E_v^+. The number of occurrences of each vertex is $1 + \Delta - \frac{\Delta+1}{2} = \frac{\Delta+1}{2}$. The lower bound equals $\lceil \frac{C+1}{C} \frac{\Delta}{2} \rceil = \frac{\Delta-1}{2} + \lceil \frac{1}{2} + \frac{\Delta}{2C} \rceil$, which equals $\frac{\Delta+1}{2}$ as $\Delta < C$.

(2) $\Delta \geq C$ and $C \geq 3$ is odd (the case $C = 1$ is trivial by Lemma 1). For each edge $\{u, v\} \in M$, build the tree with C edges consisting of $\{u, v\}$, $\frac{C-1}{2}$ edges from E_u^+, and $\frac{C-1}{2}$ edges from E_v^+. Partition the remaining $\frac{\Delta-1}{2} - \frac{C-1}{2} = \frac{\Delta-C}{2}$ edges assigned to each vertex into $\lceil \frac{\Delta-C}{2C} \rceil$ stars with at most C edges. The number of occurrences of each vertex is

$$1 + \left\lceil \frac{\Delta - C}{2C} \right\rceil + \Delta - \frac{\Delta + 1}{2} = \left\lceil \frac{C+1}{C} \frac{\Delta}{2} \right\rceil.$$

(3) $\Delta \geq C$ and $C \geq 4$ is even (the case $C = 2$ is solved by Corollary 4). Build the tree with $C - 1$ edges consisting of $\{u, v\}$, $\frac{C-2}{2}$ edges from E_u^+, and $\frac{C-2}{2}$ edges from E_v^+. Partition the remaining $\frac{\Delta-1}{2} - \frac{C-2}{2} = \frac{\Delta-C+1}{2}$ edges assigned to each vertex into stars with at most C edges. The number of occurrences of each vertex is

$$1 + \left\lceil \frac{\Delta - C + 1}{2C} \right\rceil + \frac{\Delta - 1}{2} = \left\lceil \frac{\Delta(C+1) + 1}{2C} \right\rceil = \left\lceil \frac{C+1}{C} \frac{\Delta}{2} \right\rceil,$$

where the last equality holds because both Δ and $(C+1)$ are odd. □

5.3 Improved Lower Bounds

In this section we prove a new lower bound which strictly improves on Proposition 1 when $\Delta \equiv C \pmod{2C}$. The idea is to generalize the counterexample given in [12, Proposition 4] to prove that $M(3,3) = 3$.

Theorem 7. *Let $\Delta \geq 5$ be odd and let $LB(C, \Delta) = \lceil \frac{C+1}{C} \frac{\Delta}{2} \rceil$, the lower bound of Proposition 1. If $\Delta \equiv C \pmod{2C}$, then $M(C, \Delta) = LB(C, \Delta) + 1$.*

Proof: We prove that if $\Delta = kC$ with k odd, then $M(C, \Delta) \geq LB(C, \Delta) + 1$ and thus, by Proposition 2, $M(C, \Delta)$ is equal to $LB(C, \Delta) + 1$. Since both Δ and k are odd, so is C, and therefore $LB(C, \Delta) = k \cdot \frac{C+1}{2}$.

We proceed to build a Δ-regular graph G with no C-edge-partition where each vertex is incident to at most $LB(C, \Delta)$ subgraphs, hence implying that $M(C, \Delta) > LB(C, \Delta)$. First, we construct a graph H where all vertices have degree Δ except one which has degree $\Delta - 1$. Furthermore, we build H so that it has girth strictly greater than C. H exists by [6]. Make Δ copies of H and add a cut-vertex v joined to all vertices of degree $\Delta - 1$ to make our Δ-regular graph G.

Now suppose for the sake of contradiction that there is a C-edge-partition $\mathcal{B}$ of G where each vertex is incident to at most $LB(C, \Delta)$ subgraphs. Since the girth of G is greater than C, all the subgraphs in $\mathcal{B}$ are trees. Since $LB(C, \Delta) < \Delta$, v must have degree at least 2 in some subgraph $T' \in \mathcal{B}$. Since $|E(T')| \leq C$, the tree T' contains at most $\lfloor \frac{C-2}{2} \rfloor = \frac{C-3}{2}$ edges of a copy H' of H intersecting T'. Now we only work in H'. Let $\alpha = |E(T' \cap H')| \leq \frac{C-3}{2}$.

Let $\mathcal{B}' = \{B \cap H'\}_{B \in (\mathcal{B} - \{T'\})}$, with the empty subgraphs removed. That is, $\mathcal{B}'$ contains the subgraphs in $\mathcal{B}$ that partition the edges in H' that are not in T'. Let $n = |V(H')|$, which is odd as in H' there is one vertex of degree $\Delta - 1$ and all the others have degree Δ. Therefore, the total number of edges of the trees in $\mathcal{B}'$ is

$$\sum_{T \in \mathcal{B}'} |E(T)| = |E(H')| - \alpha = \frac{n\Delta - 1}{2} - \alpha = \frac{nkC - 1}{2} - \alpha. \qquad (2)$$

As $\alpha \leq \frac{C-3}{2}$, from Equation (2) we get

$$\sum_{T \in \mathcal{B}'} |E(T)| \geq \frac{nkC - 1}{2} - \frac{C-3}{2} = \left(\frac{nk-1}{2}\right) \cdot C + 1. \qquad (3)$$

As each tree in $\mathcal{B}'$ has at most C edges, from Equation (3) we get that $|\mathcal{B}'|$, the number of trees in $\mathcal{B}'$, satisfies

$$|\mathcal{B}'| \geq \left\lceil \frac{nk-1}{2} + \frac{1}{C} \right\rceil = \frac{nk-1}{2} + \left\lceil \frac{1}{C} \right\rceil = \frac{nk-1}{2} + 1. \qquad (4)$$

Clearly, the total number of vertices in the trees in $\mathcal{B}'$ is exactly the total number of edges in the trees in $\mathcal{B}'$ plus the number of trees in $\mathcal{B}'$, that is, $\sum_{T \in \mathcal{B}'} |V(T)| = \sum_{T \in \mathcal{B}'} |E(T)| + |\mathcal{B}'|$. On the other hand, the tree T' contains

$\alpha + 1$ vertices of H', that is, $|V(T' \cap H')| = \alpha + 1$. Therefore, using Equations (2) and (4), we get that the total number of occurrences of the vertices in H' in some tree of $\mathcal{B}$ is

$$\sum_{v \in V(H')} |\{T \in \mathcal{B} : v \in T\}| = \sum_{T \in \mathcal{B}'} |V(T)| + |V(T' \cap H')| = \sum_{T \in \mathcal{B}'} |E(T)| + |\mathcal{B}'| + \alpha + 1$$

$$= \frac{nkC - 1}{2} - \alpha + |\mathcal{B}'| + \alpha + 1 \geq \frac{nkC - 1}{2} + \frac{nk - 1}{2} + 1 + 1$$

$$= nk \cdot \frac{C + 1}{2} + 1 = n \cdot \text{LB}(C, \Delta) + 1 ,$$

which implies that at least one vertex of H' appears in at least $\text{LB}(C, \Delta) + 1$ subgraphs, which is a contradiction to $\mathcal{B}$ being a C-edge-partition of G in which each vertex appears in at most $\text{LB}(C, \Delta)$ subgraphs. The theorem follows. □

It turns out that Theorem 7 allows us to find the value of $M(3, \Delta)$ for any $\Delta \geq 5$ odd.

Corollary 5. *For any $\Delta \geq 5$ odd, $M(3, \Delta) = \lceil \frac{2\Delta + 1}{3} \rceil$.*

Proof: If $\Delta \equiv 1 \pmod{6}$ or $\Delta \equiv 5 \pmod{6}$, then by Corollary 3, $M(3, \Delta) = \lceil \frac{2\Delta}{3} \rceil = \lceil \frac{2\Delta + 1}{3} \rceil$. Otherwise, if $\Delta \equiv 3 \pmod{6}$, then by Theorem 7, $M(3, \Delta) = \lceil \frac{2\Delta}{3} \rceil + 1 = \lceil \frac{2\Delta + 1}{3} \rceil$. □

6 Conclusions

We considered the traffic grooming problem in unidirectional WDM rings when the request graph belongs to the class of graphs with maximum degree Δ. This problem is essentially equivalent to finding the least integer $M(C, \Delta)$ such that

Table 1. Known values of $M(C, \Delta)$. The **bold** cases remain open. The cases in brackets only hold if the graph has a perfect matching. The symbol "(=)" means that the corresponding lower bound is attained.

$C\|\Delta$	1	2	3	4	5	6	7	8	9	...	Δ even	Δ odd	
1	1	2	3	4	5	6	7	8	9	...	Δ	Δ	
2	1	2	3	3	4	5	6	6	7	...	$\lceil \frac{3\Delta}{4} \rceil$	$\lceil \frac{3\Delta}{4} \rceil$	
3	1	2	3 (2)	3	4	5 (4)	5	6	7 (6)	...	$\lceil \frac{2\Delta}{3} \rceil$	$\lceil \frac{2\Delta + 1}{3} \rceil$	$(\lceil \frac{2\Delta}{3} \rceil)$
4	1	2	2	3	4	4	5	5	6	...	$\lceil \frac{5\Delta}{8} \rceil$	$\geq \lceil \frac{5\Delta}{8} \rceil$	(=)
5	1	2	2	3	4 (3)	4	5	5	6	...	$\lceil \frac{3\Delta}{5} \rceil$	$\geq \lceil \frac{3\Delta}{5} \rceil$	(=)
6	1	2	2	3	≥ 3 (=)	4	5	5	6	...	$\lceil \frac{7\Delta}{12} \rceil$	$\geq \lceil \frac{7\Delta}{12} \rceil$	(=)
7	1	2	2	3	≥ 3 (=)	4	5 (4)	5	6	...	$\lceil \frac{4\Delta}{7} \rceil$	$\geq \lceil \frac{4\Delta}{7} \rceil$	(=)
8	1	2	2	3	≥ 3 (=)	4	≥ 4 (=)	5	6	...	$\lceil \frac{9\Delta}{16} \rceil$	$\geq \lceil \frac{9\Delta}{16} \rceil$	(=)
9	1	2	2	3	≥ 3 (=)	4	≥ 4 (=)	5	6 (5)	...	$\lceil \frac{5\Delta}{9} \rceil$	$\geq \lceil \frac{5\Delta}{9} \rceil$	(=)
...	...	...	...	...	...	...	...	...	...	...			
C	1	2	2	3	≥ 3 (=)	4	≥ 4 (=)	5	≥ 5 (=)	...	$\lceil \frac{C+1}{C} \frac{\Delta}{2} \rceil$	$\geq \lceil \frac{C+1}{C} \frac{\Delta}{2} \rceil$	(=)

the edges of any graph with maximum degree at most Δ can be partitioned into subgraphs with at most C edges and each vertex appears in at most $M(C, \Delta)$ subgraphs. We established the value of $M(C, \Delta)$ for many cases, leaving open only the case where $\Delta \geq 5$ is odd, Δ (mod $2C$) is between 3 and $C-1$, $C \geq 4$, and the graph does not contain a perfect matching. Table 1 summarizes what is known about $M(C, \Delta)$, including the case where the graph has a perfect matching. For the remaining cases, we hope to either extend the counterexample given in Section 5.3 or strengthen Proposition 3 in order to meet the lower bound.

References

1. Alon, N., Teague, V., Wormald, N.C.: Linear Arboricity and Linear k-Arboricity of Regular Graphs. Graphs and Combinatorics 17(1), 11–16 (2001)
2. Amini, O., Pérennes, S., Sau, I.: Hardness and Approximation of Traffic Grooming. In: Tokuyama, T. (ed.) ISAAC 2007. LNCS, vol. 4835, pp. 561–573. Springer, Heidelberg (2007)
3. Bermond, J.-C., Coudert, D.: Traffic Grooming in Unidirectional WDM Ring Networks using Design Theory. In: IEEE ICC, vol. 2, pp. 1402–1406 (2003)
4. Bermond, J.-C., Coudert, D.: Grooming. In: Colbourn, C.J., Dinitz, J.H. (eds.) The CRC Handbook of Combinatorial Designs, ch. VI.27, 2nd edn. Discrete Mathematics and its Applications, vol. 42, pp. 493–496. CRC Press, Boca Raton (2006)
5. Bermond, J.-C., Fouquet, J.-L., Habib, M., Péroche, B.: On linear k-arboricity. Discrete Mathematics 52(2-3), 123–132 (1984)
6. Chandran, L.S.: A high girth graph construction. SIAM J. Discrete Math. 16(3), 366–370 (2003)
7. Dutta, R., Rouskas, N.: Traffic grooming in WDM networks: Past and future. IEEE Network 16(6), 46–56 (2002)
8. Flammini, M., Monaco, G., Moscardelli, L., Shalom, M., Zaks, S.: Approximating the traffic grooming problem in tree and star networks. Journal of Parallel and Distributed Computing 68(7), 939–948 (2008)
9. Flammini, M., Moscardelli, L., Shalom, M., Zaks, S.: Approximating the traffic grooming problem. Journal of Discrete Algorithms 6(3), 472–479 (2008)
10. Goldschmidt, O., Hochbaum, D., Levin, A., Olinick, E.: The SONET edge-partition problem. Networks 41(1), 13–23 (2003)
11. Modiano, E., Lin, P.: Traffic grooming in WDM networks. IEEE Communications Magazine 39(7), 124–129 (2001)
12. Muñoz, X., Sau, I.: Traffic Grooming in Unidirectional WDM Rings with Bounded Degree Request Graph. In: Broersma, H., Erlebach, T., Friedetzky, T., Paulusma, D. (eds.) WG 2008. LNCS, vol. 5344, pp. 300–311. Springer, Heidelberg (2008)
13. Petersen, J.P.: Die Theorie der Regulären Graphs. Acta Mathematica 15, 193–220 (1891)
14. Thomassen, C.: Two-coloring the edges of a cubic graph such that each monochromatic component is a path of length at most 5. J. Comb. Theory Ser. B 75(1), 100–109 (1999)
15. Zhu, K., Mukherjee, B.: A review of traffic grooming in WDM optical networks: Architectures and challenges. Optical Networks Magazine 4(2), 55–64 (2003)

Injective Oriented Colourings

Gary MacGillivray[1,*], André Raspaud[2], and Jacobus Swarts[1]

[1] Mathematics and Statistics, University of Victoria, P.O. BOX 3060 STN CSC, Victoria, B.C., Canada V8W 3R4
[2] LaBRI UMR CNRS 5800, Université Bordeaux I, 33405 Talence Cedex, France

Abstract. We develop the theory of oriented colourings which are injective on in-neighbourhoods. The complexity of deciding if the minimum number of colours is at most the fixed integer k is determined, as is Brooks-Theorem type bound. The latter relies, in part, on a characterization of the oriented graphs in which each vertex must be assigned a different colour. A better, tight bound is determined for oriented trees, and a linear algorithm that decides if a given tree can be coloured with at most k colours, where k is fixed, is described.

Keywords: oriented colouring, injective colouring, digraph. homomorphisms.

1 Introduction

We study oriented colourings which are injective on in-neighbourhoods. These first arose in work of Courcelle [2] under the name "good" and "semi-strong". Raspaud and Sopena [14] improved a result of Courcelle's by showing that every planar graph can be oriented to have such a colouring with at most 320 colours. Related topics that are studied in the literature include injective homomorphisms [12,16] injective colourings and homomorphisms of undirected graphs (e.g. see [5,6,7,8]), and oriented colourings (e.g. see [3,15,17]).

The parameter in which we are interested is the injective oriented chromatic number of an oriented graph D. Informally, it is the minimum number of colours needed in an oriented colouring of D having the additional property that no two in-neighbours of any vertex are assigned the same colour. Our goal is to develop a comprehensive theory of injective oriented colourings of digraphs.

When considering a colouring parameter, the following questions naturally arise:

- Is there a homomorphism model?
- What is the complexity of deciding if it is no more than a fixed integer k?
- Can obstructions and critical digraphs be identified in the polynomial cases?
- What bounds are available?
- Is there anything that can be said for special classes of digraphs?
- How many such colourings exist?

---
* Research supported by NSERC.

A homomorphism model for injective oriented colourings is described in Section Two. It leads to the complexity results presented in the following section. The obstructions and critical digraphs can be identified, and algorithms that either return an injective oriented colouring, or an obstruction that certifies that there is no such colouring exist. These are described elsewhere [11]. Before giving bounds for the injective oriented chromatic number in terms of vertex degrees in Section Five, in Section Four we give a structural characterization of the oriented graphs whose injective oriented chromatic number equals the number of vertices. The results imply a polynomial time recognition algorithm. Finally, Section Six presents a polynomial time algorithm that tests whether a given oriented tree has an injective homomorphism to a fixed directed graph H. A consequence is that it is possible to decide in polynomial time if an oriented tree has an injective oriented colouring with k colours, and find a colouring when one exists. We have obtained similar results to those presented in this paper in the case where the colourings need not be proper [11]. Finding results on the enumeration of injective oriented colourings remains an open problem.

2 Preliminaries

For basic results and notation concerning graphs and digraphs we follow the text of Bondy and Murty [1]. We consider *oriented graphs*: directed graphs such that, for every pair of distinct vertices x and y, at most one of the arcs xy and yx exists. A vertex x of digraph D is a *dominating vertex* for a $X \subseteq V$ (respectively, subgraph H) if it is adjacent to every vertex of X (respectively, H). Similarly, a vertex x is *dominated* by $X \subseteq V$ (respectively, subgraph H) if it is adjacent from every vertex of X (respectively, H).

Let G and H be directed graphs. A *homomorphism of G to H* is a function $f : V(G) \to V(H)$ such that $f(x)f(y)$ is an arc of H whenever xy is an arc of D. A homomorphism of G to H is *injective* if, for each vertex x of D, no two in-neighbours of x have the same image. Because homomorphisms generalize colourings, the problem of deciding if a given digraph has a homomorphism to a fixed digraph H has been called *H-colouring*. If the mapping is required to be injective, it is called *injective H-colouring*. The book by Hell and Nešetřil [9] is an excellent introduction to the theory of homomorphisms of graphs and digraphs.

Let D be an oriented graph and k be a positive integer. An *oriented k-colouring* of D is an assignment of the colours $1, 2, \ldots, k$ to the vertices of D so that adjacent vertices get different colours and if some arc of D joins a vertex of colour i to a vertex of colour j, then no arc of D joins a vertex of colour j to a vertex of colour i. The *oriented chromatic number of D* is the smallest k for which there is an oriented k-colouring of D. Equivalently, an oriented k-colouring of D is a homomorphism to an oriented graph T on k vertices, and the oriented chromatic number of D is the least k for which such a homomorphism exists. Since arcs can be added joining non-adjacent vertices of T without destroying the existence of a homomorphism of D to T, the oriented graph T can be taken to be a tournament.

An *injective oriented k-colouring* of an oriented graph D is an oriented colouring of D such that, for each vertex x of D, no two in-neighbours of x are assigned the same colour. The *injective oriented chromatic number* of an oriented graph D, denoted $\chi_i(D)$, is the smallest k for which there is an injective oriented k-colouring of D. It is clear from comparing the definitions that an injective oriented k-colouring of an oriented graph D is an injective homomorphism of D to a digraph T on k vertices, and $\chi_i(D)$ is the least k for which such a homomorphism exists. As for oriented colourings, the digraph T can be assumed to be a tournament.

The directed graph that consists of two arcs meeting at a vertex plays an important role in our work. We define the *hat* $H_3(v_0, s, v_1)$ with *point* s and *ends* v_0 and v_1 to be the digraph with vertices v_0, s, v_1 and arcs $v_0 s, v_1 s$. It will be denoted by H_3 when it is not necessary to emphasize the vertices. All three vertices belonging to a hat must be assigned different colours in an injective oriented colouring.

3 Complexity

In this section we determine, for each fixed positive integer k, the complexity of deciding whether a given oriented graph D has an injective oriented k-colouring. The dividing line is the same as for oriented colouring (see [10]).

Theorem 1. *[12,16] Let T be a tournament. Then the injective T-colouring problem is NP-complete, except when T has at most three vertices or consists of 3-cycle dominated by a single vertex. If T has at most three vertices or consists of 3-cycle dominated by a single vertex, then the injective T-colouring problem is polynomial.*

The polynomial algorithms are via reduction to 2-SAT.

The proof of the following theorem can be greatly simplified if the restriction to connected inputs is dropped. On the one hand, the restriction seems unnecessary because connected components can always be considered separately. Our motivation for doing some extra work in the proof is a subtlety that arises in oriented colourings of disconnected digraphs: one has to take care that all of the components of the given digraph are mapping to the same tournament on k vertices. The proof shows how to handle that small concern when the target tournament is strongly connected.

Theorem 2. *Let k be a fixed positive integer. If $k \leq 3$ then oriented k-colouring is polynomial. If $k \geq 4$ then oriented k-colouring is NP-complete when restricted to inputs whose underlying graph is connected.*

Proof. When $k \leq 3$ the result follows from Theorem 1.

Let T_4 denote the unique strong tournament on four vertices. Let z be a vertex with in-degree two, and let x and y be its two in-neighbours. Every vertex of T_4 is joined to y by a directed walk of length nine.

Let P be the oriented path $p_0, p_1, \ldots, p_{12}$, where $p_0 p_1 \in E(P)$ and $p_{i+1} p_i \in E(P)$ for $i = 1, 2, \ldots, 11$. For any vertex v of T_4 there is an injective homomorphism of P to T_4 that maps p_0 to x and p_{12} to v.

We first show that injective oriented 4-colouring is NP-complete for connected inputs. The transformation is from injective T_4-colouring. Given an oriented graph D, construct D' from the disjoint union of D and T_4 by, for each component C of the underlying graph of D, adding a new copy of P and identifying p_0 with x and p_{12} with some vertex c of C. We claim that there is an injective homomorphism of D to T_4 if and only if D' has an injective oriented 4-colouring.

Suppose first that there is an injective homomorphism of D to T_4. It can be extended to D' by mapping the copy of T_4 identically to itself and extend the mapping to each of the oriented paths.

Suppose D' has an injective 4-colouring. Then there is an injective homomorphism of D' to a tournament T on four vertices. Since T_4 is a subdigraph of D', it follows that $T = T_4$. The result follows on restricting the mapping to the vertices of D.

This argument generalizes to larger values of k. For a fixed integer $k > 4$, let T_k be any tournament obtained from T_4 by adding $k - 4$ dominating vertices and arbitrarily orienting the arcs among them. Then injective T_k-colouring is NP-complete. The argument is identical to that for T_4 except that T_4 is replaced by T_k. □

4 Cliques

An *io-clique* is a oriented graph G such that $\chi_i(G) = |V(G)|$. For a given oriented graph D, if $\omega_i(D)$ denotes the maximum number of vertices in a subgraph of D which is an io-clique, then $\chi_i(D) \geq \omega_i(D)$. In this section we give a structural characterization of io-cliques. It is similar to the characterization of *ocliques*: oriented graphs for which the oriented chromatic number equals the number of vertices [10].

Theorem 3. *An oriented graph G is an io-clique if and only if every two vertices x and y either have a common out-neighbour or are joined by a directed path of length at most two.*

Proof. Suppose first that the condition holds. Let $x, y \in V(G)$. Since adjacent vertices must be assigned different colours, suppose that x and y are non-adjacent. If x and y have a common out-neighbour, then they must be assigned different colours because of injectivity, and if they are joined by a directed path of length two then they must be assigned different colours in order to have an oriented colouring.

We prove the contrapositive of the converse implication. Suppose that the condition does not hold. Then there are non-adjacent vertices x and y which have no common out-neighbour and are not joined by a directed path of length two. Consider the colouring in which x and y are assigned the same colour and every other vertex is assigned its own unique colour. Since x and y are not joined by a directed path of length two, this is an oriented colouring, and since they have no common out-neighbour, it is an injective colouring. Hence G is not an io-clique. □

Theorem 3 implies that io-cliques can be recognized in polynomial time. It also implies that a tournament is an io-clique, as is any o-clique (because any two non-adjacent vertices are joined by a directed path of length two). More examples of io-cliques can be obtained from the following corollary.

Informally, the *wreath product*, $D[H]$, of two digraphs D and H is obtained by replacing each vertex of D by a copy of H and, if uv is an arc of D, putting all possible arcs from the vertices of the copy of H that replaced u to the vertices of the copy of H that replaced v. Formally, $D[H]$ is the digraph with vertex set $V(D) \times V(H)$ and (d_1, h_1) adjacent to (d_2, h_2) if and only if $d_1 d_2 \in E(D)$ or $d_1 = d_2$ and $h_1 h_2 \in E(H)$.

Corollary 4. *If G and H are both io-cliques, so is the wreath product $G[H]$.*

Proof. Let x and y be two vertices of $G[H]$. If they belong to the same copy of H then, since H is an io-clique, they are either joined by a directed path of length at most two, or they have a common out-neighbour. Otherwise, these two vertices belong to different copies of H corresponding to to different vertices of G. Since G is an io-clique, these vertices are joined by a directed path of length at most two, or they have a common out-neighbour. By the definition of wreath product, the same is true of x and y. Therefore, by Theorem 3, the digraph $G[H]$ is an io-clique. □

By the corollary, the wreath products $T[H_3]$, where T is a transitive tournament and H_3 denotes the hat, are acyclic io-cliques (which are not o-cliques).

5 Bounds on the Oriented Injective Chromatic Number

The quantity $\Delta^- + 1$ is a lower bound for χ_i. It is natural to wonder whether there is a theorem giving upper bounds on χ_i in terms of Δ^-. To see that there is no such bound, let $k \geq \Delta^-$ be a fixed positive integer. Start with a set L of k vertices, and for each Δ^--subset $X \subseteq L$ add a new vertex adjacent from all vertices in X. The maximum in-degree of the resulting graph is Δ^-, but since no two vertices in L can be assigned the same colour, $\chi_i \geq k$.

It is possible to bound the injective oriented chromatic number if one takes into account the maximum in-degree, maximum out-degree and the maximum degree of the underlying undirected graph. To see that such a bound is necessarily exponential, let $D = T_1 \cup T_2 \cup \cdots \cup T_m$ be the disjoint union of all tournaments on $k+1$ vertices. Then the maximum in-degree of D is k. Any tournament T for which there is an injective homomorphism of D to T has to contain each tournament on $k+1$ vertices as a subgraph. Therefore T is a k-universal tournament, and must have at least $2^{k/2}$ vertices [13].

Next, we present an upper bound on $\chi_i(D)$. It is derived by constructing and colouring a suitable undirected graph, and then refining the (undirected) colouring into an injective oriented colouring.

We define the *undirected io-square* of an oriented graph D to be the undirected graph $D^\bullet$ with vertex $V(D)$ and x adjacent to y whenever they have a common out-neighbour in D, or are joined by a directed path of length at most two in D. Any two vertices which are adjacent in $D^\bullet$ must be assigned different colours in an injective oriented colouring of D.

Theorem 5. *Let D be an oriented graph. If $D^\bullet$ is k-colourable, then $\chi_i(D) \leq 2^k - 1$.*

Proof. Consider a colouring of $D^\bullet$ with colours $1, 2, \ldots, k$. By definition of $D^\bullet$, no two vertices with a common out-neighbour in D are assigned the same colour. We will obtain an injective oriented colouring of D by refining this partition of $V(D)$ into independent sets.

Let the colour of v be j. For any colour $i \neq j$, the vertex v has at most one in-neighbour of colour i because the in-neighbours of v are adjacent in $D^\bullet$. Furthermore, v does not have both an in-neighbour and an out-neighbour of colour i because vertices joined by a directed path of length two in D are adjacent in $D^\bullet$. Therefore, v either has exactly one in-neighbour in colour class i or only out-neighbours in colour class i.

Define the *signature* of a vertex v to be a k-tuple where the jth entry is "·" and the ith entry is either "+" or "−" depending on whether v has out-neighbours or in-neighbours respectively in colour class i, $1 \leq i \leq k$, $i \neq j$.

We now use the information encoded in the signatures of each vertex to decide how to recolour it so that we have an oriented colouring. Since the collection of vertices of colour 1 is an independent set, we assign them all the same colour. Next, consider the possible arcs between vertices of colour 1 and vertices of colour 2: every vertex of colour 2 either has an out-neighbour of colour 1 (so its signature starts with $(+ \cdot \ldots)$) or an in-neighbour of colour 1 (so its signature starts with $(- \cdot \ldots)$). These vertices must receive different colours in an oriented colouring. If we now consider the arcs between the vertices of colour 3 and those of colour i, $i = 1, 2$, we see that the signatures start in one of four ways: $(- - \cdot \ldots)$, $(- + \cdot \ldots)$, $(+ - \cdot \ldots)$ or $(+ + \cdot \ldots)$. The vertices of colour 3 with different signatures on the first two coordinates must be assigned different colours in an oriented colouring. In general, the vertices in colour class j can be partitioned into 2^{j-1} independent sets depending on their signatures on the first $j-1$ coordinates. As before, vertices of colour j in different blocks of the partition need different colours in an oriented colouring. In total we therefore need $1 + 2 + 2^2 + \cdots + 2^{k-1} = 2^k - 1$ colours in an injective oriented colouring of D. Colouring according to the partition just described yields an injective oriented colouring of D. □

Corollary 6. *Let D be an oriented graph. Then $\chi_i(D) \leq 2^{\chi(D^\bullet)} - 1$.*

Lemma 7. *Let D be an oriented graph. The graph $D^\bullet$ is complete if and only if D is an io-clique. The graph $D^\bullet$ can not be an undirected odd cycle of length greater than three.*

Proof. It is clear that $D^\bullet$ is complete if and only if D is an io-clique.

In order for $D^\bullet$ to be an undirected odd cycle, the underlying undirected graph of D must be connected and have maximum degree two. Thus it is either a path or a cycle. If $D^\bullet$ contains no 3-cycles, then D can have no vertices of in-degree two and no directed path of length two. But any orientation of a path or cycle with more than three vertices contains either a vertex of in-degree two or a directed path of length two. This completes the proof. □

Corollary 8. *Let D be an oriented graph which is not an io-clique Then*

$$\chi_i(D) \leq 2^{(\Delta+(\Delta^--1)\Delta^++(\Delta^+)^2)} - 1.$$

Proof. Each vertex in $D^\bullet$ has at most Δ neighbours from the underlying graph of D, at most $(\Delta^- - 1)\Delta^+$ neighbours that are derived from joining common in-neighbours by an edge and at most $(\Delta^+)^2$ vertices that are at a distance two from it. Therefore the maximum degree of $D^\bullet$ is at most $k = \Delta + (\Delta^- - 1)\Delta^+ + (\Delta^+)^2$. By Brooks' Theorem, $D^\bullet$ has a k-colouring, so the result follows from Corollary 6. □

A tight bound which is linear in Δ^-, and depends only on Δ^-, is available for oriented trees.

Proposition 9. *Let T be an oriented tree. Then the injective chromatic number of T satisfies $\chi_i \leq 2\Delta^- + 1$, and the bound is best possible.*

Proof. Let $k = \Delta^-(T)$, and let H be a k-regular tournament. It is easy to prove by induction on $|V(T)|$ that there is an injective homomorphism of any tree T with maximum in-degree to H. This is clear if T has only one vertex. Suppose any tree with $n-1$ vertices and maximum in-degree at most k has an injective homomorphism to H. Let x be a leaf of (the underlying graph of) T. By the induction hypothesis, there is an injective homomorphism of $T - x$ to H. Since H is k-regular, and the vertex of T adjacent to x has in-degree at most $k-1$ in $T - x$, it is possible to extend this mapping to T.

Suppose that the injective oriented chromatic number of every complete k-ary tree (a tree in which every internal vertex has in-degree k and out-degree 1 except the root, which as out-degree 0) is at most $b \leq 2k$. Let T be such a tree that requires b colours. We claim that T can be chosen so that every colour is used on an internal vertex of T. Suppose not. Then some colour appears only on a leaf of T. Let T' be obtained from T by adding $k = \Delta^-$ in-neighbours to every leaf. If T' does not have the desired property then some colour is only used on a leaf. Restricting this colouring of T' to T gives a colouring of T with fewer than $b = \chi_i(T)$ colours, a contradiction.

By the claim, an optimal colouring of T is a homomorphism to a tournament H on $b \leq 2k$ vertices such that every vertex of H has in-degree at least k, and therefore out-degree at most $k-1$. Therefore, the sum of the in-degrees can not equal the sum of the out-degrees, a contradiction.

It follows that there exist k-ary trees with injective chromatic number $2k+1$. □

6 Oriented Injective Colourings of Trees

Let T be an oriented tree. In this section we give an algorithm that tests whether there is an injective homomorphism of T to a fixed digraph H, and finds one if it exists. In turn, this implies a polynomial time algorithm to find an injective oriented k-colouring of T (for fixed k), if one exists. Our algorithm is straightforward and involves a small amount of local graph information and consistency checking. A different algorithm follows from the results of Courcelle [2].

If T is an oriented tree rooted at a vertex x, we may arrange the vertices of T into its level sets based on the distance from x (in the underlying undirected tree T'): $V_i = \{v \mid d_{T'}(v,x) = i\}$, $0 \leq i \leq \ell$, where ℓ is the eccentricity of x. V_ℓ is considered to be the bottom of the tree and V_0 the top of the tree.

The algorithm first assigns lists, $L(v) \subseteq V(H)$, $v \in V(T)$, to the vertices of T. These lists are to be thought of as possible images (or colours) for the vertices of T. The assignment is based on the fact that if f is an injective homomorphism of T to H, then $d_H^-(f(v)) \geq d_T^-(v)$ since the image of v has to accommodate the in-neighbours of v in H. The lists are then processed from the bottom-up and the eventual colouring is from the top-down.

Since the colouring will be from the top-down, the processing of the lists has to ensure that once a possible image for a vertex has been decided upon (i.e. we've made a choice from $L(v)$), this choice can be extended downwards. "Extending downwards" means that we can make choices for vertices lower down in the tree (from their lists) that will preserve arcs as well as respect injectivity on in-neighbours.

To preserve arcs we essentially do a one-sided consistency check. That is if $u \in V_i$ and $v \in V_{i+1}$ such that uv (vu) is an arc of T, we remove $a \in L(u)$ if there does not exist a $b \in L(v)$ such that ab (ba) is an arc of H. This is done for all $u \in V_i$, for i from $\ell - 1$ to 0. In this way if we reach the root, x, and $L(x) \neq \emptyset$, we can make a choice for the image of x and extend all the way down preserving arcs in the process.

Let $A = \{v \in V(T) \mid |N^-(v)| \geq 2\}$. If $u \in A$, with $N^-(u) = \{u_1, u_2, \ldots, u_k\}$ and $y \in L(u)$, then there exists an injective mapping of $N^-(u)$ with $u \mapsto y$, if and only if there exists a system of distinct representatives (SDR) for the sets

$$L(u_1) \cap N^-(y), \; L(u_2) \cap N^-(y), \; \ldots, \; L(u_k) \cap N^-(y).$$

Therefore to process $L(u)$ we remove y from $L(u)$ if there does not exist an SDR as shown above. Note that the lists have to be intersected with $N^-(y)$ in order to "localize" them around y in H.

If $u \in A \cap V_i$, for some $0 \leq i \leq \ell - 1$, is such that $N^-(u) \subseteq V_{i+1}$, then the processing above is sufficient since a top-down colouring will colour u first (make a choice from $L(u)$) and if the SDR checks succeeded we will then be able to extend downwards (injectively). On the other hand it could happen that $N^-(v) \cap V_{i-1} \neq \emptyset$ for some $v \in A \cap V_i$. In this case $|N^-(v) \cap V_{i-1}| = 1$. Let u be the vertex in $|N^-(v) \cap V_{i-1}|$. Here u will be coloured (in a top down colouring) before any of $\{v\} \cup (N^-(v) - \{u\})$ have been coloured. For such a vertex u we have to ensure that a choice $a \in L(u)$ can be extended downwards. Of course, u may be an in-neighbour of more than one vertex in V_i. Let $A^* = \{v \in V(T) \mid v \in V_i$ and $N^+(v) \cap V_{i+1} \neq \emptyset,\ 0 \leq i \leq \ell - 1\}$. Suppose that $u \in A^* \cap V_i$. That is, u is an in-neighbour of some vertices $v_1, v_2, \ldots, v_k \in V_{i+1}$. We remove $a \in L(u)$ if there exists $v_j \in \{v_1, v_2, \ldots, v_k\} \cap A$, such that for every $z \in L(v_j)$ there does not exist an SDR for the sets

$$\{a\} \cap N^-(z),\ L(u_{j_1}) \cap N^-(z),\ L(u_{j_2}) \cap N^-(z),\ \ldots,\ L(u_{j_k}) \cap N^-(z),$$

where $(N^-(v_j) - \{u\}) = \{u_{j_1}, u_{j_2}, \ldots, u_{j_k}\}$. In essence, we are checking here whether $u \mapsto a$ can be extended to an injective mapping on all the in-neighbourhoods of vertices in $\{v_1, v_2, \ldots, v_k\} \cap A$ — we only need to consider vertices with more than one in-neighbour, hence the intersection with A.

The algorithm is shown below (Algorithm 10).

Algorithm 10. *Injective homomorhism of an oriented tree T to H.*
INPUT: An oriented tree T rooted at a vertex x.
        Level sets $V_0 = \{x\}, V_1, \ldots, V_\ell$.
        The sets A and A^* as defined as above.
        A target digraph H.

TASK: Find an injective homomorphism of T to H if one exists.
        Assign lists to $V(T)$ as follows:
           $L(v) = \{y \in V(H) \mid d_H^-(y) \geq d_T^-(v)\}$.
        For $i = \ell - 1$ to 0 perform the following for all $u \in V_i$.
           For each arc uv (vu) with $v \in V_{i+1}$, remove $a \in L(u)$ if there does not exist a $b \in L(v)$ with ab (ba) an arc of H.
           If $u \in A$, remove $y \in L(u)$ if there does not exist an SDR for the sets:
               $L(u_1) \cap N^-(y),\ L(u_2) \cap N^-(y),\ \ldots,\ L(u_k) \cap N^-(y)$,
              where $N^-(u) = \{u_1, u_2, \ldots, u_k\}$.
           If $u \in A^*$ and $N^+(u) \cap V_{i+1} = \{v_1, v_2, \ldots, v_k\}$, remove $a \in L(u)$ if there exists a $v_j \in \{v_1, v_2, \ldots, v_k\} \cap A$, such that for every $z \in L(v_j)$ there does not exist an SDR for the sets:
               $\{a\} \cap N^-(z),\ L(u_{j_1}) \cap N^-(z),\ L(u_{j_2}) \cap N^-(z),\ \ldots,$
               $L(u_{j_k}) \cap N^-(z)$, where $(N^-(v_j) - \{u\}) = \{u_{j_1}, u_{j_2}, \ldots, u_{j_k}\}$.

Let T be an oriented tree rooted at a vertex x together with the corresponding level sets $V_0 = \{x\}, V_1, \ldots, V_\ell$. For $v \in V_i$, consider the forest induced by $V(T) - (V_0 \cup V_1 \cup \cdots \cup V_{i-1})$. Let T_v be the sub-tree of T that contains v in the aforementioned forest. This is the sub-tree of T rooted at v, relative to x. The height of T_v is at most $\ell - i$.

Theorem 11. *Let H be a digraph and T an oriented tree rooted at a vertex x. If Algorithm 10 terminates with $L(x) \neq \emptyset$, then for every $v \in V(T)$ and for every $a \in L(v)$, there exists an injective homomorphism f of T_v to H such that $f(v) = a$.*

Proof. If $L(x) \neq \emptyset$, then $L(v) \neq \emptyset$ for every $v \in V(T)$ (an empty list in V_i would lead to empty lists in V_j, $j \leq i$, in particular $V_0 = \{x\}$ would have an empty list).

Let $v \in V_i$. The proof is by induction on the height of T_v: $h = \ell - i$, $0 \leq i \leq \ell$. When $h = 0$, $i = \ell$, v is a leaf and $T_v = \{v\}$. Therefore any element in $L(v)$ defines an injective homomorphism f of T_v to H.

Assume that the statement is true for all $0 \leq h < k \leq \ell$. That is, the statement is true for all $v \in V_j$ with $\ell - k + 1 \leq j \leq \ell$. Let $v \in V_{\ell-k}$ so that T_v has height $h = k$. Since $L(v) \neq \emptyset$, there exists an $a \in L(v)$ such that we can define an injective homomorphism f of $T[(N(v) \cap V_{\ell-k+1}) \cup \{v\}]$ to H with $f(v) = a$ (we might have to re-compute some of the SDRs to do this). Let $u \in N(v) \cap V_{\ell-k+1}$, with $f(u) = b \in L(u)$. The height of $(T_v)_u$ is $k-1$ and by the induction hypothesis there exists an injective homomorphism f_u of $:(T_v)_u$ to H such that $f_u(u) = f(u) = b$. Since this applies to every $u \in N(v) \cap V_{k-1}$ we can extend f to all of T_v. □

Corollary 12. *Let H be a digraph and T an oriented tree rooted at a vertex x. Then there exists an injective homomorphism of T to H if and only if Algorithm 10 terminates with $L(x) \neq \emptyset$.*

Proof. If there is an injective homomorphism f of T to H, then $d_H^-(f(v)) \geq d_T^-(v)$ and so $f(v) \in L(v)$. Furthermore, $f(v)$ is never removed from $L(v)$ during the execution of the algorithm. In particular $f(x) \in L(x)$, and so $L(x) \neq \emptyset$.

The converse follows from Theorem 11. □

The algorithm has a running time that is proportional to the number of vertices in T. This follows from the fact that each vertex of T is processed only once and the processing (one-sided consistency check and SDR computation) at each vertex is a function of $|V(H)|$, which is fixed.

Corollary 13. *For each fixed positive integer k there exists a linear time algorithm that determines if a given oriented tree has an injective oriented k-colouring, and finds one if it exists.*

Proof. The oriented tree T has an injective oriented k-colouring if and only if it has an injective homomorphism to a tournament on k vertices. The number of such tournaments is a function of k alone (i.e. it does not depend on the number of vertices of T). The existence of an injective homomorphism to each candidate tournament can be tested in linear time by Algorithm 10. □

References

1. Bondy, J.A., Murty, U.S.R.: Graph Theory. Springer, Berlin (2007)
2. Courcelle, B.: The monadic second order logic of graphs VI: On several reseresentations of graphs by relational structures. Discrete Appl. Math. 54, 117–149 (1994); Erratum: Discrete Appl. Math. 63, 199–200 (1995)
3. Esperet, L., Ochem, P.: Oriented colorings of 2-outerplanar graphs. Inform. Process. Lett. 101, 215–219 (2007)

4. Feder, T., Vardi, M.: The computational structure of monotone monadic SNP and constraint satisfaction: A study through Datalog and group theory. SIAM J. Comput 28, 57–104 (1998)
5. Fertin, G., Rizzi, R., Vialette, S.: Finding exact and maximum occurrences of protein complexes in protein-protein interaction graphs. In: Jedrzejowicz, J., Szepietowski, A. (eds.) MFCS 2005. LNCS, vol. 3618, pp. 328–339. Springer, Heidelberg (2005)
6. Fiala, J., Kratochvil, J.: Locally injective graph homomorphism: lists guarantee dichotomy. In: Fomin, F.V. (ed.) WG 2006. LNCS, vol. 4271, pp. 15–26. Springer, Heidelberg (2006)
7. Fiala, J., Kratochvil, J., Por, A.: On the computational complexity of partial covers of theta graphs. Discrete Appl. Math. 156, 1143–1149 (2008)
8. Hahn, G., Kratochvil, J., Siřan, J., Sotteau, D.: On the injective chromatic number of graphs. Discrete Math. 256, 179–192 (2002)
9. Hell, P., Nešetřil, J.: Graphs and Homomorphisms. Oxford University Press, London (2004)
10. Klostermeyer, W., MacGillivray, G.: Analogues of cliques for oriented colouring. Discussions Mathematicae Graph Theory 24, 373–387 (2004)
11. MacGillivray, G. , Raspaud, A. and Swarts, J.: Injective Oriented Colourings II (Manuscript) (2009)
12. MacGillivray, G., Swarts, J.: The complexity of injective homomorphisms (Submitted) (2008)
13. Moon, J.W.: Topics on Tournaments. Holt, Rinehart and Winston, New York (1968)
14. Raspaud, A., Sopena, E.: Good and semi-strong colorings of oriented planar graphs. Information Processing Letters 51, 171–174 (1994)
15. Sopena, E.: Oriented graph coloring. Discrete Math. 229, 359–369 (2001)
16. Swarts, J.: The complexity of digraph homomorphisms: Local tournaments, Injective Homomorphisms and Polymorphisms. Ph.D. Thesis, Department of Mathematics and Statistics, University of Victoria, Victoria, BC, Canada (2008)
17. Wood, D.: Acyclic, star and oriented colourings of graph subdivisions. Discrete Math. Theor. Comput. Sci. 7, 37–50 (2005)

Chordal Digraphs*

Daniel Meister and Jan Arne Telle

Department of Informatics, University of Bergen, Norway
daniel.meister@ii.uib.no, jan.arne.telle@ii.uib.no

Abstract. Chordal graphs, also called triangulated graphs, are important in algorithmic graph theory. In this paper we generalise the definition of chordal graphs to the class of directed graphs. Several structural properties of chordal graphs that are crucial for algorithmic applications carry over to the directed setting, including notions like simplicial vertices, perfect elimination orderings, and characterisation by forbidden subgraphs resembling chordless cycles. Moreover, just as chordal graphs are related to treewidth, the chordal digraphs will be related to Kellywidth. Chordal digraphs coincide with the perfect elimination digraphs arising in the study of Gaussian elimination on sparse linear systems [Haskins and Rose, 1973].

1 Introduction

Chordal graphs have many applications in algorithmic graph theory. In some cases the input graph itself is chordal, in other cases we work on a (minimal) triangulation of the input graph, with edges added so that we have a chordal graph. The algorithmic interest in triangulations and chordality is based on the many structural properties associated with chordal graphs. Since graphs can be regarded as a subclass of directed graphs (digraphs) a basic question is which algorithmic properties of graphs extend to digraphs. In this paper we generalise the definition of chordal graphs to digraphs and show that many of the properties that hold for chordal graphs carry over to the directed setting. Such properties are essential for algorithmic applications of chordal digraphs. Many problems that are NP-hard on general graphs become polynomial-time solvable on chordal graphs of bounded clique-size, and also on their subgraphs. The corresponding graph parameter, called treewidth, is in this way related to chordal graphs. The chordal digraphs will in an analogous way be related to the digraph parameter called Kelly-width, recently introduced by Hunter and Kreutzer [7]. The amount of research devoted to algorithms for digraphs is steadily increasing, see e.g. the monograph of Bang-Jensen and Gutin [1]. The structural properties we would like the chordal digraphs to capture include the following equivalent characterisations of chordal graphs:

(a) iteratively constructed by adding a new vertex adjacent to a clique
(b) have a perfect elimination ordering [10]

---
* This work is supported by the Research Council of Norway.

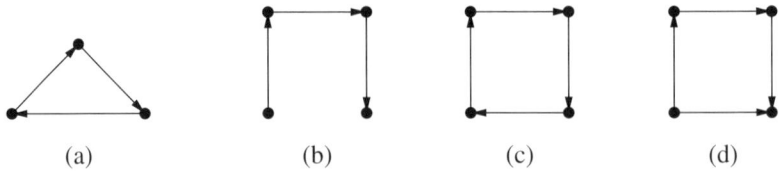

Fig. 1. The forbidden induced subgraphs of uni(G) for G a chordal digraph

(c) have vertex layout such that if $u \prec v \prec w$, and $\{v,u\}, \{u,w\}$ edges then $\{v,w\}$ also an edge [10]
(d) contain no chordless (or induced) cycle of length at least 4
(e) every minimal separator is a clique [3]
(f) are the intersection graphs of families of subtrees of trees [2,4,13].

Our definition of chordal digraphs in Section 3 is based on a generalisation of the iterative construction (a), which also serves to define Kelly-width [7,9]. We then show that chordal digraphs allow equivalent characterisations in terms of natural generalisations of perfect elimination orderings (b) and also vertex layouts (c). In fact, chordal digraphs coincide with the perfect elimination digraphs introduced by Haskins and Rose [6], see also [8]. Their interest stemmed from the study of Gaussian elimination of (unsymmetric) sparse matrices. Chordal digraphs also satisfy nice properties like closure under taking induced subgraphs and closure under reversal of all arcs. To generalise the characterisation by forbidden subgraphs (d) to chordal digraphs requires more work. Let us describe our results informally.

Partition the arcs of digraph G to define two digraphs uni(G) and bi(G), with uni(G) containing the arcs (u,v) for which (v,u) is not an arc, and bi(G) containing the arcs (u,v) for which (v,u) is also an arc. Our first result states that if bi(G) is empty (equivalently, if G is an orientation) then G is a chordal digraph if and only if G is acyclic (equivalently, contains no chordless directed cycle of length at least 3). On the other hand, if uni(G) is empty then G is chordal if and only if G contains no chordless directed cycle of length at least 4. Note that G could be viewed as an undirected graph precisely when uni(G) is empty, and hence the latter result shows that indeed the chordal digraphs are a generalisation of chordal (undirected) graphs. We also show that if G is a chordal digraph then bi(G) contains no chordless directed cycle of length at least 4. An important and large class of digraphs are the *semi-complete* digraphs where every pair of vertices have at least one arc. We show that a semi-complete digraph G is chordal if and only if bi(G) is chordal and uni(G) does not contain an induced subgraph isomorphic to a digraph in Figure 1.

Regarding the characterisation of chordal graphs by minimal separators (e) and by intersections of subtrees of a tree (f) it is an interesting open problem to provide a nice generalisation that will work for chordal digraphs. We discuss these and other open problems in the Conclusion section. For example, perfect elimination digraphs and thus chordal digraphs can be recognised in time $\mathcal{O}(nm)$ [11]. Rose and Tarjan showed that recognising perfect elimination digraphs is closely related to transitivity check [11]. Thus, perfect elimination digraph recognition falls into a well-studied class of problems.

2 Preliminaries

We consider directed and undirected graphs. Directed graphs are called *digraphs*. All graphs are simple and finite. In particular, digraphs contain no loops. By "graph", we may refer to a directed or undirected graphs. Let G be a digraph. The vertex and arc set of G are denoted as respectively $V(G)$ and $A(G)$. To vertices u, v of G can be connected by one or two arcs, namely (u, v) or (v, u). If u and v are not connected by any of these arcs, we call u and v *non-adjacent*; otherwise, u and v are *adjacent*. If $(u, v) \in A(G)$ then u is *in-neighbour* of v and v is *out-neighbour* of u. By $N^{\text{in}}(u)$ and $N^{\text{out}}(u)$, we denote the respectively *in-neighbourhood* and *out-neighbourhood* of u. For a set $X \subseteq V(G)$, $G[X]$ denotes the *subgraph* of G *induced* by X, i.e., the digraph on vertex set X and $(u, v) \in A(G[X])$ if and only if $u, v \in X$ and $(u, v) \in A(G)$ for all $u, v \in V(G)$. For x a vertex of G, we denote $G[V(G) \setminus \{x\}]$ as $G - x$. A *directed path* in G is a sequence $(x_1, \ldots, x_k)$ of pairwise different vertices of G where $(x_i, x_{i+1}) \in A(G)$ for all $1 \leq i < k$. A *directed cycle* in G is a sequence $(x_1, \ldots, x_k)$ of pairwise different vertices of G that is a directed path and $(x_k, x_1) \in A(G)$. Note that $k \geq 2$, since G contains no loops.

Let H be an undirected graph. The vertex and edge set of H are denoted as $V(H)$ and $E(H)$. Edges of H are denoted as $\{u, v\}$. For undirected graphs, we use definitions analogous to the definitions for digraphs. For a set $E' \subseteq E(H)$, the undirected graph $H \setminus E'$ has vertex set $V(H)$ and edge set $E(H) \setminus E'$. A *cycle* C in H is a sequence $(x_1, \ldots, x_k)$ of pairwise different vertices of H with $k \geq 3$ where $\{x_i, x_{i+1}\} \in E(H)$ for all $1 \leq i < k$ and $\{x_1, x_k\} \in E(H)$; k denotes the *length* of C. If $\{x_i, x_j\} \in E(H)$ for some $i, j \in \{1, \ldots, k\}$ and $1 < |j - i| < k - 1$ then $\{x_i, x_j\}$ is a *chord* in C. A cycle without chord is called *chordless*. A *vertex layout* for a graph F is a linear order of its vertices, denoted as $\beta = \langle x_1, \ldots, x_n \rangle$. For a pair u, v of vertices of F, we write $u \prec_\beta v$ if $u = x_i$ and $v = x_j$ and $i < j$.

An undirected graph without chordless cycles of length at least 4 is called *chordal*. Chordal graphs have a large number of different characterisations, such as by properties of minimal separators [3] or as intersection graphs [2,4,13]. Another characterisation is by vertex layouts.

Theorem 1 ([10]). *An undirected graph H is chordal if and only if there is a vertex layout β for H such that for all vertex triples u, v, w of H with $u \prec_\beta v \prec_\beta w$, $\{v, u\}, \{u, w\} \in E(H)$ implies $\{v, w\} \in E(H)$.*

A digraph is called *acyclic* if it contains no directed cycle. Acyclic digraphs have at most one arc between every pair of vertices. The following characterisation is folklore: a digraph G is acyclic if and only if there is a vertex layout β for G such that for all arcs (u, v) of G, $u \prec_\beta v$.

The structure of digraphs is often studied by looking only at the adjacency relation. For a digraph G, the *underlying graph* is the undirected graph H on vertex set $V(G)$, and for every pair u,v of vertices of G, $\{u,v\} \in E(H)$ if and only if $(u,v) \in A(G)$ or $(v,u) \in A(G)$.

3 Definition of Chordal Digraphs and First Results

We define a new class of digraphs. This class is defined inductively, in a manner similar to k-trees. We call our digraphs *chordal* and show by several characterisation results that the chordal digraphs and the chordal undirected graphs have analogous properties.

Definition 1
1. A d-*clique of a digraph* G *is a pair* (A, B) *with* $A, B \subseteq V(G)$ *where for all* $a \in A$ *and* $b \in B$ *with* $a \neq b$, $(a, b) \in A(G)$.
2. *The class of* chordal digraphs *is inductively defined as follows:*
    – *a graph on a single vertex is a chordal digraph*
    – *let* G *be a chordal digraph and let* u *be a vertex that does not appear in* G. *Let* (A, B) *be a d-clique of* G. *The graph is also a chordal digraph that is obtained from* G *by adding* u *and the set of arcs* $\{(a, u) : a \in A\} \cup \{(u, b) : b \in B\}$.

In other words, a chordal digraph is extended by a new vertex u by chosing a d-clique and making the one set the in-neighbours of u and the other set the out-neighbours of u. In case of k-trees or general chordal undirected graphs, the chosen d-clique corresponds to the chosen clique and there is no distinction between in- and out-neighbours. For a digraph G that is constructed according to Definition 1, the vertex layout that lists the vertices in order they are added to build G where the leftmost vertex is the last added vertex is called *construction sequence*. Note that a chordal digraph can have different construction sequences, since the construction of a chordal digraph is no unique process.

The class of chordal digraphs is closely related to the notion of Kelly-width [7]. We can introduce a parameter k that bounds the size of the second component in a d-clique. A chordal digraph has *width* k if it can be constructed according to Definition 1 by only choosing d-cliques with size of the second component at most k. With the results in [7] and [9], it is easy to verify that a digraph has Kelly-width at most $k+1$ if and only if it is subgraph of a chordal digraph of width at most k.

Chordal digraphs can be understood as a generalisation of chordal undirected graphs, since every chordal undirected graph can be seen as a chordal digraph in a natural sense: from an undirected graph H, obtain a digraph $\overline{H}$ by replacing every edge by the two arcs between the connected vertices. If we only use the adjacency matrix as graph representation, we can call H and $\overline{G}$ equivalent. We will see in the next section that G is chordal if and only if H is chordal. For chordal digraphs with all pairs of adjacent vertices connected by two arcs, we can make the following observation about the construction process: for every added vertex x, the in-neighbourhood and out-neighbourhood have to be the same, which means that the chosen d-clique is of the form (A, A). Such d-cliques induce complete subgraphs.

Chordal undirected graphs have many equivalent characterisations. We show that some of these characterisations have analogues for chordal digraphs. We begin with a vertex layout characterisation for chordal digraphs, that is an analogue of Theorem 1.

Definition 2. *Let G be a digraph with vertex layout β. We say that β is* directed transitive *if for every triple u, v, w of pairwise different vertices of G with $u \prec_\beta v$, $u \prec_\beta w$, $(v, u) \in A$ and $(u, w) \in A$, it holds $(v, w) \in A$.*

Theorem 2. *Let G be a digraph with vertex layout β. G is a chordal digraph with construction sequence β if and only if the reverse of β is directed transitive.*

Proof. We show the theorem by induction over the number of vertices in a digraph. The statement is obviously true for digraphs on a single vertex. Now, let G be a digraph on at least two vertices, and let $\beta = \langle x_1, \ldots, x_n \rangle$ be a vertex layout for G. Let $G' =_{\text{def}} G[\{x_1, \ldots, x_{n-1}\}]$ and $\beta' =_{\text{def}} \langle x_1, \ldots, x_{n-1} \rangle$. Let G be a chordal digraph with construction sequence β. Then, G' is a chordal digraph with construction sequence β' according to Definition 1. Applying the induction hypothesis, β' is directed transitive for G'. It remains to prove that $(a, b) \in A$ for every pair a, b of vertices where $a \in N_G^{\text{in}}(x_n)$ and $b \in N_G^{\text{out}}(x_n)$. Let x_n be added to G' by chosing d-clique (A, B). By definition, every vertex in A is in-neighbour of every vertex in B. Let $a \in N_G^{\text{in}}(x_n)$ and $b \in N_G^{\text{out}}(x_n)$ with $a \neq b$, i.e., $a \in A$ and $b \in B$. Then, $(a, b) \in A(G)$. For the converse, let β be directed transitive for G. Then, β' is directed transitive for G', and by induction hypothesis, G' is a chordal digraph with construction sequence β'. Let $A =_{\text{def}} N_G^{\text{in}}(x_n)$ and $B =_{\text{def}} N_G^{\text{out}}(x_n)$. By definition of directed transitive, $(a, b) \in A(G)$ for every $a \in A$ and $b \in B$ and $a \neq b$. Hence, (A, B) is a d-clique in G', and G can be obtained from G' by adding x_n in the sense of Definition 1. Hence, G is a chordal digraph with construction sequence β. □

Since "directed transitive" generalises "topological ordering", it directly follows from Theorem 2 that acyclic digraphs are chordal. We will see later that acyclic digraphs are the only chordal digraphs that have at most one arc between every pair of vertices.

Lemma 1. *Every induced subgraph of a chordal digraph is chordal.*

For a digraph G, denote by $\text{rev}(G)$ the digraph on vertex set $V(G)$ and with arc set $A(\text{rev}(G))$ where for all $u, v \in V(G)$, $(u, v) \in A(\text{rev}(G))$ if and only if $(v, u) \in A(G)$. We call $\text{rev}(G)$ the *reverse graph* of G.

Lemma 2. *A digraph G is chordal if and only if $\text{rev}(G)$ is chordal.*

It is a natural question to ask whether chordality of a digraph can be determined by looking at the connected components separately. With the characterisation of Theorem 2, it is an easy observation that a digraph is chordal if and only if all its

Fig. 2. The depicted digraph is chordal and has exactly one di-simplicial vertex

weakly connected components are chordal. A directed transitive vertex layout for the whole digraph can be constructed by concatenating directed transitive vertex layouts for the weakly connected components. However, the same is generally not true for strongly connected components. In fact, there is a non-chordal digraph on four vertices with only chordal strongly connected components.

Our first type of characterisations of chordal digraphs was by vertex layouts. The second type of characterisations involves vertices of special properties. In fact, we define and consider analogues of simplicial vertices for undirected chordal graphs.

Definition 3. *A vertex u of a digraph G is* di-simplicial *if $(N_G^{in}(u), N_G^{out}(u))$ is a d-clique of G.*

A vertex of an undirected graph is *simplicial* if its neighbourhood is a clique. It is an easy result that a vertex u of an undirected graph G is simplicial if and only if for every path P of G of length at least 1, $P-u$ is a path in G. An analogue characterisation holds for digraphs and di-simplicial vertices where paths are directed paths. In this sense, di-simplicial vertices **are** the directed analogue of simplicial vertices.

Lemma 3. *The first vertex of a directed transitive vertex layout is di-simplicial. In particular, every chordal digraph has a di-simplicial vertex.*

Non-complete chordal undirected graph have two non-adjacent simplicial vertices [3]. The same is not generally true for chordal digraphs. The digraph in Figure 2 has exactly one di-simplicial vertex.

Theorem 3. *A digraph G is chordal if and only if G can be reduced to a digraph on a single vertex by repeatedly deleting an arbitrary di-simplicial vertex.*

Theorem 3 also shows that chordal digraphs are exactly the perfect elimination digraphs, introduced by Haskins and Rose [6], see also [8]. Rose and Tarjan gave an $\mathcal{O}(nm)$-time algorithm for recognising perfect elimination digraphs [11], and therefore for chordal digraphs.

It is not difficult to see that there is a 1-to-1 correspondence between the directed transitive vertex layouts of a chordal digraph and the orderings defined by the elimination process in Theorem 3. Note that a result analogous to Theorem 3 exists for chordal undirected graphs [10].

4 Two Classes of Chordal Digraphs

We consider two classes of digraphs and characterise their chordal digraphs. First, we consider digraphs that have at most one arc between every pair of vertices, and second, we consider digraphs that have two arcs between every pair of adjacent vertices. For a digraph G, we denote by uni(G) the digraph on vertex set $V(G)$, and for all $u, v \in V(G)$, $(u,v) \in A(\text{uni}(G))$ if and only if $(u,v) \in A(G)$ and $(v,u) \notin A(G)$. In other words, uni(G) is the restriction of G to the arcs that uniquely connected two vertices. We call uni(G) the uni-restriction of G. As examples, if G is a tournament graph or an acyclic digraph then uni(G) = G.

Theorem 4. *A digraph G with* uni(G) = G *is chordal if and only if G contains no chordless directed cycle of length at least 3.*

Proof. If G contains no chordless directed cycle of length at least 3 then G is acyclic. Any topological ordering for G is directed transitive for G. Let G be chordal with directed transitive vertex layout β. Suppose that $(x_1, \ldots, x_k)$ is a shortest chordless directed cycle in G. We can assume $x_1 \prec_\beta x_2$ and $x_1 \prec_\beta x_k$. Then, $(x_k, x_1), (x_1, x_2) \in A(G)$ implies $(x_k, x_2) \in A(G)$, which contradicts the choice of the cycle. □

Corollary 1. *A digraph G with* uni(G) = G *is chordal if and only if G is acyclic.*

Every undirected graph is underlying graph of an acyclic digraph. So, every undirected graph is underlying graph of a chordal digraph. This also means that looking at underlying graphs of chordal digraphs cannot provide any insight into the structure of chordal digraphs.

The second class of digraphs in this section consists of the digraphs with always two arcs between every pair of adjacent vertices. For a digraph G, we denote by bi(G) the digraph on vertex set $V(G)$, and for all $u, v \in V(G)$, $(u,v) \in A(\text{bi}(G))$ if and only if $(u,v) \in A(G)$ and $(v,u) \in A(G)$. We call bi(G) the bi-restriction of G. Clearly, uni(G) and bi(G) are complementary to each other. The digraphs G with bi(G) are the ones that are obtained from undirected graphs by replacing every edge $\{u,v\}$ by the two arcs (u,v) and (v,u). Note that for digraphs G with bi(G) = G, the adjacency matrices of G and the underlying graph of G are equal.

Lemma 4. *If a digraph G is chordal then* bi(G) *contains no chordless directed cycle of length at least 4.*

Proof. Let G be chordal with β directed transitive vertex layout for G. Suppose that bi(G) contains a chordless directed cycle $(x_1, \ldots, x_k)$ where $k \geq 4$. We can assume that $x_1 \prec_\beta x_2, \ldots, x_k$. Then, $(x_k, x_1), (x_1, x_2) \in A(G)$ implies $(x_k, x_2) \in A(G)$, and $(x_2, x_1), (x_1, x_k) \in A(G)$ implies $(x_2, x_k) \in A(G)$. Thus, x_2 and x_k are adjacent in bi(G), contradicting $k \geq 4$. □

In other words, Lemma 4 shows that for every chordal digraph G, the underlying graph of bi(G) is chordal.

Theorem 5. *A digraph G with $\operatorname{bi}(G) = G$ is chordal if and only if G contains no chordless directed cycle of length at least 4.*

For every chordal undirected graph H, there is a digraph G with $\operatorname{bi}(G) = G$ such that H is the underlying graph of G. Since H is chordal, G contains no chordless directed cycle of length at least 4, and thus G is chordal due to Theorem 5. Hence, the chordal undirected graphs are exactly the underlying graphs of the bi-restriction of chordal digraphs.

5 Chordal Semi-complete Digraphs

A digraph is called *semi-complete* if every pair of vertices is adjacent. Pairs of vertices can be connected by one or two arcs. We give a complete characterisation of chordal semi-complete digraphs by forbidden induced subgraphs. We will obtain this result by mainly studying simplicial vertices in the underlying undirected graph of $\operatorname{bi}(G)$. Our approach to the forbidden induced subgraphs characterisation is to give a characterisation of semi-complete digraphs without di-simplicial vertices. Remember from Lemma 3 that every chordal digraph has a di-simplicial vertex.

We use the notion of uni- and bi-restriction defined in the previous section. The uni-restrictions will be our main study objects. Let F be a semi-complete digraph and let $G =_{\text{def}} \operatorname{uni}(F)$. Let u, v, w be a vertex triple of F. We call (u, v, w) a *witness triple for u in G* if one of the following three conditions is satisfied:

- $(u, v), (v, u), (u, w), (w, u) \notin A(G)$ and $(v, w) \in A(G)$
- $(u, v), (v, w) \in A(G)$ and $(u, w), (w, u) \notin A(G)$, or
  $(w, v), (v, u) \in A(G)$ and $(u, w), (w, u) \notin A(G)$
- $(v, u), (u, w), (w, v) \in A(G)$.

We refer to the different schemes as "witness triple of the first, second or third type".

Lemma 5. *1) A vertex u of a semi-complete digraph F is di-simplicial if and only if there is no witness triple for u in $\operatorname{uni}(F)$.*
2) If a vertex u of a digraph F is di-simplicial in F then u is simplicial in the underlying graph of $\operatorname{bi}(F)$.

For the proof of our main result, we need two properties of chordal undirected graphs. Let x and y be adjacent simplicial vertices of an undirected graph H. Then, $N_H[x] = N_H[y]$. Equivalently, x and y have the same non-neighbours. The second tool property is the following: for H a chordal undirected graph u a vertex of H that is not universal, every connected component of $H \setminus N_H[u]$ contains a vertex that is simplicial in H. We are ready for the main result of this section.

Lemma 6. *Let F be a semi-complete digraph with the underlying graph of $\operatorname{bi}(F)$ being chordal. If F is not chordal then $\operatorname{uni}(F)$ contains one of the digraphs of Figure 1 as induced subgraph.*

Proof. Let F not be chordal. We first consider the case that F contains no di-simplicial vertex; the other case is discussed at the end of the proof. Let $G =_{\text{def}} \text{uni}(F)$ and let H be the underlying graph of $\text{bi}(F)$. By assumption, H is chordal. Due to Lemma 5, a vertex that is not simplicial in H is not di-simplicial in F. If a simplicial vertex of H is not di-simplicial in F then it is not di-simplicial because of orientations of arcs in G. Applying Lemma 5, the assumption that F contains no di-simplicial vertex means that every vertex of G has a witness triple in G. In particular, every simplicial vertex of H has a witness triple in G. If there is a vertex with a witness triple of the third type then G contains a copy of digraph (a) of Figure 1 as induced subgraph. Now, assume that all witness triples in G are of the first or second type. Let u be a simplicial vertex of H. Then, u cannot have a witness triple of the first type. Hence, all simplicial vertices of H have only witness triples of the second type in G. We construct an auxiliary digraph based on these witness triples and show the existence of a digraph of Figure 1 as induced subgraph.

Let S be the set of vertices of F that are simplicial in H. We construct a digraph D that has vertex set S, and there is an arc (u, v) if and only if there is a vertex w of G such that (u, v, w) is a witness triple for u in G. We show that every vertex of D has an out-neighbour. Let u be a vertex of S. By the above considerations, there is a witness triple (u, y, z) for u in G. Since u and z are adjacent to y in G, u and z are non-adjacent to y in H. Furthermore, u and z are non-adjacent in G, thus adjacent in H. So, y is vertex in a connected component K of $H \setminus (N_H(u) \cup N_H(z))$. Since u is simplicial and z is a neighbour of u in H, $N_H(u) \subseteq N_H[z]$. There exists a vertex v from S in K. We want to show that (u, v, z) is a witness triple for u in G or G contains a copy of a digraph of Figure 1 as induced subgraph. Since K is connected, there is a spanning tree T for K. Let all vertices of T be unmarked. We show by induction that (u, x, z) is a witness triple for u in G for all vertices x of T. Mark y in T. By assumption, the claim is true for every marked vertex of T. Let x' be an unmarked vertex of T that has a marked neighbour x'' in T. Mark x'. We consider the triple (u, x', z). Remember that u and z are non-adjacent in G and x' is adjacent to both u and z in G. Assume that (u, x', z) is not a witness triple for u in G. Then, $(u, x'), (z, x') \in A(G)$ or $(x', u), (x', z) \in A(G)$. This implies that $\{u, z, x', x''\}$ induces a copy of digraph (d) of Figure 1 in G. If no copy of digraph (d) has been detected, (u, v, z) is a witness triple for u in G, since all vertices of T have been marked, and thus D contains arc (u, v). We conclude that every vertex of D has an out-neighbour or G contains a copy of digraph (d) as induced subgraph. This completes the construction of D.

Assume that every vertex of D has an out-neighbour in D, i.e., D is not acyclic and contains a directed cycle. Let $C = (u_1, \ldots, u_k)$ be a directed cycle in D of shortest length. It can be shown that in case $k = 2$, G contains a copy of digraph (b), (c) or (d) of Figure 1 as induced subgraph. Henceforth, let $k \geq 3$. If $(u_1, u_2), \ldots, (u_{k-1}, u_k), (u_k, u_1) \in A(G)$ or $(u_1, u_k), (u_k, u_{k-1}), \ldots, (u_2, u_1) \in A(G)$ then G contains a directed cycle as subgraph, and therefore, G contains digraph (a) or (c) of Figure 1 as induced subgraph. Now, assume that $(u_1, \ldots, u_k)$

does not define a directed cycle in G. We can assume that $(u_1, u_2), (u_1, u_k) \in A(G)$. It can be shown that no witness triple has all its vertices on C.

Suppose that $k \geq 4$. We consider the vertices u_1, u_2, u_k. Let a, b be vertices such that (u_1, u_2, b) is a witness triple for u_1 and (u_k, u_1, a) is a witness triple for u_k in G. Remember that $(u_1, u_2), (u_1, u_k) \in A(G)$. Thus, $(u_1, u_2), (u_2, b) \in A(G)$ and $(a, u_1), (u_1, u_k) \in A(G)$. By the considerations above, a and b are not vertices on C. Furthermore, $a \neq b$. If a and u_2 are non-adjacent in G then (u_2, u_1, a) is a witness triple for u_2 in G, which means $(u_2, u_1) \in A(D)$, and D has a cycle of length 2. This contradicts our assumption $k \geq 4$. Thus, a and u_2 are adjacent in G. If $(u_2, a) \in A(G)$ then $\{u_1, u_2, a\}$ induces a copy of digraph (a) of Figure 1 in G, and we are done. Otherwise, let $(a, u_2) \in A(G)$. Since u_2 and u_k are simplicial in H and have different neighbourhoods, they are adjacent in G. If $(u_2, u_k) \in A(G)$ then (u_k, u_2, a) is a witness triple for u_k in G, which means $(u_k, u_2) \in A(D)$, and C has a chord. This contradicts the above results. Hence, $(u_k, u_2) \in A(G)$. If u_k and b are non-adjacent in G then (u_k, u_2, b) is a witness triple for u_k in G, and $(u_k, u_2) \in A(G)$. Since this yields a contradiction, u_k and b must be adjacent. If $(b, u_k) \in A(G)$ then $\{u_k, u_2, b\}$ induces a copy of digraph (a) of Figure 1 in G. Otherwise, if $(u_k, b) \in A(G)$ then (u_1, u_k, b) is a witness triple for u_1 in G, which means $(u_1, u_k) \in A(D)$, thus a contradiction. We conclude that $k \not\geq 4$, i.e., $k = 3$.

We distinguish between two cases: $(u_2, u_3) \in A(G)$ and $(u_3, u_2) \in A(G)$. Let a, b, c be vertices such that $(u_1, u_2, b), (u_2, u_3, c)$ and (u_3, u_1, a) are witness triples for respectively u_1, u_2, u_3 in G. Note that a, b, c are pairwise different. And since u_1, u_2, u_3 are pairwise adjacent in G, $\{a, b, c\} \cap \{u_1, u_2, u_3\} = \emptyset$. Hence, u_1, u_2, u_3, a, b, c are pairwise different. For the first case, let $(u_2, u_3) \in A(G)$. This means that $(u_3, c) \in A(G)$. We consider a and u_2. If a and u_2 are non-adjacent in G then $\{a, u_1, u_2, b\}$ induces a copy of digraph (b) or (c) or (d) of Figure 1 in G. If $(u_2, a) \in A(G)$ then $\{a, u_1, u_2\}$ induces a copy of digraph (a) of Figure 1 in G. And if $(a, u_2) \in A(G)$ then $\{a, u_2, u_3, c\}$ induces a copy of digraph (b) or (c) or (d) of Figure 1 in G. For the second case, let $(u_3, u_2) \in A(G)$. This means that $(c, u_3) \in A(G)$. We consider b and u_3 and conclude in the above manner that $\{c, u_3, u_2, b\}$ or $\{u_3, u_2, b\}$ or $\{a, u_1, u_3, b\}$ induces a copy of a digraph of Figure 1 in G. Hence, we have found a copy of a digraph of Figure 1 as induced subgraph in G in every case, so that we can conclude the case when G has no di-simplicial vertex.

Now, we consider the case that F contains di-simplicial vertices. We apply Theorem 3 and conclude that F contains an induced subgraph F' on at least two vertices without di-simplicial vertex. Then, F' is not chordal and satisfies the conditions of the above case. Note that the underlying graph of $\mathrm{bi}(F')$ is chordal, since it is an induced subgraph of a chordal undirected graph. We apply the above case and conclude that $\mathrm{uni}(F')$ contains a copy of a digraph of Figure 1 as induced subgraph. This completes the proof. □

Theorem 6. *A semi-complete digraph F is chordal if and only if the underlying graph of $\mathrm{bi}(F)$ is chordal and $\mathrm{uni}(F)$ does not contain a copy of any of the digraphs of Figure 1 as induced subgraph.*

Proof. Let H be the underlying graph of bi(F). If H is not chordal then bi(F) contains a chordless directed cycle of length at least 4. With Lemma 4, F is not chordal. Now, let uni(F) contain a copy C of a digraph of Figure 1 as induced subgraph. Then, F contains a non-chordal induced subgraph, thus is not chordal due to Lemma 1. For the converse, let F not be chordal and let H be chordal. Then, uni(F) contains a copy of one of the digraphs of Figure 1 as induced subgraph due to Lemma 6. ☐

The actual set of minimal forbidden induced subgraphs for chordal semi-complete digraphs is the following: all semi-complete digraphs F with the underlying graph of bi(F) a chordless cycle of length at least 4 and the four digraphs that are obtained from the digraphs of Figure 1 by adding the two arcs between every pair of non-adjacent vertices (of each digraph). We want to conclude with two remarks. Firstly, note that even though the actual set of minimal forbidden induced subgraphs for chordal semi-complete digraphs is much bigger than the set of minimal forbidden induced subgraphs for chordal undirected graphs (due to the many different orientations), the structure of chordal semi-complete digraph is already much richer than the structure of the whole class of chordal undirected graphs. This can give a first impression of the significant difference between directed and undirected graphs. Secondly, it is an interesting observation that each digraph of Figure 1 is isomorphic to its own reverse graph.

6 Conclusion

In Section 3, we have given two characterisations of chordal digraphs (Theorem 2 and Theorem 3), which are analogues of characterisations of chordal undirected graphs. We have also introduced the notion of *d-clique* as a directed analogue or even generalisation of the undirected notion of *clique*. A famous characterisation of chordal undirected graphs gives a connection between cliques and minimal separators [3]. Is there a directed notion of minimal separator that is connected to d-cliques in a similar way for chordal digraphs?

Another famous characterisation of chordal undirected graphs is as intersection graphs of subtrees of a tree. Can this be generalised to chordal digraphs? Let us remark that there is a generalisation of intersection graphs to 'intersection digraphs' that results in an interesting directed analogue of interval graphs [12,14]. However, if using this definition of 'intersection digraphs' then *all digraphs* become representable by 'intersection subtrees' [5], see also Bang-Jensen and Gutin [1] [Proposition 4.13.2]. Thus, a different approach is needed to define the proper directed analogue of 'intersection graph of subtrees of a tree'.

On the structural side, the main open problem for chordal digraphs is a characterisation by forbidden induced subgraphs. We have given such characterisations for large subclasses of digraphs, such as digraphs G with bi(G) empty and semi-complete digraphs. The case of semi-complete digraphs shows that a forbidden induced subgraphs characterisation for the whole class of chordal digraphs is challenging.

On the algorithmic side, there are many interesting problems. A di-simplicial vertex can trivially be found in $\mathcal{O}(nm)$ time. Can this be reduced to $\mathcal{O}(n^2)$ or even linear time? There is an $\mathcal{O}(nm)$-time algorithm for recognising chordal digraphs [6,8,11]. Rose and Tarjan showed that this problem is closely related to other well-studied problems like transitivity test [11]. Improvements on the chordal digraph recognition directly carry over to other problems, and thus are of great interest. The study of chordal digraph recognition for digraph classes may be a way of implicitly attacking also other problems. Finally, we want to ask whether there exists an $\mathcal{O}(n^2)$-time algorithm for verifying whether a given vertex ordering is directed transitive.

One motivation for the study of chordal digraphs is their connection to Kelly-width. Can the structural properties exhibited by chordal digraphs be exploited algorithmically for graphs of bounded Kelly-width?

References

1. Bang-Jensen, J., Gutin, G.: Digraphs. Theory, Algorithms and Applications. Springer, Heidelberg (2000)
2. Buneman, P.: A Characterisation of Rigid Circuit Graphs. Discrete Mathematics 9, 205–212 (1974)
3. Dirac, G.A.: On rigid circuit graphs. Abhandlungen aus dem Mathematischen Seminar der Universität Hamburg 25, 71–76 (1961)
4. Gavril, F.: The Intersection Graphs of Subtrees in Trees Are Exactly the Chordal Graphs. Journal of Combinatorial Theory (B) 16, 47–56 (1974)
5. Harary, F., Kabell, J.A., McMorris, F.R.: Bipartite intersection graphs. Commentationes Mathematicae Universitatis Carolinae 23, 739–745 (1982)
6. Haskins, L., Rose, D.J.: Toward characterization of perfect elimination digraphs. SIAM Journal on Computing 2, 217–224 (1973)
7. Hunter, P., Kreutzer, S.: Digraph measures: Kelly decompositions, games, and orderings. Theoretical Computer Science 399, 206–219 (2008)
8. Kleitman, D.J.: A note on perfect elimination digraphs. SIAM Journal on Computing 3, 280–282 (1974)
9. Meister, D., Telle, J.A., Vatshelle, M.: Characterization and Recognition of Digraphs of Bounded Kelly-width. In: Brandstädt, A., Kratsch, D., Müller, H. (eds.) WG 2007. LNCS, vol. 4769, pp. 270–279. Springer, Heidelberg (2007)
10. Rose, D.J.: Triangulated Graphs and the Elimination Process. Journal of Mathematical Analysis and Applications 32, 597–609 (1970)
11. Rose, D.J., Tarjan, R.E.: Algorithmic Aspects of Vertex Elimination on Directed Graphs. SIAM Journal on Applied Mathematics 34, 176–197 (1978)
12. Sanyal, B.K., Sen, M.K.: New characterization of digraphs represented by intervals. Journal of Graph Theory 22, 297–303 (1996)
13. Walter, J.R.: Representations of Chordal Graphs as Subtrees of a Tree. Journal of Graph Theory 2, 265–267 (1978)
14. West, D.B.: Short proofs for interval digraphs. Discrete Mathematics 178, 287–292 (1998)

A New Intersection Model and Improved Algorithms for Tolerance Graphs

George B. Mertzios[1], Ignasi Sau[2], and Shmuel Zaks[3]

[1] Department of Computer Science, RWTH Aachen University, Germany
mertzios@cs.rwth-aachen.de

[2] Mascotte joint Project of INRIA/CNRS/UNSA, Sophia-Antipolis, France; and Graph Theory and Combinatorics Group, Applied Maths. IV Dept. of UPC, Barcelona, Spain
ignasi.sau@sophia.inria.fr

[3] Department of Computer Science, Technion, Haifa, Israel
zaks@cs.technion.ac.il

Abstract. Tolerance graphs model interval relations in such a way that intervals can tolerate a certain degree of overlap without being in conflict. This class of graphs, which generalizes in a natural way both interval and permutation graphs, has attracted many research efforts since their introduction in [9], as it finds many important applications in constraint-based temporal reasoning, resource allocation, and scheduling problems, among others. In this article we propose the first non-trivial intersection model for general tolerance graphs, given by three-dimensional parallelepipeds, which extends the widely known intersection model of parallelograms in the plane that characterizes the class of bounded tolerance graphs. Apart from being important on its own, this new representation also enables us to improve the time complexity of three problems on tolerance graphs. Namely, we present optimal $\mathcal{O}(n \log n)$ algorithms for computing a minimum coloring and a maximum clique, and an $\mathcal{O}(n^2)$ algorithm for computing a maximum weight independent set in a tolerance graph with n vertices, thus improving the best known running times $\mathcal{O}(n^2)$ and $\mathcal{O}(n^3)$ for these problems, respectively.

Keywords: Tolerance graphs, parallelogram graphs, intersection model, minimum coloring, maximum clique, maximum weight independent set.

1 Introduction

A graph $G = (V, E)$ on n vertices is a *tolerance graph* if there is a set $I = \{I_i \mid i = 1, \ldots, n\}$ of closed intervals on the real line and a set $T = \{t_i \mid i = 1, \ldots, n\}$ of positive real numbers, called *tolerances*, such that for any two vertices $v_i, v_j \in V$, $v_i v_j \in E$ if and only if $|I_i \cap I_j| \geq \min\{t_i, t_j\}$, where $|I|$ denotes the length of the interval I. These sets of intervals and tolerances form a *tolerance representation* of G. If G has a tolerance representation such that $t_i \leq |I_i|$ for $i = 1, \ldots, n$, then G is called a *bounded tolerance graph* and its representation is a *bounded tolerance representation*.

Tolerance graphs were introduced in [9], mainly motivated by the need to solve scheduling problems in which resources that would be normally used exclusively, like rooms or vehicles, can tolerate some sharing among users. Since then, tolerance graphs have been widely studied in the literature [1,2,5,10,11,14,16,20], as they naturally generalize both interval graphs (when all tolerances are equal) and permutation graphs (when $|I_i| = t_i$ for $i = 1, \ldots, n$) [9]. For more details, see [12].

Notation. All the graphs considered in this paper are finite, simple, and undirected. Given a graph $G = (V, E)$, we denote by n the cardinality of V. An edge between vertices u and v is denoted by uv, and in this case vertices u and v are said to be *adjacent*. $\overline{G}$ denotes the *complement* of G, i.e. $\overline{G} = (V, \overline{E})$, where $uv \in \overline{E}$ if and only if $uv \notin E$. Given a subset of vertices $S \subseteq V$, the graph $G[S]$ denotes the graph *induced* by the vertices in S, i.e. $G[S] = (S, F)$, where for any two vertices $u, v \in S$, $uv \in F$ if and only if $uv \in E$. A subset $S \subseteq V$ is an *independent set* in G if the graph $G[S]$ has no edges. For a subset $K \subseteq V$, the induced subgraph $G[K]$ is a *complete subgraph* of G, or a *clique*, if each two of its vertices are adjacent (equivalently, K is an independent set in $\overline{G}$). The maximum cardinality of a clique in G is denoted by $\omega(G)$ and is termed the *clique number* of G. A *proper coloring* of G is an assignment of different colors to adjacent vertices, which results in a partition of V into independent sets. The minimum number of colors for which there exists a proper coloring is denoted by $\chi(G)$ and is termed the *chromatic number* of G. A partition of V into $\chi(G)$ independent sets is a *minimum coloring* of G.

Motivation and previous work. Besides generalizing interval and permutation graphs in a natural way, the class of tolerance graphs has other important subclasses and superclasses. Let us briefly survey some of them.

A graph is *perfect* if the chromatic number of every induced subgraph equals the clique number of that subgraph. Perfect graphs include many important families of graphs, and serve to unify results relating colorings and cliques in those families. For instance, in all perfect graphs, the graph coloring problem, maximum clique problem, and maximum independent set problem can all be solved in polynomial time using the Ellipsoid method [13]. Since tolerance graphs were shown to be perfect [10], there exist polynomial time algorithms for these problems. However, these algorithms are not very efficient and therefore, as it happens for most known subclasses of perfect graphs, it makes sense to devise specific fast algorithms for these problems on tolerance graphs.

A *comparability* graph is a graph which can be transitively oriented. A *co-comparability* graph is a graph whose complement is a comparability graph. Bounded tolerance graphs are co-comparability graphs [9], and therefore all known polynomial time algorithms for co-comparability graphs apply to bounded tolerance graphs. This is one of the main reasons why for many problems the existing algorithms have better running time in bounded tolerance graphs than in general tolerance graphs.

A graph $G = (V, E)$ is the *intersection graph* of a family $F = \{S_1, \ldots, S_n\}$ of distinct nonempty subsets of a set S if there exists a bijection $\mu : V \to F$ such that for any two distinct vertices $u, v \in V$, $uv \in E$ if and only if $\mu(u) \cap \mu(v) \neq \emptyset$. In that case, we say that F is an *intersection model* of G. It is easy to see that each graph has a trivial intersection model based on adjacency relations [18]. Some intersection models provide a natural and intuitive understanding of the structure of a class of graphs, and turn out to be very helpful to find efficient algorithms to solve optimization problems [18]. Therefore, it is of great importance to establish non-trivial intersection models for families of graphs. A graph G on n vertices is a *parallelogram graph* if we can fix two parallel lines L_1 and L_2, and for each vertex $v_i \in V(G)$ we can assign a parallelogram $\overline{P}_i$ with parallel sides along L_1 and L_2 so that G is the intersection graph of $\{\overline{P}_i \mid i = 1, \ldots, n\}$. It was proved in [1, 17] that a graph is a bounded tolerance graph if and only if it is a parallelogram graph. This characterization provides a useful way to think about bounded tolerance graphs. However, this intersection model cannot cope with general tolerance graphs, in which the tolerance of an interval can be greater than its length.

Our contribution. In this article we present the first non-trivial intersection model for general tolerance graphs, which generalizes the widely known parallelogram representation of bounded tolerance graphs. The main idea is to exploit the third dimension to capture the information given by unbounded tolerances, and as a result parallelograms are replaced with parallelepipeds. The proposed intersection model is very intuitive and can be efficiently constructed from a tolerance representation (actually, we show that it can be constructed in linear time).

Apart from being important on its own, this new representation proves to be a powerful tool for designing efficient algorithms for general tolerance graphs. Indeed, using our intersection model we improve the best existing running times of three problems on tolerance graphs. We present algorithms to find a minimum coloring and a maximum clique in $\mathcal{O}(n \log n)$ time, which turns out to be optimal. The best existing algorithm was $\mathcal{O}(n^2)$ [11, 12]. We also present an algorithm to find a maximum weight independent set in $\mathcal{O}(n^2)$ time, whereas the best known algorithm was $\mathcal{O}(n^3)$ [12]. We note that [20] proposes an $\mathcal{O}(n^2 \log n)$ algorithm to find a maximum *cardinality* independent set on a general tolerance graph, and that [12] refers to an algorithm transmitted by personal communication with running time $\mathcal{O}(n^2 \log n)$ to find a maximum weight independent set on a general tolerance graph; to the best of our knowledge, this algorithm has not been published.

It is important to note that the complexity of recognizing bounded and general tolerance graphs is a challenging open problem [3, 12, 20], and this is the reason why we assume throughout this paper that along with the input tolerance graph we are also given a tolerance representation of it. The only "positive" result in the literature concerning recognition of tolerance graphs is a linear time algorithm for the recognition of bipartite tolerance graphs [3].

Organization of the paper. We provide the new intersection model of general tolerance graphs in Section 2. In Section 3 we present a canonical representation of tolerance graphs, and then show how it can be used in order to obtain optimal $\mathcal{O}(n \log n)$ algorithms for finding a minimum coloring and a maximum clique in a tolerance graph. In Section 4 we present an $\mathcal{O}(n^2)$ algorithm for finding a maximum weight independent set. Finally, Section 5 is devoted to conclusions and open problems. Some proofs have been omitted due to space limitations; a full version can be found in [19].

2 A New Intersection Model for Tolerance Graphs

One of the most natural representations of bounded tolerance graphs is given by parallelograms between two parallel lines in the Euclidean plane [1,12,17]. In this section we extend this representation to a three-dimensional representation of general tolerance graphs.

Given a tolerance graph $G = (V, E)$ along with a tolerance representation of it, recall that vertex $v_i \in V$ corresponds to an interval $I_i = [a_i, b_i]$ on the real line with a tolerance $t_i \geq 0$. W.l.o.g. we may assume that $t_i > 0$ for every vertex v_i [12].

Definition 1. *Given a tolerance representation of a tolerance graph $G = (V, E)$, vertex v_i is bounded if $t_i \leq |I_i|$. Otherwise, v_i is unbounded. V_B and V_U are the sets of bounded and unbounded vertices in V, respectively. Clearly $V = V_B \cup V_U$.*

We can also assume w.l.o.g. that $t_i = \infty$ for any unbounded vertex v_i, since if v_i is unbounded, then the intersection of any other interval with I_i is strictly smaller than t_i. Let L_1 and L_2 be two parallel lines at distance 1 in the Euclidean plane.

Definition 2. *Given an interval $I_i = [a_i, b_i]$ with tolerance t_i, $\overline{P}_i$ is the parallelogram defined by the points c_i, b_i in L_1 and a_i, d_i in L_2, where $c_i = \min\{b_i, a_i + t_i\}$ and $d_i = \max\{a_i, b_i - t_i\}$. The slope ϕ_i of $\overline{P}_i$ is $\phi_i = \arctan\left(\frac{1}{c_i - a_i}\right)$.*

An example is depicted in Figure 1, where $\overline{P}_i$ and $\overline{P}_j$ correspond to bounded vertices v_i and v_j, and $\overline{P}_k$ corresponds to an unbounded vertex v_k. Observe that when vertex v_i is bounded, the values c_i and d_i coincide with the *tolerance points* defined in [7, 12, 15], and $\phi_i = \arctan\left(\frac{1}{t_i}\right)$. On the other hand, when vertex v_i is unbounded, the values c_i and d_i coincide with the endpoints b_i and a_i of I_i, respectively, and $\phi_i = \arctan\left(\frac{1}{|I_i|}\right)$. Observe also that in both cases $t_i = b_i - a_i$ and $t_i = \infty$, parallelogram $\overline{P}_i$ is reduced to a line segment (c.f. $\overline{P}_j$ and $\overline{P}_k$ in Figure 1). Since $t_i > 0$ for every vertex v_i, it follows that $0 < \phi_i < \frac{\pi}{2}$. Furthermore, we can assume w.l.o.g. that all points a_i, b_i, c_i, d_i and all slopes ϕ_i are distinct [7, 12, 15].

Observation 1. *Let $v_i \in V_U, v_j \in V_B$. Then $|I_i| < t_j$ if and only if $\phi_i > \phi_j$.*

We are ready to give the main definition of this article.

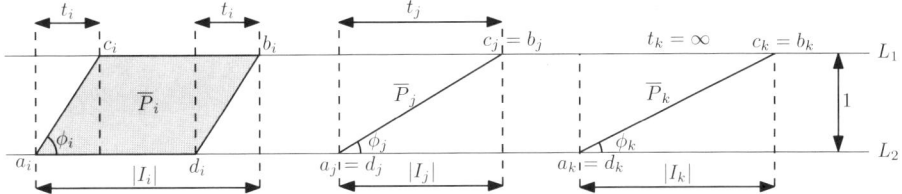

Fig. 1. Parallelograms $\overline{P}_i$ and $\overline{P}_j$ correspond to bounded vertices v_i and v_j, respectively, whereas $\overline{P}_k$ corresponds to an unbounded vertex v_k

Definition 3. *Let $G = (V, E)$ be a tolerance graph with a tolerance representation $\{I_i = [a_i, b_i], t_i \mid i = 1, \ldots, n\}$. For every $i = 1 \ldots, n$, P_i is the parallelepiped in $\mathbb{R}^3$ defined as follows:*

(a) *If $t_i \leq b_i - a_i$ (that is, v_i is bounded), then $P_i = \{(x, y, z) \in \mathbb{R}^3 \mid (x, y) \in \overline{P}_i,\ 0 \leq z \leq \phi_i\}$.*
(b) *If $t_i > b_i - a_i$ (v_i is unbounded), then $P_i = \{(x, y, z) \in \mathbb{R}^3 \mid (x, y) \in \overline{P}_i,\ z = \phi_i\}$.*

The set of parallelepipeds $\{P_i \mid i = 1, \ldots, n\}$ is a parallelepiped representation *of G.*

Observe that for each interval I_i, the parallelogram $\overline{P}_i$ of Definition 2 (see also Figure 1) coincides with the projection of the parallelepiped P_i on the plane $z = 0$. An example of the construction of these parallelepipeds is given in Figure 2, where a set of eight intervals with their associated tolerances is given in Figure 2(a). The corresponding tolerance graph G is depicted in Figure 2(b), while the parallelepiped representation is illustrated in Figure 2(c). In the case $t_i < b_i - a_i$, the parallelepiped P_i is three-dimensional, c.f. P_1, P_3, and P_5, while in the border case $t_i = b_i - a_i$ it degenerates to a two-dimensional rectangle, c.f. P_7. In these two cases, each P_i corresponds to a bounded vertex v_i. In the remaining case $t_i = \infty$ (that is, v_i is unbounded), the parallelepiped P_i degenerates to a one-dimensional line segment above plane $z = 0$, c.f. P_2, P_4, P_6, and P_8.

We prove now that these parallelepipeds form a three-dimensional intersection model for the class of tolerance graphs (namely, that every tolerance graph G can be viewed as the intersection graph of the corresponding parallelepipeds P_i).

Theorem 1. *Let $G = (V, E)$ be a tolerance graph with a tolerance representation $\{I_i = [a_i, b_i], t_i \mid i = 1, \ldots, n\}$. Then for every $i \neq j$, $v_i v_j \in E$ if and only if $P_i \cap P_j \neq \emptyset$.*

Proof. We distinguish three cases according to whether vertices v_i and v_j are bounded or unbounded:

(a) Both vertices are bounded, that is $t_i \leq b_i - a_i$ and $t_j \leq b_j - a_j$. It follows that $v_i v_j \in E(G)$ if and only if $\overline{P}_i \cap \overline{P}_j \neq \emptyset$ [12]. However, due to the definition

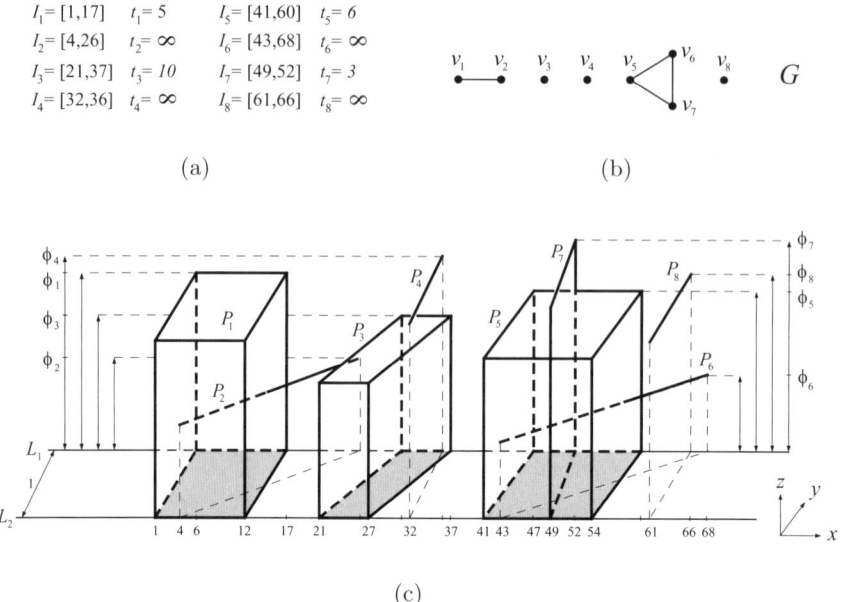

Fig. 2. The intersection model for tolerance graphs: (a) a set of intervals $I_i = [a_i, b_i]$ and tolerances t_i, $i = 1, \ldots, 8$, (b) the corresponding tolerance graph G and (c) a parallelepiped representation of G

of the parallelepipeds P_i and P_j, in this case $P_i \cap P_j \neq \emptyset$ if and only if $\overline{P}_i \cap \overline{P}_j \neq \emptyset$ (c.f. P_1 and P_3, or P_5 and P_7, in Figure 2).

(b) Both vertices are unbounded, that is $t_i = t_j = \infty$. Since no two unbounded vertices are adjacent, $v_i v_j \notin E(G)$. On the other hand, the line segments P_i and P_j lie on the disjoint planes $z = \phi_i$ and $z = \phi_j$ of $\mathbb{R}^3$, respectively, since we assumed that the slopes ϕ_i and ϕ_j are distinct. Thus, $P_i \cap P_j = \emptyset$ (c.f. P_2 and P_4).

(c) One vertex is unbounded (that is, $t_i = \infty$) and the other is bounded (that is, $t_j \leq b_j - a_j$). If $\overline{P}_i \cap \overline{P}_j = \emptyset$, then $v_i v_j \notin E$ and $P_i \cap P_j = \emptyset$ (c.f. P_1 and P_6). Suppose that $\overline{P}_i \cap \overline{P}_j \neq \emptyset$. We distinguish two cases:

  (i) $\phi_i < \phi_j$. It is easy to check that $|I_i \cap I_j| \geq t_j$ and thus $v_i v_j \in E$. Since $\overline{P}_i \cap \overline{P}_j \neq \emptyset$ and $\phi_i < \phi_j$, then necessarily the line segment P_i intersects with the parallelepiped P_j on the plane $z = \phi_i$, and thus $P_i \cap P_j \neq \emptyset$ (c.f. P_1 and P_2).
  (ii) $\phi_i > \phi_j$. Clearly $|I_i \cap I_j| < t_i = \infty$. Furthermore, since $\phi_i > \phi_j$, Observation 1 implies that $|I_i \cap I_j| \leq |I_i| < t_j$. It follows that $|I_i \cap I_j| < \min\{t_i, t_j\}$, and thus $v_i v_j \notin E$. On the other hand, $z = \phi_i$ for all points $(x, y, z) \in P_i$, while $z \leq \phi_j < \phi_i$ for all points $(x, y, z) \in P_j$, and therefore $P_i \cap P_j = \emptyset$ (c.f. P_3 and P_4). ∎

Clearly, for each $v_i \in V$ the parallelepiped P_i can be constructed in constant time. Therefore, given a tolerance representation of a tolerance graph G with n vertices, a parallelepiped representation of G can be constructed in $\mathcal{O}(n)$ time.

3 Coloring and Clique Algorithms in $\mathcal{O}(n \log n)$

In this section we present optimal $\mathcal{O}(n \log n)$ algorithms for constructing a minimum coloring and a maximum clique in a tolerance graph $G = (V, E)$ with n vertices, given a parallelepiped representation of G. These algorithms improve the best known running time $\mathcal{O}(n^2)$ of these problems on tolerance graphs [11, 12]. First, we introduce a canonical representation of tolerance graphs in Section 3.1, and then we use it to obtain the algorithms for the minimum coloring and the maximum clique problems in Section 3.2.

3.1 A Canonical Representation of Tolerance Graphs

We associate with every vertex v_i of G the point $p_i = (x_i, y_i)$ in the Euclidean plane, where $x_i = b_i$ and $y_i = \frac{\pi}{2} - \phi_i$. Since all endpoints of the parallelograms $\overline{P}_i$ and all slopes ϕ_i are distinct, all coordinates of the points p_i are distinct as well. Similarly to [11, 12], we state the following two definitions.

Definition 4. *An unbounded vertex $v_i \in V_U$ of a tolerance graph G is called* inevitable *(for a certain parallelepiped representation), if replacing P_i with $\{(x, y, z) \mid (x, y) \in P_i, 0 \leq z \leq \phi_i\}$ creates a new edge in G. Otherwise, v_i is called evitable.*

Definition 5. *Let $v_i \in V_U$ be an inevitable unbounded vertex of a tolerance graph G (for a certain parallelepiped representation). A vertex v_j is called a* hovering vertex *of v_i if $a_j < a_i$, $b_i < b_j$, and $\phi_i > \phi_j$.*

It is now easy to see that, by Definition 5, if v_j is a hovering vertex of v_i, then $v_i v_j \notin E$. Note that, in contrast to [11], in Definition 4, an isolated vertex v_i might be also inevitable unbounded, while in Definition 5, a hovering vertex might be also unbounded. Definitions 4 and 5 imply the following lemma:

Lemma 1. *Let $v_i \in V_U$ be an inevitable unbounded vertex of the tolerance graph G (for a certain parallelepiped representation). Then, there exists a hovering vertex v_j of v_i.*

Definition 6. *A parallelepiped representation of a tolerance graph G is called* canonical *if every unbounded vertex is inevitable.*

For example, in the tolerance graph depicted in Figure 2, v_4 and v_8 are inevitable unbounded vertices, v_3 and v_6 are hovering vertices of v_4 and v_8, respectively, while v_2 and v_6 are evitable unbounded vertices. Therefore, this representation is not canonical for the graph G. However, if we replace P_i with

$\{(x, y, z) \mid (x, y) \in P_i, 0 \leq z \leq \phi_i\}$ for $i = 2, 6$, we get a canonical representation for G.

In the following, we present an algorithm that constructs a canonical representation of a given tolerance graph G.

Definition 7. *Let $\alpha = (x_\alpha, y_\alpha)$ and $\beta = (x_\beta, y_\beta)$ be two points in the plane. Then α dominates β if $x_\alpha > x_\beta$ and $y_\alpha > y_\beta$. Given a set A of points, the point $\gamma \in A$ is called an* extreme point *of A if there is no point $\delta \in A$ that dominates γ. $Ex(A)$ is the set of the extreme points of A.*

Algorithm 1. Construction of a canonical representation of a tolerance graph G

Input: A parallelepiped representation R of a given tolerance graph G with n vertices
Output: A canonical representation R' of G

  Sort the vertices of G, such that $a_i < a_j$ whenever $i < j$
  $\ell_0 \leftarrow \min\{x_i : 1 \leq i \leq n\}$; $r_0 \leftarrow \max\{x_i : 1 \leq i \leq n\}$
  $p_s \leftarrow (\ell_0 - 1, \frac{\pi}{2})$; $p_t \leftarrow (r_0 + 1, 0)$
  $P \leftarrow (p_s, p_t)$; $R' \leftarrow R$
  for $i = 1$ to n **do**
    Find the point p_j having the smallest x_j with $x_j > x_i$
    if $y_j < y_i$ **then** {no point of P dominates p_i}
      Find the point p_k having the greatest x_k with $x_k < x_i$
      Find the point p_ℓ having the greatest y_ℓ with $y_\ell < y_i$
      if $x_k \geq x_\ell$ **then**
        Replace points $p_\ell, p_{\ell+1} \ldots, p_k$ with point p_i in the list P
      else
        Insert point p_i between points p_k and p_ℓ in the list P
      if $v_i \in V_U$ **then** {v_i is an evitable unbounded vertex}
        Replace P_i with $\{(x, y, z) \mid (x, y) \in P_i, 0 \leq z \leq \phi_i\}$ in R'
    else {$y_j > y_i$; p_j dominates p_i}
      if $v_i \in V_U$ **then** {v_i is an inevitable unbounded vertex}
        v_j is a hovering vertex of v_i
  return R'

Given a tolerance graph $G = (V, E)$ with the set $V = \{v_1, v_2, \ldots, v_n\}$ of vertices (and its parallelepiped representation), we can assume w.l.o.g. that $a_i < a_j$ whenever $i < j$. Recall that with every vertex v_i we associated the point $p_i = (x_i, y_i)$, where $x_i = b_i$ and $y_i = \frac{\pi}{2} - \phi_i$, respectively. The following theorem shows that, given a parallelepiped representation of a tolerance graph G, we can construct in $\mathcal{O}(n \log n)$ a canonical representation of G. This result is crucial for the time complexity analysis of the algorithms of Section 3.2.

Theorem 2. *Every parallelepiped representation of a tolerance graph G with n vertices can be transformed by Algorithm 1 to a canonical representation of G in $\mathcal{O}(n \log n)$ time.*

3.2 Minimum Coloring and Maximum Clique

In the next theorem we present an optimal $\mathcal{O}(n \log n)$ algorithm for computing a minimum coloring of a tolerance graph G with n vertices, given a parallelepiped representation of G. The informal description of the algorithm is identical to the one in [11], which has running time $\mathcal{O}(n^2)$; the difference is in the fact that we use our new representation, in order to improve the time complexity.

Theorem 3. *A minimum coloring of a tolerance graph G with n vertices can be computed in $\mathcal{O}(n \log n)$ time.*

In the next theorem we prove that a maximum clique of a tolerance graph G with n vertices can be computed in optimal $\mathcal{O}(n \log n)$ time, given a parallelepiped representation of G. This theorem follows from Theorem 2 and from the clique algorithm presented in [6], and it improves the best known $\mathcal{O}(n^2)$ running time mentioned in [11].

Theorem 4. *A maximum clique of a tolerance graph G with n vertices can be computed in $\mathcal{O}(n \log n)$ time.*

Based on a lower time bound of $\Omega(n \log n)$ for computing the length of a longest increasing subsequence in a permutation [6,8], it turns out that the time complexity $\mathcal{O}(n \log n)$ of the presented algorithms for the minimum coloring and the maximum clique problems presented in Theorems 3 and 4 are oprimal.

4 Weighted Independent Set Algorithm in $\mathcal{O}(n^2)$

In this section we present an algorithm for computing a maximum weight independent set in a tolerance graph $G = (V, E)$ with n vertices in $\mathcal{O}(n^2)$ time, given a parallelepiped representation of G, and a weight $w(v_i) > 0$ for every vertex v_i of G. The proposed algorithm improves the running time $\mathcal{O}(n^3)$ of the one presented in [12]. In the following, consider as above the partition of the vertex set V into the sets V_B and V_U of bounded and unbounded vertices of G, respectively.

Similarly to [12], we add two isolated bounded vertices v_s and v_t to G with weights $w(v_s) = w(v_t) = 0$, such that the corresponding parallelepipeds P_s and P_t lie completely to the left and to the right of all other parallelepipeds of G, respectively. Since both v_s and v_t are bounded vertices, we augment the set V_B by the vertices v_s and v_t. In particular, we define the set of vertices $V_B' = V_B \cup \{v_s, v_t\}$ and the tolerance graph $G' = (V', E)$, where $V' = V_B' \cup V_U$. Since $G'[V_B']$ is a bounded tolerance graph, it is a co-comparability graph as well [10,12]. A transitive orientation of the comparability graph $\overline{G'[V_B']}$ can be obtained by directing each edge according to the upper left endpoints of the parallelograms $\overline{P}_i$. Formally, let $(V_B', \prec)$ be the partial order defined on the bounded vertices V_B', such that $v_i \prec v_j$ if and only if $v_i v_j \notin E$ and $c_i < c_j$. Recall that a *chain* of elements in a partial order is a set of mutually comparable elements in this order [4].

Observation 2 ([12]). *The independent sets of $G[V_B]$ are in one-to-one correspondence with the chains in the partial order $(V'_B, \prec)$ from v_s to v_t.*

Using a dynamic programming algorithm that exploits the properties of the new parallelepiped representation of tolerance graphs, we derive the next theorem. The details can be found in [19].

Theorem 5. *A maximum weight independent set of a tolerance graph G with n vertices can be computed in $\mathcal{O}(n^2)$ time.*

5 Conclusions and Further Research

In this article we proposed the first non-trivial intersection model for general tolerance graphs, given by parallelepipeds in the three-dimensional space. This representation generalizes the parallelogram representation of bounded tolerance graphs. Using this representation, we presented improved algorithms for computing a minimum coloring, a maximum clique, and a maximum weight independent set on a tolerance graph. The complexity of the recognition problem for tolerance and bounded tolerance graphs is the main open problem in this class of graphs. Even when the input graph is known to be a tolerance graph, it is not known how to obtain a tolerance representation for it [20].

References

1. Bogart, K.P., Fishburn, P.C., Isaak, G., Langley, L.: Proper and unit tolerance graphs. Discrete Applied Mathematics 60(1-3), 99–117 (1995)
2. Busch, A.H.: A characterization of triangle-free tolerance graphs. Discrete Applied Mathematics 154(3), 471–477 (2006)
3. Busch, A.H., Isaak, G.: Recognizing bipartite tolerance graphs in linear time. In: Brandstädt, A., Kratsch, D., Müller, H. (eds.) WG 2007. LNCS, vol. 4769, pp. 12–20. Springer, Heidelberg (2007)
4. Diestel, R.: Graph Theory, 3rd edn. Springer, Berlin (2005)
5. Felsner, S.: Tolerance graphs and orders. Journal of Graph Theory 28, 129–140 (1998)
6. Felsner, S., Müller, R., Wernisch, L.: Trapezoid graphs and generalizations, geometry and algorithms. Discrete Applied Mathematics 74, 13–32 (1997)
7. Fishburn, P.C., Trotter, W.T.: Split semiorders. Discrete Mathematics 195, 111–126 (1999)
8. Fredman, M.L.: On computing the length of longest increasing subsequences. Discrete Mathematics 11, 29–35 (1975)
9. Golumbic, M.C., Monma, C.L.: A generalization of interval graphs with tolerances. In: Proceedings of the 13th Southeastern Conference on Combinatorics, Graph Theory and Computing, Congressus Numerantium, vol. 35, pp. 321–331 (1982)
10. Golumbic, M.C., Monma, C.L., Trotter, W.T.: Tolerance graphs. Discrete Applied Mathematics 9(2), 157–170 (1984)
11. Golumbic, M.C., Siani, A.: Coloring algorithms for tolerance graphs: Reasoning and scheduling with interval constraints. In: Joint International Conferences on Artificial Intelligence, Automated Reasoning, and Symbolic Computation (AISC/Calculemus), pp. 196–207 (2002)

12. Golumbic, M., Trenk, A.: Tolerance Graphs. Cambridge Studies in Advanced Mathematics (2004)
13. Grötshcel, M., Lovász, L., Schrijver, A.: The Ellipsoid Method and its Consequences in Combinatorial Optimization. Combinatorica 1, 169–197 (1981)
14. Hayward, R.B., Shamir, R.: A note on tolerance graph recognition. Discrete Applied Mathematics 143(1-3), 307–311 (2004)
15. Isaak, G., Nyman, K., Trenk, A.: A hierarchy of classes of bounded bitolerance orders. Ars Combinatoria 69 (2003)
16. Keil, J.M., Belleville, P.: Dominating the complements of bounded tolerance graphs and the complements of trapezoid graphs. Discrete Applied Mathematics 140(1-3), 73–89 (2004)
17. Langley, L.: Interval tolerance orders and dimension. PhD thesis, Dartmouth College (June 1993)
18. McKee, T., McMorris, F.: Topics in Intersection Graph Theory. Society for Industrial and Applied Mathematics. SIAM, Philadelphia (1999)
19. Mertzios, G.B., Sau, I., Zaks, S.: A New Intersection Model and Improved Algorithms for Tolerance Graphs. Technical report, RWTH Aachen University (March 2009)
20. Narasimhan, G., Manber, R.: Stability and chromatic number of tolerance graphs. Discrete Applied Mathematics 36, 47–56 (1992)

Counting the Number of Matchings in Chordal and Chordal Bipartite Graph Classes

Yoshio Okamoto[1,*], Ryuhei Uehara[2,**], and Takeaki Uno[3,***]

[1] Graduate School of Information Science and Engineering, Tokyo Institute of Technology, Ookayama 2-12-1-W8-88, Meguro-ku, Tokyo 152-8552, Japan
okamoto@is.titech.ac.jp
[2] School of Information Science, JAIST, Asahidai 1-1 , Nomi, Ishikawa 923-1292, Japan
uehara@jaist.ac.jp
[3] National Institute of Informatics, Hitotsubashi 2-1-2, Chiyoda-ku, Tokyo 101-8430, Japan
uno@nii.jp

Abstract. We provide polynomial-time algorithms for counting the number of perfect matchings in chain graphs, cochain graphs, and threshold graphs. These algorithms are based on newly developed subdivision schemes that we call a recursive decomposition. On the other hand, we show the #P-completeness for counting the number of perfect matchings in chordal graphs, split graphs and chordal bipartite graphs. This is in an interesting contrast with the fact that counting the number of independent sets in chordal graphs can be done in linear time.

1 Introduction

The study of graph classes has been motivated by the fact that a lot of NP-hard problems can be solved in polynomial time when the input is restricted. While this research direction leads to many polynomial-time algorithms for decision problems and optimization problems, such results for counting problems seem rare. With this motivation, the authors studied problems to count the independent sets in chordal graphs [16], and refinement for interval graphs has been proposed by Lin [14] and Lin and Chen [15]. However, the current understanding for counting problems in graph classes is still poor. Counting algorithms may require properties of graphs that are not needed for solving decision and optimization problems.

This paper is concerned with perfect matchings. A perfect matching of a graph is one of the fundamental objects when we study counting problems.

---

* Supported by Global COE Program "Computationism as a Foundation of the Sciences" and Grant-in-Aid for Scientific Research from Ministry of Education, Science and Culture, Japan, and Japan Society for the Promotion of Science.
** Supported by Ministry of Education, Science and Culture, Japan, and Japan Society for the Promotion of Science.
*** Supported by Ministry of Education, Science and Culture, Japan, and Japan Society for the Promotion of Science.

When Valiant [21] introduced the complexity class #P, he already proved that counting the perfect matchings in a bipartite graph is #P-complete. In another paper [22], he also proved that counting all matchings in a bipartite graph is #P-complete. His results were refined by Dagum and Luby [4] showing that counting the perfect matchings in a 3-regular bipartite graph is #P-complete, and by Vadhan [19] showing that counting all matchings in a bipartite graph of maximum degree 4 and in a planar bipartite graph of maximum degree 6 is #P-complete. There are also some results on the positive side, namely for polynomial-time algorithms. The perfect matchings in a planar graph can be counted in polynomial time [6,11,17] via the so-called Pfaffian orientations. A generalization of this approach yields a polynomial-time algorithm for graphs of bounded genus [8,18]. Furthermore, we can count the perfect matchings in a graph of bounded treewidth in polynomial time [1]. Basically, these positive results are concerned with sparse graphs.

This paper concentrates on classes of chordal graphs and chordal bipartite graphs. An interesting phenomenon to be proven here is the #P-completeness for the counting problem of perfect matchings in chordal graphs, while we can count the number of independent sets in chordal graphs in linear time [16]. We also prove that the matching counting is #P-complete even for chordal bipartite graphs. Therefore, we seek for subclasses of these graph classes for which the perfect matchings can be counted in polynomial time. We give $O(n^2 \log n)$-time algorithms for the following classes of graphs on n vertices: Chain graphs, cochain graphs, and threshold graphs. The definitions will be given later, but there is a relation among these classes as depicted in Fig. 1. In the figure, we also have the classes of bipartite permutation graphs, proper interval graphs, interval graphs and interval bigraphs. For these four classes, complexity of counting the matchings is unsettled. This is a main open problem this paper leaves for us.

We should note here that there exists an $O(n^{2k+1})$-time algorithm to count the perfect matchings in a graph of cliquewidth k [13]. Since a threshold graph is a cograph and the cliquewidth of a cograph is at most 2, this immediately yields an $O(n^5)$-time algorithm for threshold graphs. Similarly, since a chain graph is distance-hereditary and the cliquewidth of a distance-hereditary graph is at most 3 [9], we obtain an $O(n^7)$-time algorithm for chain graphs. Furthermore, the complement of a graph of cliquewidth k has cliquewidth at most $2k$ [3] and a cochain graph is the complement of some chain graph. Therefore, we obtain an $O(n^{13})$-time algorithm for cochain graphs. However, these algorithms are less efficient than ours.

Due to the page limitation, some proofs are omitted. They can be found in the journal version.

2 Preliminaries

We assume the reader is familiar with basic terminology on graphs. A graph is denoted by $G = (V, E)$ when V is the vertex set and E is the edge set of G. The *neighborhood* of a vertex $v \in V$ is the set $N_G(v) = \{u \in V \mid \{u, v\} \in E\}$.

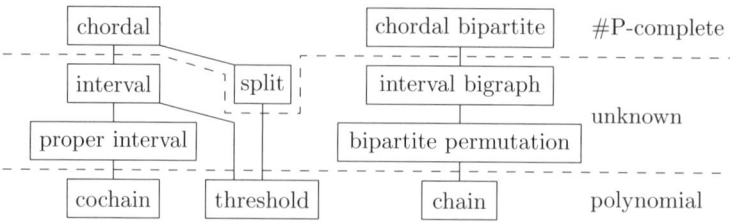

Fig. 1. Inclusion relationship among the graph classes in this paper and the summary of our results

For a subset $U \subseteq V$, the *subgraph of G induced by U* is the graph (U, F), where $F = \{\{u, v\} \in E \mid u, v \in U\}$, and denoted by $G[U]$. For given two graphs $G = (V, E)$ and $G' = (V', E')$, we denote the graph $(V \cup V', E \cup E')$ by $G \cup G'$. A vertex set C is a *clique* if all pairs of vertices in C are joined by an edge. A vertex set I is *independent* if no pair of vertices in I is joined by an edge. An edge set M is a *matching* if no pair of edges in M shares an endpoint. Note that the empty set is a matching of size zero in any graph. An endpoint v of an edge e in a matching M is said to be *matched* by M. A matching is *perfect* if all vertices are matched by the matching.

A graph $G = (V, E)$ is *bipartite* if V can be partitioned into two sets X and Y such that every edge joins a vertex in X and the other vertex in Y. We denote a bipartite graph by $G = (X, Y, E)$ when the partition is given. A bipartite graph G is *complete* if every vertex in X is adjacent to all vertices in Y.

For a graph G, the number of matchings in G is denoted by $\mu(G)$, the number of matchings of size i in G is denoted by $\mu_i(G)$, and the number of perfect matchings in G is denoted by $\pi(G)$.

3 Polynomial-Time Algorithm for Chain Graphs

In this section, we study chain graphs. A chain graph is also called a difference graph [10], a bisplit graph [7], and nonseparable bipartite graph [5].

To define a chain graph, we need to define monotonicity on vertex sets. Let $G = (X, Y, E)$ be a bipartite graph. An order $<$ on X in G is *increasing* if $x < x'$ implies $N(x) \subseteq N(x')$. Similarly, $<$ on X is *decreasing* if $x < x'$ implies $N(x) \supseteq N(x')$. An order is *monotone* if it is increasing or decreasing. A bipartite graph $G = (X, Y, E)$ is a *chain graph* if there exist monotone orders $<_X, <_Y$ on X, Y respectively [12,23]. We assume that $<_X$ is decreasing and $<_Y$ is increasing. It is not hard to observe the following.

Proposition 1. *Let $G = (X, Y, E)$ be a connected chain graph with $|X| = n_x$ and $|Y| = n_y$. Then there exist a decreasing order $<_X$ and an increasing order $<_Y$ such that $y_{n_y} \in N(x_i)$ for every $i \in \{1, \ldots, n_x\}$ and $x_1 \in N(y_j)$ for every $j \in \{1, \ldots, n_y\}$, where $x_1 <_X x_2 <_X \cdots <_X x_{n_x}$ and $y_1 <_Y y_2 <_Y \cdots <_Y y_{n_y}$.*

For a given chain graph, monotone orders on X and Y can be found in linear time (e.g., using a PQ-tree), and from them, the orderings as in Proposition 1 can

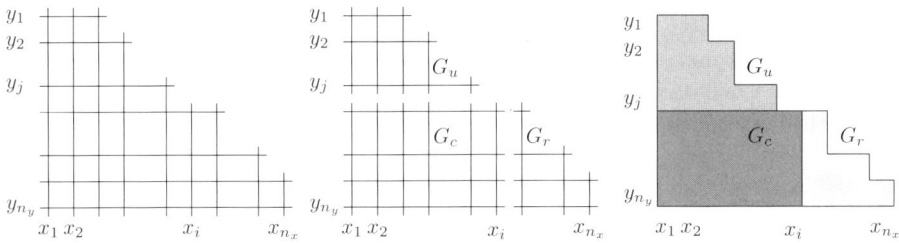

Fig. 2. The intersection model of a chain graph (left), a derived decomposition (middle) and its schematic representation (right)

be computed in linear time (e.g., [20]). Hence, hereafter, we assume that a chain graph is given with two ordered vertex sets stated as in Proposition 1. Without loss of generality, we assume that a chain graph $G = (X, Y, E)$ is connected, and the vertex sets X and Y are ordered as stated in Proposition 1. That is, we have $N(x_i) = \{y_{j_i}, \ldots, y_{n_y}\}$ with $1 \leq i \leq n_x$ and $j_1 \leq j_2 \leq \cdots \leq j_{n_x}$, and $N(y_j) = \{x_1, \ldots, x_{i_j}\}$ with $1 \leq j \leq n_y$ and $i_1 \geq i_2 \geq \cdots \geq i_{n_y}$, where $|X| = n_x$ and $|Y| = n_y$. The main theorem in this section is as follows.

Theorem 1. *Given a connected chain graph $G = (X, Y, E)$, the number of perfect matchings in G can be computed in $O(n^2 \log n)$ time, where $n = |X \cup Y|$.*

We prove Theorem 1 by providing an algorithm. This is based on the following recursive subdivision structure $\mathcal{T}(G)$. The structure $\mathcal{T}(G)$ is a rooted tree, where each node possesses an induced subgraph of G and a node G' is a descendant of a node G''' only if G' is a subgraph of G'''.

The structure $\mathcal{T}(G)$ is inspired by an intersection model of a chain graph (Fig. 2 (left)): The vertices x_i in X correspond to vertical line segments I_{x_i} which are located from left to right according to the ordering, and the bottom of the segments are on a horizontal line. The vertices y_j in Y correspond to horizontal line segments J_{y_j} from top to bottom and the left endpoints of the segments are on a vertical line. From Proposition 1, it is easy to see that G is a chain graph if and only if G can be represented by the intersection model of those horizontal line segments and vertical line segments such that I_{x_i} is longer than or equal to $I_{x_{i+1}}$ with $1 \leq i < n_x$ and $J_{y_{j+1}}$ is longer than or equal to J_{y_j} with $1 \leq j < n_y$.

Let $G = (X, Y, E)$ be a chain graph with two ordered vertex sets X, Y as in Proposition 1. An edge $e = \{x_i, y_j\} \in E$, $x_i \in X, y_j \in Y$, is *extremal* if $\{x_{i'}, y_j\} \notin E$ for any $i' > i$ or $\{x_i, y_{j'}\} \notin E$ for any $j' < j$. Fix an extremal edge $e = \{x_i, y_j\} \in E$. Then we partition X and Y into X_u, X_r, Y_u, and Y_r as follows; $X_u := \{x_{i'} \mid i' \leq i\}$, $X_r := \{x_{i'} \mid i' > i\}$, $Y_u := \{y_{j'} \mid j' \leq j\}$, and $Y_r := \{y_{j'} \mid j' > j\}$. Using these vertex sets, we define three graphs G_c, G_u, and G_r as follows; $G_c = G[X_u \cup Y_r]$, $G_r = G[X_r \cup Y_r]$, and $G_u = G[X_u \cup Y_u]$. Note that the edge sets of these three graphs form a partition of E, and furthermore, G_u and G_r are connected chain graphs (unless empty), and G_c is a complete bipartite graph. See Fig. 2 (middle, right).

Algorithm 1. RD(G, d)

Input : A connected chain graph $G = (X, Y, E)$ and a non-negative integer d, where $n_x = |X|, n_y = |Y|$;
Output: A recursive decomposition $\mathcal{T}(G)$ of G;
1 add G as the root;
2 **if** G is complete bipartite **then return**;
3 **if** d is even **then** set $k_x := n_x/2$ and $k_y := \min\{j \mid \{x_{k_x}, y_j\} \in E\}$;
4 **if** d is odd **then** set $k_y := n_y/2$ and $k_x := \max\{i \mid \{x_i, y_{k_y}\} \in E\}$;
5 find G_u, G_r, G_c from k_x and k_y;
6 add RD($G_u, d+1$) as the left subtree rooted at G;
7 add RD($G_r, d+1$) as the right subtree rooted at G;
8 **return**;

We are now ready for defining $\mathcal{T}(G)$ for a connected chain graph G. The root of $\mathcal{T}(G)$ is G. Let G' be a node of $\mathcal{T}(G)$. If G' is complete bipartite, then G' has no child and hence it is a leaf of $\mathcal{T}(G)$. Otherwise, G' has two children G'_u and G'_r constructed by an appropriate choice of an extremal edge of G'. We call $\mathcal{T}(G)$ a *recursive decomposition* of G.

Let us describe how to choose an appropriate extremal edge when we construct a recursive decomposition. To do this, we look at the depth of each node in $\mathcal{T}(G)$. Namely, the depth of a root node is zero, and if a node has depth d, then its children have depth $d+1$. According to the parity of the depth, we make the choice. If d is even, then we let $k_x = n_x/2$ and $k_y = \min\{j' \mid \{x_i, y_{j'}\} \in E\}$. If d is odd, then we let $k_y = n_y/2$ and $k_x = \max\{i' \mid \{x_{i'}, y_j\} \in E\}$. In both cases, we see that $\{x_{k_x}, y_{k_y}\}$ is an extremal edge of G. Algorithm 1 computes a recursive decomposition $\mathcal{T}(G)$ of G according to this choice of an extremal edge when RD($G, 0$) is called. We omit the proof of the following lemma.

Lemma 1. *Let $G = (X, Y, E)$ be a connected chain graph. Then, Algorithm 1 finds a recursive decomposition $\mathcal{T}(G)$ with at most $O(n)$ nodes and height at most $\frac{1}{2}\log_2 n$ in $O(n)$ time, where $n = |X|+|Y|$.* □

To describe the number of matchings, we denote by $\pi(G; a, b)$ the number of perfect matchings in a chain graph $G = (X, Y, E)$ with a vertices from X and b vertices from Y deleted. Namely, $\pi(G; a, b) = |\{M \subseteq E \mid M$ is a perfect matching of $G - (A \cup B)$, where $A \subseteq X, |A| = a, B \subseteq Y, |B| = b\}|$. Intuitively, we will count the number of perfect matchings in G such that a vertices in X and b vertices in Y have been matched in the previous level. Since $\pi(G; 0, 0)$ is the number of perfect matchings of G, it suffices to compute $\pi(G; a, b)$ for all possible a and b.

Let us look at how we can decompose $\pi(G; a, b)$ into several independent parts. This gives a fundamental idea for our algorithm.

Lemma 2. *Let $G = (X, Y, E)$ be a connected chain graph. For $0 \le a < |X|$ and $0 \le b < |Y|$, it holds that $\pi(G; a, b) = \sum_{a_u, b_r} \pi(G_u; a_u+i_c, b_u) \cdot \pi(G_r; a_r, b_r+i_c) \cdot \pi(G_c; k_x-i_c, n_y-k_y-i_c)$, where the sum is taken over the ranges $0 \le a_u \le$*

$\min\{a, k_x\}$ and $0 \le b_r \le \min\{b, n_y-k_y\}$, and other symbols are defined as $b_u = b-b_r$, $a_r = a-a_u$, and $i_c = k_x-k_y-a_u+b-b_r$.

Proof. We first show the inequality "the left-hand side $\le$ the right-hand side." Every matching M counted in the left-hand side can be partitioned into three parts $M = M_u \cup M_r \cup M_c$, where M_u, M_r, M_c is a matching of G_u, G_r, G_c, respectively. Since M is a perfect matching of $G-(A\cup B)$ for some $A \subseteq X, B \subseteq Y$ with $|A|=a, |B|=b$, it holds that $|M|=n_x-a=n_y-b$. Let $|M_u|=i_u, |M_r|=i_r, |M_c|=i_c$. Let $A_c \subseteq X_u$ and $B_c \subseteq Y_r$ be the set of vertices matched by M_c. Note that $|A_c|=i_c=|B_c|$. Then, M_u is a perfect matching of $G_u-(A_u\cup A_c\cup B_u)$ for some $A_u \subseteq X_u \setminus A_c$ and $B_u \subseteq Y_u$, and M_r is a perfect matching of $G_r - (A_r \cup B_r \cup B_c)$ for some $A_r \subseteq X_r$ and $B_r \subseteq Y_r \setminus B_c$. It is important to observe that such A_u, B_u, A_r, B_r are unique. For example, A_u is determined as the set of vertices in $X_u \setminus A_c$ that are not matched by M_u. Therefore, if we let $|A_u|=a_u$ and $|B_r|=b_r$, then M_u is counted exactly once in $\pi(G_u; a_u+i_c, b-b_r)$ and similarly M_r is counted exactly once in $\pi(G_r; a-a_u, b_u+i_c)$. Then, we see that $k_x-(a_u+i_c)=i_u=k_y-(b-b_r)$ and $n_x-k_x-(a-a_u)=i_r=n_y-k_y-(b_u+i_c)$, and therefore, $i_c=k_x-k_y-a_u+b-b_r$. Then, it suffices to note that M_c is counted exactly once in $\pi(G_c; k_x-i_c, n_y-k_y-i_c)$.

Conversely, we show the inequality "the left-hand side $\ge$ the right-hand side." Let M_u and M_r be perfect matchings counted in $\pi(G_u; a_u+i_c, b_u)$ and $\pi(G_r; a_r, b_r+i_c)$ respectively, for some a_u, b_r in the appropriate ranges and a_r, b_u, i_c as defined in the statement of the lemma. Specifically, let M_u be a perfect matching of $G_u-(A_u \cup A_c \cup B_u)$ for some $A_u \cup A_c \subseteq X_u$ and $B_u \subseteq Y_u$, and M_r is a perfect matching of $G_r-(A_r \cup B_r \cup B_c)$ for some $A_r \subseteq X_r$ and $B_r \cup B_c \subseteq Y_r$, where $|A_u|=a_u, |A_r|=a_r, |B_u|=b_u, |B_r|=b_r, |A_c|=|B_c|=i_c$. Consider constructing a perfect matching M of $G-(A\cup B)$ for some A, B with $|A|=a, |B|=b$ as $M=M_u\cup M_r\cup M_c$ with some matching M_c of G_c. Then, such M_c should be a perfect matching of $G_c-((X_u-A_c)\cup(Y_r-B_r))$. This completes the proof. □

Since G_c is a complete bipartite graph, it is not difficult to see that $\pi(G_c; k_x-i_c, n_y-k_y-i_c) = i_c! \binom{k_x}{i_c}\binom{n_y-k_y}{i_c}$. Hence Lemma 2 readily gives Algorithm 2. To compute the number of perfect matchings in a given chain graph G, we first call #M$(G, 0, 0)$.

Proof (of Theorem 1). Algorithm 2 can be implemented by dynamic programming on $\mathcal{T}(G)$. For each node G' of $\mathcal{T}(G)$, we store the values returned by calls #M(G', a, b) for all possible a and b. If the depth of G' is d, then the number of such possibilities is at most $n_x/2^{\lceil d/2 \rceil} \cdot n_y/2^{\lfloor d/2 \rfloor} \le n_x n_y/2^d$. Therefore, the number of values stored for each node of $\mathcal{T}(G)$ is $O(n^2/2^d)$. At the call to #M(G', a, b) we need to look up at most $n_x n_y/2^d$ values. Note that the values of factorials and binomial coefficients can be computed beforehand and stored as well in $O(n^2)$ time and space. Since the number of nodes at depth d is at most 2^d, the overall running time is at most $\sum_{d=0}^{\frac{1}{2}\log_2 n} 2^d O(n^2/2^d) = O(n^2 \log n)$. The space requirement is also $O(n^2 \log n)$. □

Algorithm 2. $\#M(G, a, b)$

Input : A connected chain graph $G = (X, Y, E)$ and two integers a, b together with a recursive decomposition $\mathcal{T}(G)$;
Output: $\pi(G; a, b)$;

1 **if** $n_x - a \neq n_y - b$ **then return** 0;
2 **else if** G is complete bipartite **then return** $(n_x - a)!$;
3 **else**
4       $sum := 0$; $a_r := a - a_u$; $b_u = b - b_r$; $i_c = k_x - k_y - a_u + b - b_r$;
5   **foreach** $a_u = 0, 1, \ldots, \min\{a, k_x\}$ and $b_r = 0, 1, \ldots, \min\{b, n_y - k_y\}$ **do**
6            $sum := sum + i_c! \cdot \binom{k_x}{i_c} \cdot \binom{n_y - k_y}{i_c} \cdot \#M(G_u, a_u + i_c, b_u) \cdot \#M(G_r, a_r, b_r + i_c)$;
7   **end**
8   **return** sum.
9 **end**

Note that if $n_x - a = n_y - b$, then $\pi(G; a, b)$ is the number of matchings of size $n_x - a$. Hence, Algorithm 2 computes the number of matchings of each possible size. This implies the following corollary.

Corollary 1. *Given a chain graph $G = (X, Y, E)$, the number of matchings and the number of matchings of fixed size in G can be computed in $O(n^2 \log n)$ time, where $n = |X \cup Y|$.* □

4 Polynomial-Time Algorithm for Cochain Graphs and Threshold Graphs

Similarity among chain graphs, cochain graphs and threshold graphs allows us to provide polynomial-time algorithms to count the number of perfect matchings in cochain graphs and threshold graphs.

A *cochain graph* is simply defined as the complement of a chain graph. From Proposition 1, we can immediately see a cochain graph has the following property.

Proposition 2. *Let $G = (V, E)$ be a cochain graph and $\overline{G} = (X, Y, \overline{E})$ be a chain graph that is the complement of G with $|X| = n_x$ and $|Y| = n_y$. Then X and Y are cliques of G, and there exist a decreasing order $<_X$ and an increasing order $<_Y$ such that $y_{n_y} \in N(x_i) \setminus X$ for every $i \in \{1, \ldots, n_x\}$ and $x_1 \in N(y_j) \setminus Y$ for every $j \in \{1, \ldots, n_y\}$, where $x_1 <_X x_2 <_X \cdots <_X x_{n_x}$ and $y_1 <_Y y_2 <_Y \cdots <_Y y_{n_y}$.* □

Namely, a cochain graph can be constructed from a chain graph by filling up both of the color classes to cliques. See Fig. 3.

A graph $G = (V, E)$ is a *threshold* graph if there exist a weight assignment $w : V \to \mathbb{R}$ such that $\{u, v\} \in E$ if and only if $w(u) + w(v) > 0$. The following is a well-known property (or actually a characterization) of threshold graphs.

Proposition 3 (Chvátal and Hammer [2]). *For a threshold graph $G = (V, E)$, a partition $\{X, Y\}$ of V with the following properties can be found in*

$O(|V|+|E|)$ time. First, $X = \{x_1, \ldots, x_{n_x}\}$ is a clique of G, $Y = \{y_1, \ldots, y_{n_y}\}$ is an independent set of G, and there exist a decreasing order $<_X$ and an increasing order $<_Y$ such that $y_{n_y} \in N(x_i) \setminus X$ for every $i \in \{1, \ldots, n_x\}$ and $x_1 \in N(y_j)$ for every $j \in \{1, \ldots, n_y\}$, where $x_1 <_X x_2 <_X \cdots <_X x_{n_x}$ and $y_1 <_Y y_2 <_Y \cdots <_Y y_{n_y}$.

Namely, a threshold graph can be constructed from a chain graph by filling up one of the color classes to a clique. See Fig. 3.

Since a cochain graph and a threshold graph possess a structure similar to a chain graph, we may define a recursive decomposition for them analogously. An important difference is that G_c is not an induced subgraph of G, but G_c will be a subgraph of G with vertex set $X_u \cup Y_r$ and edge set consisting of those edges between X_u and Y_r. Namely, G_c is complete bipartite. Then, the equation

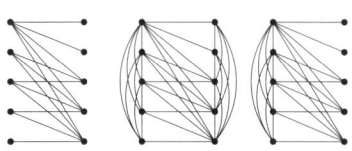

Fig. 3. Comparison of three classes. (Left) A chain graph. (Middle) A cochain graph. (Right) A threshold graph.

similar to one in Lemma 2 holds. The whole arguments are verbatim. Hence we obtain the following theorem.

Theorem 2. *The number of perfect matchings in a cochain graph and a threshold graph with n vertices can be computed in $O(n^2 \log n)$ time.* □

5 Hardness Results

In this section, we prove the #P-completeness of counting the perfect matchings in split graphs and chordal bipartite graphs. The #P-completeness for split graphs immediately implies that for chordal graphs.

A graph is a *split graph* if the vertex set can be partitioned into two parts such that one part is a clique and the other is an independent set. In other words, a split graph is constructed from a bipartite graph by filling up one color class to a clique. A graph is *chordal* if every induced cycle has length three. It is easy to see that every split graph is chordal.

We omit the proof of the following theorem.

Theorem 3. *Counting the number of perfect matchings in a split graph is #P-complete.* □

By utilizing an interpolation technique, we are able to show the following.

Theorem 4. *Counting the number of matchings in a split graph is #P-complete.*

Proof. For our reduction, we use the problem to count the number of perfect matchings in a bipartite graph, which is known to be #P-complete [21].

Let $G = (U, V, E)$ be a bipartite graph with $|U| = |V| = n$. We construct a graph $G_i = (V_i, E_i)$ for every $i \in \{1, \ldots, n+1\}$ out of G as follows. Let $V := \{v_1, \ldots, v_n\}$. For each vertex $v_\ell \in V$, we use a set $V_i^{(\ell)}$ of i vertices for G_i

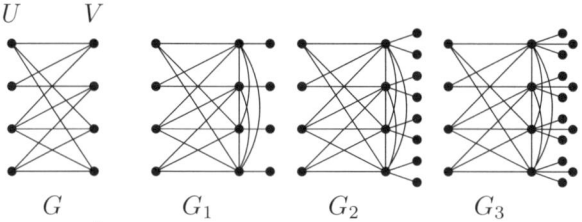

Fig. 4. Construction in the proof of Theorem 4

where $V_i^{(1)}, \ldots, V_i^{(n)}$ are all disjoint. We set $V_i := U \cup V \cup \bigcup_{\ell=1}^n V_i^{(\ell)}$, and $E_i := E \cup \binom{V}{2} \cup \bigcup_{\ell=1}^n F_i^{(\ell)}$, where $F_i^\ell := \{\{v_\ell, v\} \mid v \in V_i^{(\ell)}\}$ for every $\ell \in \{1, \ldots, n\}$. Fig. 4 illustrates the construction.

For a graph H with n vertices, remember that we denote by $\mu_j(H)$ the number of matchings in H of size j, by $\mu(H)$ the number of matchings in H. Furthermore, let $F_i := \bigcup_{\ell=1}^n F_i^{(\ell)}$.

Let us consider $\mu(G_i)$. Each matching M of G_i potentially uses some edges from E and some edges from F_i. Let M use j edges from E and k edges from F_i. Consider constructing M by first choosing j edges from E, then k edges from F_i, and finally the rest of edges from $\binom{V}{2}$. Since M is a matching, we have $\mu_j(G)$ ways to choose j edges from E for M. Then j vertices in V are already matched, so there are $n-j$ vertices left unmatched in V. Therefore, the number of ways to choose k edges from F_i for M is $\binom{n-j}{k}i^k$. Then there are $n-j-k$ vertices left unmatched in V. Among them we choose some edges for M. Therefore, the number of choices is $\mu(K_{n-j-k})$, where K_{n-j-k} is a complete graph with $n-j-k$ vertices. This way, we obtain the following formula: for every $i \in \{1, \ldots, n+1\}$ it holds that $\mu(G_i) = \sum_{j=0}^n \mu_j(G) \left(\sum_{k=0}^{n-j} \left(\binom{n-j}{k} i^k \right) \mu(K_{n-j-k}) \right)$. In a matrix form, this can be written as

$$\begin{bmatrix} \mu(G_1) \\ \mu(G_2) \\ \vdots \\ \mu(G_{n+1}) \end{bmatrix} = A \begin{bmatrix} \mu_0(G) \\ \mu_1(G) \\ \vdots \\ \mu_n(G) \end{bmatrix},$$

where A is a matrix with row index set $\{1, \ldots, n+1\}$ and column index set $\{0, \ldots, n\}$ defined as $A_{i,j} := \sum_{k=0}^{n-j} \left(\binom{n-j}{k} i^k \right) \mu(K_{n-j-k})$ for each $i \in \{1, \ldots, n+1\}$ and $j \in \{0, \ldots, n\}$.

Now we claim that A defined above is non-singular. We would like to notice that the claim finishes the proof of the theorem. If we are able to know $\mu(G_i)$ for every $i \in \{1, \ldots, n+1\}$, then by computing the inverse of A, we are also able to know $\mu_j(G)$ for all $j \in \{0, \ldots, n\}$. (Note that each entry of A can be computed efficiently since $\mu(K_m) = \sum_{j=0}^{\lfloor m/2 \rfloor} \mu_j(K_m) = \sum_{j=0}^{\lfloor m/2 \rfloor} \binom{m}{2j} \frac{(2j)!}{j! 2^j}$ holds.) In particular we obtain $\mu_n(G)$, the number of perfect matchings in G. This completes the reduction.

Therefore, it suffices to prove the claim. To do that, first observe that for each row index $i \in \{1, \ldots, n+1\}$ and each column index $j \in \{0, \ldots, n\}$ the i, j-entry $A_{i,j}$ can be written as $A_{i,j} = \sum_{k=0}^{n-j} i^k \left(\binom{n-j}{k} \mu(K_{n-j-k})\right)$. Let B be a matrix with row index set $\{1, \ldots, n+1\}$ and column index set $\{0, \ldots, n\}$ defined as $B_{i,k} := i^k$ for each $i \in \{1, \ldots, n+1\}$ and $k \in \{0, \ldots, n\}$, and C be a matrix with row index set $\{0, \ldots, n\}$ and column index set $\{0, \ldots, n\}$ defined as

$$C_{k,j} := \begin{cases} \binom{n-j}{k} \mu(K_{n-j-k}) & \text{if } 0 \le k \le n-j, \\ 0 & \text{otherwise,} \end{cases}$$

for each $k \in \{0, \ldots, n\}$ and $j \in \{0, \ldots, n\}$. Then, we can see that for every $i \in \{1, \ldots, n+1\}$ and $j \in \{0, \ldots, n\}$ it holds that $A_{i,j} = \sum_{k=0}^{n} B_{i,k} C_{k,j}$. In other words, $A = BC$ as a matrix. The matrix B is a famous Vandermonde matrix, which is known to be non-singular. How about the non-singularity of C? Since $\binom{n-j}{k} \mu(K_{n-j-k}) \neq 0$ when $0 \le k \le n-j$, the upper-left half of C is occupied with non-zero entries, and the lower-right half of C is occupied with zero entries. So, the matrix C is also non-singular. Thus, A is non-singular. □

A modification of this proof shows the following. We omit the proof.

Theorem 5. *Counting the number of maximal matchings in a split graph is #P-complete.* □

Next, we switch to chordal bipartite graphs. A bipartite graph is *chordal bipartite* if every induced cycle is of length four. The #P-completeness for chordal bipartite graphs will be proven via the interpolation technique.

Theorem 6. *The problem to count the number of perfect matchings in a chordal bipartite graph is #P-complete.*

Proof. We again use a reduction from the problem to count the number of perfect matchings in a bipartite graph.

Given a bipartite graph $G = (X, Y, E)$ with $|X| = |Y| = n$, we construct the following chordal bipartite graph $cb_i(G)$ for each $i \in \{1, \ldots, n+1\}$. The vertex set of $cb_i(G)$ is defined as $V(G) = X \cup Y \cup \{p_{j,v,e} \mid 1 \le j \le i, v \in X, e \in E\} \cup \{q_{j,v,e} \mid 1 \le j \le i, v \in Y, e \in E\}$. The edge set of $cb_i(G)$ is defined as $E(G) = \{\{x, y\} \mid x \in X, y \in Y\} \cup \{\{x, p_{j,x,e}\} \mid x \in X, e \in E, x \in e, 1 \le j \le i\} \cup \{\{y, q_{j,y,e}\} \mid y \in Y, e \in E, y \in e, 1 \le j \le i\} \cup \{\{p_{j,x,e}, q_{j,y,e}\} \mid x \in X, y \in Y, e = \{x, y\} \in E\}$. Namely, to construct $cb_i(G)$ from G, we replace each edge of G by i paths of length three, and join the vertices of X and Y by edges to make them complete bipartite. It is not difficult to see that $cb_i(G)$ is chordal bipartite. Fig. 5 shows an example.

Consider a perfect matching M of $cb_i(G)$. We map M to a matching M' of G if and only if the following conditions are satisfied.

– When $e = \{x, y\} \notin M'$, it holds that $\{p_{j,x,e}, q_{j,y,e}\} \in M$ for all $j \in \{1, \ldots, i\}$.
– When $e = \{x, y\} \in M'$, it holds that $\{x, p_{j,x,e}\}, \{y, q_{j,y,e}\} \in M$ for exactly one $j \in \{1, \ldots, i\}$. (Then, it must hold that $\{p_{j,x,e}, q_{j,y,e}\} \in M$ for all other $j \in \{1, \ldots, i\}$.)

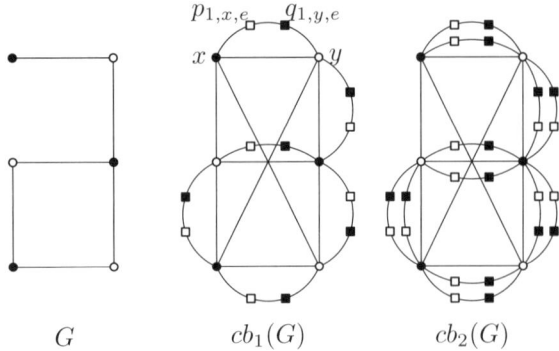

Fig. 5. Hardness for chordal bipartite graphs

There are several perfect matchings M that corresponds to M'. We can count the number of such matchings M from M'. In the second condition, we have i choices for each edge of M'. Let $|M'| = k$. Then, this gives rise to i^k choices. Moreover, there are $n-k$ vertices in both X and Y that do not appear in these conditions, and they are supposed to be matched. Since these vertices induce a complete bipartite subgraph of G, the number of ways to match them is exactly $(n-k)!$.

In this way, we obtain $\pi(cb_i(G)) = \sum_{k=0}^{n} \mu_k(G) i^k (n-k)!$ for each $i \in \{1, \ldots, n+1\}$, where $\pi(cb_i(G))$ means the number of perfect matchings in $cb_i(G)$. When we write it down in the matrix form as we did before, the coefficient matrix A can be defined as $A_{i,k} := i^k(n-k)!$ for every row index $i \in \{1, \ldots, n+1\}$ and every column index $k \in \{0, \ldots, n\}$. We can see that the determinant of A is the determinant of a non-singular Vandermonde matrix times $\prod_{k=0}^{n}(n-k)!$, thus non-zero. Therefore, from the equality above, we can recover $\mu_n(G)$, the number of perfect matchings of the given bipartite graph G, in polynomial time. □

Similar arguments show the following theorem. We omit the proofs.

Theorem 7. *Counting the number of matchings in chordal bipartite graphs is #P-complete. Similarly, counting the number of maximal matchings in chordal bipartite graphs is #P-complete.* □

Acknowledgment. We thank anonymous referees for their valuable comments.

References

1. Arnborg, S., Lagergren, J., Seese, D.: Easy problems for tree-decomposable graphs. J. Algor. 12, 308–340 (1991)
2. Chvátal, V., Hammer, P.L.: Set-packing and threshold graphs. Res. Rep., Comput. Sci. Dept., Univ. Waterloo, CORR 73-21 (1973)
3. Courcelle, B., Olariu, S.: Upper bounds to the clique width of graphs. Discr. Appl. Math. 101, 77–114 (2000)

4. Dagum, P., Luby, M.: Approximating the permanent of graphs with large factors. Theor. Comput. Sci. 102, 283–305 (1992)
5. Ding, D.: Covering the edges with consecutive sets. J. Graph Theory 15, 559–562 (1991)
6. Fisher, M.E.: Statistical mechanics of dimers on a plane lattice. Phys. Rev. Ser. 2 124, 1664–1672 (1961)
7. Frost, H., Jacobson, M., Kabell, J., Morris, F.R.: Bipartite analogues of split graphs and related topics. Ars Comb. 29, 283–288 (1990)
8. Galluccio, A., Loebl, M.: On the theory of Pfaffian orientations I. Perfect matchings and permanents. Electr. J. Comb. 6, Research Paper 6, 19 pages (1999)
9. Golumbic, M.C., Rotics, U.: On the clique-width of some perfect graph classes. Internat. J. Found. Comput. Sci. 11, 423–443 (2000)
10. Hammer, P.L., Peled, U.N., Sun, X.: Difference graphs. Discr. Appl. Math. 28, 35–44 (1990)
11. Kasteleyn, P.W.: Dimer statistics and phase transitions. J. Math. Phys. 4, 287–293 (1963)
12. Kloks, T., Kratsch, D., Müller, H.: Bandwidth of chain graphs. Infor. Proc. Lett. 68, 313–315 (1998)
13. Makowsky, J.A., Rotics, U., Averbouch, I., Godlin, B.: Computing graph polynomials on graphs of bounded clique-width. In: Fomin, F.V. (ed.) WG 2006. LNCS, vol. 4271, pp. 191–204. Springer, Heidelberg (2006)
14. Lin, M.-S.: Fast and simple algorithms to count the number of vertex covers in an interval graph. Infor. Proc. Lett. 102, 143–146 (2007)
15. Lin, M.-S., Chen, Y.-J.: Linear time algorithms for counting the number of minimal vertex covers with minimum/maximum size in an interval graph. Infor. Proc. Lett. 107, 257–264 (2008)
16. Okamoto, Y., Uno, T., Uehara, R.: Counting the independent sets in a chordal graph. J. Discr. Algor. 6, 229–242 (2008)
17. Temperley, H.N.V., Fisher, M.E.: Dimer problem in statistical mechanics — an exact result. Phil. Magazine, Ser. 8 6, 1061–1063 (1961)
18. Tesler, G.: Matchings in graphs on non-orientable surfaces. J. Comb. Theory, Ser. B 78, 198–231 (2000)
19. Vadhan, S.P.: The complexity of counting in sparse, regular, and planar graphs. SIAM J. Comput. 31, 398–427 (2001)
20. Uehara, R., Uno, Y.: On computing longest paths in small graph classes. Internat. J. Found. Comput. Sci. 18, 911–930 (2007)
21. Valiant, L.G.: The complexity of computing the permanent. Theor. Comput. Sci. 8, 189–201 (1979)
22. Valiant, L.G.: The complexity of enumeration and reliability problems. SIAM J. Comput. 8, 410–421 (1979)
23. Yannakakis, M.: Node-deletion problems on bipartite graphs. SIAM J. Comput. 10, 310–327 (1981)

Distance d-Domination Games

Stephan Kreutzer and Sebastian Ordyniak

Oxford University Computing Laboratory
{kreutzer,ordyniak}@comlab.ox.ac.uk

Abstract. We study graph searching games where a number of cops try to capture a robber that is hiding in a system of tunnels modelled as a graph. While the current position of the robber is unknown to the cops, each cop can see a certain radius d around his position. For the case $d = 1$ these games have been studied by Fomin, Kratsch and Müller [7] under the name domination games.

We are primarily interested in questions concerning the complexity and monotonicity of these games. We show that dominating games are computationally much harder than standard graph searching games where the cops only see their own vertex and establish strong non-monotonicity results for various notions of monotonicity which arise naturally in the context of domination games. Answering a question of [7], we show that there exists graphs for which the shortest winning strategy for a minimal number of cops must necessarily be of exponential length. On the positive side, we establish tractability results for graph classes of bounded degree.

1 Introduction

Graph searching games are a form of two-player games played on graphs. A wide range of such games have been studied in the literature but they all share the common scheme that a number of cops tries to catch a robber who is hiding in the graph. The problem is to guide a party of as few cops as possible so that the robber is guaranteed to be captured regardless of his moves. In the model of graph searching games known as node searching, the cops and the robber occupy vertices of the graph. At each step of the play, the player controlling the cops can lift some of the cops from the graph and place them somewhere else. While they are in transit, the robber can move in the graph following any path from his current to his new position as long as this path does not go through a vertex occupied or "blocked" by a remaining cop (in which case the robber would be have been captured).

Variants of this game are obtained by varying the abilities of the cops, for instance, whether or not they know the current position of the robber, and by the precise definition of "blocking". The minimal number of cops needed to catch a robber on a graph yields an interesting graph invariant related to the global connectivity of the graph. See [6] for a recent survey on the subject.

Graph searching games have found a wide range of applications in Computer Science in seemingly unrelated areas: there is a strong resemblance of graph searching games to pebble games modelling sequential computation as described in [10]. In [8], graph searching games have been employed as a model for privacy in distributed systems, where the cops model eavesdroppers or intruders in networks. Furthermore, applications of graph searching games can be found in VLSI design as the game theoretical approach to important graph layout parameters providing valuable tools for the design of efficient algorithms. Of particular importance is the connection between graph searching games and well-known graph parameters such as tree-width and path-width (see e.g. [5,3,4]). For instance, Seymour and Thomas [12] characterised the tree-width of a graph in terms of a variant of graph searching games where the robber is visible and hides on vertices of the graph.

An important concept in the theory of graph searching games is monotonicity. Intuitively, a strategy for the cops is monotone if they can catch the robber without allowing him to revisit vertices from which he has previously been exspelled. Monotonicity has featured highly in research on graph searching games for a number of reasons. For instance, monotone strategies correspond directly to graph decompositions such as tree- or path-decompositions. Also, for many game variants, winning strategies for the robber can often be characterised by simple combinatorial structures, such as *brambles* for the case of games corresponding to tree-width, and hence provide natural and intuitive obstructions for tree-width and similar measures. However, these structures usually provide a winning strategy even against cops following a non-monotone strategy. Hence, showing for a game variant that the number of cops needed to win against a robber is always the same as the number of cops needed for a monotone strategy brings all these concepts together and establishes a smooth theory of decompositions and games in terms of min-max or duality theorems.

From an algorithmic perspective, an important property of monotone strategies is that their length is usually linearly bounded in the order of the graph, whereas non-monotone strategies can have up to exponential length, although almost no game variant actually requires such long strategies. Hence, monotone strategies often provide polynomial certificates and thereby yield NP-algorithms for deciding the number of cops needed to catch a robber.

Originally, graph searching games were introduced to model the chivvy for a robber that is hiding in a system of tunnels. While the cops do not know the current position of the robber they do have knowledge of the graph modelling the system of tunnels. In this paper we follow this idea of catching an invisible robber but consider games, which we call d-*domination games*, where the cops do not only see their current vertex but have a radius d of visibility. That is, a cop placed on a vertex v can see any other vertex within distance d of v and if this vertex is occupied by the robber then the cop can see the robber and capture him. We are primarily interested in complexity and monotonicity questions related to these games.

For the case $d = 1$ these games correspond to *domination games* as introduced by Fomin, Kratsch and Müller [7]. This variant is related to the notion of "see-catch" games studied in Computational Geometry and Robotics, for instance motivated by applications in robotics such as surveillance with a mobile robot equipped with a camera. In their paper, the authors develop the fundamental theory of domination games and establish a relationship between domination games and the size of a minimum dominating set of a graph and an interesting connection between these games and a graph parameter called *domination target number* introduced in [11]. The focus of [7] is on establishing bounds on the domination search number – the minimal number of cops that are required to guarantee capture of the robber – for various classes of graphs such as k-dimensional cubes, *asteroidal-triple free* graphs, *claw-free* graphs, and graphs with certain types of spanning trees and caterpillars. They also exhibit an example showing that domination games are non-monotone.

In this paper we study d-domination games with a focus on complexity and monotonicity. Following the initial results on monotonicity of graph searching games mentioned above, monotonicity proofs for a large number of graph searching games and also non-monotonicity proofs for some games have been obtained (see e.g. [6]). Most variants of graph searching games are either monotone or, if not, at least a bound on the difference between the number of cops needed for arbitrary or monotone strategies can be established. As it turns out, d-domination games exhibit a completely different behaviour in this respect.

Organisation and results. In Section 4, we establish very strong non-monotonicity results by exhibiting classes of graphs on which two cops can win on any graph in this class but the number of cops required for monotone winning strategies is unbounded. Hence, domination games are one of only very few types of games for which such a difference has been proved.

In [7, Problem 7], Fomin et al. raise the question whether any polynomial bound could be proved for the length of winning strategies in domination games. We give a negative answer to this question by exhibiting a class of graphs where two cops have a winning strategy but only with an exponential number of steps. To the best of our knowledge, this is the first type of graph search games for which such a lower bound has been proved.

In terms of complexity, domination games are also much harder than standard cops and robber games. In particular, we show that deciding if two cops have a (non-monotone) winning strategy is PSPACE-complete. Again, to the best of our knowledge, this is the first type of graph searching games exhibiting this worst-case complexity. This result is in sharp contrast to other variants of graph searching games on undirected graphs, which often are in polynomial time for a fixed number of cops and often even fixed-parameter tractable with the numbers of cops being the parameter. For monotone strategies we also prove that it is NP-hard to decide whether two cops have a monotone winning strategy in domination games. The complexity results are the focus of Section 5.

Finally, we establish a relation between domination games and Robber and Marshal games played on hypergraphs. Robber and Marshal games were introduced in [9] to provide a game theoretical characterisation of hypertree-width. In particular, we show that every Robber and Marshal game on a hypergraph can be translated into a domination game on an undirected graph and derive interesting consequences from this fact.

2 Preliminaries

We use standard notation from graph theory as can be found in, e.g., [5]. In particular, we write $V(G)$ for the vertex set of a graph G and $E(G)$ for its edge set. All graphs in this paper are simple and undirected and all graphs and hypergraphs are finite. Let G be a graph and $d \geq 1$. The (open) d-neighbourhood of a vertex v in G is $N_d^G(v) := \{u : 0 < \mathrm{dist}_G(u,v) \leq d\}$, where $\mathrm{dist}_G(u,v)$ is the distance between u and v in G.

The closed d-neighbourhood of v is $N_d^G[v] := N_d^G(v) \cup \{v\}$. If X is a set, we define $N_d^G[X] := \bigcup_{v \in X} N_d^G[v]$. For the case $d = 1$, we omit the index d and e.g. write $N^G(v)$ for $N_1^G(v)$. Also, we omit the index G whenever G is clear from the context.

The notions of tree-width and path-width were introduced by Robertson and Seymour as part of their work on graph minors. We refer to [3,5] for definitions and further information. We write $\mathrm{pw}(G)$ for the path-width of a graph G and $\mathrm{tw}(G)$ for its tree-width.

3 d-Domination Games

In this section we introduce d-*domination games* and present basic results.

A d-*domination game* on a graph G is played between two players, the cop and the robber, where the goal of the cops is to capture the robber. At each step of the play, the robber occupies a vertex of the graph and the cop player controls a finite number of cops each occupying vertices. A play starts by the robber choosing an initial position. In each step of the game, the cop either places a new cop on a vertex or removes an already placed cop from the graph. Suppose X is the set of vertices currently occupied by the cops and they want to place a new cop on vertex v. They first have to announce this to the robber. The robber can then run away, but is not allowed to run through a vertex that is in the d-neighbourhood of a vertex occupied by a cop, i.e. he can pick a new position u anywhere on the graph as long as there is a path from his current position to u that contains no vertex in $N_d[X]$.

After the robber has chosen his new position, the new cop is placed on v and the play continues. The cops win a play if they can capture the robber, i.e. if they can place a cop occupying or dominating the vertex occupied by the robber so that the robber is not able to escape. If the robber can escape forever, he wins.

d-domination games are a variant of the well-known cops and robber games used to characterise graph parameters such as tree-width or path-width (see e.g. [12]). The difference is that in a cops and robber game, a cop only occupies his current position but does not block the d-neighbourhood of this position.

We will distinguish between two variants of d-domination games, i.e. the visible and invisible variant. In the *visible* case, the cops can see the robber and can adapt their strategy accordingly. In the *invisible* case, the cops do not see the robber and hence have to search the graph independently of the robbers current position. In this case, we are essentially dealing with a one player game and in describing the game, we can discard the robber positions. In both cases, the aim of the cop player is to capture the robber using as few cops as possible. In this paper we primarily consider the invisible case and will therefore present the relevant notation and definitions in terms of the invisible domination game. We briefly comment on the visible case in Section 6.

In the invisible domination game, the cops have to capture the robber without being able to see him – and hence without being able to react to his actions. We can therefore represent any cop strategy on a graph G in the invisible d-domination game by a sequence $\mathcal{S} := (S_1, \ldots, S_n)$, where, for $1 \le i \le n$, $S_i \subseteq V(G)$ is the cop position after step i. With any strategy $\mathcal{S} := (S_1, \ldots, S_n)$ we associate the corresponding sequence $R_0, \ldots, R_n$ of *robber spaces* as follows: $R_0 := V(G)$ and for all

$i > 0$, $R_i := \{v \in V(G) \setminus N_d[S_i] :$ there is $u \in R_{i-1}$ and a path from u to v in $G \setminus N_d[S_{i-1} \cap S_i]\}$, where we take $S_0 := \emptyset$. Hence, R_i is the set of vertices available to the robber after i steps of the play. Vertices in $V(G) \setminus R_i$ are called *clear at stage i*.

Definition 3.1. *Let $\mathcal{S} := (S_1, \ldots, S_n)$ be a strategy and $(R_0, \ldots, R_n)$ be the corresponding robber spaces.*
1. *$\mathcal{S}$ is a winning strategy if it is finite and $R_n = \emptyset$.*
2. *The width $w(\mathcal{S})$ of $\mathcal{S}$ is defined as $w(\mathcal{S}) := \max\{|S_i| : 1 \leq i \leq n\}$.*
3. *The d-domination search number $ds_d(G) := \min\{w(\mathcal{S}) : \mathcal{S}$ is a winning strategy on $G\}$ of G is the minimal number of cops required to win the invisible d-domination game on G.*

Clearly, every graph of order n can be searched by n cops. Hence $ds_d(G)$ is well-defined. We next introduce a general construction that will be used frequently throughout the paper. As a first application of this we show that questions about complexity and monotonicity of d-domination games for $d > 1$ can be reduced to the corresponding questions for the case of $d = 1$.

For $k > 0$, let K_k be the k-clique, i.e. the complete graph on k vertices. Further, if X is a set, we write $K[X]$ for the complete graph with vertex set X. For each $k > 0$ and $d > 0$, we define S_k^d as the graph (up to isomorphism) obtained from K_k by subdividing each edge $2d$ times, i.e. replacing each edge by a path of length $2d + 1$. We call S_k^d a *d-subdivided k-clique*. Note that S_k^d contains more than k vertices but in the rest of the paper the vertices in the paths replacing edges will usually not play a role. We say that S is the *d-subdivided clique over* a set X if S is obtained from $K[X]$ by subdividing each edge $2d$ times. We write $S^d[X]$ for this graph and call X the *original vertices* of $S^d[X]$. As before, we omit the indices in case $d = 1$. The following lemma, whose proof is straightforward, will be used frequently in the sequel.

Lemma 3.2. *For all $k > 0$ and $d > 0$, $ds_d(S_k^d) = k$.*

For a graph G, $k > 0$ and a function $f : V(G) \to 2^{V(G)}$ we define the *subdivided k-clique graph* of G, denoted by $SC(G, k, f)$, to be the graph obtained from G by 1) replacing each vertex $v \in V(G)$ by a disjoint copy of S_k^1, denoted $SC(v)$, and 2) replacing each edge $\{u, v\} \in E(G)$ by a perfect matching between the original vertices in $SC(u)$ and the original vertices in $SC(v)$ and 3) for each $v \in V(G)$ we add a new vertex denoted $c(v)$ so that $\{c(v) : v \in V(G)\}$ induces a clique in $SC(G, k, f)$ and for each $v \in V(G)$, $SC(G, k, f)$ contains edges between $c(v)$ and all vertices in every $SC(u)$ for $u \in f(v)$.

Now it is easily seen that k cops have a winning strategy in the d-domination game on G if, and only if, k cops have a winning strategy in the 1-domination game on $SC(G, k, N_d^G[])$, where in addition they only play on the new extra vertices $c(v)$, for $v \in V(G)$. The same holds for monotone winning strategies as defined in Section 4 below. Here, $N_d^G[]$ denotes the function $f(v) := N_d^G[v]$. By setting $k := |V(G)|$ we obtain the following corollary.

Corollary 3.3. *Fix $d > 0$. There is a polynomial time algorithm which constructs for each graph G a graph G' such that for all $k > 0$, k cops win the d-domination game on G if, and only if, k cops win the 1-domination game on G'. The analogous statement holds for monotone winning strategies.*

The converse direction is also true. By subdividing each edge $2d$-times, we can construct for each graph G a graph G' so that k cops win the 1-domination game on G if, and only if, k cops win the d-domination game on G'. This construction follows essentially from [7] and also shows that the cops and robber game underlying tree-width can be reduced to the 1-domination game. It follows that all questions concerning monotonicity and complexity about d-domination games can be reduced to the case of $d = 1$. We will therefore only consider this case in the sequel. As described in the introduction, this case was already studied under the name of domination games by Fomin et al. [7]. We will therefore follow their terminology and refer to these games as domination games and write $ds(G)$ for the minimal number of cops required to win the domination game on a graph G.

4 Monotonicity of Domination Games

In this section we study monotone strategies of invisible domination games. In particular, we establish strong non-monotonicity results for common notions of monotonicity – *cop*- and *robber-monotonicity* – in showing that in general more cops are needed to catch a robber with a monotone strategy than with an unrestricted strategy and that the ratio between the monotone and the non-monotone case is unbounded. We then consider a third type of monotonicity specific to domination games.

Definition 4.1. *Let* $\mathcal{S} := (S_1, \ldots, S_n)$ *be a strategy and* $(R_0, \ldots, R_n)$ *be the corresponding robber spaces (see Section 3).*

1. $\mathcal{S}$ *is* robber-monotone, *if* $R_i \supseteq R_j$ *for all* $i < j$.
2. $\mathcal{S}$ *is* cop-monotone *if for all* $i < j < l$ *and all* $v \in V(G)$, *if* $v \in S_i \setminus S_j$ *then* $v \notin S_l$.
3. *The* cop-monotone domination search number *is defined as* $c\text{-}ds(G) := \min\{w(\mathcal{S}) : \mathcal{S} \text{ is a cop-monotone winning strategy on } G\}$. *The* robber-monotone domination search number $r\text{-}ds(G)$ *is defined analogously.*

In a non-monotone strategy, a vertex $v \in R_j \setminus R_i$, *for* $j > i$, *is called* recontaminated.

Note that, unlike cops and robber games, in domination games cop-monotone strategies might not be robber-monotone and vice versa. In [7], Stefan Dobrev exhibited an example where three cops can win the domination game but four cops are needed to search the graph using a monotone strategy. We now strengthen this result considerably by showing that the ratio between the (robber- or cop-) monotone and the non-monotone search numbers is unbounded.

Lemma 4.2. *For every* $k > 2$, *there is a graph* G_k *such that* $ds(G_k) = 2$ *but* $r\text{-}ds(G_k) = c\text{-}ds(G_k) = k$.

Proof. For $k \in \mathbb{N}$ we define G_k as follows. Let $U := \{u_1, \ldots, u_k\}$ be a set of size k. For all permutations ρ of $(1, \ldots, k)$ and all $1 \leq i \leq k$, let P_i^ρ be a subdivided clique on k vertices and let H_ρ be the graph obtained from the disjoint union $\dot\bigcup_i P_i^\rho$ of these subdivided cliques by adding edges forming a perfect matching of the original vertices in P_i^ρ and P_{i+1}^ρ, for $1 \leq i < k$. Then G_k is defined as $K[X] \dot\cup \bigcup_\rho H_\rho$ augmented by edges

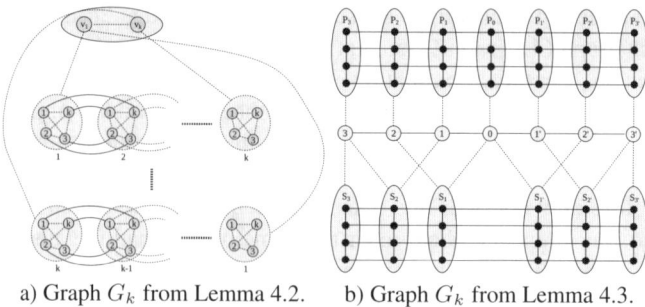

a) Graph G_k from Lemma 4.2. **b)** Graph G_k from Lemma 4.3.

Fig. 1. Examples for non-monotonicity in domination games

$\{\{v_i, v\} : v \in P^\rho_{\rho(i)}, 1 \leq i \leq k$ and ρ is a permutation of $(1, \ldots, k)\}$. The construction is illustrated in Figure 1 $a)$. Here, dashed lines represent edges from a vertex u_i to all vertices in a subdivided clique whereas solid lines represent actual edges.

It is easily seen that two cops can search G_k as follows: for each permutation ρ of $(1, \ldots, k)$ they play $\mathcal{S}_\rho := (\{u_{\rho(1)}, u_{\rho(2)}\}, \{u_\rho(2), u_\rho(3)\}, \ldots, \{u_{\rho(k-1)}, u_{\rho(k)}\})$, i.e. they search the "path" P_ρ by going through $u_1, \ldots, u_k$ using the ordering given by ρ. As the only connection between H_ρ and $H_{\rho'}$ is through the vertices in U and these form a clique, they can search the H_ρ independently.

It remains to show that $k - 1$ cops do not have a cop-monotone or a robber-monotone strategy on G_k. We can assume that the cops are only playing on the vertices in $u_1, \cdots, u_k$ as otherwise they need at least k cops to clear a subdivided k-clique.

Suppose the cops start by occupying all but one vertex u_i in U. Then in each H_ρ, the clique $P^\rho_{\rho(i)}$ is still contaminated. Furthermore, in the next step the cops have to remove a cop from a vertex u_j. But then, there is a permutation ρ such that $\rho(i)$ and $\rho(j)$ are consecutive numbers and thus in H_ρ the subdivided clique $P^\rho_{\rho(j)}$ becomes recontaminated. This shows that the strategy is not robber-monotone. As $P^\rho_{\rho(j)}$ can only be cleared again by playing on v_j the strategy for the cops can not be cop-monotone. This concludes the proof.

Considering again the example above exhibiting non-monotone strategies for the cops, the main source for non-monotonicity appears to be that while clearing some parts of the graph, the cops accidentally and unintentionally clear other parts of the graph also – which later on they have to allow to be recontaminated. For instance, in the example above, while clearing a sub-graph H_ρ they also clear parts of other sub-graphs $H_{\rho'}$ but in the wrong order. If we gave the cops the power to choose which vertices in the neighbourhood of a cop they really want to dominate, then they could easily search the graphs G_k with a robber- and cop-monotone strategy. We call this *selective monotonicity*. It seems conceivable, thus, that such *selective* strategies are always sufficient, i.e. whenever k cops can win in any form, they can do so with a selective monotone strategy. Such a result would be extremely interesting as it would imply a linear upper bound for the length of minimal winning strategies for the cop player. This hope is dashed, though, by the following theorem.

Theorem 4.3. *There exists a graph G with $ds(G) = 2$ but 3 cops are needed for any selective monotone winning strategy.*

Proof. The graph G is shown in Figure 1 b). Here, solid lines represent actual edges whereas a dashed line such as between 3 and S_3 indicates that there is an edge between 3 and every vertex in S_3.

Now, $ds(G) = 2$ as witnessed by the following two strategies: $\mathcal{S}_1 := (\{3, 2\}, \{2, 1\}, \{1, 0\}, \{0, 1'\}, \{1', 2'\}, \{2', 3'\})$ and $\mathcal{S}_2 := (\{3', 2'\}, \{2', 1'\}, \{1', 0\}, \{0, 1\}, \{1, 2\}, \{2, 3\})$. Note that both strategies are not robber monotone. For instance, in $\mathcal{S}_1$ the vertices in S_1 are recontaminated in the step from $\{2, 1\}$ to $\{1, 0\}$ and similarly in the symmetric strategy $\mathcal{S}_2$. Further, observe that in order for these strategies to work, at each step all neighbours of every vertex occupied by a cop need to be dominated. Hence, none of the two strategies can be turned into a selective monotone strategy.

We claim that there is no selective monotone strategy with only two cops. For the sake of contradiction let $\mathcal{S}$ be a selective monotone winning strategy with two cops using a minimal number of steps. We first show that $\mathcal{S}$ cannot use any vertex other than those in $X := \{3, 2, 1, 0, 1', 2', 3'\}$. For, if $v \in S_i$ or $v \in P_i$ is occupied by a cop then at the first step where this cop is lifted from v, v will be recontaminated unless it is dominated by the other cop. Hence, placing a cop on v either can be avoided, as v is dominated anyway, or it leads to non-monotonicity.

Thus, a selective monotone strategy with two cops essentially searches the path $3, 2, 1, 0, 1', 2', 3'$. However, it is easily seen that a path of length 7 can be searched in only two ways by two cops using a monotone strategy: left to right or right to left. If follows that the only possible strategies are $\mathcal{S}_1$ or $\mathcal{S}_2$ and neither is selective monotone. This yields the contradiction.

As argued above, an important aspect of monotonicity for a variant of graph searching games is that in this way a bound on the maximal number of steps in a strategy is obtained. As domination games are strongly non-monotone, no such bound can be achieved using this approach. In Corollary 5.3 below we show that there exist graphs such that the number of steps needed by a strategy in the domination game is exponential in the size of the graph and thus cannot be bounded bounded by a polynomial.

5 Complexity of Domination Games

In this section we study the complexity of deciding whether k cops have a (monotone) winning strategy in the domination game on a graph G. We measure the complexity of this problem in different ways – classically and in the context of parametrised complexity. Let DOMINATION SEARCH be the problem of deciding for a given graph G and $k \in \mathbb{N}$ whether k cops have a winning strategy on G. In [7], Fomin et al. study this problem and show that it is NP-hard.

Theorem 5.1 ([7]). DOMINATION SEARCH *is* NP-*hard.*

No upper bound for the complexity of the problem was given. We settle this problem by giving precise complexity bounds for DOMINATION SEARCH.

Theorem 5.2. DOMINATION SEARCH *is* PSPACE-*complete. More precisely, we show that even deciding whether two cops have a winning strategy on a graph is* PSPACE-*complete.*

In [7, Problem 7], Fomin et al. raise the question whether for every graph G there is a winning strategy of length $\mathcal{O}(n)$ using $ds(G)$ cops in the invisible domination search game. As a consequence of the proof of the previous theorem we answer this question negatively by showing that there exist graphs on which the number of steps needed by a strategy in the domination game is at least exponential in the size of the graph and thus can not be bounded bounded by a polynomial. Clearly, exponential length of strategies is also the worst possible.

Corollary 5.3. *There exists a family $\mathcal{C}$ of graphs such that two cops have a winning strategy in the invisible domination game on each $G \in \mathcal{C}$ but any such strategy is at least of exponential length, i.e. there is no polynomial $p(n)$ so that the length of these strategies is bounded by $p(|G|)$.*

We now consider the problem to decide for a given graph G whether k cops have a monotone winning strategy in the invisible domination game, where we consider cop- and selective-monotonicity. Clearly, as the length of monotone strategies is polynomially bounded in the size of the graph, these problems are necessarily in NP. We again give tight complexity bounds by showing that even deciding whether two (or three, respectively) cops have monotone winning strategies is NP-hard.

Theorem 5.4. *Let G be a graph. Deciding whether two cops have a cop-monotone winning strategy in the domination search game on G is NP-complete.*

Theorem 5.5. *Let G be a graph. Deciding whether three cops have a selective monotone winning strategy in the domination game on G is NP-complete.*

We do not know corresponding results for robber-monotone strategies and leave this as an open problem.

The previous results settle the classical complexity of the domination game problem. We now study the parametrised complexity of this problem. The parametrised domination search problem p-DOMINATION SEARCH is defined as the problem, given a graph G and $k \in \mathbb{N}$ as input, to decide if k cops have a winning strategy in the invisible domination game on G. We take k as the parameter. The problem is in the parametrised complexity class XP if it can be solved in time $|G|^{f(k)}$ for some computable function $f : \mathbb{N} \to \mathbb{N}$. It is *fixed-parameter tractable*, or in FPT, if it can be solved in time $f(k) \cdot |G|^c$, for some $c \in \mathbb{N}$ and computable $f : \mathbb{N} \to \mathbb{N}$. The following is an immediate consequence of Theorem 5.2, 5.4 and 5.5.

Corollary 5.6. p-DOMINATION SEARCH *is not in XP. This holds true even for the cop- or selective monotone version of the problem.*

The previous results establish fixed-parameter intractability for domination games. Hence, domination games are considerably more complex than standard cops and robber games, which are NP-complete and fixed-parameter tractable. The latter follows from the parametrised tractability of tree-width and path-width and the monotonicity of the games.

We now turn to special cases where tractability can be obtained. A natural choice of graph classes where the problem might be easier are classes of bounded tree- or path-width. One is tempted to think that fixed-parameter tractability of domination search on classes $\mathcal{C}$ of graphs of tree-width at most d could be established along the following lines: given $G \in \mathcal{C}$ and $k \in \mathbb{N}$, we first compute a tree-decomposition of G of width d and then use dynamic programming to decide whether there is a winning strategy of width k. This is the approach taken to show that the analogous questions for cops and robber games (visible and invisible) can be solved by linear time parametrised algorithms. Typically, one proceeds bottom-up along the tree-decomposition and for each node in the decomposition tree one computes a constant size data structure containing information about the sub-graph induced by the vertices in the sub-tree rooted at this node. For domination games, however, this approach fails as a vertex in a bag can be dominated by vertices not contained in this bag. The ways in which this happens can be rather complex and hence a constant size data structure seems difficult to obtain. It is still possible, though, that domination search is fpt on classes of bounded tree-width and we leave this for future work.

We are, however, able to obtain parametrised algorithms for classes of graphs of bounded degree (recall that the problem is already NP-hard on the class of graphs of degree at most 3).

Lemma 5.7. *For $d > 0$ let $\mathcal{C}_d$ be the class of graphs of maximum degree at most d. Then the problem, given $G \in \mathcal{C}_d$ and $k \in \mathbb{N}$, to decide whether k cops have a cop-monotone winning strategy on G is fixed-parameter tractable with parameter $d + k$.*

Furthermore, if k cops have a winning strategy on any $G \in \mathcal{C}_d$, then at most $dk + 1$ cops have a cop- and a selective-monotone winning strategy.

6 Games on Hypergraphs and Visible Robbers

In this section we briefly explore the relation between domination games and Robber and Marshal games on hypergraphs and comment on domination games with a visible robber.

Robber and Marshal games, with a visible robber, have been introduced in [9] as a game-theoretical approach to hypertree-width and have, since then, been studied intensively. Essentially, a Robber and Marshal game is a Cops and Robber game on a hypergraph where the robber occupies a vertex whereas each marshal (= cop) occupies a hyperedge and blocks all vertices contained in it.

We will show next that every hypergraph game can be translated into a domination game – in the visible and the invisible case. There is a small difference between the Robber and Marshal game we use here and the original robber and marshal game in [9]. In the original game the marshals *slide* along edges in the sense that if a marshal moves from hyperedge e to e' then the vertices in $e \cap e'$ remain blocked (an equivalent notion for domination games could easily be defined). Here, we consider the variant of Robber and Marshal games where only the vertices in edges on which a marshal remains are blocked. It is easy to see that both variants are within a constant factor of each other.

Lemma 6.1. *Let H be a hypergraph and $k \geq 1$ be an integer. Then there exists a graph H_{k+1}^{dom}, such that k marshalls have a (marshal-/robber-monotone) winning strategy in* the (visible) robber and marshals game on H, if and only if, k cops have a (cop-/robber-monotone) winning strategy in the (visible) domination game on H_{k+1}^{dom} and H_{k+1}^{dom} can be constructed from H in polynomial time.

The lemma allows us to translate Robber and Marshal games to domination games. It follows immediately from Lemma 6.2 below that there is no translation in the converse direction.

So far, we have primarily considered domination games with an invisible robber. Here, we briefly summarise our knowledge of the visible case. Clearly, notions such as monotonicity and the domination search number translate easily.

In [1], Adler showed that the visible robber and marshall game mentioned above is not robber-monotone. Together with Lemma 6.1, this implies that the visible domination game is also not robber-monotone. However, the robber-monotone and non-monotone variant of the visible robber and marshall Game are within a constant factor of each other (see [2]). We show next that no such bound can be obtained for domination games.

Lemma 6.2. *For every $k > 2$, there is a graph G_k such that 2 cops have a non-monotone but k cops are needed for a robber-monotone winning strategy in the visible domination game on G_k.*

Finally, we consider the complexity of visible domination games. In terms of classical complexity, we can show the following.

Theorem 6.3. *Let G be a graph. Deciding whether three cops have a selective monotone winning strategy in the visible domination game on G is NP-complete.*

It is easily seen that all visible game variants except for the selective monotone variant are in XP, as the current cop and robber position completely determine the current state of the play and there are only $n^{\mathcal{O}(k)}$ such positions. We show next that the problem is not in FPT unless FPT=W[2].

As observed in [7], domination search is closely related to dominating sets in graphs. A *dominating set* of a graph G is a set X such that for all $v \in V(G)$ either $v \in X$ or there is a $u \in X$ such that $\{u, v\} \in E(G)$. The *domination number* of G, denoted by $\gamma(G)$, is the minimal size of a dominating set of G.

Lemma 6.4 ([7]). *Let G be a graph and H be the graph obtained from G by connecting every pair of non-adjacent vertices in G by a path of length three. Then $\gamma(G) \leq \mathrm{ds}(H) \leq \gamma(G) + 1$.*

We establish a similar but exact correspondence using a slightly different construction.

Theorem 6.5. *For all graphs G, there exists a graph G' such that $\gamma(G) + 1 = \mathrm{ds}(G')$ and G' is constructable in polynomial time.*

The theorem immediately gives a parametrised reduction from the dominating set problem, parametrised by the size of the solution, to the domination search problem, parametrised by the number k of cops. The following result follows from the W[2]-hardness of the dominating set problem, where W[2] is a parametrised complexity class strongly believed to be different from FPT.

Theorem 6.6. *The problem* p-DOMINATION SEARCH*: "given a graph G and* $k \in \mathbb{N}$*, with parameter* k*, decide whether* k *cops have a winning strategy in the (in-)visible domination game on G" is* $W[2]$*-hard.*

However, Lemma 5.7 also applies to the visible case and thus calculating the visible domination search number for graphs of bounded degree is fixed-parameter tractable.

Acknowledgements. We are grateful to Fedor Fomin for bringing domination games to our attention and thereby stimulating the research reported in this paper and to Fedor Fomin, Paul Hunter and Paul Dorbec for valuable discussions on the subject.

References

1. Adler, I.: Marshals, monotone marshals, and hypertree-width. JGT 47(4), 275–296 (2004)
2. Adler, I., Gottlob, G., Grohe, M.: Hypertree width and related hypergraph invariants. Journal of Combinatorics 28(8), 2167–2181 (2007)
3. Bodlaender, H.: A partial k-aboretum of graphs with bounded tree-width. TCS 209, 1–45 (1998)
4. Bodlaender, H.L.: Treewidth: Algorithmic techniques and results. In: Privara, I., Ružička, P. (eds.) MFCS 1997. LNCS, vol. 1295, pp. 19–36. Springer, Heidelberg (1997)
5. Diestel, R.: Graph Theory, 3rd edn. Springer, Heidelberg (2005)
6. Fedor, F., Thilikos, D.M.: An annotated bibliography on guaranteed graph searching. TCS 399(3), 236–245 (2008)
7. Fomin, F., Kratsch, D., Müller, H.: On the domination search number. Discrete Applied Mathematics 127(3), 565–580 (2003)
8. Franklin, M.K., Galil, Z., Yung, M.: Eavesdropping games: a graph-theoretic approach to privacy in distributed systems. Journal of the ACM 47(2), 225–243 (2000)
9. Gottlob, G., Leone, N., Scarcello, F.: Robbers, marshals, and guards: game theoretic and logical characterizations of hypertree width. Journal of Computer and Systems Science 66(4), 775–808 (2003)
10. Kirousis, L.M., Papadimitriou, C.H.: Searching and pebbling. TCS 47(3), 205–218 (1986)
11. Kloks, T., Kratsch, D., Müller, H.: On the structure of graphs with bounded asteroidal number. Graphs and Combinatorics 17(2), 295–306 (2001)
12. Seymour, P.D., Thomas, R.: Graph searching and a min-max theorem for tree-width. Journal of Combinatorial Theory, Series B 58(1), 22–33 (1993)

Cycles, Paths, Connectivity and Diameter in Distance Graphs

Lucia Draque Penso[1], Dieter Rautenbach[2], and Jayme Luiz Szwarcfiter[3]

[1] Institut für Theoretische Informatik, Technische Universität Ilmenau, Postfach 100565, D-98684 Ilmenau, Germany
lucia.penso@tu-ilmenau.de
[2] Institut für Mathematik, Technische Universität Ilmenau, Postfach 100565, D-98684 Ilmenau, Germany
dieter.rautenbach@tu-ilmenau.de
[3] Universidade Federal do Rio de Janeiro, Instituto de Matemática, NCE and COPPE, Caixa Postal 2324, 20001-970 Rio de Janeiro, RJ, Brasil
jayme@nce.ufrj.br

Abstract. For $n \in \mathbb{N}$ and $D \subseteq \mathbb{N}$, the distance graph P_n^D has vertex set $\{0, 1, \ldots, n-1\}$ and edge set $\{ij \mid 0 \leq i, j \leq n-1, |j-i| \in D\}$. The class of distance graphs generalizes the important and very well-studied class of circulant graphs which have been proposed for numerous applications concerning networks, distributed systems and chip design.

We prove that the class of circulant graphs coincides with the class of regular distance graphs. Extending some of the fundamental results concerning circulant graphs, we study the existence of long cycles and paths in distance graphs and analyse the computational complexity of problems related to their connectivity and diameter.

Keywords: Circulant graph; distance graph; multiple loop networks; connectivity; diameter; Hamiltonian cycle; Hamiltonian path.

1 Introduction

Circulant graphs form an important and very well-studied class of graph [1, 12, 13, 16, 18]. They are Cayley graphs of cyclic groups and have been proposed for numerous applications such as local area computer networks, large area communication networks, parallel processing architectures, distributed computing, and VLSI design. Their connectivity and diameter [2, 1, 12, 13], cycle and path structure, and further graph-theoretical properties [3, 11, 22] have been studied in great detail. Polynomial time algorithms for isomorphism testing and recognition of circulant graphs have been long-standing open problems which were completely solved only recently [5, 8, 17].

For $n \in \mathbb{N}$ and $D \subseteq \mathbb{N}$, the *circulant graph* C_n^D has vertex set $[0, n-1] = \{0, 1, \ldots, n-1\}$ and the neighbourhood $N_{C_n^D}(i)$ of a vertex $i \in [0, n-1]$ in C_n^D is given by

$$N_{C_n^D}(i) = \{(i+d) \bmod n \mid d \in D\} \cup \{(i-d) \bmod n \mid d \in D\}.$$

Clearly, we may assume $\max(D) \leq \frac{n}{2}$ for every circulant graph C_n^D.

Our goal here is to extend some of the fundamental results concerning circulant graphs to the similarly defined yet more general class of distance graphs: For $n \in \mathbb{N}$ and $D \subseteq \mathbb{N}$, the *distance graph* P_n^D has vertex set $[0, n-1]$ and

$$N_{P_n^D}(i) = \{i + d \mid d \in D \text{ and } (i + d) \in [0, n-1]\}$$
$$\cup \{i - d \mid d \in D \text{ and } (i - d) \in [0, n-1]\}$$

for all $i \in [0, n-1]$. Clearly, we may assume $\max(D) \leq n - 1$ for every distance graph P_n^D.

Every distance graph P_n^D is an induced subgraph of the circulant graph $C_{n+\max(D)}^D$. More specifically, distance graphs are the subgraphs of sufficiently large circulant graphs induced by sets of consecutive vertices. Conversely, the following simple observation shows that every circulant graph is in fact a distance graph.

Proposition 1. *A graph is a circulant graph if and only if it is a regular distance graph.*

Proof: Clearly, every circulant graph C_n^D is regular and isomorphic to the distance graph $P_n^{D'}$ for $D' = D \cup \{n - d \mid d \in D\}$.

Now let P_n^D be a regular distance graph. Let $D = \{d_1, d_2, \ldots, d_k\}$ with $d_1 < d_2 < \ldots < d_k \leq n - 1$. Since the vertex 0 has exactly k neighbours D, P_n^D is k-regular.

Let $i \in [1, k]$. The vertex $d_i - 1$ has exactly $i - 1$ neighbours j with $j < d_i - 1$. Hence $d_i - 1$ has exactly $k + 1 - i$ neighbours j with $j > d_i - 1$ which implies $(d_i - 1) + d_{k+1-i} \leq n - 1$. The vertex d_i has exactly i neighbours j with $j < d_i$. Hence d_i has exactly $k - i$ neighbours j with $j > d_i$ which implies $d_i + d_{k+1-i} > n - 1$.

We obtain $d_i + d_{k+1-i} = n$ for every $i \in [1, k]$ which immediately implies that P_n^D is isomorphic to the circulant graph $C_n^{D'}$ for $D' = \{d \in D \mid d \leq \frac{n}{2}\}$. □

Distance graphs lack the symmetry and algebraic interpretation of circulant graphs and the algebraic methods used in [8, 17] will not apply to them. In view of Proposition 1, the recognition of distance graphs will be at least as difficult as the recognition of circulant graphs. Originally motivated by coloring problems for infinite distance graphs studied by Eggleton, Erdős, and Skilton [7], most research on distance graphs focused on colorings (cf. eg. [6, 14, 21]).

One of the most fundamental results for circulant graphs is the following beautiful equivalence.

Theorem 1 (Boesch and Tindell [2], Burkard and Sandholzer [4], Garfinkel [9]). *For $n \in \mathbb{N}$ and a finite set $D \subseteq \mathbb{N}$, the following statements are equivalent.*

(i) C_n^D *is connected.*
(ii) The greatest common divisor $\gcd(\{n\} \cup D)$ *of the integers in* $\{n\} \cup D$ *equals* 1.

(iii) C_n^D *has a Hamiltonian cycle.*

Our contributions split into two parts. In the first part, we establish an analogue of the above equivalence for distance graphs. While connectivity and hamiltonicity of circulants are equivalent to a simple necessary gcd-condition, we prove that a similar condition for distance graphs only ensures the existence of a large component and a long cycle. We also discuss consequences and variants of our result. In the second part, we study the complexity of various connectivity and distance problems for distance graphs. Whereas deciding the connectivity of circulants only requires a simple gcd-computation, several related problems become hard for distance graphs.

2 Cycles and Paths in Distance Graphs

We immediately proceed to our main result in this section. The residue of an integer $n \in \mathbb{Z}$ modulo $d \in \mathbb{N}$ will be denoted by $n \bmod d$.

Theorem 2. *For a finite set $D \subseteq \mathbb{N}$, the following statements are equivalent.*

(i) There is a constant $c_1(D)$ such that for every $n \in \mathbb{N}$, the distance graph P_n^D has a component of order at least $n - c_1(D)$.
(ii) $\gcd(D) = 1$.
(iii) There is a constant $c_2(D)$ such that for every $n \in \mathbb{N}$, the distance graph P_n^D has a cycle of order at least $n - c_2(D)$.

Proof (i) $\Rightarrow$ (ii): Let n be such that n is even and $n > 2c_1(D)$. By (i), more than half the vertices are in the same component of P_n^D. By the pigeonhole principle, there is some $i \in [0, n-2]$ such that the two vertices i and $i+1$ are in the same component of P_n^D. This implies that there is a path in P_n^D from i to $i+1$. Hence 1 is an integral linear combination of the elements in D. It is a well-known consequence of the Euclidean algorithm that this is equivalent to (ii).

(ii) $\Rightarrow$ (iii): The essential idea in order to obtain a cycle which contains almost all vertices of P_n^D is to use increasing and decreasing paths which only use edges uv such that $v - u$ is one fixed element d^* of D. Because the vertices on these paths always remain in the same residue class modulo d^*, such paths can be overlayed without intersecting. In order to connect these paths to a cycle, we use short paths which are close to 0 or $n-1$ and whose end vertices are in different residue classes modulo d^*. In this way the cycle can collect all vertices of P_n^D in some middle section and only misses vertices close to 0 or $n-1$ in terms of D (cf. Figure 1).

Due to space limitations, we will only discuss the case that $\max(D)$ is even in detail.

In view of the constant $c_2(D)$ in (iii), we may tacitly assume in the following that n is sufficiently large in terms of D.

Let $d_{\max} = \max(D)$ and $D^- = D \setminus \{d_{\max}\}$. Since $\gcd(D) = 1$, 1 is an integral linear combination of the elements of D. Hence there are integers n_d

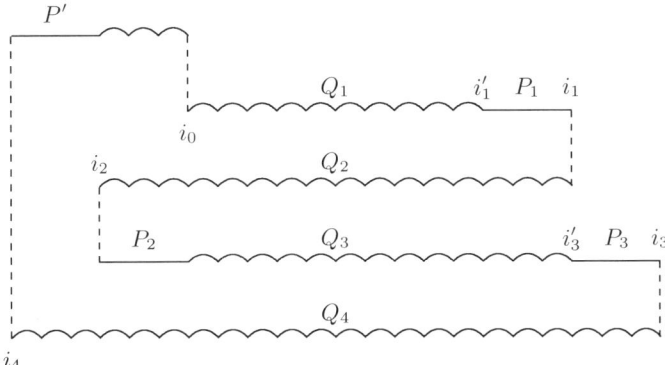

Fig. 1.

for $d \in D^-$ such that $1 = \left(\sum_{d \in D^-} n_d d \right) \bmod d_{\max}$. This implies the existence of a path $P : v_0 v_1 \ldots v_k$ in P_n^D such that $v_i - v_{i-1} \in D^-$ for all $i \in [1, k]$ and $\{v_i \bmod d_{\max} \mid i \in [0, k]\} = [0, d_{\max} - 1]$, i.e. P is a monotonously increasing path which only uses edges uv with $v - u \in D^-$ and contains a vertex from every residue class modulo $d_{\max}$.

We assume that P is chosen so as to be shortest possible. This implies that the residues modulo $d_{\max}$ of the vertices v_0 and v_k appear exactly once on P. Let $r_1, r_2, \ldots, r_{d_{\max}}$ denote the residues modulo $d_{\max}$ in the order in which they appear for the first time when traversing P from v_0 to v_l. Clearly, $r_1 = v_0 \bmod d_{\max}$, $r_2 = v_1 \bmod d_{\max}$, and $r_{d_{\max}} = v_k \bmod d_{\max}$. Furthermore, P is the concatenation of $(d_{\max} - 1)$ edge-disjoint paths $P = P_1 P_2 \ldots P_{d_{\max}-1}$ such that for $i \in [1, d_{\max} - 1]$, the path P_i begins at the smallest vertex v_j on P with $r_i = v_j \bmod d_{\max}$ and ends at the smallest vertex $v_{j'}$ on P with $r_{i+1} = v_{j'} \bmod d_{\max}$; let $l_i = v_{j'} - v_j$ for these indices. By the choice of P, for $i \in [1, d_{\max} - 1]$, all internal vertices of P_i have residues modulo $d_{\max}$ in $\{r_j \mid 2 \leq j \leq i - 1\}$. Let $l = l_1 + l_2 + \ldots + l_{d_{\max}-1}$. Note that $l = v_k - v_0$.

We now describe a long cycle C in P_n^D. The general structure of C is illustrated in Figure 1. For simplicity we will first assume that $d_{\max}$ is even.

Let i_0 be the smallest integer at least $(d_{\max} + l_2) + (d_{\max} + l_4) + \ldots + (d_{\max} + l_{d_{\max}-2}) + (d_{\max} + l)$ of residue r_1 modulo $d_{\max}$. Furthermore, let i'_1 be the largest integer at most $(n - 1) - l_1 - (d_{\max} + l_3) - (d_{\max} + l_5) - \ldots - (d_{\max} + l_{d_{\max}-1})$ of residue r_1 modulo $d_{\max}$. (Note that i_0 has to be chosen large enough in order to fit the paths $P_2, P_4, \ldots, P_{d_{\max}-2}$ and a path P' of length at most l - all starting at vertices with a specified residue - within $[0, i_0]$ as illustrated in Figure 1. A similar reasoning motivates the choice of i'_1.)

We start C with an increasing path Q_1 only using edges uv with $v - u = d_{\max}$ which begins at i_0 and ends at i'_1. We continue C with P_1 shifted by a multiple of $d_{\max}$ such that it begins at i'_1 and ends at a vertex i_1.

For j from 1 up to $\frac{d_{\max}}{2} - 1$, we proceed as follows: We assume that we have already constructed C until the end of a shifted path P_{2j-1} which ends at a vertex i_{2j-1}. We continue C with a decreasing path Q_{2j} which only uses edges uv with $v - u = d_{\max}$ and ends at the largest integer i_{2j} at most $i_{2j-2} - l_{2j}$ with residue r_{2j} modulo $d_{\max}$. We continue C with P_{2j} shifted by a multiple of $d_{\max}$ such that it begins at i_{2j}. We continue C with an increasing path Q_{2j+1} which only uses edges uv with $v - u = d_{\max}$ and ends at the smallest integer i'_{2j+1} at least i_{2j-1} with residue r_{2j+1} modulo $d_{\max}$. We continue C with P_{2j+1} shifted by a multiple of $d_{\max}$ such that it begins at i'_{2j+1} and ends at a vertex i_{2j+1}. At this point, we increase j until it reaches $\frac{d_{\max}}{2} - 1$.

To complete C, we may assume now that we have already constructed C until the end of the shifted path $P_{d_{\max}-1}$ which ends at a vertex $i_{d_{\max}-1}$. We continue C with a decreasing path $Q_{d_{\max}}$ which only uses edges uv with $v - u = d_{\max}$ and ends at the largest integer $i_{d_{\max}}$ at most $i_{d_{\max}-2} - l$ with residue $r_{d_{\max}}$ modulo $d_{\max}$.

Let $P' : u_0 u_1 \ldots u_{k'}$ be a path in P_n^D such that $u_0 = i_{d_{\max}}$, $u_i - u_{i-1} \in D^-$ for all $i \in [1, k']$, $r_1 = u_{k'} \bmod d_{\max}$, and $l' = u_{k'} - u_0$ is minimum possible. Clearly, $l' \leq l$. Furthermore, no internal vertex of P' has residue $r_{\max}$ modulo $d_{\max}$. We continue C with P'. Finally, we complete C with an increasing path which only uses edges uv with $v - u = d_{\max}$, begins at $u_{k'}$ and ends at i_0.

At this point we have completely described C as the concatenation of paths. Clearly, the choices of i_0 and i'_1 imply that C never leaves $[0, n-1]$, i.e. C is in fact a closed walk within P_n^D. In order to show that C is a cycle, it remains to prove that it visits no vertex twice. This follows easily from the facts that

- the vertices on Q_i all have residue r_i modulo $d_{\max}$ for all $i \in [1, d_{\max}]$,
- the end vertices of the shifted paths P_i are the first vertices on C - traversed as constructed above - which have residue r_{i+1} modulo $d_{\max}$ for all $i \in [1, d_{\max} - 1]$,
- all internal vertices of P_i have residues modulo $d_{\max}$ in $\{r_j \mid 2 \leq j \leq i - 1\}$ for all $i \in [1, d_{\max} - 1]$, and
- no internal vertex of P' has residue $r_{\max}$ modulo $d_{\max}$.

Since C contains all vertices between i_0 and i'_1, it misses at most $2d_{\max} + l_1 + (d_{\max} + l_2) + \ldots + (d_{\max} + l_{d_{\max}-1}) + (d_{\max} + l) = d_{\max}(d_{\max} + 1) + 2l$ many vertices of P_n^D. Since this expression is bounded in terms of D, the proof of (iii) in the case that $d_{\max}$ is even is complete.

In the case that $d_{\max}$ is odd a very similar construction with a small modification yields a long cycle.

(iii) $\Rightarrow$ (i): Since this implication is trivial, the proof is complete. □

We add some comments concerning Theorem 2.

It is easy to see that a distance graph P_n^D with $\gcd(D) = 1$ and $n \geq 2\max(D) + 1$ is actually connected. Therefore, the constant $c_1(D)$ in (i) of Theorem 2 can be eliminated for sufficiently large n. For (iii) in Theorem 2, a similar change is not possible, i.e. no lower bound on the order n would imply that P_n^D has a Hamiltonian cycle. If n as well as all elements of D are odd for instance, then P_n^D is bipartite and every cycle misses at least one vertex. In this sense, Theorem 2 is best-possible.

As observed in the proof of Theorem 2, there are integers n_d for $d \in D^-$ with $1 = \left(\sum_{d \in D^-} (n_d \bmod d_{\max}) d \right) \bmod d_{\max}$. Repeating the path corresponding to the last sum $d_{\max} - 1$ times yields a monotonously increasing path which contains vertices of all residues modulo $d_{\max}$. This implies $l \leq (d_{\max} - 1) \sum_{d \in D^-} (n_d \bmod d_{\max}) d \leq O(d_{\max}^3 |D|)$ for the value l considered in the proof of Theorem 2 and yields the estimate $c_2(D) = O(\max(D)^2 + l) = O(\max(D)^3 |D|)$.

Theorem 2 (iii) trivially implies the existence of a path of order at least $n - c_2(D)$ which traverses $[0, n-1]$ several times back and forth just like the cycle does. We believe that there is also always a path containing almost all vertices of P_n^D which is *essentially monotonic*, i.e. it traverses $[0, n-1]$ once. The following conjecture makes this precise.

Conjecture 1. *For a finite set $D \subseteq \mathbb{N}$, the following statements are equivalent.*

(i) $\gcd(D) = 1$.
(ii) There are two constants $c_3(D)$ and $c_4(D)$ such that for every $n \in \mathbb{N}$, the distance graph P_n^D has a path $u_0 u_1 \ldots u_l$ of order at least $n - c_3(D)$ such that $u_j > u_i$ for all $0 \leq i, j \leq l$ with $j - i \geq c_4(D)$.

A simple modification of the construction used in the proof of Theorem 2 implies the following weak version of Conjecture 1.

Theorem 3. *If $D \subseteq \mathbb{N}$ is a finite set with $\gcd(D) = 1$ and $\epsilon > 0$, then there are constants $c_5(D, \epsilon)$ and $c_6(D, \epsilon)$ such that for every $n \in \mathbb{N}$, the distance graph P_n^D has a path $u_0 u_1 \ldots u_l$ of order at least $(1 - \epsilon)n - c_5(D, \epsilon)$ such that $u_j > u_i$ for all $0 \leq i, j \leq l$ with $j - i \geq c_6(D, \epsilon)$.*

Note that Conjecture 1 is trivial, if D contains only one element. If D contains exactly two elements, then Conjecture 1 easily follows from the following result.

Proposition 2. *If $d_1, d_2 \in \mathbb{N}$ are such that $d_1 > d_2$ and $\gcd(\{d_1, d_2\}) = 1$, then $P_{d_1+d_2+1}^{\{d_1,d_2\}}$ has a Hamiltonian path which begins at 0 and ends at $d_1 + d_2$.*

3 Connectivity and Diameter in Distance Graphs

The most fundamental connectivity problem for distance graphs is the following.

CONNECTIVITY OF P_n^D
Instance: $n \in \mathbb{N}$ and $D \subseteq \mathbb{N}$.
Question: Is P_n^D connected?

We have not been able to determine the complexity of CONNECTIVITY OF P_n^D and pose the following conjecture.

Conjecture 2. CONNECTIVITY OF P_n^D *is NP-hard.*

Clearly, P_n^D is connected if and only if for every $i \in [0, n-2]$, there is a path in P_n^D from i to $i+1$. Equivalently, for every $i \in [0, n-2]$, there are integers $x_1, x_2, \ldots, x_l$ such that

$$|x_i| \in D \text{ for all } i \in [1, l], \tag{1}$$

$$1 = \sum_{j=1}^{l} x_j, \text{ and} \tag{2}$$

$$i + \sum_{j=1}^{k} x_j \in [0, n-1] \text{ for all } k \in [0, l]. \tag{3}$$

It is a well-known consequence of the Euclidean Algorithm that 1 is an integral linear combination of the elements of D if and only if $\gcd(D) = 1$. Hence the existence of integers x_i which satisfy (1) and (2) can be decided in polynomial time. Unfortunately, these integers are by far not unique. Furthermore, given integers x_i which satisfy (1) and (2), deciding the existence of an ordering of them which satisfies (3) is in general a hard problem as we show next.

BOUNDED PARTIAL SUMS
Instance: $x_0, x_1, x_2, \ldots, x_l \in \mathbb{Z}$ and $n \in \mathbb{N}$.
Question: Is there a permutation $\pi \in S_l$ such that

$$x_0 + \sum_{j=1}^{k} x_{\pi(j)} \in [0, n-1] \text{ for all } k \in [0, l]?$$

Proposition 3. BOUNDED PARTIAL SUMS *is NP-complete.*

Proof: Clearly, BOUNDED PARTIAL SUMS is in NP. We will reduce the classical NP-complete problem PARTITION to BOUNDED PARTIAL SUMS. In order to relate to the preceding discussion we will reduce to instances of BOUNDED PARTIAL SUMS which satisfy (2). Let $x_1, x_2, \ldots, x_{l-2} \in \mathbb{N}$ be an instance of PARTITION. Let $X = \sum_{i=1}^{l-2} x_i$.

Let $x_0 = 0$, $x_{l-1} = -X$, $x_l = -X + 1$, and $n = X + 1$. It is easy to see that the instance $x_1, x_2, \ldots, x_{l-2}$ of PARTITION is "yes"-instance if and only if the instance of BOUNDED PARTIAL SUMS defined by $x_0, 2x_1, 2x_2, \ldots, 2x_{l-2}, x_{l-1}, x_l$ and n is a "yes"-instance. This completes the proof. □

Clearly, if $|D| = 1$, then P_n^D is connected if and only if $D = \{1\}$. Already for $|D| = 2$, the following characterization of the pairs (n, D) for which P_n^D is connected is not simple.

Theorem 4. *Let $n, d_1, d_2 \in \mathbb{N}$ be such that $d_1 < d_2$. For $i \in [0, d_1 - 1]$, let $r_i = (id_2) \bmod d_1$ and $s_i = (n - 1 - r_i) \bmod d_1$. Furthermore, for $i^* \in [1, d_1 - 1]$, let*

$$d_{i^*}^+ = \max\{r_i \mid i \in [0, i^* - 1]\} \text{ and}$$

$$d_{i^*}^- = \max\{s_{-i \bmod d_1} \mid i \in [0, d_1 - i^* - 1]\}.$$

Finally, let $d^* = \max_{i^* \in [1, d_1 - 1]} \min\{d_{i^*}^+, d_{i^*}^-\}$.
$P_n^{\{d_1, d_2\}}$ *is connected if and only if* $\gcd(\{d_1, d_2\}) = 1$ *and* $d^* + d_2 \leq n - 1$.

While deciding connectivity is easy for circulant graphs, the exact calculation and minimization of the diameter of C_n^D are difficult and well-studied problems even for the case $|D| = 2$ [1,12,13]. Many of the general upper and lower bounds on the diameter of circulant graphs easily generalize to distance graphs. The arguments used by Wong and Coppersmith [23] to obtain their classical estimates (cf. Theorems 4.6 and 4.7 in [12]) imply

$$\operatorname{diam}\left(P_n^D\right) \geq \frac{1}{2}(|D|!n)^{\frac{1}{|D|}} - |D| \text{ and}$$
$$\operatorname{diam}\left(P_{d^k}^{\{1,d,\dots,d^{k-1}\}}\right) \leq k(d-1).$$

For our final hardness result, we consider the following decision problem which closely relates to the diameter of distance graphs.

SHORT PATH IN P_n^D
Instance: $n \in \mathbb{N}$, $D \subseteq \mathbb{N}$ and $l \in \mathbb{N}$.
Question: Is there some $u \in [0, n-2]$ such that P_n^D contains a path of length at most l between u and $u + 1$?

The hardness of BOUNDED PARTIAL SUMS implies that an encoding of a certificate for a "yes"-instance of SHORT PATH IN P_n^D which can be checked in polynomial time would most likely have to use at least $\Omega(l)$ bits which would not be polynomially bounded in the encoding length of the triple (n, D, l).

The construction used in the proof of the following result is inspired by van Emde Boas's proof [20] that WEAK PARTITION is NP-complete.

Theorem 5. SHORT PATH IN P_n^D *is NP-hard.*

As a final remark, we note that the existence of a monotonic path between two vertices of P_n^D is equivalent to the feasibility of an integer linear program in $|D|$ dimensions which can be decided in polynomial time for bounded $|D|$ [15].

References

1. Bermond, J.-C., Comellas, F., Hsu, D.F.: Distributed Loop Computer Networks: A Survey. J. of Parallel and Distributed Computing 24, 2–10 (1985)
2. Boesch, F., Tindell, R.: Circulants and their connectivities. J. Graph Theory 8, 487–499 (1984)
3. Brimkov, V.E., Barneva, R.P., Klette, R., Straight, J.: Efficient computation of the lovász theta function for a class of circulant graphs. In: Hromkovič, J., Nagl, M., Westfechtel, B. (eds.) WG 2004. LNCS, vol. 3353, pp. 285–295. Springer, Heidelberg (2004)
4. Burkard, R.E., Sandholzer, W.: Efficiently solvable special cases of bottleneck travelling salesman problems. Discrete Appl. Math. 32, 61–76 (1991)

5. Cohen, J., Fraigniaud, P., Gavoille, C.: Recognizing Bipartite Incident-Graphs of Circulant Digraphs. In: Widmayer, P., Neyer, G., Eidenbenz, S. (eds.) WG 1999. LNCS, vol. 1665, pp. 215–227. Springer, Heidelberg (1999)
6. Deuber, W., Zhu, X.: The chromatic number of distance graphs. Discrete Math. 165/166, 195–204 (1997)
7. Eggleton, R.B., Erdős, P., Skilton, D.K.: Coloring the real line. J. Combin. Theory Ser. B 39, 86–100 (1985)
8. Evdokimov, S.A., Ponomarenko, I.N.: Circulant graphs: recognizing and isomorphism testing in polynomial time (English. Russian original). St. Petersbg. Math. J. 15, 813–835 (2004); translation from Algebra Anal. 15, 1–34 (2003)
9. Garfinkel, R.S.: Minimizing wallpaper waste, part I: A class of traveling salesman problems. Oper. Res. 25, 741–751 (1977)
10. Gauyacq, G., Micheneau, C., Raspaud, A.: Routing in Recursive Circulant Graphs: Edge Forwarding Index and Hamiltonian Decomposition. In: Hromkovič, J., Sýkora, O. (eds.) WG 1998. LNCS, vol. 1517, pp. 227–241. Springer, Heidelberg (1998)
11. Golin, M.J., Leung, Y.-C.: Unhooking Circulant Graphs: A Combinatorial Method for Counting Spanning Trees and Other Parameters. In: Hromkovič, J., Nagl, M., Westfechtel, B. (eds.) WG 2004. LNCS, vol. 3353, pp. 296–307. Springer, Heidelberg (2004)
12. Hwang, F.K.: A complementary survey on double-loop networks. Theoret. Comput. Sci. 263, 211–229 (2001)
13. Hwang, F.K.: A survey on multi-loop networks. Theoret. Comput. Sci. 299, 107–121 (2003)
14. Kemnitz, A., Marangio, M.: Colorings and list colorings of integer distance graphs. Congr. Numerantium 151, 75–84 (2001)
15. Lenstra, H.W.: Integer programming with a fixed number of variables. Mathematics of Operations Research 8, 538–548 (1983)
16. Liu, M.T.: Distributed Loop Computer Networks. Advances in Computers, vol. 17, pp. 163–221. Academic Press, New York (1981)
17. Muzychuk, M.: A solution of the isomorphism problem for circulant graphs. Proc. Lond. Math. Soc., III. Ser. 88, 1–41 (2004)
18. Raghavendra, C.S., Sylvester, J.A.: A survey of multi-connected loop topologies for local computer networks. Comput. Network ISDN Syst. 11, 29–42 (1986)
19. van Dorne, E.A.: Connectivity of circulant graphs. J. Graph Theory 10, 9–14 (1986)
20. van Emde Boas, P.: Another NP-complete partition problem and the complexity of computing short vectors in a lattice, Technical Report 81-04, Mathematisch Instituut, Amsterdam, Netherlands (1981)
21. Voigt, M.: Colouring of distance graphs. Ars Combin. 52, 3–12 (1999)
22. Wanka, R.: Any Load-Balancing Regimen for Evolving Tree Computations on Circulant Graphs Is Asymptotically Optimal. In: Kučera, L. (ed.) WG 2002. LNCS, vol. 2573, pp. 413–420. Springer, Heidelberg (2002)
23. Wong, C.K., Coppersmith, D.: A combinatorial problem related to multimode memory organizations. J. ACM 21, 392–402 (1974)

Smallest Odd Holes in Claw-Free Graphs (Extended Abstract)

Shimon Shrem[1], Michal Stern[2], and Martin Charles Golumbic[1]

[1] Caesarea Rothschild Institute and Department of Computer Science, University of Haifa
[2] The Academic College of Tel-Aviv Yaffo

Abstract. In this paper, we give general structure properties of a smallest odd hole in a claw-free graph that lead to a polynomial time algorithm. The algorithm is based on a modified BFS we call Γ-BFS. For a graph G with n vertices and m edges, the time complexity of the algorithm is $O(nm^2)$. The algorithm is very easy to implement. We conclude the paper with a suggestion for an extension of our approach in order to detect an odd hole in a general graph.

Keywords: claw-free graphs, triangle-free graphs, odd holes, holes, polynomial time algorithms.

1 Introduction

Throughout this paper we consider simple finite undirected graphs with n vertices and m edges, no loops and no multiple edges.

Let G be a graph and let $v_0, v_1, \ldots, v_{k-1}$ be a sequence of k distinct vertices such that there is an edge from v_i to $v_{(i+1) \bmod k}$ (for all $i = 0, \ldots, k-1$), and no other edge between any two of these vertices; we say that this is a *chordless cycle* on k vertices. A *hole* is a chordless cycle on five or more vertices; an *antihole* is the complement of a hole. The chordless cycle of length k is denoted by C_k; in particular, C_5 is the chordless cycle on 5 vertices. A graph G is C_k-*free*, for any k, when it does not contain C_k as an induced subgraph.

The *chromatic number* of a graph G, denoted by $\chi(G)$, is the minimum number of colors needed to color the vertices of G in such a way that no two adjacent vertices receive the same color. Clearly, $\chi(G)$ is bounded from below by the size of a largest clique in G, denoted by $\omega(G)$. In 1960, Berge introduced the notion of a perfect graph. A graph G is *perfect* if for every induced subgraph H of G, $\chi(H) = \omega(H)$.

It is easy to see that odd holes and odd antiholes are not perfect. Berge [1] conjectured that these are the only minimal imperfect graphs, i.e., a graph is perfect if and only if it does not contain an odd hole nor an odd antihole. When we say that a graph G *contains* a graph H, we mean as an induced subgraph. Berge's conjecture was known as the Strong Perfect Graph Conjecture (SPGC), which was proved by Chudnovsky et al.[3]. Thus, efficient algorithms for detecting induced holes or antiholes on an odd number of vertices will imply fast recognition of perfect graphs; the currently fastest algorithms for these problems run in $O(n^9)$ time [4,7].

Chvátal and Sbihi [6] showed a polynomial-time algorithm to recognize claw-free perfect graphs. The algorithm is based on a decomposition theorem of the structure of these graphs.

Nikolopoulos and Palios [13] presented two algorithms, one for the detection of holes and another for the detection of antiholes in arbitrary graphs. Both algorithms run in $O(n+m^2)$ time and require $O(nm)$ space. Thus, they provide a solution to the open problem posed by Hayward, Spinrad and Sritharan in [12]. They generalized their approach so that, for a fixed constant $k \geq 5$, they obtain an $O(n^{k-1})$-time algorithm for the detection of a hole (antihole, respectively) on at least k vertices. Additionally, they describe a different approach to detect antiholes on graphs that do not contain chordless cycles on 5 vertices in $O(n+m^2)$ time requiring $O(n+m)$ space. Again, for a fixed constant $k \geq 6$, the approach can be extended to yield $O(n^{k-2})$-time and $O(n^2)$-space algorithms for detecting holes (antiholes, respectively) on at least k vertices in graphs which do not contain holes (antiholes, respectively) on $k-1$ vertices.

The problem of determining whether a given graph on n vertices and m edges contains a chordless cycle of k or more vertices, for some fixed value of $k \geq 4$, was originally solved in $O(n^k)$ time (Hayward [11]). Spinrad [16] reduced the time complexity of the problem to $O(n^{k-3}M)$, where $M \simeq n^{2.376}$ is the time required to multiply two $n \times n$ matrices. Note that the problem of determining whether a graph contains a chordless cycle of four or more vertices, i.e., the well-known chordal graph recognition problem, can be solved in $O(n+m)$ time [9,14,18].

Holes and antiholes have been extensively studied in many different contexts in algorithmic graph theory. A typical example is that of the weakly chordal graphs (also known as weakly triangulated graphs) [9,10], which contain neither holes nor antiholes.

The algorithms of Hayward [11] and Spinrad [16] can be used for the recognition of weakly chordal graphs in $O(n^5)$ and $O(n^{4.376})$ time, respectively. Further progress on the weakly chordal graph recognition problem includes the $O(n^4)$ time and $O(nm)$ space algorithm of Spinrad and Sritharan [17], and the $O(m^2)$ time and $O(n+m)$ space algorithms of Hayward, Spinrad and Sritharan [12] and of Berry, Bordat and Heggernes [2]. It is interesting to note that the algorithm of Hayward et al. [12] produces a hole or an antihole certificate whenever the given graph is not weakly chordal. In the same paper, the authors posed as an open problem the designing of an $O(n^4)$ algorithm to find a hole in an arbitrary graph.

The detection of *odd length* holes and antiholes appears to be a much more difficult problem, see also [15]. It is this area which we address in this paper. In this paper, we present an algorithm to detect a smallest odd hole in claw-free graphs.

2 Preliminaries

Let $G = (V, E)$ be a simple finite undirected graph with no loops and no multiple edges. The length of a path is the number of its edges. We denote a subpath

$p_i p_{i+1} \ldots p_j$ of a path P by $P_{i,j}$. The chordless cycle of length k is denoted by C_k. A *hole* is a chordless cycle of five or more vertices. The hole of length k is also denoted by H_k.

From now on, when we use the phrase *hole*, we will assume that its vertices are ordered clockwise. Let $H = u_0 u_1 \ldots u_{k-1} u_0$ be a hole in a graph G with vertices ordered clockwise. The *arc* $A_{i,j}$ is the path going around clockwise from u_i to u_j, i.e., $A_{i,j}$ is the path $u_i u_{i+1} u_{i+2} \ldots u_j$ and $A_{j,i}$ is the path $u_j u_{j+1} u_{j+2} \ldots u_i$. Let H be an odd hole, then with respect to u_i and u_j, the path with the shorter length among $A_{i,j}$ and $A_{j,i}$ is called the *short arc* and the path with the longer length is called the *long arc*. An arc $A_{i,i}$ of length 0 is called a *trivial arc*. Let $(u_i, u_{i+1}), (u_j, u_{j+1}) \in E(H)$ we define the *arcs between* (u_i, u_{i+1}) and (u_j, u_{j+1}) to be $A_{i+1,j}$ and $A_{j+1,i}$. Furthermore, if H is an odd hole then, with respect to $(i, i+1)$ and $(j, j+1)$, H contains one arc with shorter length, called the *short arc* and the path with the longer length is called the *long arc*. An arc $A_{i,i}$ of length 0 is called a *trivial arc*.

A *claw* is the graph consisting of vertices w, x, y, z and only the edges (w, x), $(w, y), (w, z)$. Seymour and Chudnovsky [5] have found a complete description of claw-free graphs. They prove that all claw-free graphs are constructed starting from line graphs, circular interval graphs, subgraphs of the Schläfli graph, and a few other basic graphs, by piecing them together in prescribed ways.

3 Tents

Definition 1. *Let H be a hole in a graph G. A vertex v is called a* tent *of H if v has four consecutive neighbors on H. Two tents v and w are called* isomorphic tents *if they share the same neighbors on H.*

Property 2. *Let H be a hole in a (claw,H_5)-free graph G, let $v \notin V(H)$ and let n be the number of neighbors of v on H. Then:*

1. *Either $n = 0$ or $2 \leq n \leq 4$.*
2. *If $n = 2$ ($n = 3$) then the neighbors of v on H lie consecutively on a P_2 (P_3).*
3. *If $n = 4$ and H is a smallest odd hole in G, then the neighbors of v on H lie consecutively on a P_4.*

Corollary 3. *Let H be a smallest odd hole in a (claw,H_5)-free graph G, let $v \notin V(H)$, v has neighbors on H and let n be the number of neighbors of v on H. Then the neighbors of v on H lie consecutively on a P_n, $2 \leq n \leq 4$.*

Lemma 4. *Let H be a smallest odd hole in a (claw,H_5)-free graph G, and let t_1 and t_2 be tents of H. Then all the neighbors of t_1 and t_2 on H lie consecutively on at most a P_6.*

Corollary 5. *Let H be a smallest odd hole in a (claw,H_5)-free graph G. Then all the tents of H have at least two common neighbors on H, and H has at most three tents that are pairwise non-isomorphic.*

4 Edge-Path with Respect to a Hole

In this section we lay the foundations of the paper. First we define and characterize an edge-path with respect to a hole (edge-path in short) and the corresponding arc of an edge-path. Then we conclude that there are only two types of edge-paths.

Definition 6. *Let $H = H_{2r+1} = u_0 u_1 \ldots u_{2r} u_0$ be an odd hole in a graph G and let $\alpha = (u_i, u_{i+1})$, $\beta = (u_j, u_{j+1})$. W.l.o.g., from now on we may assume that $A_{i+1,j}$ is the short arc between α and β on H and $A_{j+1,i}$ is the long arc between α and β on H. We say that a path $P = p_0 p_1 \ldots p_{l-1} p_l$ is an edge-path with respect to H of length l from a source α to a target β if all the following conditions are satisfied:*

1. *P is a simple chordless path.*
2. *$|P| \leq |A_{i+1,j}| < |A_{j+1,i}|$.*
3. *p_0 is one of u_i, u_{i+1} and p_l is one of u_j, u_{j+1}, such that:*
   If $p_0 = u_{i+1}$ then $(u_i, p_x) \notin E(G)$ for every $1 \leq x \leq l$.
   If $p_0 = u_i$ then $(u_{i+1}, p_x) \notin E(G)$ for every $1 \leq x \leq l$.
   If $p_l = u_j$ then $(u_{j+1}, p_y) \notin E(G)$ for every $0 \leq y \leq l-1$.
   If $p_l = u_{j+1}$ then $(u_j, p_y) \notin E(G)$ for every $0 \leq y \leq l-1$.
4. *Let $P' \subseteq P$ be the set of the inner vertices of P that have neighbors on H, then P' has neighbors only on the long arc between α and β or P' has neighbors only on the short arc between α and β, but not on both.*
5. *From all the paths that satisfy 1-4, we choose P to be with minimal length.*

Definition 7. *Let H be a hole in a graph G. Let $P = p_0 p_1 \ldots p_l$ be an edge-path with respect to H from a source $\alpha = (u_i, u_{i+1})$ to a target $\beta = (u_j, u_{j+1})$. We say that an arc A between α and β is the corresponding arc of P on H, denoted by A_P, if the neighbors of $P_{1,l-1}$ on H lie on A. The other arc between α and β on H is denoted by $\overline{A_P}$.*

Property 8. *Let $H = H_{2r+1} = u_0 u_1 \ldots u_{2r} u_0$ be a smallest odd hole in a graph G. Let $P = p_0 p_1 \ldots p_l$ be an edge-path with respect to H from a source $\alpha = (u_i, u_{i+1})$ to a target $\beta = (u_j, u_{j+1})$. Then:*

1. *P does not start at $p_0 = u_i$ and end at $p_l = u_j$.*
2. *P does not start at $p_0 = u_{i+1}$ and end at $p_l = u_{j+1}$.*
3. *There can be only the following two types of edge-paths with respect to H between α and β:*

   Type 1 *The edge-path starts at $p_0 = u_{i+1}$ and ends at $p_l = u_j$. The neighbors of $P_{1,l-1}$ on H lie on the short arc between α and β.*
   Type 2 *The edge-path starts at $p_0 = u_i$ and ends at $p_l = u_{j+1}$. The neighbors of $P_{1,l-1}$ on H lie on the long arc between α and β.*

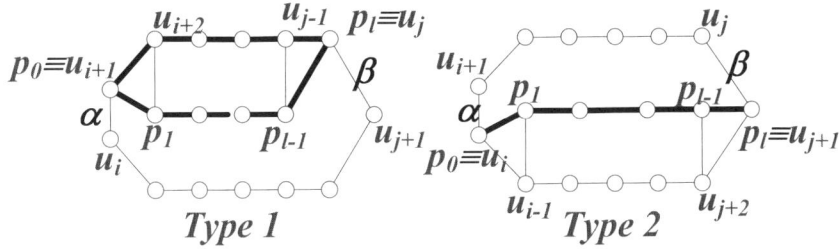

Fig. 1. Two types of edge-paths

In Figure 1 there are examples of edge paths. In Type 1 there are two edge-paths with the same length. One of them is the short arc between α and β. The length of the edge-path of Type 2 is smaller than the length of the long arc. In order to detect an odd hole, we need to avoid "walking" on edge paths shorter than the arcs.

5 Shortcuts of a Hole

This is the main theoretical section of the paper. Here we define a shortcut and a minimal shortcut of a hole. It can be prove that the length of a shortcut is smaller by 1 than its corresponding arc and we present important properties of the structure of a minimal shortcut. We conclude with Theorem 20, which is the main theorem of the paper. Furthermore, the theorem is the main foundation of the algorithm for detecting a smallest odd hole in a claw-free graph in following sections.

Throughout this section we consider $H = H_{2r+1} = u_0 u_1 \ldots u_{2r} u_0$ to be a smallest odd hole in a (claw,H_5)-free graph G.

Definition 9. *Let S be an edge-path with respect to H from a source $\alpha = (u_i, u_{i+1})$ to a target $\beta = (u_j, u_{j+1})$. We say that S is a shortcut of H if $|S| < |A_S|$. From now on we consider S to be a shortcut of H from a α to β.*

Note that, by Definition 6.2 $|S| \leq |A_{i+1,j}| < |A_{j+1,i}|$.

Definition 10. *Consider the set of all shortcuts of H. We say that $\alpha \in E(G)$ has no shortcut if there is no shortcut of H such that α is the source. We say that α has no shortcut of Type i, $i = 1, 2$, if there is no shortcut of Type i of H such that α is the source.*

Definition 11. *Let S be a shortcut of H from a source $\alpha = (u_i, u_{i+1})$ to a target $\beta = (u_j, u_{j+1})$. We say that S' is a minimal shortcut obtained from S if the following conditions hold:*
1. S' is a shortcut of H (not necessarily from α to β). 2. S' contains a minimum number of vertices from S. 3. Only the two end points of S' are in $V(H)$.

Note that, in the trivial case $S' = S$. In Figure 2 on page 334 the minimal shortcuts are in bold and with the following details. (1) $|S| = 5$, $|A_S| = 6$,

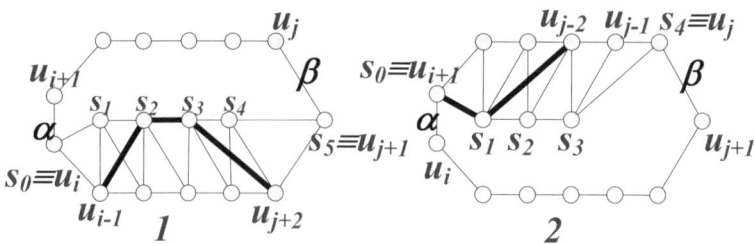

Fig. 2. Minimal Shortcuts

$S' = u_{i-1}s_2s_3u_{j+2}$ is a minimal shortcut of H from (u_{j+1}, u_{j+2}) to (u_{i-1}, u_i) of length 3 and $|A_{j+2,i-1}| = 4$. (2) $|S| = 4$, $|A_S| = 5$, $S' = s_0s_1u_{j-2}$ is a minimal shortcut of H from (u_i, u_{i+1}) to (u_{j-2}, u_{j-1}) of length 2 and $|A_{i+1,j-2}| = 3$.

Property 12. *For any shortcut $S = s_0s_1s_2$ of H, i.e., $|S| = 2$, S is a minimal shortcut of H, $|A_S| = 3$, i.e., s_1 is a tent of H.*

Theorem 13. 1. *For any shortcut $S = s_0s_1s_2$ of H, $|S| = |A_S| - 1$ and $|S| \leq r - 1$.*
2. *If S is of Type 2, then $|S| = |\overline{A_S}| = r - 1$.*

It is clear that if a hole does not have a minimal shortcut, it does not have a shortcut. Recall that we denote by S a shortcut of H from a source $\alpha = (u_i, u_{i+1})$ to a target $\beta = (u_j, u_{j+1})$. From now on, we consider $S' = s'_0 \ldots s'_l$ to be a minimal shortcut of H obtained from S.

Theorem 14. *If $|S'| \geq 3$, then:*

1. *Each one of s'_1 and s'_{l-1} has exactly three neighbors on H.*
2. *s'_{l-2} and s'_{l-1} have only one common neighbor on H.*
3. *s'_1 and s'_2 have only one common neighbor on H.*

Theorem 15. *If $|S'| \geq 4$, then every $s'_k \in S'$, $2 \leq k \leq l-2$, has exactly two neighbors on H.*

Now we introduce the definition of a shortcut substitution. A shortcut substitution defines a hole with a tent. In particular, by substituting the corresponding arc of a shortcut with a shortcut substitution we get a hole with a tent.

Definition 16. *Let $S' = s'_0 \ldots s'_l$ be a minimal shortcut of a hole H from a source $\alpha = (u_i, u_{i+1})$ to a target $\beta = (u_j, u_{j+1})$ such that $|S'| \geq 3$. Let H^1 be the hole induced as follows:*

1. *If S' is of Type 1 then $H^1 = (V(H) \setminus \{u_{i+4} \ldots u_{j-1}\}) \bigcup \{s'_2 \ldots s'_{l-1}\}$.*
2. *If S' is of Type 2 then $H^1 = (V(H) \setminus \{u_{j+2} \ldots u_{i-3}\}) \bigcup \{s'_2 \ldots s'_{l-1}\}$.*

We say that H^1 is obtained from H through a shortcut substitution.

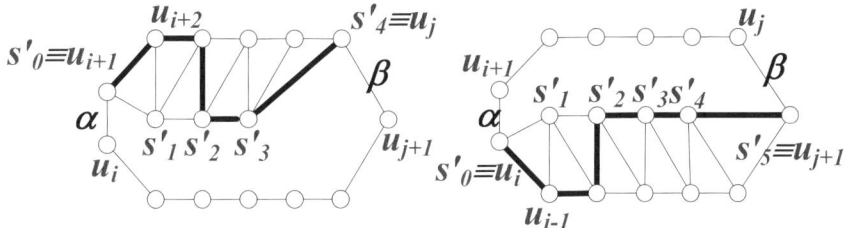

Fig. 3. Shortcut substitutions
The length of the new short arc between α and β (in bold) equals to the length of $A_{S'}$. Note that s'_1 is a tent of H^1.

Note that, s'_1 is a tent of H^1 since by Theorem 14, s'_1 has exactly three neighbors on H and these three together with s'_2 are the neighbors of s'_1 on H^1. Moreover, $|H| = |H^1|$ and $\alpha, \beta \subset H, H^1$ (See Figure 3 on page 335).

Lemma 17. *Let t_1 and t_2 be tents of H, such that the neighbors of t_1 and t_2 on H lie on a P_6. Let $\alpha = (u_i, u_{i+1}) \in E(H)$, such that u_i and u_{i+1} are common neighbors of t_1 and t_2. Then α has no shortcut of H.*

Lemma 18. *Let t_1 and t_2 be tents of H, such that the neighbors of t_1 and t_2 lie on a P_5. Then there is a smallest odd hole H^* and an edge $\gamma \in E(H^*)$, such that γ has no shortcut of H^*.*

Lemma 19. *Let t be a tent of H, then there is a smallest odd hole H^* and an edge $\gamma \in E(H^*)$, such that γ has no shortcut of H^*.*

Theorem 20. *Let $H = H_{2r+1} = u_0 u_1 \ldots u_{2r} u_0$ be a smallest odd hole in a (claw,H_5)-free graph G. Then there is a smallest odd hole H^* and an edge $\gamma \in E(H^*)$, such that γ has no shortcut of H^*.*

6 Γ-BFS

The general idea of the detection algorithm of an odd hole is to start a traversal of the graph from an edge and "walk" only on edge-paths.

The Gallai forcing graph Γ^G of a graph $G = (V, E)$, denoted by Γ^G, is the graph with vertex set $V(\Gamma^G) = E(G)$ and two vertices $\alpha = (v_1, v_2), \beta = (v_3, v_4) \in V(\Gamma^G)$ are connected in $E(\Gamma^G)$ iff either $v_1 = v_4$ and $(v_2, v_3) \notin E(G)$ or $v_2 = v_3$ and $(v_1, v_4) \notin E(G)$, see [8,9].

Definition 21. *Let $\alpha = (v_1, v_2) \in V(\Gamma)$. We define the elements of the vertex α to be v_1 and v_2.*

Property 22. *Γ^G preserves the holes of G, that is, if G contains a hole H of length k then Γ^G contains a hole H^Γ of length k such that the vertices of H^Γ are the edges of H.*

However, the opposite does not hold. There can be a hole in Γ^G whose corresponding edges of G do not form a hole. In [8], Gallai describes the list of induced subgraphs that create a hole in Γ^G.

6.1 The Algorithm

A Γ-BFS is a partial BFS on Γ^G, which uses the structure of Γ^G to discover and explore only some of the vertices of Γ^G (the edges of G). We keep track of the discovery time of the vertices of G and the vertices of Γ^G. This helps us to discover paths in G with special properties. The Γ-BFS starts from a source $\alpha = (v_1, v_2) \in V(\Gamma^G)$. The definition of Γ-BFS is given in Algorithm 1 on page 336.

Given a graph Γ^G and a distinguished *source* vertex α, Γ-BFS systematically explores Γ^G to "reveal" some of the vertices of Γ^G and ignores other vertices of Γ^G. In Γ-BFS of Γ^G, a newly discovered vertex φ is first tested for certain criteria to hold, and only if it passes the criteria it is marked "revealed" and added to the queue. Otherwise, φ is "ignored" and never explored.

Algorithm 1. Γ-BFS

Input: A graph G, $\alpha = (v_1, v_2) \in E(G)$

1 **procedure** Γ-BFS(G, α)
2   compute $\Gamma^G = (V, E)$
3   **forall** $v \in V(G)$ **do**
4       $d[v] \leftarrow \infty$
5   **forall** $\gamma \in V(\Gamma^G)$ **do**
6       $d[\gamma] \leftarrow \infty$; $p[\gamma] \leftarrow$ NIL
7     $d[\alpha] \leftarrow 0$; $d[v_1] \leftarrow 0$; $d[v_2] \leftarrow 0$
8   enqueue (α, Q)
9   **while** Q *is not empty* **do**
10       $\psi \leftarrow$ dequeue(Q)
11   **forall** *Neighbors* $\varphi = (v_4, v_5)$ *of* $\psi = (v_3, v_4)$ **do**
12   **if** $d[\varphi] = \infty$ **then**
13           $d[\varphi] \leftarrow d[\psi] + 1$
14           $p[\varphi] \leftarrow \psi$
15   **if** $d[v_5] \geq d[\varphi]$ **then**
16             $d[v_5] \leftarrow d[\varphi]$
17   **if** $d[v_4] = d[\psi]$ **then**
18   enqueue(φ, Q)

19 **end procedure**

Definition 23. *Let* $\psi = (v_3, v_4) \in V(\Gamma^G)$ *be the vertex that is now dequeued in the main loop of* Γ-*BFS. Let* $\varphi = (v_4, v_5) \in V(\Gamma^G)$ *be a neighbor of* ψ *on* Γ, *such that* $d[\varphi] = \infty$.

1. *We say that* φ *is discovered by* ψ.
2. *We say that* $v_5 \in V(G)$ *is discovered by* φ *if* $d[v_5] = \infty$. *We call* φ *the discoverer of* v_5.

3. We say that φ is revealed by ψ if φ is added to the queue, i.e., φ is discovered by ψ, $d[v_5] \geq d[\varphi]$ and $d[v_4] = d[\psi]$.
4. We say that φ is ignored by ψ if φ is not added to the queue, i.e., φ is discovered by ψ and $d[v_5] < d[\varphi]$ or $d[v_4] < d[\psi]$.
5. The discovered time of the source $\alpha = (v_1, v_2)$ is 0 and the discovered time of v_1, v_2 is 0, i.e., $d[\alpha] = d[v_1] = d[v_2] = 0$.
6. The discovered time of φ, is the discovered time of ψ plus one, i.e., $d[\varphi] = d[\psi] + 1$. The discovered time of v_5 is the discovered time of φ, where φ is the discoverer of v_5, i.e., $d[v_5] = d[\varphi] = d[\psi] + 1$.

The Γ-BFS procedure introduced above maintains several data structures. The parent of $\varphi \in V(\Gamma^G)$ is stored in the variable $p[\varphi]$. The discovered time, i.e., the distance from the source $\alpha \in V(\Gamma^G)$ to a vertex $\varphi \in V(\Gamma^G)$ computed by the algorithm, is stored in $d[\varphi]$. The discovered time in G, i.e., the distance from the source $\alpha \in V(\Gamma^G)$ to a vertex $v \in V(G)$ computed by the algorithm, is stored in $d[v]$. The algorithm also uses a first-in-first-out queue Q to manage the vertices.

6.2 Complexity

Let G be a graph with n vertices and m edges and let $\psi = (v_3, v_4) \in E(G)$. There are at most n vertices which are either neighbors of v_3 and not neighbors of v_4 or neighbors of v_4 and not neighbors of v_3. Therefore, $\psi = (v_3, v_4)$ has at most n neighbors on Γ^G. Thus Γ^G has m vertices and at most nm edges.

The loop at lines 3–4 is performed for every vertex $v \in V(G)$, therefore, runs in $O(n)$. The computation of Γ^G runs in $O(|V(\Gamma^G)| + |E(\Gamma^G)|) = O(m + nm) = O(nm)$. The loop at lines 5–7 is performed for every vertex $\gamma \in V(\Gamma^G)$, thus runs in $O(m)$. The main loop of the algorithm is at lines 10–19. The outer loop goes as long as Q is not empty, hence, runs in $O(m)$. The inner loop is performed for every neighbor φ of ψ, thus runs in $O(n)$. Therefore, the main loop runs in $O(nm)$. Thus, Algorithm 1 runs in $O(nm)$ time.

7 Γ-Path

In order to detect an odd hole in a claw-free graph G, first we define and characterize the paths that are detected by the Γ-BFS of Γ^G, and their interpretation on the edges of G.

Definition 24. *Let G be a graph, let $\alpha = (u, p_0), \beta = (p_l, v) \in E(G)$ and let $P = p_0 p_1 \ldots p_l$ be a path in G, such that $\alpha, \beta, (p_i, p_{i+1}) \in V(\Gamma^G)$, for every $0 \leq i \leq l - 1$. We say that P is a Γ-path of G from a source α to a target β if there is a Γ-BFS on Γ^G that reveals P, i.e., all the following hold:*

1. *starts from a source α.*
2. *$\alpha = (u, p_0)$ discovers and reveals (p_0, p_1).*
3. *(p_i, p_{i+1}) discovers and reveals (p_{i+1}, p_{i+2}) for every $0 \leq i \leq l - 2$.*
4. *(p_{l-1}, p_l) discovers and reveals $(p_l, v) = \beta$.*

We say that P starts at p_0 and ends at p_l. The length of P is l, the number of edges of P.

Theorem 25. Let $H = H_{2r+1} = u_0 u_1 \ldots u_{2r} u_0$ be a smallest odd hole in a (claw,H_5)-free graph G. Let $P = p_0 p_1 \ldots p_l$ be a Γ-path from a source $\alpha = (u_i, u_{i+1})$ to a target $\beta = (u_j, u_{j+1})$. Then, P is an edge-path with respect to H.

8 Detection of an Odd Hole

8.1 The Algorithm

Our approach for detecting a smallest odd hole in a claw-free graph G, is to start a Γ-BFS from all $\alpha \in V(\Gamma^G)$. If G contains an odd hole then the output of Algorithm 2 is the length of the smallest odd hole; if G is odd hole-free then the output is "infinity".

Algorithm 2. Smallest Odd Hole Detection in a Claw-Free Graph

Input: A claw-free graph G.
Output: The size of the smallest odd hole in G, if one exists.

1 **procedure** ODDHOLEDETECTION(G)
2     $minSizeOddHole \leftarrow \infty$
3   **forall** $\alpha \in V(\Gamma^G)$ **do**
4        $localMin \leftarrow \infty$
5   Start a Γ-BFS from α
6   **if** after ψ is removed from the queue, ψ has neighbor φ such that $d[\psi] = d[\varphi]$ **then**
7           $localMin \leftarrow 2d[\psi] + 1$
8   stop Γ-BFS from α
9   **if** $localMin < minSizeOddHole$ **then**
10           $minSizeOddHole \leftarrow localMin$
11   **return** $minSizeOddHole$.
12 **end procedure**

Theorem 26. Let G be a claw-free graph. G contains an odd hole $H = H_{2r+1}$ iff Algorithm 2 yields $2r + 1$.

Algorithm 2 can be easily augmented so that it provides a certificate whenever it decides the input graph G contains an odd hole.

8.2 Complexity

Let G be a graph with n vertices and m edges. Note that, there is an obvious recognition algorithm for a claw-free graph, with running time $O(n^4)$. The complexity of Γ-BFS is $O(nm)$. The main loop at lines 3–10 is performed for every vertex $\alpha \in V(\Gamma^G)$, thus runs in $O(m)$. Thus, Algorithm 2 runs in $O(nm^2)$ time.

9 Further Research

The problems we solve in this paper together with the work that has been done before, raise more questions on the subject introduced here. We summarize some of the questions below.

(1) An interesting problem is to give a better recognition algorithm for claw-free perfect graphs.
(2) Can we use Γ-BFS to detect holes and antiholes with better complexity?
(3) Define Γ_2^G and characterize the odd holes in Γ_2^G? Furthermore, generalize Γ^G and Γ_2^G to get a definition for Γ_n^G, and characterize the odd holes in Γ_n^G.
(4) In general graphs, can we find a smallest odd hole without a shortcut?

Acknowledgments

We would like to thank Maria Chudnovsky for fruitful discussions. Furthermore, we owe our gratitude to Irith Hartman and Dror Rawitz.

References

1. Berge, C.: Färbung von Graphen deren sämtliche bzw. deren ungerade Kreise starr sind, Wiss. Zeitschrift, Mathematisch-Naturwissenschaftliche Reihe, Martin-Luther-Univ. Halle-Wittenberg, pp. 114–115 (1961)
2. Berry, A., Bordat, J.P., Heggerens, P.: Recognizing weakly triangulated graphs by edge separability. Nordic J. Computing 7, 164–177 (2000)
3. Chudnovsky, M., Robertson, N., Seymour, P.D., Thomas, R.: The strong perfect graph theorem. Ann. of Math. 164(1), 51–229 (2006)
4. Chudnovsky, M., Cornuéjols, G., Liu, X., Seymour, P., Vušković, K.: Recognizing Berge Graphs. Combinatorica 25(2), 143–186 (2005)
5. Chudnovsky, M., Seymour, P.D.: The structure of claw-free graphs. In: Surveys in Combinatorics. London Math Soc. Lecture Note Series, vol. 327 (2005)
6. Chvátal, V., Sbihi, N.: Recognizing claw-free perfect graphs. Journal of Combinatorial Theory Series A 44(3), 154–176 (1987)
7. Cornuéjols, G., Liu, X., Vušković, K.: A polynomial algorithm for recognizing perfect graphs. In: Proc. 44th IEEE Symp. on Foundations of Computer Science (FOCS 2003), pp. 20–27 (2003)
8. Gallai, T.: Transitiv orientierbare Graphen. Acta Math. Acad Sci. Hungar. 18, 25–66 (1967)
9. Golumbic, M.C.: Algorithmic graph theory and perfect graphs. Annals of Discrete Mathematics, vol. 57. Elsevier, Amsterdam (2004)
10. Hayward, R.B.: Weakly triangulated graphs. J. Comb. Theory Ser. B 39, 200–208 (1985)
11. Hayward, R.B.: Two classes of perfect graphs, PhD Thesis, School of Computer Science, McGill University (1987)
12. Hayward, R.B., Spinrad, J., Sritharan, R.: Weakly chordal graph algorithms via handles. In: Proc. 11th ACM-SIAM Symp. on Discrete Algorithms (SODA 2000), pp. 42–49 (2000)
13. Nikolopoulos, S.D., Palios, L.: Holes and antihole detection in graphs. In: Proc. 15th ACM-SIAM Sympos. Discrete Algorithms, pp. 843–852 (2004)

14. Rose, D.J., Tarjan, R.E., Lueker, G.S.: Algorithmic aspects of vertex elimination on graphs. SIAM J. Computing 5, 266–283 (1976)
15. Shrem, S.: Odd hole-free Graphs, M.Sc. thesis, Department of Computer Science, University of Haifa, February 27 (2006), http://digitool.haifa.ac.il/R/?func=dbin-jump-full&object_id=175621&local_base=GEN01
16. Spinrad, J.P.: Finding large holes. Inform. Process. Letters 39, 227–229 (1991)
17. Spinrad, J.P., Sritharan, R.: Algorithms for weakly triangulated graphs. Discrete Applied Math. 59, 181–191 (1995)
18. Tarjan, R.E., Yannakakis, M.: Simple linear-time algorithms to test chordality of graphs, test acyclicity of hypergraphs, and selectively reduce acyclic hypergraphs. SIAM J. Computing 13, 566–579 (1984)

Finding Induced Paths of Given Parity in Claw-Free Graphs[*]

Pim van 't Hof[1], Marcin Kamiński[2], and Daniël Paulusma[1]

[1] Department of Computer Science, University of Durham,
Science Laboratories, South Road, Durham DH1 3LE, England
{pim.vanthof,daniel.paulusma}@durham.ac.uk
[2] Computer Science Department, Université Libre de Bruxelles,
Boulevard du Triomphe CP212, B-1050 Brussels, Belgium
marcin.kaminski@ulb.ac.be

Abstract. The PARITY PATH problem is to decide if a given graph G contains both an odd length and an even length induced path between two specified vertices s and t. In the related problems ODD INDUCED PATH and EVEN INDUCED PATH, the goal is to determine whether an induced path of odd, respectively even, length between two specified vertices exists. Although all three problems are NP-complete in general, we show that they can be solved in $\mathcal{O}(n^5)$ time for the class of claw-free graphs. Two vertices s and t form an even pair in G if every induced path from s to t in G has even length. Our results imply that the problem of deciding if two specified vertices of a claw-free graph form an even pair, as well as the problem of deciding if a given claw-free graph has an even pair, can be solved in $\mathcal{O}(n^5)$ time and $\mathcal{O}(n^7)$ time, respectively. We also show that we can decide in $\mathcal{O}(n^7)$ time whether a claw-free graph has an induced cycle of given parity through a specified vertex.

1 Introduction

Finding a shortest path, a maximum stable set or a hamiltonian cycle in a graph are just a few examples from the wide spectrum of problems dealing with finding a subgraph (or induced subgraph) with some particular property. In this context, simplest subgraphs, such as paths, trees and cycles, with some prescribed property are often studied. The following problem has been extensively studied in the context of perfect graphs. Here, the *length* of a path refers to its number of edges, and a path is said to be *odd* (respectively *even*) it has odd (respectively even) length.

PARITY PATH
Instance: A graph G and two vertices s, t of G.
Question: Does there exist both an odd and even induced path from s to t in G?

---

[*] This work has been supported by EPSRC (EP/D053633/1) and the Actions de Recherche Concertées (ARC) fund of the Communauté française de Belgique.

We focus on the closely related problem of deciding whether there exists an induced path of given parity between a pair of vertices. In particular, we study the following two problems.

ODD INDUCED PATH
Instance: A graph G and two vertices s, t of G.
Question: Does there exist an odd induced path from s to t in G?

EVEN INDUCED PATH
Instance: A graph G and two vertices s, t of G.
Question: Does there exist an even induced path from s to t in G?

The ODD INDUCED PATH problem was shown to be NP-complete by Bienstock [6]. Consequently, the EVEN INDUCED PATH problem and the PARITY PATH problem are NP-complete as well. Several authors however have identified a number of graph classes that admit polynomial-time algorithms for these problems. Below we survey those results, as well as results on related problems, before stating our contribution. Throughout the paper, we use n and m to denote the number of vertices and the number of edges of the input graph, respectively.

ODD PATH and EVEN PATH. In the ODD PATH and EVEN PATH problems the task is to find a (not necessarily induced) path of given parity between a specified pair of vertices. These problems were considered by LaPaugh and Papadimitriou [20]. They mention an $\mathcal{O}(n^3)$ algorithm for solving both problems due to Edmonds, using a reduction to matching, and propose a faster one of $\mathcal{O}(m)$ time complexity. Their algorithm also finds a *shortest* (not necessarily induced) path of given parity between two vertices in $\mathcal{O}(m)$ time, even in a weighted graph. Interestingly, as they also show in their paper, the problem of finding a directed path of given parity is NP-complete for directed graphs. Arkin, Papadimitriou and Yannakakis [1] generalized the result of [20] and designed a linear-time algorithm deciding if all (not necessarily induced) paths between two specified vertices are of length p mod q, for fixed integers p and q.

EVEN PAIR. First interest in induced paths of given parity comes from the theory of perfect graphs. Two non-adjacent vertices are called an *even pair* if every induced path between them is even. The EVEN PAIR problem is to decide if a given pair of vertices forms an even pair. The EVEN PAIR problem is co-NP-complete due to Bienstock [6], as is the problem of deciding if a graph contains an even pair. The interest in even pairs was sparked by an observation of Fonlupt and Uhry [17]: if a graph is perfect and contains an even pair, then the graph obtained by identifying the vertices that form the even pair is also perfect. Later Meyniel showed that minimal non-perfect graphs contain no even pair [23]. Those two facts triggered a series of theoretical and algorithmic results which are surveyed in [14] and its updated version [15].

There is some evidence that perfect graphs without an even pair can be generated by performing a small number of composition operations on some basic graphs. Using such a structural result could then lead to a combinatorial algorithm for coloring perfect graphs. Indeed, for coloring perfect graphs using at most three colors this approach turned out to be successful, as was shown by Chudnovsky and Seymour in [10]. Linhares Sales and Maffray [22] study even pairs in order to give characterizations of claw-free graphs that are strict quasi-parity and perfectly contractile, respectively.

PARITY PATH and GROUP PATH. Arikati, in a series of papers with different coauthors, developed polynomial-time algorithms for the PARITY PATH problem in different classes of graphs. Chordal graphs are considered in [2], where the authors present a linear-time algorithm for the GROUP PATH problem, a generalization of the ODD INDUCED PATH problem. In the GROUP PATH problem the edges of the input graph are weighted with elements of some group $\mathcal{G}$. The problem is to find an induced path of given weight between two specified vertices, where the weight of a path is defined as the product of the weights of the edges of the path. They present an $\mathcal{O}(|\mathcal{G}|m+n)$ algorithm for the GROUP PATH problem on chordal graphs using a perfect elimination ordering.

The topic of [4] is PARITY PATH on circular-arc graphs. The authors show how to reduce the problem to interval graphs by recursively applying a set of reductions. Since interval graphs are chordal, the algorithm of [2] can be used to obtain the solution. This way they obtain a polynomial-time algorithm for circular-arc graphs. In [26] polynomial-time algorithms for the PARITY PATH problem on comparability and cocomparability graphs, and a linear-time algorithm for permutation graphs are given. A polynomial-time algorithm for PARITY PATH on perfectly orientable graphs is presented in [3]. Sampaio and Sales [25] obtain a polynomial-time algorithm for planar perfect graphs. The authors of [16] characterize even and odd pairs in comparability and P_4-comparability graphs and give polynomial-time algorithms for the PARITY PATH problem in those classes. Hoàng and Le [18] show that PARITY PATH can be solved in polynomial time for the class of 2-split graphs.

Note that a set F of vertices of a line graph $G = L(H)$ form an odd (respectively even) induced path in G if and only if the set of edges corresponding to F form an even (respectively odd) path in the preimage graph H of G. It is well-known that the preimage graph of a line graph can be found in polynomial time [24]. Combining these two facts with the polynomial-time algorithm for finding (not necessarily induced) paths of given parity in [20] yields a polynomial-time algorithm for solving the PARITY PATH problem for the class of line graphs (cf. [28]).

Our Results. Our interest in the ODD INDUCED PATH problem was in part stirred by studying Bienstock's NP-completeness reduction in [6]. He builds a graph out of a 3-SAT formula and shows that the formula is satisfiable if and only if there exists an odd induced path between a certain pair of vertices. This is also shown to be equivalent to the existence of two disjoint induced paths (with no edges between the two paths) between certain pairs of vertices in the construction. Finding such two paths is then NP-hard in general but has been proved solvable in polynomial time for claw-free graphs [21]. A natural question to ask is whether the ODD INDUCED PATH problem can also be solved in polynomial time for this class of graphs. In this paper, we answer this question in the affirmative by presenting an algorithm that solves both the ODD INDUCED PATH problem and the EVEN INDUCED PATH problem in $\mathcal{O}(n^5)$ time for the class of claw-free graphs. This implies that the EVEN PAIR problem can be solved in $\mathcal{O}(n^5)$ time for claw-free graphs.

As we saw earlier in this section, the PARITY PATH problem has been extensively studied in different graph classes. However, a polynomial-time algorithm for claw-free graphs has never been proposed; somewhat surprising, since claw-free graphs form a large and important class containing, e.g., the class of line graphs and the class of complements of triangle-free graphs. Our $\mathcal{O}(n^5)$ algorithm for solving the ODD INDUCED PATH and EVEN INDUCED PATH problems for claw-free graphs immediately implies that we can solve the PARITY PATH problem for claw-free graphs in $\mathcal{O}(n^5)$ time, thus generalizing the aforementioned polynomial-time result on line graphs.

Apart from the ODD INDUCED PATH problem, Bienstock [6] mentioned two more NP-complete problems in the abstract of his paper. The first one is to decide whether a graph has an odd hole passing through a given vertex. The second one is to decide whether a graph has an odd induced path between *every* pair of vertices. We show that our polynomial-time algorithm for the ODD INDUCED PATH problem implies that both these problems are solvable in $\mathcal{O}(n^7)$ time when restricted to the class of claw-free graphs. As a result the problem of deciding whether or not a claw-free graph contains an even pair can be solved in $\mathcal{O}(n^7)$ time.

Our paper is organized as follows. In Section 2 we state our terminology and discuss some results on claw-free perfect graphs. In Section 3 we show how to solve the ODD INDUCED PATH and EVEN INDUCED PATH problem in $\mathcal{O}(n^5)$ time for claw-free graphs. There, we also show the other results mentioned above. Section 4 contains the conclusions and mentions some open problems.

2 Preliminaries

All graphs in this paper are undirected, finite, and have no loops or multiple edges. Let G be a graph. We refer to the vertex set and edge set of G by $V(G)$ and $E(G)$, respectively. The *neighborhood* of a vertex v in G is denoted by $N_G(v) = \{y \in V(G) \mid xy \in E(G)\}$. A *claw* is the graph $(\{x, a, b, c\}, \{xa, xb, xc\})$, where vertex x is called the *center* of the claw. For any set $S \subseteq V(G)$, we write $G[S]$ to denote the subgraph of G induced by S. A *hole* is an induced cycle of length at least 4, and an *antihole* is the complement of a hole. We say that a hole is *odd* (respectively *even*) if it has an odd (respectively even) number of edges. An antihole is called odd (respectively even) if its complement is an odd (respectively even) hole. The *chromatic number* of a graph is the smallest number of colors needed to color its vertices in such a way that no two adjacent vertices receive the same color. A graph G is *perfect* if for every induced subgraph H the chromatic number of H equals the size of a largest clique in H. A graph is called *Berge* if it does not contain an odd hole or an odd antihole. A little over 40 years after Berge [5] conjectured that a graph is perfect if and only if it is Berge, Chudnovsky et al. [8] confirmed his intuition by proving the following theorem.

Theorem 1 (Strong Perfect Graph Theorem, [8]). *A graph is perfect if and only if it contains no odd hole and no odd antihole.*

Shortly afterwards, Chudnovsky et al. [7] presented an $\mathcal{O}(n^9)$ algorithm for recognizing perfect graphs. We need such an algorithm as a subroutine in the algorithm presented in Section 3. We will not use their algorithm, because for claw-free perfect graphs a faster recognition algorithm exists, namely the algorithm of Chvátal and Sbihi [11]. Chvátal and Sbihi did not explicitly state the time complexity of their recognition algorithm. Theorem 2 shows it is $O(n^4)$. We postpone the required running time analysis to the journal version of our paper.

Theorem 2 ([11]). *It is possible to decide in $\mathcal{O}(n^4)$ time whether or not a claw-free graph is perfect.*

The proof of the following corollary is also postponed to the journal version.

Corollary 1. *Let G be a claw-free graph. It is possible to find an odd hole or an odd antihole of G, or conclude that such a graph does not exist, in $\mathcal{O}(n^5)$ time.*

3 Finding Induced Paths of Given Parity

We start by giving an outline of our algorithm that solves the ODD INDUCED PATH problem in $\mathcal{O}(n^5)$ time for claw-free graphs.

Algorithm solving ODD INDUCED PATH for claw-free graphs

Input : claw-free graph G, vertices s and t of G
Output : YES if G contains an odd induced path from s to t
          NO otherwise

Preprocess G to obtain graph G''
    Step 1: add edges to make s and t simplicial
    Step 2: delete irrelevant vertices
Test whether or not G'' is perfect
If G'' is not perfect, output YES
If G'' is perfect, find a shortest path P from s to t
    If P is odd, output YES
    If P is even, define graph $G^* := (V(G'') \cup \{x\}, E(G'') \cup \{sx, tx\})$
    Test whether or not G^* is perfect
        If G^* is not perfect, output YES
        If G^* is perfect, output NO

As shown in the outline, we first preprocess the input graph G in order to obtain a new graph G'' with certain desirable properties. This preprocessing procedure is described in Section 3.1. We then distinguish two cases, depending on whether or not G'' is perfect. The case that G'' is not perfect is discussed in Section 3.2, while Section 3.3 deals with the case that G'' is perfect. In Section 3.4 we prove correctness of our algorithm and show that its time complexity is $\mathcal{O}(n^5)$. We also explain in Section 3.4 how our algorithm can be slightly modified in such a way that it also solves the EVEN INDUCED PATH problem for claw-free graphs in $\mathcal{O}(n^5)$ time.

3.1 Preprocessing the Input Graph G

Let G be a claw-free graph and let s and t be two vertices of G. Note that we may without loss of generality assume that G is connected and that s and t are not adjacent. We make these assumptions throughout the paper.

Step 1. We add an edge between each pair of non-adjacent neighbors of s, and we do the same for each pair of non-adjacent neighbors of t. Then in the resulting graph G', both s and t are *simplicial* vertices, i.e., vertices whose neighborhood form a clique in G'. In general, adding edges is not a claw-freeness preserving operation. However, the following lemma states that we do not create claws in Step 1. We postpone the straightforward proof to the journal version of this paper.

Lemma 1. *The graph G' is claw-free.*

Step 2. We "clean" G' by repeatedly deleting irrelevant vertices. A vertex $v \in V(G')$ is called *irrelevant* (for vertices s and t) if v does not lie on any induced path from s to t, and we say that G' is *clean* if none of its vertices is irrelevant. Let G'' denote the graph obtained from G' by repeatedly deleting vertices that are irrelevant. Note that G'' is claw-free, as G'' is an induced subgraph of G'.

We now show that we can perform Step 2 in polynomial time by showing that we can identify irrelevant vertices in polynomial time. In general, the problem of deciding whether a vertex is irrelevant is NP-complete. This follows from a result by Derhy and Picouleau [13], who prove that the following problem is NP-complete for the class of graphs of maximum degree at most 3.

THREE-IN-A-PATH
Instance: A graph G and three vertices v_1, v_2, v_3 of G.
Question: Does there exist an induced path of G containing v_1, v_2 and v_3?

Chudnovsky and Seymour [9] study the following closely related problem.

THREE-IN-A-TREE
Instance: A graph G and three vertices v_1, v_2, v_3 of G.
Question: Does there exist an induced tree of G containing v_1, v_2 and v_3?

Theorem 3 ([9]). *The* THREE-IN-A-TREE *problem can be solved in $\mathcal{O}(n^4)$ time, and a desired tree can be found in $\mathcal{O}(n^4)$ time in case one exists.*

Observe that the THREE-IN-A-PATH problem is equivalent to the THREE-IN-A-TREE problem for the class of claw-free graphs, since every induced tree in a claw-free graph is an induced path. This, together with Theorem 3, implies the following result.

Lemma 2. *The problem of deciding whether a vertex v of a claw-free graph G is irrelevant for two simplicial vertices s and t of G can be solved in $\mathcal{O}(n^4)$ time.*

After preprocessing the input graph G we have obtained a graph G'' that satisfies the following three conditions: (1) G'' is claw-free; (2) both s and t are simplicial vertices of G''; and (3) G'' is clean for s and t. The following lemma implies that solving the ODD INDUCED PATH problem for G is equivalent to solving the problem for G''. The lemma also shows that the entire preprocessing procedure can be performed in $\mathcal{O}(n^5)$ time. The proof of the lemma is postponed to the journal version of this paper.

Lemma 3. *Every induced path from s to t in G'' is also an induced path from s to t in G, and vice versa. Moreover, G'' can be obtained from G in $\mathcal{O}(n^5)$ time.* □

We now distinguish two cases, depending on whether or not G'' is perfect.

3.2 G'' Is Not Perfect

Suppose G'' is not perfect. Then G'' contains an odd hole or an odd antihole by virtue of the Strong Perfect Graph Theorem. We consider odd antiholes and odd holes in Lemma 4 and Lemma 5, respectively. The *length* of an antihole is the number of edges in its complement.

Lemma 4. *Let H be a connected claw-free graph. If H contains a simplicial vertex, then H does not contain an odd antihole of length more than 5.*

Proof. Let s be a simplicial vertex of a connected claw-free graph H. For contradiction, suppose H contains an odd antihole X such that $\overline{X} = x_1 x_2 \ldots x_{2k+1} x_1$ is an odd induced cycle with $k \geq 3$. Vertex s does not belong to X, since s is simplicial. Let P be an induced path from s to a vertex of X such that $|V(P)|$ is minimum. Note that such a path P exists since H is connected. Without loss of generality assume that $V(P) \cap V(X) = \{x_1\}$.

Let s' be the neighbor of x_1 on P. We claim that s' is adjacent to at most one vertex of $\{x_i, x_{i+1}\}$ for $1 \leq i \leq 2k$. If $s' = s$, this claim immediately follows from the assumption that s is simplicial and the fact that x_i and x_{i+1} are not adjacent. Suppose $s' \neq s$, and let s'' be the neighbor of s' on P not equal to x_1. Note that s'' is not adjacent to any vertex of X due to the minimality of $|V(P)|$. Vertex s' cannot be adjacent to both x_i and x_{i+1}, since then the set $\{s', s'', x_i, x_{i+1}\}$ induces a claw in H with center s'. Hence s' is adjacent to at most one vertex of $\{x_i, x_{i+1}\}$ for $1 \leq i \leq 2k$.

Note that vertex s' is adjacent to at least one vertex of $\{x_i, x_{i+1}\}$ for $3 \leq i \leq 2k-1$, as otherwise $\{x_1, s', x_i, x_{i+1}\}$ induces a claw in H with center x_1. This, together with the fact that s' is adjacent to at most one vertex of $\{x_i, x_{i+1}\}$ for $1 \leq i \leq 2k$, implies that s' is adjacent to exactly one vertex of $\{x_3, x_{2k}\}$. Without loss of generality, assume that s' is adjacent to x_{2k} and not to x_3. Since s' is adjacent to x_1 and s' is adjacent to at most one vertex of $\{x_i, x_{i+1}\}$ for $1 \leq i \leq 2k$, s' is not adjacent to x_2. Note that x_3 is adjacent to x_{2k}, since $k \geq 3$. But then $\{x_{2k}, s', x_2, x_3\}$ induces a claw in H with center x_{2k}. This contradiction finishes the proof of Lemma 4. □

We point out that the arguments in the proof of Lemma 4 can also be used to prove that every odd antihole X of a connected claw-free graph H is dominating, i.e., every vertex of H either belongs to X or has a neighbor in X.

Lemma 5. *Let H be a connected claw-free graph that is clean for two simplicial vertices s and t. If H contains an odd hole, then there exists both an odd and an even induced path from s to t.*

Proof. Let C be an odd hole of H. Let P be an induced path from s to a vertex p of C and let Q be an induced path from t to a vertex q of C, such that there is no edge in H connecting a vertex in $P[V(P) \setminus \{p\}]$ to a vertex in $Q[V(Q) \setminus \{q\}]$ and such that $|V(P)| + |V(Q)|$ is minimum. Note that such paths P and Q exist since H is clean and connected. Let s' be the neighbor of p on P, and let t' be the neighbor of q on Q; we note that possibly $s' = s$ and $t' = t$.

CLAIM 1. *Both s' and t' are adjacent to exactly two adjacent vertices of C.*

Suppose p is the only vertex of C that is adjacent to s'. Let p^- (respectively p^+) denote the neighbor of p on C when we traverse C in counter-clockwise (respectively clockwise) order. The set $\{p, p^-, p^+, s'\}$ induces a claw in H with center p, contradicting the claw-freeness of H. Hence s' must be adjacent to at least one vertex of $\{p^-, p^+\}$. Suppose there exists a set $D \subseteq V(C)$ such that $|D| \geq 3$ and s' is adjacent to every vertex in D. Since C is an induced cycle, we know that D contains two vertices d_1 and d_2 that are not adjacent. Since s is simplicial and therefore does not have two non-adjacent neighbors, we must have $s' \neq s$. Let $s'' \neq p$ be a neighbor of s' on P; possibly $s'' = s$. Vertex s'' is not adjacent to any vertex of C due to the minimality of $|V(P)| + |V(Q)|$, which means the set $\{s', d_1, d_2, s''\}$ induces a claw in H with center s'. This contradiction finishes the proof of Claim 1 for vertex s'. By symmetry the claim also holds for vertex t'.

We assume, without loss of generality, that $N_H(s') \cap V(C) = \{p, p^+\}$ and $N_H(t') \cap V(C) = \{q, q^+\}$. We distinguish three cases.

Suppose $|\{p, p^+\} \cap \{q, q^+\}| = 0$. Since C is an odd hole, the induced path $s'p^+\overrightarrow{C}qt'$ and the induced path $s'p\overleftarrow{C}q^+t'$ have different parity. Hence there exists both an odd and an even induced path from s to t in H.

Suppose $|\{p, p^+\} \cap \{q, q^+\}| = 1$. Without loss of generality, suppose $p^+ = q$. Then the path $s'qt'$ is an even induced path from s' to t', and the path $s'p\overleftarrow{C}q^+t'$ is an odd induced path from s' to t'. Since by definition there is no edge connecting a vertex in $P[V(P) \setminus \{p\}]$ to a vertex in $Q[V(Q) \setminus \{q\}]$, this means there exists both an odd and an even induced path from s to t in H.

Suppose $|\{p,p^+\}\cap\{q,q^+\}| = 2$. By Claim 1, neither s' nor t' is adjacent to p^-. Since s' and t' are not adjacent by the choice of P and Q, the set $\{p,p^-,s',t'\}$ induces a claw in H with center p. This contradiction finishes the proof of Lemma 5. □

Recall that G'' is not perfect and has two simplicial vertices s and t. This, together with Lemma 4 and Lemma 5, implies that G'' contains both an odd and an even induced path from s to t. We now show that we can also *find* such paths in $\mathcal{O}(n^5)$ time.

Lemma 6. *If G'' is not perfect, then it is possible to find both an odd and an even induced path from s to t in G'' in $\mathcal{O}(n^5)$ time.*

Proof. Since G'' has two simplicial vertices s and t, G'' does not contain an odd antihole of length more than 5 by Lemma 4. Since an odd antihole of length 5 is also an odd hole of length 5, G'' contains an odd hole by virtue of the Strong Perfect Graph Theorem. We can find such a hole C in $\mathcal{O}(n^5)$ time by Corollary 1. Let c be any vertex of C, and let P be an induced path in G'' from s to t containing c. Note that such a path P exists since G'' is clean for s and t. We can find P in $\mathcal{O}(n^4)$ time as a result Theorem 3. It is clear from the proof of Lemma 5 that we can use P to find both an odd and an even induced path from s to t in G''. □

3.3 G'' Is Perfect

Suppose G'' is perfect. In the concluding remarks of their paper, Corneil and Fonlupt [12] pointed out that a polynomial-time recognition algorithm for perfect graphs implies a polynomial-time algorithm for the PARITY PATH problem for the class of perfect graphs. Interestingly, the arguments they used to prove this implication were already mentioned by Hsu [19] in the paper in which he introduced the PARITY PATH problem. Using their arguments, we can prove the following lemma.

Lemma 7. *If G'' is perfect, then it is possible to find an odd induced path from s to t in G'', or conclude that such a path does not exist, in $\mathcal{O}(n^5)$ time.*

Proof. Let P be a shortest path from s to t in G''. If P has odd length, then we are done. Suppose P has even length. Let G^* be the graph obtained from G by adding a vertex x and edges sx and tx. Note that the graph G^* is claw-free, since s and t are simplicial vertices of G''. We determine whether or not G^* is perfect, which we can do in $\mathcal{O}(n^4)$ time by Theorem 2. If G^* is perfect, then G^* does not contain an odd hole or an odd antihole by virtue of the Strong Perfect Graph Theorem. This means that all induced paths from s to t must be even, so we conclude that there does not exist an odd induced path from s to t. Suppose G^* is not perfect. Then G^* must contain an odd hole or an odd antihole, and vertex x must be in this odd hole or odd antihole since G is perfect. Since x has degree two, G^* cannot contain an odd antihole. Hence G^* contains an odd hole. We can find an odd hole C of G^* in $\mathcal{O}(n^5)$ time by Corollary 1. The graph obtained from C by removing vertex x is an odd induced path from s to t in G''. □

3.4 Finding Induced Paths of Given Parity from s to t in G

We are now ready to prove the main result of this section.

Theorem 4. *Both the* ODD INDUCED PATH *problem and the* EVEN INDUCED PATH *problem can be solved in* $\mathcal{O}(n^5)$ *time for the class of claw-free graphs. Moreover, an induced path from s to t of given parity can be found in $\mathcal{O}(n^5)$, if one exists.*

Proof. Let G be a claw-free graph, and let s and t be two vertices of G. Recall that we may without loss of generality assume that G is connected and that s and t are not adjacent. We preprocess G in $\mathcal{O}(n^5)$ time as described in Section 3.1, thus obtaining a graph G''. Recall that G'' is claw-free, that s and t are simplicial vertices in G'', and that G'' is clean for s and t. We test whether or not G'' is perfect, which we can do in $\mathcal{O}(n^4)$ time by Theorem 2. Below we show that we can find an induced path of given parity from s to t in G'', or conclude that such a path does not exist, in $\mathcal{O}(n^5)$ time. Lemma 3 implies that this suffices to prove Theorem 4.

If G'' is not perfect, then we can find both an odd and an even induced path from s to t in G'' in $\mathcal{O}(n^5)$ time by Lemma 6. If G'' is perfect, then we can find an odd induced path from s to t in G'', or conclude that such a path does not exist, in $\mathcal{O}(n^5)$ time by Lemma 7. In order to find an even induced path from s to t, we define the graph G^* as the graph obtained from G'' by adding the edge st. It is easy to verify that adding the edge st creates neither a claw nor an odd antihole. Hence the arguments used in the proof of Lemma 7 can also be used to find an even induced path from s to t in G'', or conclude that such a path does not exist, in $\mathcal{O}(n^5)$ time. □

Theorem 4 immediately implies the following.

Corollary 2. *Both the* PARITY PATH *problem and the* EVEN PAIR *problem can be solved in $\mathcal{O}(n^5)$ time for the class of claw-free graphs.*

Bienstock [6] proved that the problem of deciding if a graph contains an odd induced path between every pair of vertices, as well as the problem of deciding if a graph has an odd hole through a given vertex, is NP-complete. The following two corollaries of Theorem 4, the proofs of which are postponed to the journal version of this paper, imply that both problems can be solved in $\mathcal{O}(n^7)$ time when restricted to the class of claw-free graphs.

Corollary 3. *Deciding whether or not a claw-free graph has an even pair can be done in $\mathcal{O}(n^7)$ time.*

Corollary 4. *It is possible to find an odd hole passing through a prescribed vertex of a claw-free graph, or conclude that such a hole does not exist, in $\mathcal{O}(n^7)$ time.*

4 Conclusions and Open Problems

We have proved that both the ODD INDUCED PATH problem and the EVEN INDUCED PATH problem, and consequently the PARITY PATH problem, can be solved in $\mathcal{O}(n^5)$ time for the class of claw-free graphs. This implies that we can decide in $\mathcal{O}(n^7)$ time whether a claw-free graphs contains a hole of given parity passing through a given vertex. In the SHORTEST ODD INDUCED PATH and the SHORTEST EVEN INDUCED PATH problem, the goal is to find a *shortest* induced path between two given vertices of odd and even length, respectively. Using the structure of claw-free perfect graphs, we can show that these problems can be solved in $\mathcal{O}(n^7)$ time for this class; the details are postponed to the journal version of this paper. We conclude by mentioning some open problems. Does there exist a polynomial-time algorithm for the SHORTEST ODD INDUCED PATH and SHORTEST EVEN INDUCED PATH problems for *general* claw-free graphs? And does there exist a polynomial-time algorithm for the ODD INDUCED PATH and EVEN INDUCED PATH problems for the class of planar graphs?

Acknowledgements. The authors would like to thank Benjamin Lévêque, Luke Mathieson and Nicolas Trotignon for helpful suggestions.

References

1. Arkin, E.M., Papadimitriou, C.H., Yannakakis, M.: Modularity of cycles and paths in graphs. Journal of the ACM 38(2), 255–274 (1991)
2. Arikati, S.R., Peled, U.N.: A linear algorithm for the group path problem on chordal graphs. Discrete Applied Mathematics 44(1-3), 185–190 (1993)
3. Arikati, S.R., Peled, U.N.: A polynomial algorithm for the parity path problem on perfectly orientable graphs. Discrete Applied Mathematics 65(1), 5–20 (1996)
4. Arikati, S.R., Rangan, C.P., Manacher, G.K.: Efficient reduction for path problems on circular-arc graphs. BIT 31(2), 182–193 (1991)
5. Berge, C.: Färbung von Graphen, deren sämtliche bzw. deren ungerade Kreise starr sind. Wissenschaftliche Zeitschrift der Martin-Luther-Universität Halle-Wittenberg, Mathematisch-Naturwissenschaftliche Reihe 10, 114 (1961) (in German)
6. Bienstock, D.: On the complexity of testing for odd holes and induced odd paths. Discrete Mathematics 90(1), 85–92 (1991)
7. Chudnovsky, M., Cornuéjols, G., Liu, X., Seymour, P.D., Vušković, K.: Recognizing Berge Graphs. Combinatorica 25(2), 143–186 (2005)
8. Chudnovsky, M., Robertson, N., Seymour, P.D., Thomas, R.: The strong perfect graph theorem. Annals of Mathematics 164, 51–229 (2006)
9. Chudnovsky, M., Seymour, P.D.: The three-in-a-tree problem. Combinatorica (to appear), manuscript, http://www.columbia.edu/~mc2775/threeinatree.pdf
10. Chudnovsky, M., Seymour, P.D.: Three-colourable perfect graphs without even pairs (submitted for publication), manuscript,
    http://www.columbia.edu/~mc2775/K4evenpairs.ps
11. Chvátal, V., Sbihi, N.: Recognizing claw-free perfect graphs. Journal of Combinatorial Theory, Series B 44, 154–176 (1988)

12. Corneil, D.G., Fonlupt, J.: Stable set bonding in perfect graphs and parity graphs. Journal of Combinatorial Theory, Series B 59, 1–14 (1993)
13. Derhy, N., Picouleau, C.: Finding induced trees. Discrete Applied Mathematics 157(17), 3552–3557 (2009)
14. Everett, H., de Figueiredo, C.M.H., Sales, C.L., Maffray, F., Porto, O., Reed, B.A.: Path parity and perfection. Discrete Mathematics 165-166, 233–252 (1997)
15. Everett, H., de Figueiredo, C.M.H., Linhares Sales, C., Maffray, F., Porto, O., Reed, B.A.: In: Ramirez-Alfonsin, L., Reed, B.A. (eds.) Perfect Graphs, pp. 67–92. Wiley, Chichester (2001)
16. de Figueiredo, C.M.H., Gimbel, J.G., Mello, C.P., Szwarcfiter, J.L.: Even and odd pairs in comparability and in P_4-comparability graphs. Discrete Applied Mathematics 91(1-3), 293–297 (1999)
17. Fonlupt, J., Uhry, J.P.: Transformations which preserve perfectness and H-perfectness of graphs. Annals of Discrete Mathematics 16, 83–85 (1982)
18. Hoàng, C.T., Le, V.B.: Recognizing perfect 2-split graphs. SIAM Journal on Discrete Mathematics 13(1), 48–55 (2000)
19. Hsu, W.-L.: Recognizing planar perfect graphs. Journal of the ACM 34(2), 255–288 (1987)
20. LaPaugh, A.S., Papadimitriou, C.H.: The even-path problem for graphs and digraphs. Networks 14, 507–513 (1984)
21. Lévêque, B., Lin, D.Y., Maffray, F., Trotignon, N.: Detecting induced subgraphs. Discrete Applied Mathematics 157(17), 3540–3551 (2009)
22. Linhares Sales, C., Maffray, F.: Even pairs in claw-free perfect graphs. Journal of Combinatorial Theory, Series B 74, 169–191 (1998)
23. Meyniel, H.: A new property of critical imperfect graphs and some consequences. European Journal of Combinatorics 8, 313–316 (1987)
24. Roussopoulos, N.D.: A max {m,n} algorithm for determining the graph H from its line graph G. Information Processing Letters 2, 108–112 (1973)
25. Sampaio, R.M., Sales, C.L.: On the complexity of finding even pairs in planar perfect graphs. In: Brazilian Symposium on Graphs, Algorithms and Combinatorics 2001, Fortaleza. Electronic Notes in Discrete Mathematics, vol. 7, pp. 186–189 (2001)
26. Satyan, C.R., Pandu Rangan, C.: The parity path problem on some subclasses of perfect graphs. Discrete Applied Mathematics 68(3), 293–302 (1996)
27. Tarjan, R.E.: Decomposition by clique separators. Discrete Mathematics 55, 221–232 (1985)
28. Trotignon, N.: Graphes parfaits: Structure et algorithmes. PhD Thesis, Université Joseph Fourier - Grenoble I (2004) (in French)
29. Whitesides, S.H.: A method for solving certain graph recognition and optimization problems, with applications to perfect graphs. Annals of Discrete Mathematics 21, 281–297 (1984)

Author Index

Averbouch, Ilia   33

Broersma, Hajo   44

Călinescu, Gruia   54
Chechik, Shiri   66
Crespelle, Christophe   77

Dvořák, Zdeněk   17

Eppstein, David   1

Fellows, Michael R.   88
Fernandes, Cristina G.   54
Fernau, Henning   100
Fomin, Fedor V.   44, 112

Gandhi, Rajiv   122
Gaspers, Serge   100
Golovach, Petr A.   133
Golumbic, Martin Charles   329
Greening Jr., Bradford   122
Grigoriev, Alexander   143
Grumbach, Stéphane   154
Guo, Jiong   88
Gurski, Frank   166

Hagerup, Torben   178

Kamiński, Marcin   341
Kammer, Frank   190
Kanj, Iyad A.   88, 202
Kanté, Mamadou Moustapha   214
Katz, Bastian   226
Kaul, Hemanshu   54
Král', Daniel   17
Kratochvíl, Jan   133
Kreutzer, Stephan   308

Le, Van Bang   238
Li, Zhentao   250
Lokshtanov, Daniel   112

MacGillivray, Gary   262
Makowsky, Johann A.   33

Meister, Daniel   273
Mertzios, George B.   285

Nguyen, Ngoc Tuy   238

Okamoto, Yoshio   296
Ordyniak, Sebastian   308

Paulusma, Daniël   44, 341
Peleg, David   66
Pemmaraju, Sriram   122
Penso, Lucia Draque   320

Raible, Daniel   100
Raman, Rajiv   122
Rao, Michaël   214
Raspaud, André   262
Rautenbach, Dieter   320
Rutter, Ignaz   226

Sau, Ignasi   250, 285
Saurabh, Saket   112
Shrem, Shimon   329
Sitters, René   143
Stern, Michal   329
Suchý, Ondřej   133
Swarts, Jacobus   262
Szwarcfiter, Jayme Luiz   320

Telle, Jan Arne   273
Tholey, Torsten   190
Tittmann, Peter   33

Uehara, Ryuhei   296
Uno, Takeaki   296

van 't Hof, Pim   44, 341

Wanke, Egon   166
Wiese, Andreas   202
Woeginger, Gerhard   226
Wu, Zhilin   154

Zaks, Shmuel   285
Zhang, Fenghui   202

Printing: Mercedes-Druck, Berlin
Binding: Stein+Lehmann, Berlin